T0174242

Volume I
Infrastructure Health in Civil Engineering
Theory and Components

Volume I

Infrastructure Health in Civil Engineering

Theory and Components

Mohammed M. Ettouney
Sreenivas Alampalli

CRC Press
Taylor & Francis Group
Boca Raton London New York

CRC Press is an imprint of the
Taylor & Francis Group, an **informa** business

Cover image courtesy of New York State Department of Transportation.

CRC Press
Taylor & Francis Group
6000 Broken Sound Parkway NW, Suite 300
Boca Raton, FL 33487-2742

First issued in paperback 2019

© 2012 by Taylor & Francis Group, LLC
CRC Press is an imprint of Taylor & Francis Group, an Informa business

No claim to original U.S. Government works

ISBN-13: 978-0-8493-2040-8 (hbk)
ISBN-13: 978-0-367-38235-3 (pbk)

This book contains information obtained from authentic and highly regarded sources. Reasonable efforts have been made to publish reliable data and information, but the author and publisher cannot assume responsibility for the validity of all materials or the consequences of their use. The authors and publishers have attempted to trace the copyright holders of all material reproduced in this publication and apologize to copyright holders if permission to publish in this form has not been obtained. If any copyright material has not been acknowledged please write and let us know so we may rectify in any future reprint.

Except as permitted under U.S. Copyright Law, no part of this book may be reprinted, reproduced, transmitted, or utilized in any form by any electronic, mechanical, or other means, now known or hereafter invented, including photocopying, microfilming, and recording, or in any information storage or retrieval system, without written permission from the publishers.

For permission to photocopy or use material electronically from this work, please access www.copyright.com (http://www.copyright.com/) or contact the Copyright Clearance Center, Inc. (CCC), 222 Rosewood Drive, Danvers, MA 01923, 978-750-8400. CCC is a not-for-profit organization that provides licenses and registration for a variety of users. For organizations that have been granted a photocopy license by the CCC, a separate system of payment has been arranged.

Trademark Notice: Product or corporate names may be trademarks or registered trademarks, and are used only for identification and explanation without intent to infringe.

Library of Congress Cataloging-in-Publication Data

Ettouney, Mohammed.
   Infrastructure health in civil engineering / authors, Mohammed M. Ettouney and Sreenivas Alampalli.
      p. cm.
   "A CRC title."
   Includes bibliographical references and index.
   ISBN 978-0-8493-2040-8 (alk. paper)
   1. Structural health monitoring. 2. Structural failures--Prevention. 3. Reliability (Engineering) I. Alampalli, Sreenivas. II. Title.

TA656.6.E88 2010
624.1'71--dc22   2010013691

Visit the Taylor & Francis Web site at
http://www.taylorandfrancis.com

and the CRC Press Web site at
http://www.crcpress.com

*This book is dedicated to Mohammed A. Ettouney, Fatima A. Abaza,
and William Zacharellis, may God be merciful on their souls.*

Mohammed M. Ettouney

To my wife Sharada and son Sandeep for their love and patience

Sreenivas Alampalli

Contents

Preface

OVERALL

There is a purpose for building infrastructure, and infrastructure owners are responsible for ensuring that the intended purpose is served while achieving maximum benefit at minimal costs. Taking appropriate timely actions requires a good understanding of the infrastructure—its current and expected condition in the future. This volume and its companion volume, *Infrastructure Health in Civil Engineering: Applications and Management* (CRC Press, 2012), are dedicated to discussion of these aspects (see Figure 0.1). This volume focuses on providing an overview of the infrastructure health in civil engineering (IHCE) and associated theories followed by the description of its four components: measurements, structural identification, damage identification, and decision making. Decision making aspect is a unique feature and is introduced with an argument that any project that does not integrate decision making (or cost–benefit) ideas in all tasks cannot be successful. The companion volume builds on the ideas presented in this volume and deals with the application of the IHCE and asset management aspects.

PART I: OVERVIEW AND THEORIES OF IHCE

In the first part of this volume, consisting of four chapters, we lay the foundations of IHCE with a thorough discussion of its basic components as well as related emerging issues. The introductory chapter covers the history and importance of IHCE followed by the need for it as applied to structures in general and, in particular, to bridges. Analogies between structural and human health fields are explored. The need for knowing (or monitoring) the health of the structure throughout its life (referred to as "95% solution") is shown compared to current partial (or 5%) solution, which is applied to structures at construction and rehabilitation phases. IHCE is a complex subject that encompasses many disciplines and applications; the second chapter introduces the four basic components of IHCE to make its study more systematic and relatively easier. We introduce the *baseball analogy* concept to IHCE to further illustrate its basic components. This chapter also explores engineering paradigms and their relationships with IHCE.

Chapter 3 provides analogy between human and structural healths from a chronological viewpoint. This chapter provides a comprehensive view of the infrastructure birth (construction), life (operations), and death (failure/decommissioning), with an emphasis on bridge structures. The importance of the 95% solution as well as the potentially immense contributions of IHCE to the entire bridge life will emerge in this chapter. Chapter 4 deals with optimization of structural health monitoring projects with both technical and financial considerations. The importance of "value" in an emerging field such as IHCE is well recognized; several theories and paradigms related to IHCE are thus introduced in this chapter. General and special theories of experimentation are presented in an attempt to help stake holders obtain the maximum value from the IHCE-type experiments at reasonable costs. These theories link experiments to the concept of "value" in an objective fashion by integrating all aspects of IHCE and its goals, such as improving safety and reducing costs, in a concise manner. A third theory on sensing triangulation is presented in this chapter to maximize sensing efficiency by using different types of sensors within an IHCE experiment. Finally, the chapter introduces three principles—duality, scaling, and serendipity—needed to maximize the efficiency of IHCE efforts. Several examples from the literature that illustrate the use of these theories and principles are also included in this chapter.

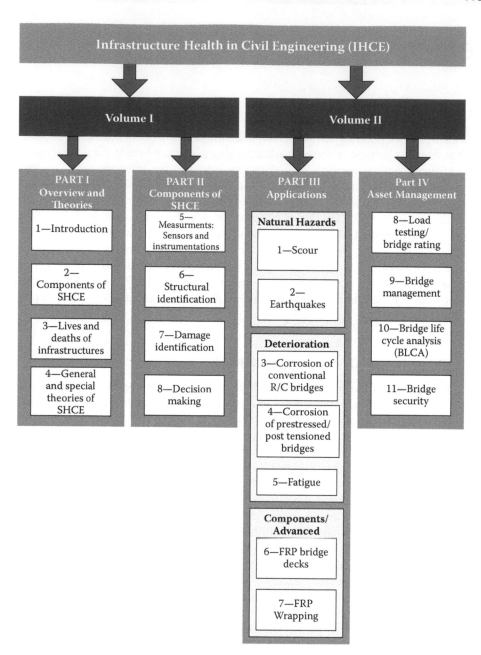

FIGURE 0.1   General layout of this volume and the companion volume, *Infrastructure Health in Civil Engineering: Applications and Management* (CRC Press, 2012).

PART II: COMPONENTS OF IHCE

The second part of this volume comprehensively reviews the four components of IHCE that were introduced in the second chapter. The first and most recognizable phase of IHCE is the measurements/sensing phase and is discussed in Chapter 5. In this chapter, we discuss sensors, sensing techniques, and instrumentation. Measurement needs for various IHCE situations are identified, and important aspect of sensors are discussed following the road map shown in Figure 0.2. This chapter also explores various sensing techniques used in IHCE experiments, including emerging topics

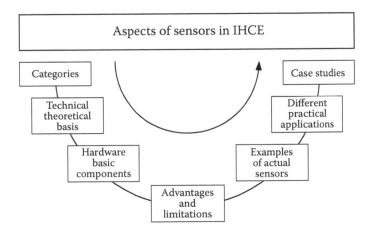

FIGURE 0.2 Sensor topics.

such as fiber optic sensors (FOS), wireless sensing, and remote sensing. Optimum sensor locations and optimum sensor selection that are intended to maximize the benefits of IHCE efforts are also explored with practical case studies from the literature.

Chapter 6 explores the second component in the IHCE field "Structural Identification (STRID)." Objectives of STRID are discussed first, followed by discussions on three STRID methodologies commonly used: modal identification, parameter identification, and functional methods. A special attention is paid to artificial neural network methods while discussing functional methods. STRID modeling techniques—static, dynamic, linear, or nonlinear analyses—as they pertain to structural health monitoring (SHM) and IHCE are then discussed, keeping in mind that the complexity of modeling varies widely, depending on the situation at hand. Even though deterministic structural modeling is predominantly discussed, some discussion of probabilistic structural modeling is also included. Scaling aspects and scale-independent methods that show promise in the STRID field are presented in the end followed by a discussion of cost–benefit issue as applied to STRID.

Chapter 7 presents an overview of damage identification (DMID). This perhaps is one of the most challenging subjects presented in this volume as the subject of damage is not very well defined in the field of SHM/IHCE. This chapter identifies damages from several view points, including damage types as they result from various hazards. We also discuss the interrelationship between STRID and DMID. We explore the vibration-based DMID methods, as well as the use of signal processing in DMID field. Widely used nondestructive testing (NDT) technologies in the field of SHM are presented, with special emphasis on their usage and applicability in the SHM field. The chapter finally explores hazard-specific DMID methods.

The main premise of this volume is the tight integration of structural health monitoring (the first three components of IHCE) and objective decision making processes that is the final component of IHCE. We advocate that without a strong decision making support, any SHM project could not fulfill its promise and thus objective decision making processes must be the basis for maintaining and improving structural health. Hence, the last chapter of this volume (Chapter 8) is devoted to decision making processes and presents numerous useful tools/techniques. This includes basics of probability and statistics tools, theory of decision making, structural reliability, and risk tools. In addition, stochastic modeling, economic tools, and probabilistic structural methods are also covered. In all situations, pertinent examples from the IHCE field are given to illustrate the use and importance of decision making tools.

It should be noted that all the opinions and views expressed in this volume are those of the authors and not necessarily of the organizations they represent.

Acknowledgments

The life of every person is measured, among other things, by the special people who kindly affect such a life. This volume is a direct result of the many beautiful interactions I have had over the years with those special people. My late mentor, Dr. Mohammed S. Aggour, taught me the very foundations of structural engineering and design. My childhood friend, Dr. Elsayed A. Elsayed, has always been with me during good and bad times. My long friendships with Dr. Raymond (Ray) Daddazio, Michael (Mike) Udvardy, Amr Aly, Dr. Loraine Whitman, Ron (Ronnie) Check, Dr. Hamid Adib, and Christina (Tina) Plink have had a direct as well as an indirect influence on my writing this volume. I am grateful for the professional help, encouragement, and support given by Antranig (Andy) Ouzoonian, Norman (Norm) Glover, Gary Higbee, Dr. Amar Chaker, and Arturo (Artie) Mendez. The special friendship of and the unparalleled technical and professional help and support given by Dr. Anil K. Agrawal will always be warmly remembered and deeply acknowledged. The advice and patience of the editors of CRC Press helped immensely in making this project materialize. For that I am deeply grateful.

Over the millennia, when darkness descended, human beings used to look up to the heavens for help and guidance. The countless stars of the Milky Way never disappointed them. The combined belts of Orion and Sirius have always guided those who needed help through the darkest of the dark hours. I am no exception: without my own belts of Orion and Sirius this volume could have never materialized. For that, I shall always be grateful to Milagros (Mila) Kennett and Dr. Sreenivas (Sreeni) Alampalli. Their friendship, kindness, help, and guidance will always be remembered as long as there are the belts of Orion and Sirius.

Mohammed M. Ettouney

It has been a great pleasure to have worked for the past 20 years with several of my hard-working and intelligent colleagues at the New York State Department of Transportation (NYSDOT), who strive to ensure that all users of transportation infrastructure are safe and secure every day. I learned a lot working with them, and several examples in this volume are the result of working with many of them. One person requiring special acknowledgment is my mentor, the late Dr. Robert J. Perry, for his encouragement and support in advancing my professional interests.

During the past 20 years, I was very fortunate to have worked with a variety of people from various state and federal agencies, universities, private industry, and professional organizations. This volume has benefitted from the practical and research experiences gained by working with them. I am specially grateful to Dr. Glenn Washer and Dr. Hamid Ghasemi for their professional help and support during these years. I thank my good friend Dr. Anil Agrawal for his support and for simply being there when needed. Several people have had an influence on me but none as much as Dr. Mohammed M. Ettouney. I met Dr. Ettouney about 15 years ago and found in him a great friend, an excellent colleague, and a mentor. I am always amazed at his professional competence and his ability to look beyond the normal. It has been a privilege writing this volume with him.

Sreenivas Alampalli

Authors

Dr Mohammed M. Ettouney, Ph.D., P.E., MBA, F. AEI, Dist. M. ASCE, was conferred the Innovators Award in 2008 by the New Jersey Inventors Hall of Fame after he was nominated to receive such a great honor by the American Society of Civil Engineers (ASCE). Dr. Ettouney also received the Homer Gage Balcom lifetime achievement award by the Metropolitan Section of ASCE (2008). He won the Project of the Year Award, Platinum Award (2008) for the New Haven Coliseum Demolition Project (ACEC, NY). He is a fellow of the Architecture Engineering Institute (AEI) and a distinguished member in American Society of Civil Engineers. Among his other recent achievements are the pioneering work on "Theory of Multihazards of Infrastructures," "Theory of Progressive Collapse," pioneering work in multihazards/multidisciplinary evaluation of risk and resiliency of buildings, tunnels, bridges, and transit stations, and an innovative green design method for protecting utilities from demolition/blasting (City of New Haven, CT). He has professional interest in diverse areas of structural engineering as demonstrated by the list of his publications, invited presentations, seminars, and sessions organized during national/international conferences, besides his membership in different professional organizations.

Dr. Ettouney is a principal with Weidlinger Associates, based in New York City, NY. He received his Doctor of Science in structural mechanics from the Massachusetts Institute of Technology (MIT), Cambridge, MA, in 1976. Since then, his interest in structural engineering has been both as a practitioner and a researcher in multihazards safety of structures; probabilistic modeling of progressive collapse of buildings; uncertainties in structural stability; blast mitigation of numerous buildings around the world; innovative concepts such as "Probabilistic Boundary Element Method," "Scale Independent Elements," and "Framework for Evaluation of Lunar-Based Structural Concepts." He is a past president and past member of the board of governors of AEI, member of the Board of Directors of the BSC, member of several technical committees on building/infrastructure security, earthquake hazards, architectural engineering, and Nondestructive Testing and Structural Health Monitoring. He was chair of the AEI National Conference 2006 and 2008. He is a member of the NIBS Advanced Materials Council. He is also Editor of the *Journal of Advanced Materials*. He serves on the Board of Multihazard Mitigation Council (MMC).

Dr. Ettouney has authored or co-authored more than 325 publications and reports and contributed to several books. He introduced several new practical and theoretical methods in the fields of earthquake engineering, acoustics, structural health monitoring, progressive collapse, blast engineering, and underwater vibrations. He co-invented the "Seismic Blast" slotted connection. More recently, he introduced the "Economic Theory of Inspection," "General and Special Theories of Instrumentation," and numerous principles and techniques in the field of infrastructure health—all pioneering efforts that can help in developing durable infrastructures at reasonable costs.

Dr. Sreenivas Alampalli, P.E., MBA, is Director of the Structures Evaluation Services Bureau at the New York State Department of Transportation (NYSDOT). His responsibilities include managing structural inspection, inventory, and safety assurance programs at the NYSDOT. Before taking up the current responsibility in 2003, Dr. Alampalli was Director of the Transportation Research and Development Bureau. In this position, he managed a targeted transportation infrastructure research and development program to enhance the quality and cost-effectiveness of transportation policies, practices, procedures, standards, and specifications. He also taught at Union College and Rensselaer Polytechnic Institute as an adjunct faculty.

Dr. Alampalli obtained his Ph.D. and MBA from Rensselaer Polytechnic Institute, his M.S. from the Indian Institute of Technology (IIT), Kharagpur, India, and his B.S. from S.V. University,

Tirupati, India. His interests include infrastructure management, innovative materials for infrastructure applications, nondestructive testing, structural health monitoring, and long-term bridge performance. He co-developed the theory of multihazards and has been a great proponent of it to integrate all vulnerabilities, including security for effective infrastructure management. Dr. Alampalli is a Fellow of the American Society of Civil Engineers (ASCE), American Society for Nondestructive Testing (ASNT), and International Society for Health Monitoring of Intelligent Infrastructure (ISHMII). He has received several awards, including the prestigious Charles Pankow Award for Innovation, from the Civil Engineering Research Foundation in 2000; ASNT Mentoring Award in 2009; and Herbert Howard Government Civil Engineer of the Year Award from ASCE Metropolitan (NYC) section in 2009. He has authored or co-authored more than 250 technical publications.

Dr. Alampalli is an active member of several technical committees in TRB, ASCE, and ASNT, and currently chairs the ASCE Technical Committee on Bridge Management, Inspection, and Rehabilitation. He served as the Transportation Research Board representative for the NYSDOT and also as a member of the National Research Advisory Committee (RAC). He is an Associate Editor of the *ASCE Journal of Bridge Engineering* and serves on the editorial board of the journal *Structure and Infrastructure Engineering: Maintenance, Management, Life-Cycle Design and Performance* and the journal *Bridge Structures: Assessment, Design and Construction.*

1 Introduction

1.1 INFRASTRUCTURE HEALTH IN CIVIL ENGINEERING (IHCE)

President Obama in his inaugural speech on January 20, 2009 said, "We will build the roads and bridges, the electric grids and digital lines that feed our commerce and bind us together" (NRP 2011). This was an acknowledgment of the importance of the role of infrastructure in modern societies. It is a multifaceted role, including commerce, communication, mobility, recreation, and comfort. We observe that there is a need to change conventional approach to infrastructure. As modern societies evolved in the past century, challenges to infrastructure stakeholders evolved, and many new challenges are emerging. Some of these challenges are increased costs, changing social expectations and demands, advent of many modern hazards, and the increased complexities of infrastructure. Take, for example, the bridge infrastructure sector. There are suspension bridges (Figures 1.1, 1.2, and 1.3) that offer grace and functionality, yet they are extremely complex constructs. Arch and truss bridges (Figures 1.4 and 1.5) and conventional multigirder, multispan bridges (Figures 1.6 and 1.7), still present immense complexities and demands. Single-span bridges (Figure 1.8) need considerable attention from bridge managers because of the potential consequences that might occur if they are damaged. Finally, historical bridges (Figure 1.9) need attention and preservation because of their function and their historic nature.

Infrastructure is aging, even newly built ones start aging as soon as the construction phase is over. The aging of infrastructure, given all other just mentioned challenges, would require diligence on part of owners to ensure that they continue operations efficiently at reasonable costs.

Fortunately, the state of the profession has also been evolving in many ways that is making it possible to meet many of these challenges. For example, computing, material, and sensing technologies have been advancing at a rapid pace. Decision making and management techniques are evolving and creating new opportunities to owners and managers. New engineering paradigms are emerging and have the potential of helping stakeholders to meet modern infrastructure challenges such as multihazards considerations, structural health monitoring (SHM), nondestructive testing (NDT), performance based engineering (PBD), life cycle analysis (LCA), and many more. Note that SHM is a subset of IHCE. SHM, as we will consider it in this volume and the companion volume (Ettouney and Alampalli 2012), contains measurements, structural identification, and damage identification. IHCE adds decision making component to SHM. Chapter 2 addresses this subject in more detail.

The new paradigms should be coordinated between stakeholders, carefully considering the interactions between them. These paradigms include, but are not limited to, economics, as built versus new construction, structural identification (STRID), damage identification (DMID), efficient management, and decision making (DM).

Perhaps an easy way to deal with all of the above complexities and potential solutions is to note the similarities of the desired infrastructure healthy performance with human health. We call this confluence IHCE (infrastructure health in civil engineering): a new and essential field that is a composition of many disciplines in civil, mechanical, and business communities. It is an inclusive way of thinking by all stakeholders of civil infrastructure. This volume aims at gathering all the pieces of this new way of addressing infrastructure and linking them at different and natural points: costs, benefits, design, analysis, sensing, hazards interaction, decision making, etc.

FIGURE 1.1   Suspension bridge (Brooklyn Bridge, NYC). (Courtesy of Francois Ghanem.)

FIGURE 1.2   Manhattan Bridge, NYC (looking from Brooklyn Bridge). (Courtesy of Francois Ghanem.)

FIGURE 1.3   Old and new Carquinez Bridges, over Suisun Bay, CA.

FIGURE 1.4   Arch bridge. (Courtesy of New York State Department of Transportation.)

FIGURE 1.5   Truss bridge. (Courtesy of New York State Department of Transportation.)

FIGURE 1.6   Conventional multigirder steel bridge. (Courtesy of New York State Department of Transportation.)

FIGURE 1.7   Multi-span overpass. (Courtesy of New York State Department of Transportation.)

FIGURE 1.8   Single-span road bridge. (Courtesy of New York State Department of Transportation.)

FIGURE 1.9   Historic Wooden bridge. (Courtesy of New York State Department of Transportation.)

This chapter gives an overview of the main points in the infrastructure health field. We first try to define what constitutes "infrastructure health." The history of and the need for structural health efforts will then be discussed. We offer an analogy between structural health and human health. The above-discussed need for IHCE is explored in depth later, including the fact that IHCE is a composition of several disciplines. We then discuss how different hazards and different disciplines need to be integrated efficiently to form important components of IHCE along with strategic challenges that require further consideration.

1.2 GENERAL CONCEPTS OF INFRASTRUCTURES HEALTH

1.2.1 WHAT ARE INFRASTRUCTURES?

Perhaps the basic question that needs to be addressed first is *what do we mean by infrastructure*? Before we attempt to answer this question, we hasten to say that we concern ourselves in this volume with civil infrastructure only. American Society of Civil Engineers (ASCE 2009) defined infrastructure as those that *may be built (such as structures, energy, water, transportation, and communication systems), natural (such as surface or ground water resources), or virtual (such as cyber, electronic data, and information systems)*. The US Department of Homeland Security (US DHS) has more specific definitions of infrastructure. As of the writing of this volume, the US DHS subdivided infrastructure into 17 sectors, US DHS (2009). See Section 1.12 for complete list of the sectors. Of the 17 sector, we can identify civil infrastructure in the transportation (bridges, tunnel, roads, etc.), dams, and commercial facilities (e.g., office buildings). Many other sectors include civil infrastructure as subsector components, such as government facilities, public health, or commercial nuclear reactors.

Many of the discussions in this volume utilize bridges as the main subject of the discussion. Most of the ideas and logics apply to almost all types of civil infrastructure.

1.2.2 WHAT IS STRUCTURAL HEALTH?

Dictionaries define health as an overall condition at any given time, freedom from abnormalities, or optimal well-being. Infrastructure is noted as an underlying base or foundation for an organization or in our context the facilities and installations needed for the functioning of transportation system. Thus, infrastructure health can be stated as its current ability to provide intended level of service in safe, secure, cost-effective manner against hazards it is designed for or expected during its life. Although this statement seems simple, it is not that simple depending on how one perceives it.

Structural health depends on the components it is made of, the network it is in, the characteristics of the individual structures as well as the network, users needs and expectations, socioeconomic aspects, expected performance, and approaches taken from the moment the structure is envisioned to its replacement before or after its expected life. Several performance indicators such as safety, mobility, reliability, and security define structural health. The physical and structural condition is an important factor as safety is the primary measure that is expected. This aspect of health is more dependent on its components, normal loading conditions, current and future environment, and expected hazards during its life. Operational conditions mostly influence other performance factors of structural health. Thus, all these factors are variables, and thus it is not easy to deterministically define the structural health. An easier, deterministic definition of structural health is the current condition or performance of the structure compared to as-built performance, assuming all other variables are the same as when it was built.

Structures are built for a purpose with certain expectations under in-service as well as extreme (limit state) loading conditions. Thus, there are two competing factors in ensuring structural health—the strength at any given time for any given loading or combination of loadings and the loading it is subjected to at any given time. These loadings can be dead loads that can be predicted with high probability, live loads that can be predicted with good probability for a long period of time, environmental loads such as wind and ice that can be predicted reasonably well, and other hazards such as blast and other manmade hazards that cannot be predicted easily. Thus, the health of infrastructure is not simple to define—as it used to be—as it varies depending on the expected level of service for a given combination of hazards and time. On the basis of improved knowledge, performance of system can vary suddenly as structural strength for a given hazard can change significantly if new failure mechanisms were discovered and current mechanism are better understood. Thus, probabilistic approaches are gaining popularity and are being implemented since the last decade.

Another way to define structural health is capacity/demand ratio. As noted above, capacity and demand vary with time (see Chapter 2). When a structure is designed, the expected demands from known hazards, based on prevalent design codes and specifications, are considered to make sure that design capacity is higher than the expected demands. Thus, the capacity is a theoretical value and can be very conservative due to several simplifications used in the analysis. Demand in its life can vary based on several factors such as change in specifications, improved knowledge on loading and hazards, change in network conditions (closure of a bridge can change traffic patterns on another bridge), expectations of users, socio-economic changes, etc. Similarly, capacity also depends on deterioration, improved knowledge, rehabilitation and replacement, deterioration, maintenance, damage, etc. Capacity can also increase due to better evaluation methods such as material testing, improved analysis methods, and test methods such as load testing that can evaluate the system in a realistic fashion. It should be noted that similar to human health, component health can vary significantly from the system health.

As noted before, infrastructure performance/health can be structural or operational, that is, a bridge may have enough capacity to handle the loads but geometry, approach pavement, and other factors can prevent vehicle capacity and mobility. A network performance can be significantly reduced due to one small link or segment unless there is redundancy built into the system.

Thus, evaluating and measuring performance is very important to make sure structural health is maintained through appropriate decision making that include operational decisions, preventive and corrective maintenance actions, and capital program. Operational performance is relatively easy to ensure when compared to structural performance. In most cases, structural health/performance or the degradation effects is measured using a direct physical parameter or an indirect parameter related to the physical parameter under a given loading. Direct parameters are those that can be correlated to structural health directly, whereas indirect parameters (symptoms) are those that give an indication of structural health. For example, strain can be directly related to the flexural capacity of a component or the bridge and thus the influence of degradation on its health. Whereas, measuring the half-cell potentials indicate that the potential to corrosion exists but does not directly indicate the extent of degradation and its effect on the structure. This is analogous to a blood test where certain counts may indicate a presence of cancer, whereas change in pulse or blood pressure may only indicate the symptom of a bigger problem but not the problem itself without further indication. Sometimes, the measurements can be passive: just observing the state of the system. Some times the measurements are active, where actuators stimulate a system response that might give indications of the state of the system (see Figures 1.10 and 1.11).

1.2.3   INFRASTRUCTURE HEALTH IN CIVIL ENGINEERING

1.2.3.1   General Examples

Numerous authors provided examples and projects related to IHCE. For example, DeWolf et al. (2002) presented examples showing the remote control/long-term SHM for different bridges. Another study that is based on the I-40 experiment in New Mexico was performed by Fritzen and Bohle (2000). They based their study on the use of finite element (FE) model of the bridge. They utilized several structural and damage identification techniques on the FE model to localize and then estimate the extent of the damage. They reported high accuracy in achieving their goals. Also Cheung and Naumoski (2002) presented a general application of SHM: a long-span bridge (smart bridge) from grounds up.

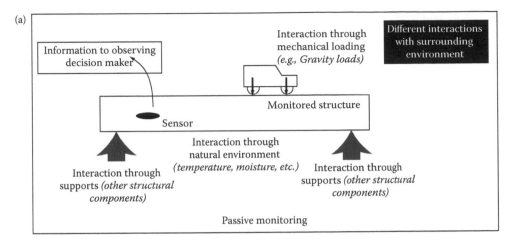

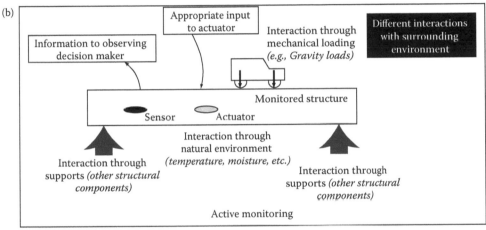

FIGURE 1.10   (a) Passive (b) active SHM experiments.

A different paradigm in sensor technology, a component of IHCE, which integrate the fiber optics sensor properties with the structural behavior, was presented by Glisic and Inaudi (2002). It is called *SOFO* and utilizes a finite-element metaphor to detect overall structural behavior with the local effect of material behavior. In other words, they contrast "long gauge sensors" versus "short range sensors" to detect structural versus material responses. The concept is investigated and generalized further in Chapter 6 of this volume via time (frequency) and space scaling. Wang et al. (2002) offered a general application of SHM to long-span bridges. They described several concepts of SHM that monitor the bridge as a whole. Oshima et al. (2002) identified three major causes for damage to aging bridges in Japan as earthquakes, fatigue, and environmental effects. The authors presented two examples: vibration monitoring and prestressed concrete monitoring.

Sikorsky et al. (2002) linked demand to damage indicators. They enumerated analytical evaluation of stiffness properties. They also offered system identification example. Rochette et al. (2002) summarized several utilizations of SHM techniques and the information that was gained from them. Taha et al. (2002) offered an overview on the use of microelectromechanical systems (MEMS), wireless sensors, and artificial intelligence (AI) in SHM context. Cheng (2004) emphasized the role of data acquisition in SHM projects. Yamada et al. (2004) related gross vehicle weight from measuring the weigh-in-motion vibrations. This is one of the needed methods for estimating demands on bridge structures. Savard and Laflamme (2004) presented generic benefits of SHM. Wenger et al. (2004) offered a general overview of the issues and topics of SHM. Yang and Newhook (2004)

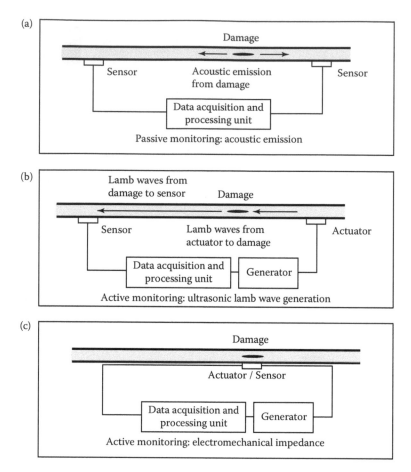

FIGURE 1.11   Examples of passive/active. SHM techniques (a) passive monitoring, (b) active monitoring I, and (c) active monitoring II.

presented the foundations of establishing a general SHM. Note that all these examples cover relatively a small portion of the IHCE and will be explored further in this book.

1.2.3.2   Objectives of IHCE

There are numerous new technologies and construction materials being developed. Obviously, each of those new technologies (materials, design procedures, analytical methods, etc.) would need to be observed more closely than conventional technologies after deployment in the field for general public use. SHM would help in such observations. Consider, for example, the research by Gangone et al. (2005). They deployed an experimental platform that can be expanded into an SHM applications for observing the newly introduced fiber-reinforced polymer (FRP) rebars for use as concrete reinforcement (in lieu of steel rebars). They used long range fiber optics strain sensors that can be attached to FRP rebars to measure strains along the rebar. As it is a long range sensor, it can measure strains over the whole length of the rebar. Since one of the design parameters of FRP rebars is concrete cracks (at a maximum width of 0.01 in, see ACI 2008, it is of importance to ensure that the FRP strains are within such a limit. The proposed usage of long range fiber optics strain sensor can be used to assess strains in the FRP rebars, hence concrete crack width along the whole rebar length.

El-Ragaby et al. (2004) presented an example of using two modern technologies to enhance safety and reduce cost: fiber optics sensing (Fabry-Perot) and glass fiber-reinforced polymer rebars. They reported a whole real-life SHM experimentation on the Cookshire-Eaton Bridge in Canada.

Another integrated project that covers all aspects of IHCE was offered by Shehata et al. (2004). The authors mention that by implementing it, the life of the bridge can be extended, thus saving costs.

1.2.4 THE 95% SOLUTION!

The performance measurement of bridges in the United States and the rest of the world is predominantly through visual inspection. The data from these visual inspections form the basis for most of the evaluation and further decision making. Detailed nondestructive testing and analyses are performed generally on a needed basis to evaluate local or component level data to better understand the structure. Monitoring the structural health on a continuous basis is done only in exceptional cases where such data is justified, generally for complex structural systems or structures with known problems where repairs could be very expensive or are not possible in a very near term and monitoring is done to closely watch the structure. This is very analogous to the way human health system works, as explained in later sections of this volume. In either periodic visual inspection or continuous health monitoring, collection of data is only a minor detail in these authors' opinion, that is, a 5% solution. Collecting the right information (data) in the right way, designed to provide enough information to make appropriate decisions within the time frame required, to make sure that the structural health is continuously maintained is the important issue, that is, termed as the rest of the story (95% solution) yielding to complete solution.

1.3 HISTORY OF STRUCTURAL HEALTH EFFORTS

1.3.1 OVERVIEW

Since the beginning of structures, structural health evaluation and monitoring existed as structures that were always visually examined by hearing to the sounds they made in operation or when tapped on them, and by the touch. Visual inspection probably is the oldest technology used for structural evaluation and has been used for centuries. It is still the most important technology used to ensure the integrity of the structure in fulfilling its mission and has come a long way in the last two decades. Thus, the New York State Department of Transportation describes their bridge visual inspection process as "Sight, Sound, and Touch." When accomplished using an educated and experienced person with commonsense, visual inspection is very important and valuable even with all the other technology used for SHM because finally a person has to make a decision based on data from any source or technology. Even though visual inspection offers advantages such as availability almost everywhere and is easily understood by everyone, it too has several disadvantages. It is subjective, that is, depends on the person and his/her capabilities, experience, and understanding of the structure. Hence, the consistency and quality depends on the entire process, manuals, training etc. Thus, changing the process based on the changing state-of-the-art and practice is very lengthy, cumbersome, and often expensive in big organizations. Reliability of the process (probability of detection) is very hard to quantify.

Nondestructive testing (NTD) is the next phase of evaluating structural health. NDT in some form has always existed as part of or as a first and foremost supplement to visual inspection. Tapping on the structure with a hand or a coin or a hammer is probably the most rudimentary and early nondestructive testing after the visual inspection. This is generally used where visual inspection is not sensitive or when a structure needs better quantifiable evaluation. As illustrated in the later sections of this volume, NDT is mostly referred to local area evaluation, and in recent years there have been several tools that are very reliable, relatively inexpensive, and easily available. These are also well accepted by the industry, owners, and public in recent years. But, most NDT applications still require manual operation, the quality of their results depends on training and experience, their applications require access to the area requiring inspection, and their results are hard to interpret and use. These limitations of NDT increases as the structure gets complicated or thicker components due to attenuation of waves. Automation is also extremely difficult in most cases. Most of these techniques are also material dependent and hence knowledge of the structural material and stress patterns is a must.

The structural health monitoring is commonly referred to sensor-based technologies that measure the structural integrity and are an extension to the common nondestructive instrumentation used for local evaluation. These are generally meant for global evaluation rather than local evaluation and in most cases are treated as continuous or periodic monitoring context. They offer new era in structural evaluation and are still in development phase. These offer to measure not only the issues with damage but also can relate them to structural stresses and strains. They also but have potential to relate them to loads causing them, environmental conditions influence, and operating conditions effect. These systems, thus, are generally referred to permanent or semipermanently attached or embedded sensors used continuously or as needed for monitoring/evaluation of structural health.

Advances in recent sensor technologies and innovation in both information technology and data transmission technologies have potential to offer opportunities that were unimaginable or unthinkable a decade ago. As described in other sections, in more detail, in the future these offer efficiencies in all aspects of managing structures—planning, design, analysis, inspection, maintenance, rehabilitation, and replacement—and thus increased optimization, innovation, and cost savings. As all these activities currently are more reactive or on schedule, in the future all these have potential to be predictive in nature and can lead to just-in-time activities, reducing unscheduled activities to a minimum.

It should be noted that all three, visual inspection, local NDT, and automated SHM, are invaluable and one can not be substituted for the other completely. Even though SHM may indicate problems in the structure they may still have to be followed up with visual inspection and local NDT to select most appropriate option/decision. All these are tools in a decision tool chest and should be used together on a need basis so that they complement each other.

The progression of health evaluation explained above, at first, is seen in human health evaluation: physical examination—visual, sound, and touch by a physician, nondestructive evaluation such as X-ray, followed by periodic or continuous monitoring that are in development. Aerospace industry has evolved in this field significantly, where all these methods have been used very effectively to meet the demands, to minimize downtimes, and to provide just-in-time maintenance schedules. Bridge engineering field still relies on visual inspection supplemented by NDT for local evaluation. See Table 1.1 for changes in bridge inspections nationally and Table 1.2 for some common NDT methods used. In recent days, there has been a push for sensor-based SHM, but there are ways to go due to limited or unknown reliability of SHM, diversity, and complexity of structures, cost-effectiveness, organizational culture, and type of workforce required.

1.3.2 FULFILLING DEFINED OBJECTIVES

Fulfilling the defined objectives is a key mission of IHCE concept. Weinmann and Lewis (2004) illustrated the structural health in civil engineering (IHCE) concept, which started with an objective and fulfilled the objective. The authors presented an example of monitoring carbon fiber composite cables (CFCC) in transverse and longitudinal direction, CFCC strands as mild reinforcement, and CFRP grid reinforcement in top slab of a bridge with 12 double-tree girder. Due to the experimental nature and the lack of design standards, several design assumptions were made during the design phase. Hence, at first, a full-scale prototype girder was lab tested to failure. Lab test results showed that the cracking moment was within 7% of the assumed value; and total ultimate mid-span moment at failure was 60% greater than the corresponding theoretical estimate.

The bridge was instrumented to monitor long-term performance in service. The objectives were well defined and instrumentation was chosen appropriately. The objectives included measuring

TABLE 1.1
Highlights of National Bridge Inspections

Before 1916	Ad hoc based inspections
1916 Act	Inspections mentioned as part of maintenance work by states and others
1967	Prompted by Silver Bridge collapse, president establishes a task force to avoid future disasters
1968	FHWA orders initial inventory and reviews of bridges
	2-year inspection interval for important structures
1971	Establishment of NBIS
	• Defines qualification of personnel
	• Defines inspection types
	• Detailed reporting format
	• Establishment of appraisal, condition, and sufficiency ratings
	• 2-year inspection cycles
1988	NBIS revisions
	• Varied inspection intervals
	• 2-year maximum inspection cycle for fracture critical bridges
	• Underwater bridge inspection requirements
2004	NBIS Revisions
	• Complex bridges requiring specialized procedures
	• QA/QC emphasis
	• Follow-up of critical findings
	• Refresher training

TABLE 1.2
NDT Methods Used in Bridge Engineering

Routine	Visual inspection
	Sounding
	Chain drag
	Ultrasonics
	Magnetic-particle
	Dye-penetrant
	Half-cell potentials
	Geophysical methods
Often	Radiography
	Ground penetrating radar
	Eddy current
	Vibration monitoring
In development	Impact echo
	Thermography
	Remote sensing

pretension load applied to CFRP tendons, concrete strain distributions, and girder camber and forces in post-tensioning CFCC tendons during fabrication/construction sequence. The integrity of longitudinal tendons was monitored during the construction sequence. The load tests indicated that all three spans exhibited similar load distribution behavior and the load distribution behavior was considerably better than what was assumed.

1.3.3 Operations and Management

Catbas et al. (2000) define bridge health monitoring as the measurement of the operating and loading environment and the critical responses of a structure to track and evaluate the symptoms of operational incidents, anomalies, and/or deterioration or damage indicators that may impact operation, serviceability, or safety reliability. It requires various strategies and specialized diagnostic tools, many of which are depicted in Figure 1.12. These strategies depend on at what point of time a facility may be in its lifespan and whether an event or symptom triggers a routine or an in-depth check-up just as in medical health management.

Authors note that effective bridge management requires accurate estimates of the actual load-carrying capacity, system reliability and failure mode rather than just a subjective measure of condition. Thus, integration of both operational and SHM of bridges is important to provide objective, accurate, and sufficiently comprehensive data to serve as a basis for reliable evaluations, projections, and forecasts. This project showed an initial attempt for a major long-span bridge with a focus to examine information management and implementation strategies.

The challenges in structural identification and health monitoring of a major cantilevered through-truss bridge were being explored in Philadelphia (Figure 1.13). The health monitoring strategy is based on various experiments with controlled input as well as continuous monitoring of ambient inputs and responses at the global, element, and localized levels. Triggered high-speed scans of traffic inputs in conjunction with global and member level responses help determine the live-load stress levels. Catbas et al. (2000) also observed that the important challenge for monitoring of large infrastructure systems as the difficulties in the integration and interpretation of data and information.

1.4   NEED FOR INFRASTRUCTURE HEALTH EFFORTS

1.4.1 Overview

The main objective of structural health efforts is to make sure that the structure is safe and secure for its users during the intended use. The second objective is to optimize the level of service and use provided the structure with minimum disruptions, minimize life cycle costs through effective decisions through just-in-time operations, avoid catastrophic failures, improve safety, evaluate new/

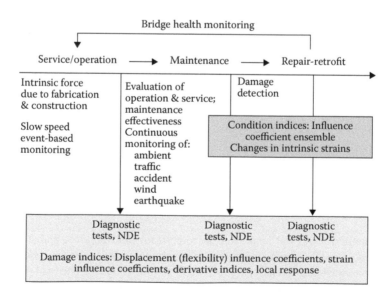

FIGURE 1.12   Infrastructure health monitoring. (With permission from CRC Press.)

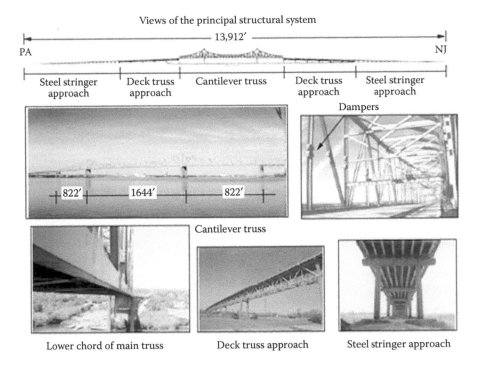

Views of the principal structural system

FIGURE 1.13 Commodore Barry Bridge. (With permission from CRC Press.)

innovative practices/materials, better understand performance, and effects of factors influencing structural/operational performance to continuously make improvements and changes as needed. In simple terms it can be stated that the need for structural health is to make better and effective management decisions. This incorporates all the others mentioned above.

Structural health evaluation, through visual inspections or using sensors and instrumentation, when effectively designed, used, and implemented can lead to improved reliability in expected performance of the structure in cost-effective manner. In the long run, as will be shown and illustrated in this volume, it can lead to system reliability approach than component reliability approach and lead to multihazard, multidisciplinary, performance-based approach. This means unnecessary maintenance, inspection, and other operations, both scheduled and periodic, can be avoided to reduce costs by better planning and execution. This will allow for making appropriate decision at all levels—component level, individual structure level, and network level. This also will lead to improving the state-of-the-art and practice through research, and to bridge the gap between both.

In essence, effective structural health evaluation can also lead to innovations both in technology and operations, can bring effective cultural change in organizations, better quality of life with increased mobility and reliability, and improve communications and coordination between all stakeholders.

1.4.2 SPECIFIC NEEDS

Structural health in civil engineering principles evaluate, periodically or continuously, the present health of the structure and compare it to baseline or expected performance to make appropriate decisions on its future. Some of the needs for IHCE activities, as described above, are given below:

A. Safety improvements
B. Assuring security

C. Improve reliability, as shown in Figure 1.14, both at component and system (network) level
D. Improving decision making on operations
   a. No action
   b. Posting
   c. Decommissioning
   d. Determination of appropriate inspection activities
      i. Inspection interval
      ii. Type and nature of inspection
      iii. Required expertise for inspection
      iv. Estimation of cost for inspection
   e. Appropriate maintenance activity
      i. Type
      ii. Interval
      iii. When and where?
      iv. Cost
   f. Repair or rehabilitation
      i. When?
      ii. Type?
      iii. Cost
      iv. Type of assessment required
   g. Replacement
      i. When?
      ii. What type of structure?
      iii. Cost
      iv. Type of assessment required
E. Research
   a. Durability evaluation
   b. Life cycle costs
   c. In-service performance
   d. Evaluation of new design concepts
   e. Evaluation of new materials
   f. Evaluation of new construction methodologies
F. Evaluation of in-service environment
   a. Hazard characteristics
   b. Vehicular characteristics
   c. Environmental characteristics

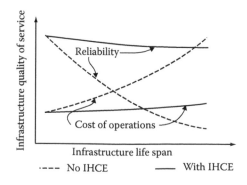

FIGURE 1.14 Qualitative SHM effects on structural reliability.

  G. Bridge management system
     a. Network level analysis
     b. Project level analysis
  H. Risk analysis: real (observed) versus conceptual (theoretical)

1.4.3 Case Studies for Uses of IHCE

1.4.3.1 Cost and Bridge Management

One of the advantages of IHCE is its potential in helping bridge managers reaching efficient decisions. Sazonov et al. (2004) discussed costs of SHM and potential help to bridge management tools. Some of the authors' observations were as follows:

1. Cost of monitoring systems, from literature, varied significantly. For example, Lynch et al. (2006) utilized 60 accelerometers in California at a reported cost of $300,000 (i.e., $5,000/sensor) while the Tsing Ma suspension bridge instrumentation with 600-channels had a reported cost of $16 million (i.e., $27,000/channel).
2. Authors note that the two popular BMS software used, PONTIS 2005 and BRIDGIT (Hawk and Small 1998), attempt to predict the remaining bridge life through generation of life cycle curves, using empirical data inputs. In PONTIS, bridge population is represented on a network level by individual bridge elements with field inspection data providing numerical condition states for each element. A probabilistic Markovian chain prediction model is applied for each element to estimate proportion of each bridge element that is expected to deteriorate in the next inspection cycle. A rank order of element condition states in any inspection cycle leads to an application at the bridge level. The BRIDGIT also applies a Markovian prediction model at the element level, but the optimization model is more bridge specific and thus addresses the element interaction more extensively than PONTIS.
3. On the basis of a comparative study of vibration signal analysis using short-time Fourier transform and Hilbert-Huang transform (HIHT) spectrum, authors note that HIHT is more sensitive to the dynamic energy-frequency distribution and is capable of capturing the difference in the structural response caused by the damage while the short-time Fourier transform indicates the natural frequency changes but fails to capture magnitude changes.

1.4.3.2 Benchmark Bridge Behavior

Four bridges in Connecticut state were instrumented by University of Connecticut, Olund et al. (2006), to monitor the effects of ambient loading on bridge behavior in a network of bridges critical for CT state's infrastructure. A steel multi girder bridge, a segmental post-tensioned box-girder bridge (see Figure 1.15), curved cast in place post-tensioned box-girder bridge, and a steel box-girder bridge were instrumented with variety of sensors with specific initial reasons and long-term monitoring. This data is also being used to create benchmarks for each bridge and then will be used as reference for future monitoring.

1.4.3.3 Inspection and Rehabilitation

Fisk et al. (2008) described a method for inspecting the Jamestown Verrazzano Bridge over Narragansett bay, RI. The bridge features a 4950-ft-long prestressed segmental box-girder main bridge with 23 spans varying in length from 109 to 636 ft, and a 2402-ft-long trestle structure. The bridge was open to traffic in 1992 and nondestructive and destructive methods were used to investigate the post-tensioning ducts for the presence of voids.

Ground penetrating radar (GPR) was first used to locate the centerline of the ducts (Figure 1.16). Then, impact echo technique was used to identify voided tendon ducts by running a four-sensor array just ahead of the impact signal device along the centerline marked out using the

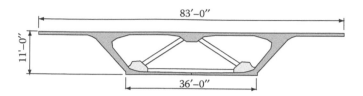

FIGURE 1.15 Typical cross-section of the segmental post-tensioned box-girder bridge. (Reprinted from ASNT Publication.)

FIGURE 1.16 GPR investigation of box bottom slab using 1500 Hz antenna. (Reprinted from ASNT Publication.)

GPR. Where voids were identified by impact echo (Figure 1.17), small diameter holes were carefully drilled through the concrete cover to the surface of the sheet metal duct, which was carefully peeled away to reveal the duct interior. Where a void was confirmed, a bore scope was used to document the size and length of the void, and the present state of the tendon. Several tendon anchorages were investigated with drilling and borescope inspections (Figure 1.18) as these anchorages are located in thicker concrete diaphragms and anchor blocks, GPR and impact echo are ineffective with these components.

Over 93,000 linear ft. of nondestructive impact-echo (sonic/ultrasonic) measurements were taken on the concrete top slab, webs, and bottom slab containing the tendons to evaluate the grouted tendon ducts for voids. Of the approximately 1520 tendon ducts tested, 7.5% or 114 tendon ducts were determined to have voids. Void lengths ranged from 1 ft to over 314 ft. In most cases the tendons were grout covered but some of the tendons were exposed and exhibited corrosion.

1.4.3.4 MONITORING SPECIFIC BRIDGE TYPES

Lloyd et al. (2004) offered some basic details of a monitoring system used, experience gained, and thoughts on how to improve the field based on this experience. The subject bridge is Kishwaukee Bridge in Rockford, IL. It is a five-span segmental post-tensioned concrete box-Girder bridge and

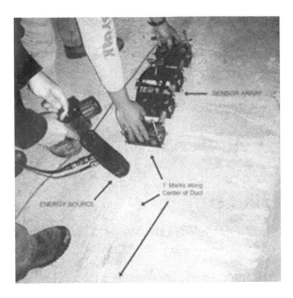

FIGURE 1.17 Impact-echo air void detection using four-sensor array. (Reprinted from ASNT Publication.)

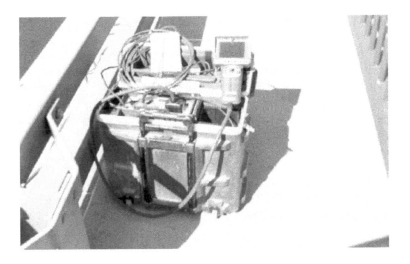

FIGURE 1.18 Borescope with video recorder and monitor. (Reprinted from ASNT Publication.)

showed extensive cracking adjacent to the piers in the web immediately after the erection. Thus the goal was to understand the crack origin and the impact of the cracks on the performance and life of the structure. Modal testing proved to be unsuccessful in understanding the relation between cracks and the structural performance due to their inability to connect to the deformational responses to the loading conditions. But this testing was useful in developing calibrated FEM models.

Monitoring system objectives were to continuously measure several hundreds of sensors to indicate long-term changes that affect load-carrying capacity, collect temperature and traffic loading conditions, statistical and expert system for postprocessing, and provide meaningful display tools. The system included a field data acquisition system (provide for continuous operation, data archiving tasks scheduling, data reduction to minimize data transfer, and allow remote configurability), Integrated Services Digital Network (ISDN) line for data transfer, and an office system (provides long-term archiving, user interface, and report generation capabilities). A bridge engineering interface was provided to access data records with appropriate information (sensor information, data type, structural location, etc.) and statistical measures as needed.

A "confidence methodology" was used for uncertainty analysis of different data collected. This scheme was used to select appropriate warning thresholds.

Analysis of data showed substantial influence of temperatures and showed the value of temperature compensation on the various data collected. On the basis of the experience, authors suggest that compensating for thermo-mechanical effects is important and should be well understood and considered in data-driven modeling techniques, load-carrying capacity methodologies need refining, and recommend development of systematic procedures to verify the response under various scenarios.

1.5   ANALOGY WITH HUMAN HEALTH

Most resources are (were) spent on health-care issues in the United States and probably in the entire world than any single subject. Significant monies are expended for R&D and monitoring to provide new resources to keep human bodies feel better, preventing measures to make sure future health-care costs are lower, and extending the overall life. This is evident if one compares any of the following data for now versus 50 or 20 years ago: the average life span, number of working days lost due to sicknesses, quality of life, and so on. One can learn quite a bit by comparing the efforts from medical/health industry to structural health in civil engineering (Figure 1.19).

Let us review the health industry first and then make some observations.

1. The care for a human being starts as soon as it is conceived in a human body. General medical practitioner (MD) is the one normally consulted first to assure that one is pregnant. As soon as it is confirmed, mothers are advised not to drink and smoke, to eat well, to take prenatal vitamins and other medications, advised to make regular visits to special doctors (gynecologists), etc. If there were any previous complications, they are referred to specialists right away. But, features and future cannot be controlled.

The structure is born in similar ways but probably requires specialists to start with—planners, surveyors, geotechnical engineers, designers, etc. The shape, size, materials, details, aesthetics, features, and so on can be controlled, but future can not be controlled like human future and depends on both controllable (humanistic) and uncontrollable factors (hazards).

2. Once a baby is born, certain data is collected and documented for identification as well as for use as baseline health data for future use as well as for statistical (census) purposes. Mandates require that these be done in certain periods. If any of the data is abnormal, more testing is done. At the same time, some medications/vaccinations are given to avoid serious problems based on probability of these problems during an expected life span. Figure 1.20 shows the analogy between human and bridge testing scopes.

Once the bridge is built, in the same fashion as a baby, the inventory data is collected for identification as well as for future use and statistical purposes. Inspections are performed within 60 days as mandated by National Bridge Inspection Standards (NBIS) to document its current condition. If any issues are noticed corrective actions may be taken. At the same time, some measures may be taken to improve its durability (such as deck sealers, etc.)

3. In the past, most people did not take any preventive medications or go to a physician unless there is a problem. In some cases, this was too late to cure or very expensive. It was realized that taking simple precautions or taking certain preventive medications help in this

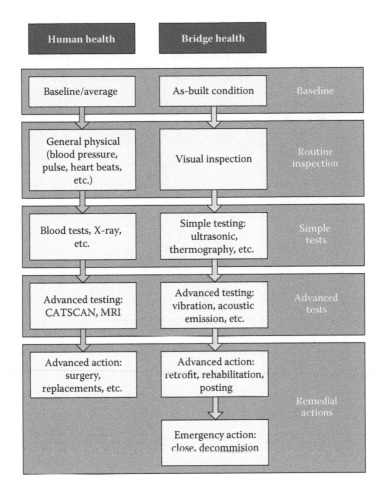

FIGURE 1.19   Analogy of human and bridge healthcare processes.

aspect. It was also realized that having regular physicals is also useful to check on changes in some routine indicators. The physical interval may change based on age, health, type of problems, and other factors.

The same can be said about structural health too. For example, until late sixties there were no national requirements for bridge inventory and inspection formats and intervals until a major bridge failure with casualties occurred. Now there is a realization that, just like human health, structures need inventory and routine inspections to make sure they are safe, did not change drastically from previous physical (inspection), and to minimize surprises (sudden costs). There are efforts under-way to have rational inspection intervals for structures too based on their age, condition, and so on. Figure 1.21 shows the age of bridges in New York State.

4. During a routine physical, doctor does a thorough physical examination, checks for some vital signs (such as BP, pulse, etc.), provides certain immunizations requiring periodic shots or booster shots, and conduct routine blood tests.

The same can be said to structures too. If routine inspection reveals issues, further actions are taken depending on the type of issue, seriousness of the issue, extent of it, and its effect on the person's life and quality of life. On the basis of an inspection report and simple tests (such as magnetic-particle

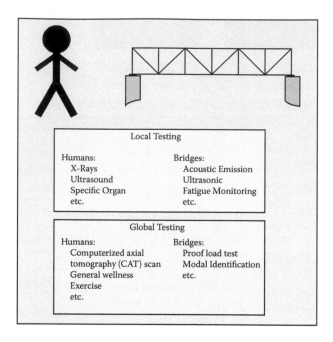

FIGURE 1.20 Analogy of human and bridge testing scopes.

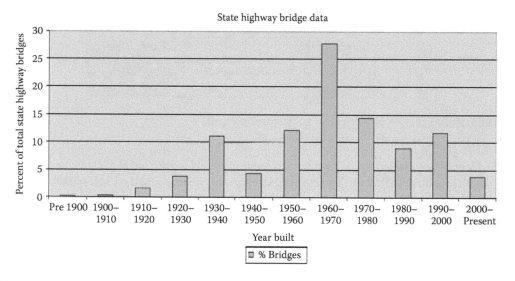

FIGURE 1.21 Age of bridges in New York State. (Courtesy of New York State Department of Transportation.)

test, sounding, or chain drag), certain recommendations are made. These include cleaning, deck sealing, joint fixing, removing loose concrete, etc. More interim inspections also may be scheduled.

5. If physical examination reveals any problems, physician may conduct some more testing or order further nondestructive testing such as X-rays, and CAT scans. Depending on diagnosis of specialists who can interpret these, further analysis or examinations may be scheduled.

If inspector and simple testing can not reveal the problem or extent, then more serious testing may be recommended. These include ultrasound testing or thermography, detailed analysis or load rating based on in-depth inspection, etc. This is an area where more work is probably required, that is, not many NDT methods are available to detect the extent of problem easily in structures and more progress probably is needed.

6. If these tests and simple trail-and-error solutions do not work, then serious testing and serious actions (such as scheduling a surgery) may be recommended.

If the structure requires extensive work, rehabilitation may be done. If not, a new structure is born. This is one area, where the structure life may be terminated due to cost–benefit analysis, whereas in human aspects cost is not an issue to save a life (in most cases).

7. With all the knowledge gained from and number of humans involved in the history and all the care taken and money spent, people do die due to natural and man made hazards. The same happens to structures too. Even though, the requirements for extending on life and quality of life are known, based on individual economy and several competing priorities, people take decisions that may not be in their best interest or have to make a choices where short-term issues or more important. Same apply to structures and their owners, that is, the choices are made based on resources at hand and hence solutions taken may not be based on long-term basis. Just like physicians knowledge and experience dictate interpretation based on several key factors affecting human health, the bridge owner's experience, and so on play a major role in a structure's life and durability. Mistakes can bring premature termination of the structure.

8. Decisions in health-care fields normally affect a single individual with effects affecting a small group of people, that is, in most cases his/her immediate family. But, in the cases of a structure, the decisions made can have far reaching effects, including affecting few people' lives to quality of life of an entire community.

9. Redundancy is another factor that differs in structure versus human life. There is little redundancy in human life/components in most cases. Any redundancy is fixed and can not be changed, that is, a second heart or a third eye. Design of human life is not in our hands; where as redundancy of a structure can be controlled. This gives a big edge to owners for future maintenance and in controlling the structure life, safety, and durability.

10. We finally note that a healthy bridge offers pleasant esthetics, in addition to the optimal safety and functionality (see Figure 1.22). In the long range, healthy bridges, and healthy infrastructure in general also will operate at an optimal life cycle cost.

1.6 STRUCTURAL HEALTH: A NEW FIELD IN CIVIL ENGINEERING?

1.6.1 WHAT CONSTITUTE A NEW FIELD?

Most civil engineering activities are reactive in nature. Even though the life span of the structure is considered during the design, at any given time, most actions are considered based on current structural and loading conditions. The structural health concept considers the time dimension and hence makes the whole field dynamic. Thus, conceptually, if fully implemented SHCE concept considers past history, present condition, and expected future condition. Hence, it gives the complete story and hence makes it a new field.

Thus, as illustrated earlier and later in this volume, SHCE concept is purely a decision making tool based on required level of service to be maintained in its life span. Thus, it has all aspects of the structure during its life built into it—hazards, performance, damage, capacity evaluation, remaining life, inspection, and preventive and corrective actions. Simply, it is a new and efficient way to make all these engineering functions making use of concepts such as risk management, benefit-cost (value) analysis, and evaluation. Let us consider couple of examples here.

FIGURE 1.22 A healthy bridge offers pleasant esthetics as well as optimal safety and functionality: Court Street Bridge over the Susquehanna River, Owego NY. (Courtesy of New York State Department of Transportation.)

Currently, the bridges are designed for specific combination of loading based on specifications. The capacity of the structure thus is not optimized for any hazard it faces. Ideally, using SHCE concept implementation, in the future, one can estimate the hazards it is expected to face, estimate probability of failure and consequence, and design the structure effectively to provide appropriate capacity. This will yield uniform reliability and the better understanding gained in this phase can be used to develop data needed to collect in the future and thus, can develop appropriate inspection procedures, and required actions.

Using the SHCE concept, the entire NDE process can be improved based on what decisions have to be made in the future. Thus, true understanding of the structure, its operating environment, and loading has to be taken into account in developing the following:

- What decisions has to be made?
- What data is required to answer these questions?
- Is NDE required?
- What methods (for local or global testing and analysis) are required?
- What kind of sensors and equipment are required?
- Where do these sensors go?
- How do these get integrated into structure/network?
- How and when the data is recorded?
- How is data transmitted?
- How and when data is analyzed?
- Processing required at structure and decisions to be taken?
- Do all the above provide answers to the questions in hand?
- Are the resources available to make the above happen?
- If the answer to the above two questions is "yes," then go to next step. If not, the whole process should not be done or reformulated.
- What is the value or benefit?
- Is it worth the cost? If the answer is "yes," then implement the evaluation plan.

This data from the evaluation system coupled with other knowledge base can be used to effectively maintain and operate the structure with the evaluation system employed. The data from individual

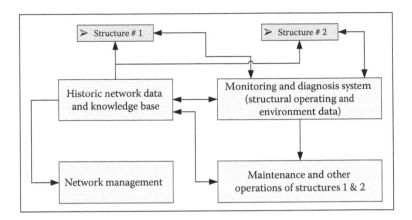

FIGURE 1.23 Network management.

structures can be generalized with historic knowledge base and expert systems to optimize the entire network management (see Figure 1.23).

By effectively implementing this at individual component/structure level, this can be generalized to and can be used for entire network as data is collected and can be used for future planning—that is, go toward predictive and proactive civil engineering than the reactive engineering. Thus, the SHCE is very different, that is, it yields a multihazard multidisciplinary, performance-based engineering centered toward constant reliability, known risk, and value-based decision making process by combining other relevant fields with civil engineering.

1.7 STRUCTURAL HEALTH: INTEGRATION OF FIELDS

As illustrated in the previous section, the structural health concept is a closed loop or an active concept, that is, it starts and ends with decision making while integrating all aspects of structural life and the environment it is part of. Thus, it is a new field in civil engineering with integration of almost all fields and all disciplines. This section briefly mentions all these fields.

Most structures are built and stay in relatively harsh environment not shielded from natural hazards (such as rain, snow, earthquakes, and wind) as well as man made hazards (such as blast). Thus, before one can design and maintain the structures, there is a need to understand the magnitude of hazards and how they behave. Even global warming needs consideration as loads expected on structures decades ago may be a lot different now and in the future. For example, temperature differences and flood loads may need reconsideration periodically. Hence, environmental engineering, risk engineering, seismic engineering, wind engineering, and blast engineering are a few that are needed to supply the data required on the environment structures are built. While lot of this data can be obtained in the literature, IHCE offers a means to verify the models used by collecting the data at the structural sites and provide a feedback to these experts to refine the models or to adopt better data for future designs and maintenance. Not only for the data but also for the analysis of structures, very good understanding of the hazard mechanisms is important to conduct optimum design any analysis. This is discussed further in multihazards sections later in this volume and in Ettouney and Alampalli (2012).

Data collection is mentioned above. Data collection again is an integral part of the SHCE. Collection of data requires understanding of the behavior of structures at system and component level. Determining system versus component performance needs require good understanding of the structural performance, deterioration aspects, etc. Material behavior is also very important to make sure appropriate sensors are used and inspection methods employed are appropriate. Data needs to be collected, transmitted, analyzed, and appropriate decisions are taken. Thus, materials engineering, mechanical engineering, sensor and electrical engineering, information technology, and decision sciences are some of the few disciplines required in this aspect. Innovations in one field can affect the entire system and need to be accounted for and communicated well. One of the authors had experience where a change in equipment in a telephone company made an entire monitoring system useless for 6 months. Thus, good project management skills also go a long way and that is again another discipline.

Construction of structures of course requires several aspects. Regulating agencies have to be dealt with, where changing role of government identities affect regulations and practices. Several industries and specialties are needed for good construction where quality control and assurance are very important. Manufacturing engineering, testing for quality control (QC)/quality assurance (QA) during fabrication and on-site construction, safety specialists, hauling industry, and traffic engineering disciplines are few that should be considered for moving toward just-in-time approach that is the goal of SHCE.

Once the structure is constructed, inspection and maintenance is the key for durability and several of the disciplines mentioned before and other expertises are required.

Resources are a big factor in any project and management. Due to the advent of internet and other innovations, the entire world has became relatively small and understanding of global economy and its implications on funding is very important for planning and other aspects. The increase in cost of construction materials due to recent Olympics in China is a good example to illustrate this fact. Since IHCE concept is based on benefit-cost aspects, finance engineering is a critical discipline in the success of this concept.

Customer expectations and demands drive the need for structures and their performance. Hence understanding the customers—demographics, urban/rural differences, cultures—and factors influencing them at present and future are very important and needs to be included. For example, a decade ago road users were happy to have a means to get to one place to other, where as now they expect reliable travel times. After the 9/11 incident, security became another aspect to be considered and has implications on all aspects of civil/bridge engineering.

A knowledgeable, competent, and adoptable workforce is very important for the well-being of future infrastructure. Thus, education and training industry along with human resource management fields cannot be forgotten. Thus, the SHCE is not a single field but integration of almost all the fields available there. For effective bridge/structural management, all these fields should work in harmony to meet the level of service expected from the bridges already built and that will be built tomorrow under SHCE umbrella.

1.8   HORIZONTAL INTEGRATION: MULTIHAZARDS

1.8.1   INTRODUCTION TO MULTIHAZARDS

1.8.1.1   General

The main responsibility of professional engineers is to provide safety of the customers/public utilizing the infrastructure. Other factors governing the designs include constructability, serviceability, durability, maintainability, inspectability, economy, and aesthetics. Most of the designs are done using the standard specifications or the guidance provided by the regulating agencies, which state, in most cases, the minimum requirements to ensure the safety of the infrastructure users. They also provide guidelines to consider the extreme events and dictate which hazards should be considered as part of the routine design along with the return periods, which may be significantly greater than the design life. In these cases the structures may be designed for survival alone or to meet different performance levels based on the importance of the structure or its use in post-hazard situations. Thus, structures in general are designed to ensure strength and stability, see Timoshenko and Gere (1961), to resist the significant load combinations they are expected to experience during their service/design life, and then checked for their survival under extreme events.

Engineers routinely deal with natural hazards such as earthquakes, floods, wind, and ice. At the same time, certain man made hazards, such as blast and impact loading, are also becoming very important and were not considered seriously until recently. The severity of these hazards can significantly increase the costs of construction and maintenance, especially if they are considered for existing infrastructure requiring rehabilitation. In some cases, some of these severe hazards have conflicting demands from the system, with the resulting increase in costs being even higher.

The primary reasons for not utilizing a comprehensive analysis under multihazard environment include (a) some of the hazards were never anticipated or demand on systems due to these hazards were not considered severe, and (b) lack of adequate computational tools and databases. Recent improvements in computational tools, database population (that include behavior of systems under several severe hazards), and SHM, all contributed to the increased consideration of multihazard approach to systems.

Multihazard analysis and design is gaining popularity in the nation. This approach considers increasing complexity of the structural systems to meet the demands of the current environment and takes advantage of the recent developments and innovations in computing, analytical, and sensing technologies. This concept requires engineers to understand the characteristics of various hazards, their interactions, and the structural response under various hazards, so that appropriate analysis can be conducted and structures can be designed to meet the expected performance to each of the hazards considered. This understanding is very important not only for design purposes but also for planning purposes, as these can affect the site selection, construction process, and future inspection requirements.

Finally, it should be noted that there are two very distinct types of multihazard considerations in the field of infrastructure: development and execution of emergency plans at local, state, and national level; and engineering and economic considerations. This section deals with the latter type, that is, engineering and economic considerations.

1.8.1.2   What Is Not Multihazards

Conventionally, infrastructure health professionals used to consider the effects of hazards on infrastructure in a serial fashion: a one hazard at a time. The potential of more than one hazard affecting the structure simultaneously is accommodated through load factors and load combinations equations. We immediately point out that this is not a multihazards consideration. This conventional method is a serial one-hazard-at-a-time approach. Multihazards considerations accommodate the effects of hazards, when they occur, on the structural response to other hazards. Several examples will explore this concept later in this section.

1.8.2   Theory of Multihazards

As noted earlier, a multihazard outlook to systems such as buildings and bridges would provide great benefits, but such an outlook is hindered by the lack of a *perspective that presents a concise and systematic view of the subject matter.* Upon reflection on how physical structures behave, we then propose the following theory of multihazards:

> *For a given system that is exposed to multihazards, there exists an inherent multihazards resiliency within the system. This multihazards resiliency implies an interrelationship between the manners that the system responds to different hazards.*

The theory, as stated above, would help in giving a perspective, which is concise and systematic. In the next section, we explore some of the ramifications of the multihazards theory. In all, we find that by adopting the perspective of the theory, we can produce safer and more economical structures.

The theory of multihazards offers two doctrines for systems: *inherent* resiliency for all hazards and the *interrelationships* between the manners in which the system responds to those hazards. By utilizing these two doctrines, the potential benefits, as illustrated earlier, would be achieved during structural analysis, design, life cycle cost analysis, risk analysis, and SHM. In what follows, we investigate each of these fields and how multihazards considerations can be used to improve accuracy, safety, and cost reduction.

1.8.3 MULTIHAZARDS CATEGORIES AND STRUCTURAL HEALTH

Our definition of hazards can be man made or natural. They can be either load-induced (earthquakes, wind, flood, collision, etc.) or durability-induced (corrosion, fatigue, freeze-thaw, wear and tear, etc.). As such, we offer that there are numerous ways to categorize those hazards as shown in Figure 1.24. The main premise of this section, and the multihazards consideration concept itself, is that there are interactions between hazards in almost every component that are shown in Figure 1.24. The only interaction that is currently employed in conventional designs is the likelihood interaction. The likelihood interaction is the probability that hazards can occur simultaneously, or as a consequence of each other. This type of hazard interaction is well studies, and will not be considered in this section.

The just-mentioned theory of multihazards deals with interaction between hazards through the structure: that is the interaction through structural vulnerabilities. This type of multihazards interaction is discussed in several parts of this volume: for example, multihazards considerations of seismic and blast hazards. Hazards also interrelate to each other through their consequences. This type of interaction is also the same as the interaction of hazards through hazard management components. For example: if a decision is made to mitigate against corrosion deterioration by using FRP wrapping solution, this decision could affect the response of the system to other hazards such as earthquakes.

Interactions of hazards can also be detected by studying how hazards are defined and how they affect the infrastructure. Any hazard affects the system in a three-parameter space as shown in Figure 1.25. The time (or frequency) distribution of the hazard is one of those parameters. The other parameters are the spatial distribution of the hazard and the amplitude of the hazard.

Figure 1.26 shows a qualitative frequency-amplitude distribution of some conventional hazards that affect infrastructure. Whenever the hazards range of effect intersects with another hazard, we expect that the two hazards would interact through the system. If the hazard zones do not intersect

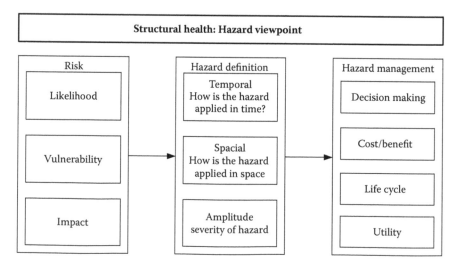

FIGURE 1.24   Elements of hazards.

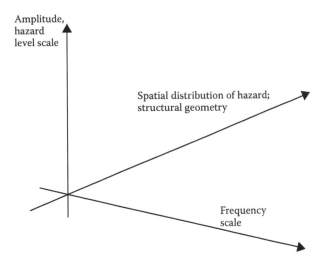

FIGURE 1.25   Three-parameter definition of hazards.

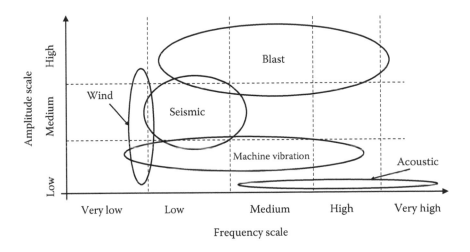

FIGURE 1.26   Frequency-amplitude distributions of hazards.

in that space, we expect the two hazards to act independently. Chapter 4 presents the spatial distribution of some popular hazards.

Figure 1.27 shows some important infrastructure hazards and a qualitative measure of the importance of their respective interactions.

1.8.4   Multihazards and Infrastructure Health

1.8.4.1   General

According to the theory of multihazards, all hazards interact through the infrastructure. These interrelationships between the hazards would require attention; it means that an action relating to one hazard can have an effect on how the infrastructure responds to other hazards. Thus, multihazards considerations are as important to infrastructure health as conventional individual hazards considerations. Tables 1.3 through 1.11 show some examples of how hazards might interact through the infrastructure.

Hazard matrix: interrelation

	Fatigue	Corrosion	Scour	Fire	Wind	Earthquakes	Impact	Overload	Wear & tear	Gravity
Fatigue	NA	H	L	M	H	H	M	M	H	H
Corrosion		NA	M	L	L	L	M	H	H	H
Scour			NA	L	L	H	H	H	H	H
Fire				NA	H	H	M	L	L	L
Wind					NA	L	M	L	L	L
Earthquakes						NA	M	M	L	L
Impact							NA	H	M	M
Overload								NA	H	H
Wear & tear									NA	H
Gravity										NA

FIGURE 1.27   Qualitative hazards interaction table.

1.8.4.2   Structural Health Monitoring

Perhaps a less obvious use of the multihazard theory of this work is the SHM field. However, recall that the theory considers that a given system comprises numerous multihazard resiliencies that are interconnected and interdependent, and that ignoring these interconnections and interdependencies might lead to erroneous consequences. It is clear that the SHM field would have several utilizations to the multihazard theory. In this volume, we present several of these utilizations such as optimum sensors locations (OSL), duality principle, and serendipity concept.

1.8.5   BENEFITS VERSUS LIMITATIONS

Given below are some of the benefits of multihazards considerations:

1. Potential for an economic design and construction.
2. A more accurate estimation of inherent resiliency of the system.
3. A more accurate treatment/estimation of life cycle cost of the system.
4. A more accurate analysis of the system
5. The multihazards design philosophy of a particular infrastructure project may affect the entire network on infrastructure systems. For example, multihazards design of an important highway bridge may improve the reliability of the entire transportation network of the region that the highway bridge serves.
6. Optimization of the SHM to increase experimental efficiency.

Such benefits seem to be fairly obvious, yet there is one limitation that must be considered; it is the lack of a concise set of tools that can accommodate multihazards considerations. This results

TABLE 1.3
Multihazards Table: Interaction with Fatigue

Corrosion	Scour	Fire	Wind	Seismic	Impact	Overload	Wear and Tear	Gravity
Stress corrosion and hydrogen embitterment (in Chapter 4 by Ettouney and Alampalli [2012]) can both have an effect on fatigue behavior of high-strength cables	Fatigue of chemical gels (in Chapter 1 by Ettouney and Alampalli [2012]) has a direct effect on scour	Response of systems to only severe fires can be affected by reduced fatigue capacity of systems	Wind-induced cyclic higher stresses have direct effects on fatigue	Seismic-induced cyclic higher stresses have direct effects on fatigue	Response of systems to demands of high impact amplitudes can be affected by reduced fatigue capacity of systems	Repeated overloading-induced cyclic higher stresses have direct effects on fatigue	Repeated environmental demands' (thermal, freeze-thaw, etc.) induced cyclic stresses have direct effects on fatigue	Repeated live loads-induced cyclic higher stresses have direct effects on fatigue

TABLE 1.4
Multihazards Table: Interaction with Corrosion

Scour	Fire	Wind	Seismic	Impact	Overload	Wear and Tear	Gravity
Corrosion in submerged foundation can be affected by soil erosion that might happen due to scouring	Limited interactions might occur between corrosion and fire	Loss of strength or ductility (in high-strength cables due to stress corrosion or hydrogen embitterment) can affect long-term resistance to high wind	Loss of strength or ductility (in high-strength cables due to stress corrosion or hydrogen embitterment) can affect long-term resistance to seismic effects	Loss of strength or ductility (in high-strength cables due to stress corrosion or hydrogen embitterment) can affect long-term resistance to impact	Loss of strength or ductility (in high-strength cables due to stress corrosion or hydrogen embitterment) can affect long-term resistance to overload	There is a direct interrelation between corrosion damage and wear and tear (in Chapter 4 by Ettouney and Alampalli 2012)	Loss of strength or ductility (in high-strength cables due to stress corrosion or hydrogen embitterment) can affect long-term resistance to gravity

TABLE 1.5
Multihazards Table: Interaction with Scour

Fire	Wind	Seismic	Impact	Overload	Wear and Tear	Gravity
Limited interactions might occur between scour and fire	Loss of foundation support due to scour might have an effect on global bridge wind resistance	Loss of foundation support due to scour might have an effect on global bridge seismic resistance	Loss of foundation support due to scour might have an effect on global bridge impact resistance	Loss of foundation support due to scour might have an effect on global bridge overload resistance	Lost of foundation support due to scour might accelerate deterioration of affected foundations	Loss of foundation support due to scour might have an effect on global bridge gravity load resistance

TABLE 1.6
Multihazards Table: Interaction with Fire

Wind	Seismic	Impact	Overload	Wear and Tear	Gravity
Fire damage, if not retrofitted properly, can reduce resistance to high wind	Fire damage, if not retrofitted properly, can reduce resistance to earthquake effects	Fire damage, if not retrofitted properly, can reduce resistance to impact hazard	Fire damage, if not retrofitted properly, can reduce resistance to overloads	Fire damage, if not retrofitted properly, can accelerate wear and tear deterioration	Fire damage, if not retrofitted properly, can reduce resistance to gravity loads

TABLE 1.7
Multihazards Table: Interaction with Wind

Seismic	Impact	Overload	Wear and Tear	Gravity
Wind and seismic loads has many similarities (including the global nature of both loads). Thus there is high degree of intersection between the two hazards	Local nature of impact loads makes their interaction with wind loads less likely. On other hand, damages of impact loads, if not treated properly, can reduce wind resistance	Differences in temporal and spatial distributions of wind and overload make interaction less likely	Wear and tear losses can affect wind resistance	There are interactions between gravity loads and wind loads

TABLE 1.8
Multihazards Table: Interaction with Seismic

Impact	Overload	Wear and Tear	Gravity
Local nature of impact loads makes their interaction with seismic loads less likely. On other hand, damages of impact loads, if not treated properly, can reduce seismic resistance	Differences in temporal and spatial distributions of seismic and overload make interaction less likely	Wear and tear losses can affect seismic resistance	There are interactions between gravity loads and seismic loads

TABLE 1.9
Multihazards Table: Interaction with Impact

Overload	Wear and Tear	Gravity
Impact effects can reduce resistance to overloads	Wear and tear losses can affect impact resistance	There are interactions between gravity loads and impact effects

TABLE 1.10
Multihazards Table: Interaction with Overload

Wear and Tear	Gravity
Wear and tear losses can affect overload resistance	There are interactions between gravity loads and overload

TABLE 1.11
Multihazards Table: Interaction with Wear and Tear

Gravity
There are interactions between gravity loads and wear and tear

in spotty applications to an otherwise increasingly important subject. Thus, even though the infrastructure community realizes this value, there are no objective roadmaps to tap on the immense potential of the multihazards philosophy.

1.9 VERTICAL INTEGRATION: MULTIDISCIPLINARY

1.9.1 DIFFERENT DISCIPLINES AND SHCE

As noted in previous sections, infrastructure or structural health in civil engineering is not simply structural monitoring but is an integrated field where it is a means to make appropriate decision from design to manufacturing to maintaining to replacement, that is, complete structural cycle to assure safety and intended level of service in a cost-effective manner. In essence, the idea is to make the entire infrastructure network smarter—that includes making structures as well as materials they are made of are smarter and intelligent. As noted, in previous section, this involves several disciplines, from several fields, dependent on each others.

These opportunities also bring challenges in making sure all these personnel and technologies works together seamlessly to improve the performance measures and decision making process to achieve increased safety at lowest costs. This may need change in organizational culture, change in thinking, and change in practices and work flow. Figure 1.28 gives a broader pictorial representation of the multidisciplinary nature of the SHCE. The broader fields include engineering, decision makers, businesses, and others.

1.9.2 TREATMENT OF SHCE MULTIDISCIPLINARY NEEDS IN INFRASTRUCTURE HEALTH ENVIRONMENT

Infrastructure process starts with identifying a need by planners, who work with network data and the level of service needed at global level. They work with finances available, using some

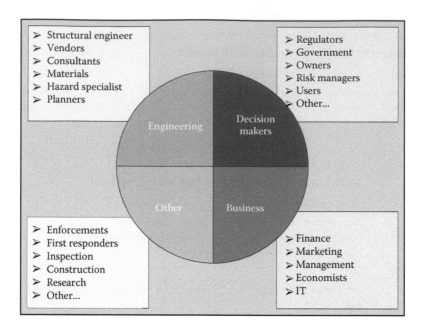

FIGURE 1.28 Interaction of multidisciplines during an SHM effort.

preliminary engineering estimates derived based on previous experience. Hence, effective SHCE is very important to provide these data that are needed to make a decision on to develop or expand or remove a new highway or a structure, replacement/rehabilitation/maintenance options, do nothing, etc. based on current and expected performance. Once several options based on social, economical, political, and engineering aspects are considered, a decision to continue with detailed studies is made. Environmental and engineering aspects based on regulations, codes, loads, and hazards the structure may face during its lifespan will shape the behavior of the structure or network along with financial and other factors. The detailed studies phase needs hazard specialist, risk evaluators, engineers, material personnel, etc. As the project continues with the construction phase, it requires quality control aspects that itself is a combination of personnel from several disciplines. After the construction, come inspection and maintenance, and eventually planning. In case of an unfortunate incident, the network or system level of service depends on informed users, enforcement, and first responders. The economy also dictates to the user needs and the finances available to all facets of this field. This shows the multidisciplinary needs in infrastructure health environment.

Effectiveness of the process depends on how effective the communication is and what data is available at every step of the process as a decision has to be made. Thus, effective SHCE process can assist this decision making process by completing the cycle, that is, start with what is an issue and end with making an optimal decision to solve the issue.

1.10   SHM AND IHCE: A QUICK PRIMER

We argued earlier that there is an analogy between human and infrastructure health. One of the major components of health of both is the need for diligent monitoring. In case of infrastructure, this means the need to have a reasonable knowledge of the status of the infrastructure. Infrastructure status would degrade as a function of passage of time and/or exposure to different hazards. In many situations, the signs of such degradations may not be visible. In addition, the functional and performance demands from infrastructure change with time. Again, in many situations those demands may need monitoring (e.g., the truck loads on roadways or bridges). The *knowledge* of the status of the infrastructure as a result of *monitoring* is generally referred to as SHM. SHM would result

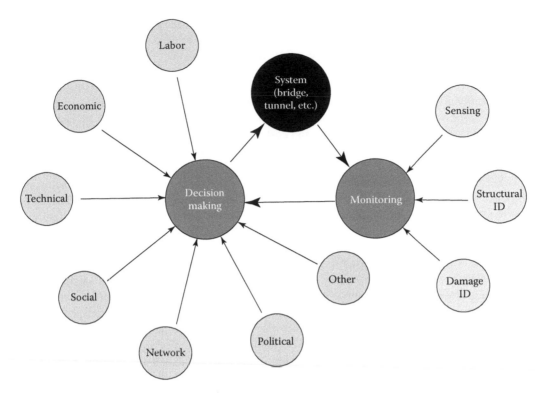

FIGURE 1.29 Combining monitoring and decision making to form IHCE/SHCE.

in estimates of structural properties such as stiffness, strength, mass, damping, or ductility. These general properties include, but are not limited to, the knowledge of damages that can affect structural performance. For our current purpose, we define damage as the degradation of a structural property from its intended state. Damages can be fatigue cracks, corrosion loss of mass, inelastic performances, or scour holes near bridge foundation. Until recently, the knowledge of structural state has been obtained on a qualitative basis. The advent of modern sensing and NDT techniques is changing the methods of monitoring structural states into more quantitative methods.

Clearly, the knowledge of structural states is only part of maintaining the health of the infrastructure. The remaining question is: what to do with such knowledge? There are many possibilities that range from doing nothing to replacing the infrastructure completely. Factors that can affect decisions are numerous. These factors can be economical, labor, social, technological, and/or political. Among other factors that can affect decisions are infrastructure network and uncertainties involved in all of these factors. After a decision is made, such a decision will affect the state of the infrastructure on hand (a null decision can also have an effect on the infrastructure). When the process is combined with SHM, we will call it SHCE/IHCE. Figure 1.29 shows the combined process.

We explore this process and the interrelationships between its components in full detail in Chapter 2. The remainder of the volume will discuss many of the complex components of SHCE.

1.11 CHALLENGES AHEAD

1.11.1 GENERAL

There are several advantages in using infrastructure health in civil engineering concepts. But, to move forward with more proactive bridge management concepts for effective decision making, there are several technological, organizational, and cultural challenges ahead.

The major challenge will be the cultural and organizational changes required in the structural engineering organizations. Most owners of structures such as bridges are public entities with no profit motivation, rightfully so, to assure safety first. These owners include state transportation agencies with better resources, experience, and expertise to small local owners, owning one or two bridges with very limited resources and expertise. Thus, type of structures built, maintenance, and other operational decision taken vary considerably. Hence, standardization and implementation of advanced processes is not only difficult but may not be affordable unless it can be justified in a rational and systematic fashion using risk and cost–benefit analyses.

The second big challenge is the diversity of infrastructure and people involved with building and maintaining them. They are in general unique as they are designed and constructed to suit the physical field settings and also to make sure mobility of the network is not affected. Manufacturing of components and field construction is not limited to very few companies around the world as the case for aero space industry, as literally thousands of organizations are involved with infrastructure (say bridges, for example) construction. Thus, standards and regulations are developed conservatively in most aspects to account for the variability associated with this fact. Most standards are prescriptive and thus are not performance based. For example, concrete mix specifications are specified by an owner in such a way that the minimum concrete strength is maintained (say 3000 psi). Thus, the real concrete strength could be anywhere from 3000 psi after 24 hours of placement to as much as 8000 psi or more. Thus, for most designs nominal conservative values of material properties are used that in most cases works well and absolves reactive based management procedures. Given, the resources are not dependent on the needs, this system works well. For a proactive and just-in-time management approach, there is a need for resources that can be planned and controlled well for the life of the network. SHCE advocates should be able to articulate this well, for making progress, by convincing not only the bridge owners but also the regulators and funding sources.

As noted, there are several stakeholders involved with SHCE that will be dealt in next sections. Understanding each others and what other stakeholders expect, and having a common understanding is another big challenge. There has also been overselling the automated SHM by several proponents of these concepts in the past and falling short on deliverable. Through effective communications, these should be avoided to make progress. Realistic goals should be set with common understanding with all stake holders involved to moving forward with these concepts. All should realize that automated SHM concepts are complementary to existing tools and optimizing the use of all tools based on the decisions to be made in a cost-effective manner should be emphasized and understood by all the stakeholders. Risk and cost–benefit analyses may seem like rudimentary concepts but are not routinely used in change management when people try to push the SHM technology to owners. These should be emphasized more along with the value of these advances more and forms the fundamental reason for SHCE.

Besides these challenges, there are several other technological challenges that have to be dealt with. Unlike, aerospace industry, the civil structures are spread around geographically and cannot be taken to a site where maintenance and other operations can be performed. This may seem simple but poses several logistical and technological challenges. Logistically, since these structures are not reproduced in mass like aeroplanes, one should have specific understanding of not only the material behavior but also the global and local structural behavior before appropriate proactive activities can be performed. There may not be power or telephone or internet connectivity at many of these sites; accessibility maybe an issue due to the required work zone control, disruptions it may cause to users, or due to the features under or over the structure. This limits type of technology and resources that can be used. Hence, for example, there is a big push for passive technology. There is also a need for portability and other aspects for sensors and other equipment specifications. Replacement or rehabilitation of components is also not as simple as in aeronautical fields unless future design and construction practices change significantly as we move forward with IHCE concepts.

Another major challenge is the type of workforce required and thus the change in educational institutions. At present, most civil engineers leaving the 4-year colleges are only aware of design aspects and not other areas of civil engineering. But, in reality, even though the structural understanding is a must, very small percentage of graduates will work in design areas. All other areas such as maintenance, inspection, analysis, planning, management, instrumentation, and other new technologies are not emphasized or made aware in the colleges. For effective progress, a radical change in the entire educational system is needed and should be well thought. All stakeholders should pay considerable attention to this aspect to make good progress in positive direction. Owners also should consider and change their hiring practices to accommodate the current state of the field. For example, in most transportation departments, there are no engineers specialized in corrosion even though corrosion is a major contributor to most problems faced by the bridges. In most cases, the civil engineers with no background in corrosion principles end up solving corrosion problems solely because they have some experience in dealing with it as part of their job. The same can be said about maintenance, research, constructions, risk analysis, and other aspects too. Even many organizations that have significant authority on the practice and specifications select their members based on the position irrespective of the knowledge and experience required to make such decisions.

In addition to all above challenges, there are other global challenges that can affect infrastructure health in numerous ways. For example, in the field of energy, energy consumption of (a) structural health technologies, and (b) manufacturing of construction materials, and construction methods need to be optimized and addressed. The sustainability issues of infrastructure are of major concern. Lately, efforts of making infrastructure green and environmentally friendly have just started. Another major challenge is infrastructure security. Security has emerged lately as a concern to infrastructure decision makers. This subject is discussed in more detail in Chapter 11 by Ettouney and Alampalli (2012).

The above are a few challenges to consider as we move forward with proactive multihazard performance-based bridge management philosophy using SHCE concepts. But in the authors' opinion, all these can be achieved with improved communications between all stakeholders. Infrastructure owners have a key role to play in this by taking the lead and working with all other stake holders in an open format.

In the remainder of this section we discuss two major challenges to infrastructure health field: manual versus automatic inspection and climate change.

1.11.2 MANUAL VERSUS AUTOMATIC INSPECTION

Manual inspection of infrastructure has been the mainstay of inspection processes since humans started building infrastructure. As sensing and other technological methods emerge, there has been slow utilizations of automatic inspection. These utilizations will increase as technology improves. For example, there are many potential advantages to using automated ultrasonic testing (AUT) for inspecting welds versus performing manual weld inspection. These include permanent record of inspection data, higher probability of detecting flaws, faster inspection times, less operator dependence on the results, and very accurate sizing. Disadvantages to AUT inspection are higher equipment costs, increased setup and calibration time, requires special calibration blocks, additional operator training and experience, and lack of specifications. Johnson and Bell (2008) attempted to verify the ability of AUT to perform acceptably on a typical AWS type component through a special qualification block fabricated with intentional weld related defects in the weld and heat-affected zone. This block contained both longitudinal and circumferential welds (Figure 1.30).

A technique combining TOFD (time of flight diffraction) and pulse echo UT methods were employed to obtain 100% volumetric inspection of the weld area. Lack of fusion, porosity, inclusions, cracks, lack of penetration, and transverse defects were put in the areas of the weld where they would be typically generated, including several defects in areas that are known to be difficult to inspect. This was done to test the limitations of the inspection techniques. According to the authors, all of the flaws in the weld area were detected with the AUT inspection. And sizing of the

FIGURE 1.30 Qualification block on rotating stand. (Reprinted from ASNT Publication.)

AUT detected flaws was in close agreement with the manual UT measurements and the intended flaw size.

The main challenge in the field of infrastructure health is to obtain a reasonable balance between automatic and manual inspections. Such a balance would include cost, technology, and accuracy balances. This volume will discuss those issues whenever appropriate.

1.11.3 CLIMATE CHANGE

There is a strong relationship between infrastructure health and climate. Since early times, the humans designed and built their infrastructure with the prevailing climate in mind. Notice the differences between early building properties in the arctic region and the equator. In addition to building properties, humans learned the hard lessons of accommodating, or ignoring, locations and material properties of infrastructure as they relate to climate changes. Two obvious examples come to attention from classical Egypt. The ancients choose location of the great pyramids on high grounds in the Nile Valley: they survived for more than five millennia. When the ancients chose to build the magnificent classical Alexandria without regards to location, sizable section of ancient Alexandria lies now a few feet below the Mediterranean Sea as a result.

More recently, two important changes have occurred in the field of civil infrastructure. First: there has been a revolution of construction materials: the utilization of modern concrete and steel has been predominant in the last two centuries. More recently, the expectations of longer life spans of modern infrastructure became predominant. A parallel ominous change occurred on a global scale: it is becoming clearer now that earth climate is changing. As a direct consequence of the confluence of those three factors, the effects of climate change on longer-lived infrastructure is quickly emerging as a major challenge in the field of infrastructure health. Generally speaking climate change demands would increase as the capacities of infrastructure decrease (as they age). The concept is shown in Figure 1.31.

There are numerous manners climate change can affect infrastructure. Some of those are summarized below:

Rising sea water: Melting of polar caps and other environmental effects would result in rising sea water levels. The effects on bridges and coastal structures can be immense. Those effects can be structural and architectural.
Corrosion: Increased rates of corrosion of reinforced concrete in submerged columns/foundation (effects can be most noticeable in causeways) in low coastal areas: low areas that are dry now will be attacked by sea water, with potential acceleration in deterioration rates.

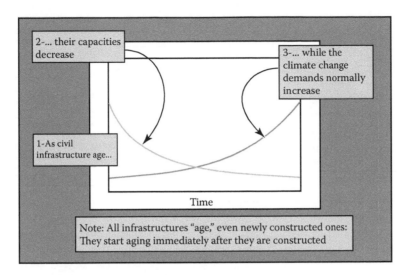

FIGURE 1.31 Climate change and aging infrastructure dilemma.

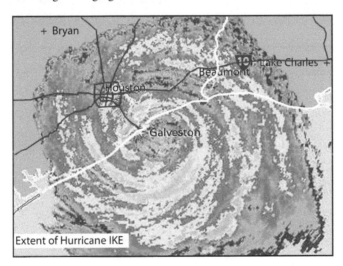

FIGURE 1.32 Extent of Hurricane Ike (2008).

Direct pressure loads: These can result in added hydrostatic pressures on retaining walls. This might result in the need for redesigns of existing systems.

Buoyancy problems: Changes in water levels can result in emerging buoyancy problems on structural foundations.

Clearance problems: Increased water levels can result in head clearance problems under low level water crossings.

Direct forces of flood and winds: Climate changes can result in an increased severity of floods and winds (Figure 1.32). Redesign of systems might be needed. Scour problems needs to be reevaluated. Storm surge can have negative effects on bridges (Figure 1.33). Debris impacts during floods might become more severe. Similarly, wind borne objects can be more severe.

Temperature rise and increased fluctuation ranges: Effects on expansion joints, and integral bridge designs might have major effects.

Deterioration issues: Potential increased deterioration rates due to sulfur attacks, corrosion, and so on. Prestressed systems can have additional shrinkage, creep: resulting in loss of prestress.

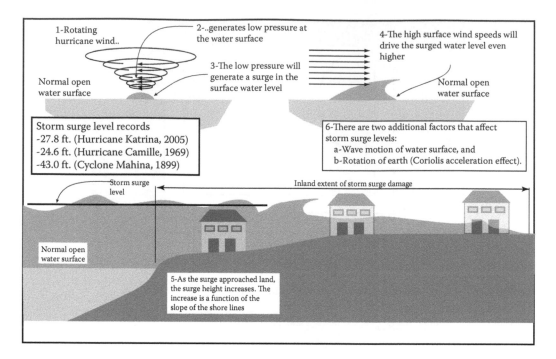

FIGURE 1.33   Anatomy of storm surge.

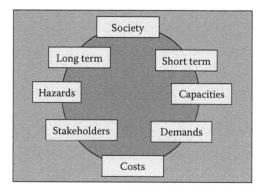

FIGURE 1.34   Integrated approach in meeting climate change challenges.

Detailed discussion of the above climate change effects on infrastructure are beyond the scope of this volume. However, many of the subjects we discuss later can be applied directly when the decision makers try to address climate change effects on the health of civil infrastructure. We note that one of the basic tenets of this volume is the integrated approach in solving infrastructure health challenges. Same can be said about climate change challenges for infrastructure, as shown in Figure 1.34.

1.12   APPENDIX I: DHS INFRASTRUCTURE SECTORS (AS OF 2009)

For ease of managing the vast infrastructures inventory of the United States, the U.S. Department of Homeland Security (DHS) categorized infrastructures into several sectors. Table 1.12 shows those sectors and the U.S. agency lead for each sector.

TABLE 1.12
DHS Infrastructure Sectors, US DHS (2009)

Critical Infrastructure/Key Resources Sectors	Federal Sector-Specific Agency (SSA) Lead
Agriculture and food	Department of Agriculture and Department of Health and Human Services
Banking and finance	Department of the Treasury
Chemical	Department of Homeland Security
Commercial facilities	Department of Homeland Security
Commercial nuclear reactors, materials and waste	Department of Homeland Security
Dams	Department of Homeland Security
Defense industrial base	Department of Defense
Drinking water and water treatment systems	Environmental Protection Agency
Emergency services	Department of Homeland Security
Energy	Department of Energy
Government facilities	Department of Homeland Security
Information technology	Department of Homeland Security
National monuments and icons	Department of the Interior
Postal and shipping	Department of Homeland Security
Public health and health care	Department of Health and Human Services
Telecommunications	Department of Homeland Security
Transportation systems	Department of Homeland Security

REFERENCES

ACI (2008). Building Code Requirements for Structural Concrete and Commentary, American Concrete Institute, Farmington Hills, MI.

ASCE, (2009), http://ciasce.asce.org/working-definitions, site accessed on March 20, 2009.

Catbas, F. N., Grimmelsman, K. A., and Aktan, A. E., (2000) "Bridge Health-Monitoring for Structural Performance," *Structural Materials Technology: An NDT Conference*, ASNT, Atlantic City, NJ.

Cheng, J., (2004) "Potential and Challenges of Using SHM in the Canadian Energy Industry," Proceedings of 2nd International Workshop on Structural Health Monitoring of Innovative Civil Engineering Structures, ISIS Canada Corporation, Manitoba, Canada.

Cheung, M. and Naumoski, N., (2002) "The First Smart Long-Span Bridge in Canada," Proceedings of 1st International Workshop on Structural Health Monitoring of Innovative Civil Engineering Structures, ISIS Canada Corporation, Manitoba, Canada.

DeWolf, J., Mao, J., and Virkler, C., (2002) "Non-Destructive Monitorinh of Bridges in Connecticut," Proceedings, NDE Conference on Civil Engineering, ASNT, Cincinnati, OH.

El-Ragaby, A., El-Salakawy, E., and Benmokrane, B., (2004) "Field Performance of a Concrete Bridge Deck Reinforced with GFRP Bars: Cookshire-Eaton Bridge," Proceedings of 2nd International Workshop on Structural Health Monitoring of Innovative Civil Engineering Structures, ISIS Canada Corporation, Manitoba, Canada.

Ettouney, M. and Alampalli, S., (2012) *Infrastructure Health in Civil Engineering: Applications and Management*, CRC Press, Boca Raton, FL.

Fisk, P., Abrams, M., Kaufman, S., (2008) "Inspection and Rehabilitation of Jamestown - Verrazzano Segmental Concrete Bridge," *NDE/NDT for Highway and Bridges: Structural Materials Technology (SMT)*, ASNT, Oakland, CA.

Fritzen, C. and Bohle, K., (2000) "Parameter Selection Stratefies in Model-Based Damage Detection," Proceedings of 2nd International Workshop on Structural Health Monitoring, Stanford University, Stanford, CA.

Gangone, M., Kroening, R., Minnetyan, L., Janoyan, K., and Grimmke, W., (2005) "Evaluation of FRP Rebar Reinforced Concrete Bridge Deck Superstructure," Proceedings of the 2005 ASNT Fall Conference, Columbus, OH.

Glisic, B. and Inaudi, D., (2002) "Long-Gauge Fibre Optic Sensors for Global Structural monitoring," Proceedings of 1st International Workshop on Structural Health Monitoring of Innovative Civil Engineering Structures, ISIS Canada Corporation, Manitoba, Canada.

Hawk, H., and Small, E. P., (1998) "The BRIDGIT bridge management system." *Structural Engineering International*, (4), 309–314. International Association of Bridge and Structural Engineering (IABSE), Zurich, Switzerland.

Johnson, W., and Bell, R., (2008) "Comparing Manual AWS UT Inspection to Automated TOFD and Pulse Echo Inspection," *NDE/NDT for Highway and Bridges: Structural Materials Technology (SMT)*, ASNT, Oakland, CA.

Lloyd, G. M., Wang, M. L., Wang, X., and Chen, H., (2004) "Components of a Real-Time Monitoring System for a Segmental Pre-Cast Concrete Box Girder Bridge," Proceedings, NDE Conference on Civil Engineering, ASNT, Buffalo, NY.

Lynch, J., and Loh, K., (2006) "A Summary Review of Wireless Sensors and Sensor Networks for Structural Health Monitoring," *Shock and Vibration Digest*, 38(2), 91–128.

NPR (2011) "Transcript: Obama's State of the Union Address," http://www.npr.org/2011/01/26/133224933/transcript-obamas-state-of-union-address, site accessed June 10, 2011.

Olund, J., Cardini, A. J., D'Attilio, P., Feldblum, E., and DeWolf, J., (2006) "Connecticut's Bridge Monitoring Systems," *NDE Conference on Civil Engineering*, ASNT, St. Louis, MO.

Oshima, T., Mikami, S., Yamazaki, T., and Sato, M., (2002) "Application of Intelligent Monitoring System to Bridge Health Diagnosis," Proceedings of 1st International Workshop on Structural Health Monitoring of Innovative Civil Engineering Structures, ISIS Canada Corporation, Manitoba, Canada.

Pontis 4.4. (2005) "Pontis Release 4.4: User's Manual," Prepared for AASHTO by Cambridge Systematics, Inc., Cambridge, MA.

Rochette, P., Neale, K., Pierre, L., and Demers, M., (2002) "Monitoring FRP-Strengthened Structures : Lessons from Applications in Quebec," Proceedings of 1st International Workshop on Structural Health Monitoring of Innovative Civil Engineering Structures, ISIS Canada Corporation, Manitoba, Canada.

Savard, M. and Laflamme, J-F., (2004) "Long-Term Monitoring of a Reinforced Concrete Bridge," Proceedings of 2nd International Workshop on Structural Health Monitoring of Innovative Civil Engineering Structures, ISIS Canada Corporation, Manitoba, Canada.

Sazonov, E., Janoyan, K., and Jha, R. (2004) "Sensor Network Application Framework for Autonomous Structural Health Monitoring of Bridges," Proceedings, NDE Conference on Civil Engineering, ASNT, Buffalo, NY.

Shehata, E., Haldane-Wilson, R., Stewart, D., Mufti, A., Tadros, G., Bakht, B., and Ebenspanger, B., (2004) "Structural Health Monitoring of the Esplanade Riel Pedestrian Bridge," Proceedings of 2nd International Workshop on Structural Health Monitoring of Innovative Civil Engineering Structures, ISIS Canada Corporation, Manitoba, Canada.

Sikorsky, C., Stubbs, N., Bolton, R., and Karbharti, V., (2002) "The Application of Structural Health Monitoring to Evaluate Bridge Strength," Proceedings of 1st International Workshop on Structural Health Monitoring of Innovative Civil Engineering Structures, ISIS Canada Corporation, Manitoba, Canada.

Taha, M, Kinawi, H., and El-Sheimy, N., (2002) "The Realization of Commercial Structural Health Monitoring Using Information Technology Based Techniques," Proceedings of 1st International Workshop on Structural Health Monitoring of Innovative Civil Engineering Structures, ISIS Canada Corporation, Manitoba, Canada.

Timoshenko, S., and Gere, J., (1961) *Theory of Elastic Stability*, McGraw Hill, New York, NY.

US DHS, (2009), http://www.dhs.gov/xnews/gc_1179776352521.shtm, site accessed on March 20, 2009.

Wang, J., Ni, Y., Ko, J., and Chan, T. (2002) "Damage Detection of Long-Span Cable-Supported Bridges," Proceedings of 1st International Workshop on Structural Health Monitoring of Innovative Civil Engineering Structures, ISIS Canada Corporation, Manitoba, Canada.

Weinmann, T., and Lewis, A. (2004) "Structural Testing and Health Monitoring for the Bridge Street Bridge Deployment Project," Proceedings, NDE Conference on Civil Engineering, ASNT, Buffalo, NY.

Wenger, L., Zhou, Z., Alwash, M., Siddique, A., and Sparling, B., (2004) "Vibration-Based Damage Detection on Bridge Superstructures," Proceedings of 2nd International Workshop on Structural Health Monitoring of Innovative Civil Engineering Structures, ISIS Canada Corporation, Manitoba, Canada.

Yamada, K., Ojio, T., Fukada, S., and Kajikawa, Y., (2004) "Use of Bridge Weigh-IN-Motion System With Environmental Monitoring of Viaducts," Proceedings of 2nd International Workshop on Structural Health Monitoring of Innovative Civil Engineering Structures, ISIS Canada Corporation, Manitoba, Canada.

Yang, C., and Newhook, J., (2004) "Load Distribution as a Damage Detection Tool for Steel-Free Bridge Deck," Proceedings of 2nd International Workshop on Structural Health Monitoring of Innovative Civil Engineering Structures, ISIS Canada Corporation, Manitoba, Canada.

2 Elements of Structural Health in Civil Engineering (SHCE)

2.1 SHCE: A NEW FIELD IN CIVIL ENGINEERING?

The field of structural health is a complex field. It includes many components, and cover numerous disciplines, subjects, and concerns. The complexities of the field are also due to the fact that many of its components and methodologies are fairly new to the civil engineering practice. For example: wide use of nondestructive testing (NDT), though known in civil engineering applications for long time, has not been an integral part of the practice. For proper and efficient execution of SHCE, the methodologies of NDT must be utilized and applied. This would require ingenious applications and perhaps some changes in NDT applications as they have been practiced in other fields such as aeronautics or mechanical engineering. Similarly, structural identification (STRID) or damage identification (DMID) methods have been utilized in other fields, such as mechanical or aerospace fields. Both STRID and DMID methodologies and tools must be retooled for efficient and successful applications in the SHCE field. In addition, there are emerging design paradigms and concepts in the civil engineering field such as performance based engineering, multihazards considerations, life cycle analysis (LCA) and component and system resiliency. All of those paradigms and concepts need to be included in the SHCE field. Their inclusion and interaction with SHCE field adds to the complexity and wide reach and coverage of the field.

Triumvirate Concept and SHCE: It is clear that for successful implementation of a healthy structure strategy, over the whole length of its life span, all of the above issues, and more, must be considered. Moreover, the interactions between all of those issues through both the stakeholders and the structures itself must be adapted properly. The triumvirate concept is shown in Figure 2.1. We offer that this triumvirate is the basis of the SHCE field. This volume aims at studying the different SHCE/triumvirate components and the interaction between them.

2.1.1 THIS CHAPTER

Since this is a first attempt to address the new field of SHCE, our first task is naturally to try to define its components or categories. This is needed to ensure adequate, efficient, and successful coverage of the subject. There are numerous ways to categorize or compartmentalize the a field of SHCE. We try to use the general components as shown in Figure 2.2, as follows:

Human Health/Baseball Analogy: Chapter 1 introduced an analogy between human and structural health. Such an analogy can be restated in terms that are more pertinent to structural health issues: the baseball analogy. The baseball analogy is surprisingly simple, yet very convenient and encompassing to all the activities of SHCE.

Design Paradigms: Other essential components in the field of SHCE are the different civil engineering design paradigms. We note that the conventional design paradigm: capacity/demand (CD) paradigm has been in use from early times. More recently, the performance-based design (PBD) paradigm is emerging. Both paradigms are related directly to SHCE.

NDT Methods: One of the bases of SHCE is the knowledge of the state of the structure in real, or near real, time. To achieve this knowledge, field and laboratory experiments

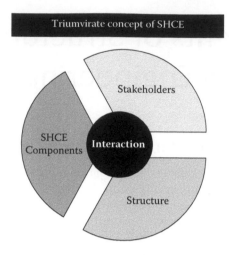

FIGURE 2.1   Triumvirate concept of SHCE.

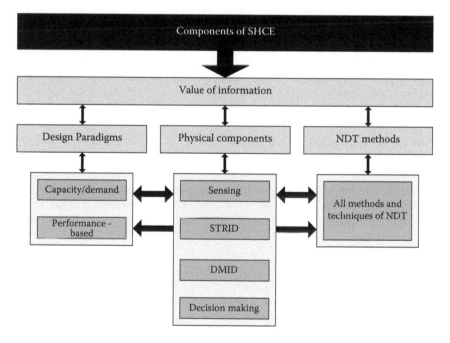

FIGURE 2.2   Components of SHCE/SHM.

are needed. The well-developed field of NDT must play an integral role in gaining this knowledge.

Value of Information: The execution of all needed activities pertaining to the field of SHCE will require additional costs and efforts that are not conventionally incurred. Of course, there are benefits that will be gained as a result of those costs. The question is whether those benefits outweigh the costs? In other words: how much value an asset manager will gain from SHCE?

This chapter will discuss the above components of SHCE in detail. Subsequent chapters explore different topics related to each component and potential interactions between these components.

2.2 SHCE: BASEBALL ANALOGY

2.2.1 HEALTH MONITORING FROM START TO FINISH

Monitoring structures such as bridges, dams, and buildings has received significant attention in the last decade. Several owners and researchers are actively pursuing development of monitoring systems, with reliable sensor technologies and remote monitoring systems for decision making. Most of the system identification process thus far is concentrated in three phases, namely data acquisition, structural identification (SI), and damage detection, and lacks decision making phase. This section presents a comprehensive view on the engineering of structural health process integrating the decision making process.

2.2.2 HOME PLATE: PROBLEM DEFINITION

Both existing and new civil engineering structures and constructed facilities throughout the world are subjected to continuous deterioration due to several reasons such as corrosion resulting from exposure to different environmental conditions. Day-to-day service loads can initiate fatigue problems over the time. Moreover, infrequent actions such as earthquake, fire, and improper use or maintenance result in acceleration of the structure deterioration. This deterioration causes significant degradation of a structure's mechanical properties and its ability to perform intended functions without failure. Due to the immense importance and cost of infrastructure, the subject of structural health and performance during the life span of the structure has emerged lately as a major issue for the engineering community.

Health monitoring of structures is perhaps the most serious attempt to address this problem. In a general sense, it involves three distinct phases: the measurement of performance phase, the STRID phase, and the damage detection phase. Considerable research and several methods in the literature address each of these three phases. We offer a detailed overview of these phases in Chapters 5, 6, and 7.

Unfortunately, health monitoring of structures in the present form lacks a major component, namely the decision making phase. As an example: assume that a certain damage profile has been detected during a health-monitoring project for a specific structure. With the damage profile information, the decision maker is faced with the question of what to do next. Thus, important factors and necessary tools that could help that decision maker should be readily available so that proper decision can be made.

In this volume, we propose adding the decision making process to the three health monitoring phases to form an integrated SHCE field. This concept was first proposed by Alampalli and Ettouney (2004). This field will then be composed of four phases: measurements, identification, damage detection, and decision making. Figure 2.3 shows a schematic representation of this process. Moreover, an integrated approach is proposed to the SHCE phases, meaning that all four phases of the SHCE process should be considered simultaneously. We will show that failing to do so may yield inefficient or even erroneous results in many situations.

A more general, and more realistic, paradigm of SHCE is shown in Figure 2.4. This general paradigm allows for base stealing, that is, it allows for cross-communications between different SHCE components, instead of the serial paradigm shown in Figure 2.3. Note that in the more general paradigm, the decision making process is always available at any step of the process.

2.2.2.1 First Base: Measurements

2.2.2.1.1 General

The first phase in SHCE is the measurement phase and forms the basis for the entire SHCE process (Figure 2.4). Measuring structural behavior for SHCE will have two main goals. First, it is desirable to numerically identify the structural system, and estimate the state of structural damage, if any.

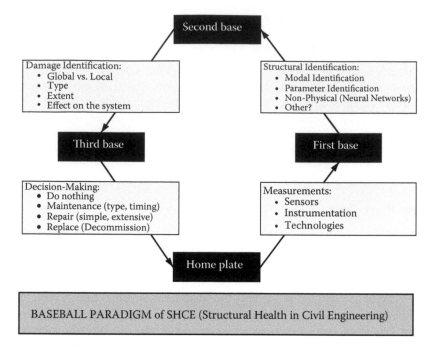

FIGURE 2.3   SHCM baseball analogy.

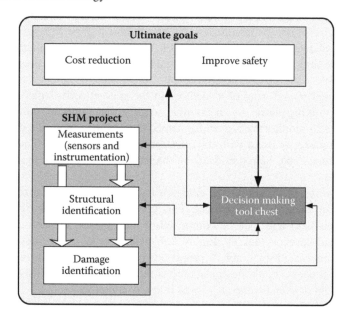

FIGURE 2.4   General SHCE paradigm.

Often this requires structural measurements. The most obvious is selecting structural parameters, which represent important structural response measures of interest. These include the following:

- Direct measurements: These relate to the structural performance and give an estimate of structural strength or failure condition directly. These include, but are not limited to, the following
  - Structural displacements
  - Velocities

- Accelerations
- Structural deformation, for example, strains
- Inclinations
- Indirect measurements: These include parameters that give the necessary information, on structural parameters, indirectly through some correlations
  - Measuring the acoustics of cable snapping in suspension bridges, and then finding the location of the breaks in the cables by tracing the time of arrival of acoustic waves.
  - Half-cell potential data from concrete columns can be correlated to corrosion rate.

In all situations, complete knowledge of the structural system is mandatory to collect successful and efficient measurements. This includes all the structural functions, intended use, design loading, required service life span, and economics of maintenance.

All of the above factors will affect the measurement phase in SHCE. In any experiment, number of sensors n_c is directly related to the economics of maintaining the structure as well as the structural function and the intended use. In addition, number of sensors and their locations are directly related to the type of structure, loading levels, and environment conditions. Any structural measurement scheme not accounting for these interrelationships can lead to inefficient/insufficient data, or possibly measurement information that is not suited for the intended SHCE use. A brief discussion of these interrelationships is presented in next sections.

2.2.2.1.2 Number of Sensors and Cost

In any SHCE measurements, the minimum number of sensors n_o that is needed to insure the required quality of measurement information q_o are related by the *number* curve as shown in Figure 2.5. In addition, the cost of obtaining q_o is C_o. They are related by the *cost* curve shown in Figure 2.5. Thus, for a required information quality q_o, both cost and number of sensors are uniquely correlated. If the actual number of sensors in an experiment n_c falls below n_o then the measurement information can be rendered insufficient. In addition, an upper bound sensor cost ceiling $C_c \geq C_o$ also exists. It is determined by economic factors that relate to the structure's intended use and importance. Both C_o and n_o requirements need to be consistent for optimum SHCE process. If this proves to be infeasible, either the cost ceiling need to be raised, or the technical scope of the measurements has to be reduced. Figure 2.5 illustrates this process.

2.2.2.1.3 Sensor Locations and Number

Another important step in the SHCE measurements is the optimum sensor location problem (OSLP). See Cobb (1996) for comprehensive review of this important issue. It can be stated as follows: For

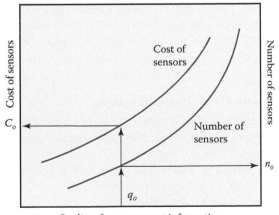

FIGURE 2.5 Cost, number of sensors, and quality of information relationships.

any given structural measurement problem, find n_o locations of n_c sensors such that the information gathered is optimum. Several methods are available for the solution of the OSLP. However, it was argued (Ettouney et al. 1999) that any OSLP solution has to account for the usage and the environment of the structure in an integral fashion. Failing to do this may result in sensor locations that do not yield optimum information. The same authors proposed goal-programming techniques to solve the general OSLP. Ettouney et al. (1999) studied OSLP for seismic, dead, and live load conditions. They showed that all loading conditions have to be considered in deciding upon sensor locations; otherwise, erroneous or inefficient sensor locations might result.

2.2.2.2 Second Base: Structural ID

Once data from measurement phase is obtained, transformation of this data to understand the structural properties is necessary. This is widely known as SI problem. The structure can be identified by several modes of information. For example, the structure can be identified by the distribution of stiffness throughout the structure, that is, the stiffness matrix of the structure. Identifying natural frequencies and mode shapes of a structure (modal analysis) is one of the popular SI methods. Detailed discussion of SI methods is beyond the scope of this work. For more information, the reader is referred to Cobb (1996).

Even with the existence of the large number of studies and methods for SI, it is wise to place SI in the context of the larger issue of SHCE. A successful SI technique should be capable of addressing the needs of the SHCE problem. For example, if the SHCE problem on hand is the corrosion of reinforcing steel in a reinforced concrete bridge, the employed SI method should be capable of addressing the initiation and propagation of corrosion, and structural deterioration due to corrosion.

This discussion also shows that the choice of a particular method of SI should also depend on the type and degree of damage expected during a structure's life span.

2.2.2.3 Third Base: Damage Detection

The third phase in the engineering of structural health is detection of damage that can occur in the structural system. For example, offshore oil platforms need continuous monitoring for structural damage that might occur below the waterline from extreme sea conditions and ship impact. Damage monitoring of bridges is increasingly necessary as the effects of corrosion, ship traffic, earthquakes and, sadly, even terrorist actions continuously threaten the soundness of these vital structures. During the last two decades, research has been focused on using the vibration characteristics of a structure for structural damage identification (Faarar 1996; Ettouney et al. 1998). The vibration characteristics of a structure may be defined by its modal parameters (natural frequencies and mode shapes), which generally depend only on characteristics of the structure and not the excitation, and may be determined from taking measurements at one or more locations on the structure. Vibration signatures obtained before and after damages may be utilized to locate the damage and estimate its severity (Alampalli 1998). Chapter 7 explores damage identification issues from defining damage to various analytical and experimental methods available in the infrastructure field.

2.2.2.4 Fourth Base (Home Plate): Decision Making

All data from the above three phases should answer the main reason why the structural health monitoring (SHM) was considered in the first place. These includes questions such as

- For a given damaged pattern in a structural component or a complete structural system, what is the reliability of such a component or a structure?
- What is the remaining service life before failure?
- Is the failure imminent?
- Are the assumptions used in design, construction, and so on are valid?
- What loads are needed to cause failure?

- What are the predominant failure mechanisms under the expected loads of a structure's life?
- What type of actions (such as maintenance, rehabilitation, or replacement) are necessary and when?

Let us explore how structures degrade to illustrate the point. When an engineering structure is loaded it will respond in a manner that depends on the type and magnitude of the load as well as the strength and stiffness of the structure. Whether the response is considered acceptable, depends on the requirements that must be satisfied (Melchers 1987). These requirements might include safety of the structure against collapse, limitation of damage, magnitude of deflection, or any other such criteria. Not meeting these requirements is considered a limiting state violation. Hence, the reliability of a structure is defined as the probability of occurrence of the limit state violation at any stage during its lifetime. This probability can be obtained from measurements of the long-term occurrence of the violations on similar structures, subjective estimation, or by using small-scale prototypes and testing them under different conditions. Alternatively, a measure of the structural degradation with time, such as rate of corrosion, can be used to determine the time at which a structure will cease to function as desired. The level of degradation at that time is referred to as threshold degradation level. Elsayed (1996) classifies degradation models as physics-based and statistics-based models. The physics-based degradation models are those in which the degradation phenomenon is described by a physics-based relationship such as Arrhenius law, the corrosion initiation equation (Enright and Frangopol 1998) or experimentally based results such as crack propagation or crack growth model (Oswald and Schuëller 1983). The statistics-based degradation models are those in which the degradation phenomenon is described by a statistical model such as regression. The description of the advantages and limitations of the two types of models are beyond the scope of this study. An example of the use of a degradation model in SHCE is as follows.

Suppose that a degradation/damaged state database for a structure, such as that shown in Figure 2.6 can be assembled. The damaged state database of Figure 2.6 was assembled for a reinforced concrete bridge that is subjected to corrosion. It was obtained using combined analytical, experimental, and statistical methods (Ettouney et al. 1999). Using the degradation model that was

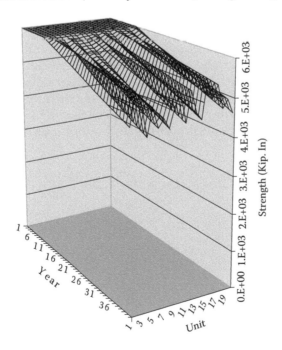

FIGURE 2.6 Deterioration data.

described by (Eghbali and Elsayed 1997), the reliability of the structure can then be estimated as a function of time, as shown in Figure 2.7. Note that this damage model relies heavily on structural function and usage.

This structural reliability information can then be used to make proper decisions concerning the optimum course of action.

2.2.2.4.1 Cost Analysis

Perhaps the most important step in SHCE is the cost analysis phase. After reviewing reliability and degradation data (similar to those of Figures 2.6 and 2.7), the question facing the owners of the structure will be; what should be done next? The answer can be obtained by employing a general cost analysis as was employed by Ettouney and Elsayed (1999). Some cost analysis tools are available for specific SHCE problems, such as corrosion cost analysis. However, cost analysis should include not only monetary costs, but also possible social and economical costs. When cost analysis of an SHCE is performed in such generalized manner, it becomes more specific to the structure under study. This shows the importance of integration and interdependability of all aspects of SHCE.

2.2.3 EXAMPLE

As an example of the baseball analogy in an SHM project that can be found in a paper by Helmicki et al. (2002). The strategy of this SHM was to monitor load rating (global) and several loacalized issues (ultrasonic pin testing), tesnile coupon testing, and Charpy V-Notch testing. This led the authors to an estimation of remaining fatigue life. A combination of localized and global SHM approach is one of the features of this case study. One result was to modify cold weather maintenance procedures for the bridge.

The study involved the Ironton–Russell Bridge, built in 1922, a 2401' long steel thru truss with steel plate girders and concrete deck approach slabs. It has gone through various rehabs during its life and was rated low in 1997 during the biennial inspection due to superstructure deficiencies. Hence it was decided by the owner to conduct instrumentation-based field test to evaluate the structure further. The objectives are to evaluate required reduced load posting, near-term retrofits, and long-term management and planning.

A finite elements model was developed. On the basis of the model, visual inspection data, and discussions with owners, the sensor locations were chosen sufficient to capture the critical bridge elements' response. Strain gages with known truck loading was used for field tests. The field tests and subsequent analysis was used to select inventory and operating capacity of the structure.

The bridge was monitored for about 20 months for its in-service long-term operating behavior (to study the effect of its operating environment). The data revealed fixity of the pin and hanger connections, which were supported by physical observations such as pack rust.

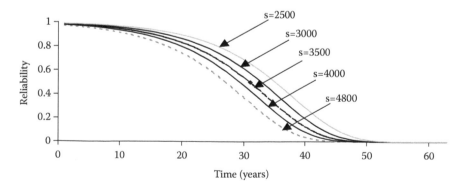

FIGURE 2.7 Reliability versus Time for different strengths (S) of a reinforced concrete beam. Courtesy of Dr. E. A. Elsayed.

On the basis of above observations, ultrasonic tests were scheduled to obtain more quantitative information on pin and hanger connections. On the basis of this testing and determination of section loss, retrofit options were considered at select locations to increase structural redundancy.

Selected material testing was conducted to make sure appropriate properties (Elastic Modulus and yield strength) were used in the analysis. Charpy V-notch testing was done to evaluate temperature effect on ductility of steel. On the basis of the testing, it was decided that the material became brittle at temperatures below 0°F. The bridge used to be closed below 20°F. On the basis of these tests, it was decided to close the bridge, when temperatures fall below –5°F.

Due to the presence of low category weld details, appropriate members were instrumented to measure continuous traffic data to observe actual stress ranges these members are subjected to. On the basis of this data, it was decided that these members had infinite fatigue life due to very low stress ranges observed.

2.2.4 Bringing All Together

In the above sections, SHCE is compared to a baseball game. Each component of SHCE is similar to a base in a baseball game. A runner in a baseball game starts at the home plate with a goal to return back there through all three bases in between to score a run. In SHCE efforts, the investigator starts with a decision to make and then go through all four components to make the final decision. If he/she obtains the required information after going through the entire SHCE cycle, then the goal is fulfilled and a run is scored.

If the player is stranded at intermediate bases and thus cannot return to home base, then the efforts are of limited or no use. In similar manner, if the SHCE study does not conclude with the information that cannot aid in making the decision, for which the study is intended for, all the efforts are of no use or limited value.

The efforts required to achieve or complete each of the components of SHCE can vary significantly depending on study at hand. In some cases, amount of time required for sensor selection, system identification, and damage identification can be relatively less due to a simple or well studied structure, previous experience, and so on. In such cases, it can be compared to a home run in a baseball game, where all bases are run through with a single shot. Such an example could be finding an impact factor associated with an easily accessible simple supported single short-span bridge with couple of girders in a rural environment. This requires installing one or two gages on the structure and measuring strain under traffic loads. The effort required for all SHCE components is very limited in such cases and one can obtain the impact factor with relative ease. As structure and service environment becomes complex, the efforts required for each SHCE component will vary significantly.

If multiple decisions have to be made using the same SHCE study, information required to make each decision can be compared to a run. Number of decisions that have to be made can be thought as number of runs required to win the game and thus the study can be concluded if all decisions the study is intended to assist are successfully made. If the study produces more information to enhance the state of the practice or state of the art besides providing the minimal information needed from it, then the study value is significant. Such studies and the teams involved in planning such studies can be considered better teams.

Several comparisons such as above, where SHCE is compared to base ball game, can be made. But, the bottom line is that every SHCE study has to be initiated based on the probability of success and when benefits outweigh costs (i.e., produces value to the decision making process). If probability of success or the probable value accomplished is not significant then study should not be initiated, that is, it is better not to play the game at all. This is one place, probably; the SHCE differs with the baseball game!

2.3   SHCE AND DESIGN PARADIGMS

2.3.1   GENERAL

The dominant civil/structural engineering paradigm has been traditionally the CD paradigm. Another engineering paradigm has emerged lately, which is the performance-based design (PBD). It is obvious that SHCE/SHM should be directly related to the paradigms of engineering practices. See, for example Figure 2.8. The figure shows the relationships between time-dependent hazards and demands from a given structure.

2.3.1.1   Statistical Pattern Recognition

An integrated SHM paradigm that was called *statistical pattern recognition* was introduced by Farrar et al. (2000). The authors considered vibration-based damage detection as the main basis for SHM. They also recognized that conventional vibration-based damage detection has two main drawbacks (a) damage is typically a local phenomenon that needs accurate recognition of higher vibration modes to be detected properly, and (b) the need for vibration-based damage detection to be performed in an unsupervised mode. The statistical pattern recognition paradigm includes several topics that aim at utilizing the advantages of vibration-based SHM, while addressing the above two

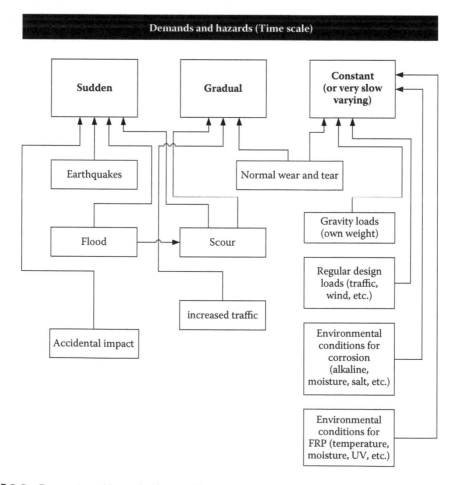

FIGURE 2.8   Demands and hazards (time scale).

drawbacks. Some of the topics of the paradigm are as follows:

Operational Evaluation: Any SHM must accommodate different operational needs of the system on hand. These operational needs include type and significance of damage, operational and environmental conditions, and limitations of data collection, if any.

Data Acquisition and Cleansing: The method(s) of acquiring and preparing data for use must be compatible with both operational conditions, above, and selected features, below.

Feature Selection: The authors of the statistical pattern recognition describe in great detail the different categories of vibration-based damage recognitions. The categories are (a) basic modal properties, (b) dynamically measured flexibility, (c) updated structural model parameters, (d) time-history and spectral pattern methods, (e) time domain methods, (f) frequency domain methods, (g) time-frequency methods, and (h) nonlinear methods. It should be noted that the above categories were discussed at even greater details by Farrar et al. (2000).

Statistical Model Development: The final topic is statistical development. The authors stated that there are five subjects that need to be defined in the damage state in a system: (a) existence of damage, (b) location of damage, (c) type of damage, (d) severity, or extent, of damage, and (e) remaining useful life.

Three statistical modeling approaches were recommended by the authors of the paradigm. They are as follows:

1. Group classification in a supervised learning mode. Group classification would try to classify results of observations into groups that can be utilized by decision makers; an example of grouping is damaged/undamaged grouping.
2. Regression analysis in a supervised learning mode. As the name implies, the authors of this model recognized that utilization of regression analysis tools can help in damage detection.
3. Density information in an unsupervised learning mode. This technique tries to handle situations when damage information is not complete. By utilizing probabilistic methods the authors contend that it is possible to identify damage information.

The proposed statistical pattern recognition paradigm is a powerful paradigm that can be of help in numerous situations. As such, it is an essential approach in the field of SHM.

2.3.2 CAPACITY/DEMAND DESIGN

2.3.2.1 General

Monitoring large civil structures such as bridges, dams, and buildings has received significant attention in the last decade. Several owners and researchers are actively pursuing development of long-term monitoring systems, with reliable sensor technologies and remote monitoring systems for system and/or damage identification. This work emphasizes that there are several other aspects to structural health in addition to structural monitoring. Considerations of all aspects are necessary before any decision or strategy is made to ensure safe, economic, and reliable structural performance.

The safety and adequate performance of any structural or nonstructural component can be assured by the well-known CD equation, as follows:

$$\frac{C}{D} \geq \alpha$$

(2.1)

Where C and D are measures of the capacity of the system and the demand from the system, respectively. The factor α is usually taken as unity. It is referred to as the CD equation.

The process of analysis and design of structures involve the evaluation of the CD equation. In addition, there is an interconnection between the subject of health monitoring and the CD equation. Understanding this interconnection is important for the decision making process in the SHM subject.

2.3.2.2   Capacity/Demand Ratio as the True Measure of Structural Health

2.3.2.2.1   General

We offer a general discussion of some characteristics of the *capacity* of a system and the *demand* from the system as they relate to SHCE/SHM in this section.

2.3.2.2.2   Evaluating Structural Capacity

Capacity of a structure, or a smaller component in the structure, can take several forms. These forms of capacity include structural strength, structural strains, or structural displacements. One of the most important tasks of any structural design code is describing the process of evaluating the structural capacity.

We argue here that structural capacity is not a constant entity. It is a time-dependent entity. From the instant the structure is built, the capacity of this structure, and its components, changes. The time-dependent capacity can be either gradual, or abrupt. Among the reasons for gradual changes in structural capacities are the following:

1. Natural deterioration of material properties. Rusting and corrosion are the best known reasons for this phenomenon.
2. Suddenly applied additional structural demands that might push the structural capacity beyond the linear limit to a nonlinear limit. Thus, after the removal of the additional demands, the structural capacity will not revert back to its original state, but to a lower state of capacity. Earthquake events are an example to such a situation.
3. Changes in structural geometry. Spalling of concrete can lead to gradual reduction of capacity. On the other hand, if a retrofit, or addition, is added to the structure, the resulting capacity might be increased.

Figure 2.9 shows schematics of the different possible modes of time-dependent capacity of a structure.

Health monitoring of any structure or a structural component is only a single step in the multi-steps evaluation of the time-dependent structural capacity as shown in Figure 2.10. The figure also

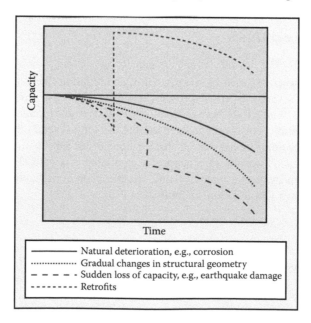

FIGURE 2.9   Time-dependent capacity of a system.

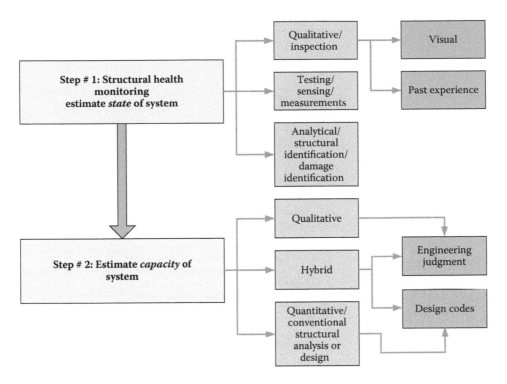

FIGURE 2.10 Steps for evaluating structural capacity.

shows the different steps and tasks that are needed to evaluate the time-dependent structural (or component) capacity. Figure 2.10 shows that to evaluate the state of the structure, at any instant of time, the health monitoring techniques might be used. Alternatively, purely analytical techniques might also be used. Each of these two alternatives has its own strong and weak points. Unfortunately, knowing the *state* of the system is only a necessary but not sufficient part of the capacity-evaluation process. An evaluation of the capacity of the system, utilizing the knowledge of the structural state, is still needed. Figure 2.10 shows that this evaluation can be either qualitative, quantitative, or a qualitative-quantitative hybrid process. In what follows, some examples of capacity-evaluation processes and their relationship with health monitoring techniques are presented.

2.3.2.2.3 Evaluating Demands on Structures

Evaluating the time-dependent demand on a structure, or a structural component, is as important as evaluating the capacity. The demand on any system can change with time. This change can be either gradual or sudden. It can either increase or decrease, depending of the functional changes of the structure. Some examples of changes in demands are as follows:

1. Changes in traffic loads on a highway bridge. This can be either increasing or decreasing demands. They usually occur gradually and imperceptibly.
2. Changes in design code specified demands. These demand changes tend to be abrupt. They can either increase or decrease the demand. For example, seismic demands on structures, as required by seismic codes, have fluctuated over the past several years. See UBC-97 (1997) and IBC-2006 (2006), for example.

Figure 2.11 shows schematically the different modes of time-dependent demand on structural systems. Note that demands on structures are decreed by the pertinent design codes for the structure of interest. Unfortunately, most of the current design codes are written for new construction. This

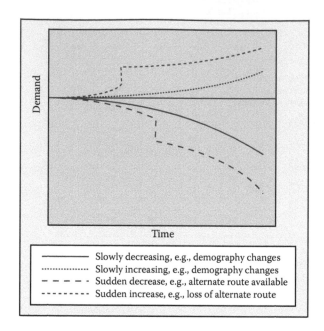

FIGURE 2.11 Time-dependent demand from a structure.

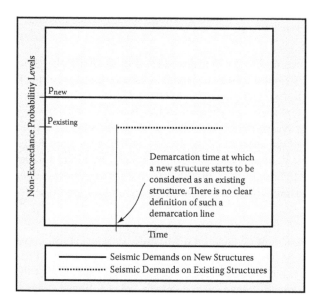

FIGURE 2.12 Changes of seismic demands for existing structures (as prescribed by seismic design codes).

means that when an engineer is evaluating the demands an existing building, he will have to use the same demands on a new building. This could lead to overly conservative and unrealistic demands. Some earthquake design codes have tried to mitigate this conservatism. Figure 2.12 shows schematically the earthquake demand reduction as a function of time as mandated by an existing building seismic design code. Note that such a code has not specified the needed elapsed time for the building designation to change from a *new* to an *existing* building. Is it one day, one week, one year, or more? The problem that this hypothetical question signifies will be discussed in more details in later sections.

2.3.2.2.4 Setting the Structural CD Ratio

The previous section discussed in some detail the concepts of capacity and demand in relationship to the current application of health monitoring of structures. It was shown that health monitoring, as it is practiced now, occupies a small part in the time-dependent structural evaluation process. This process is actually dependent on the CD ratio, as shown in Equation 2.1, not the individual values of the capacity and demand. In this section, we investigate the evaluation process as it relates to SHM.

2.3.2.2.4.1 "Optimum" CD Ratio

For new structures, it is desired to have $\alpha = 1.0$ for all components of the structure, and for the structure as a whole. Unfortunately, this is not always possible. For practical reasons, the ratio α is generally designed to be as close to unity as possible but not less than unity such that $\alpha \geq 1.0$. The closer α to unity the more efficient the structure is. In some cases, the design engineer may design a component such that $\alpha < 1.0$. This situation happens when his *engineering judgment* makes him/her conclude that a CD ratio that is slightly less than unity is still acceptable. There is no quantification of engineering judgment; it is a qualitative measure that is dependent on the situation on hand, and the personal experience of the design engineer.

Following the above discussion, it becomes clear that, for new buildings, there is a range of design for α. This can be stated as $\alpha_{min} \leq \alpha \leq \alpha_{max}$. This inequality is such that $\alpha_{min} \leq 1.0$ and $1.0 \leq \alpha_{max}$. Both α_{min} and α_{max} are desired to be as close to 1.0 as possible. The optimum structural design is when $\alpha = \alpha_{min} = \alpha_{max} = 1.0$ for the structure and all its components.

Figure 2.13 shows the time-dependent CD ratio after the structure goes into service. Figure 2.13 shows a case when CD changes abruptly. This can happen when either (or both) the capacity or the demand changes abruptly. It also shows the case when CD changes slowly. This can happen when both the capacity and the demand changes slowly.

Figure 2.13 also shows schematically the range α_{min} and α_{max} of the original design. For the purposes of this study, we will investigate only the situations when α decreases with time.

Studying Figure 2.13 further it appears that there are two main phases that can be identified. The first phase of the life span of the structure occurs while $\alpha_{min} \leq$ CD. This is where the structure, and its components, is still safe within the bounds of the *engineering judgment* of the design engineer.

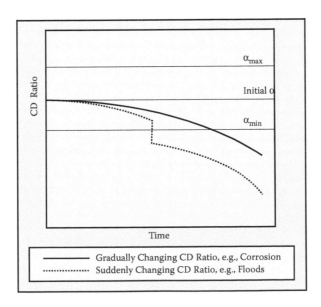

FIGURE 2.13 Time-dependency of actual versus acceptable CD levels.

The second phase of the life span of the structure occurs while $CD \leq \alpha_{min}$. This is where the CD ratio for the structure, or its components, starts dipping below the bounds of the *engineering judgment* of the design engineer.

The above discussion leads to an intriguing observation; if the structure was designed such that no engineering judgment were exercised, that is, $\alpha_{min} = 1.0$, the structure would be rendered unsafe almost from the instant it is built. This shows the importance of *engineering judgment*, either during the design phase, or during the maintenance phase of the structure.

2.3.2.2.4.2 Target CD Ratio

As was described earlier, some design codes have recognized this dilemma, and tried to pre-scribe a demand (hence CD ratio) for existing structures. We generalize this approach by identifying a time-dependent target CD ratio $\bar{\alpha}$. A safe and properly performing structure is operating such that $\alpha \geq \bar{\alpha}$, as shown in Figure 2.14. Note that $\bar{\alpha} = \alpha_{min}$ for new structures. As time progress $\bar{\alpha}$ varies in a realistic manner. This variation should reflect engineering judg-ment, probabilistic considerations, structural service time already spent, and expected/desired life of the structure.

2.3.2.2.5 Decision Making Process

Let us assume that the time-dependent CD ratio α is known for a given structure, or a structural component. This knowledge can be based partly on a health monitoring system that identifies the *state* of the structure. This knowledge is also dependent on translating the state of the structure into structural capacity. Also, the time-dependent demands on the structure on hand is needed for the full knowledge of α. Let us assume that the target CD ratio α_1 is known. Again, this knowledge can be based on design codes, engineering judgment, or any other mean of evaluation. Figure 2.14 shows the ensuing decision making process. The decision making process would lead finally to one of the three possible decisions (a) do nothing, (b) retrofit the system, or (c) replace the system. Deciding to do nothing is a simple decision. However, the decision of retrofitting, and the levels of retrofit, or replacement can be complex process. Ultimately, cost, social and environmental pro-cesses would control such a decision.

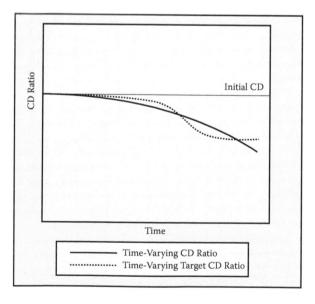

FIGURE 2.14   Time-dependency of actual versus acceptable CD levels.

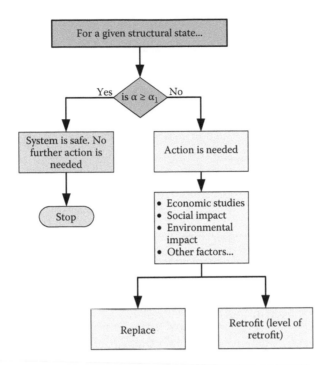

FIGURE 2.15 Decision making process when both actual and target structural states are known.

2.3.3 PERFORMANCE-BASED DESIGN

2.3.3.1 Overview

Performance-based design of buildings has been under development in the earthquake engineering community for more than a decade. It has been in use in the blast community for much longer. Communities of other hazards (progressive collapse) and infrastructure (bridges) are considering the usage of PBD. This article explores different aspects of PBD as it applies to seismic designs. The concepts are generally applicable to almost any other hazard, or any other infrastructure. We first compare prescriptive and PBD. The steps of PBD are presented next. Finally, we explore the interrelations between PBD and other emerging paradigms in the design/infrastructure community.

2.3.3.2 Prescriptive versus Performance Design Paradigms

There are several types of differences between common prescriptive and emerging PBD paradigms. The differences can be classified as (a) design objectives, (b) computational underpinnings, and (c) practical steps of the methods. Figure 2.16 illustrates the differences in design objectives between these two approaches. Basically, prescriptive design methods call for ensuring that the prescribed capacity C is larger than prescribed demand D. Design iterations continue until the capacity to demand ratios are within an acceptable range. In PBD, the objective of the design is to achieve a level of performance, as correlated to appropriate consequences (measured monetarily, for example). The design iteration involves reaching a design that would achieve the desired performance. The computational underpinnings of prescriptive design are related to the relationship between capacity and demand, and are based on structural reliability methods. PBD is based on risk methods that consider hazards (demands), vulnerabilities (capacity), and consequences. The differences between prescriptive and PBD steps are shown in Table 2.1.

Clearly, PBD involves more efforts by different building stakeholders, and in return it offer many advantages, such as the potential cost savings in the long run, options of continued operations and

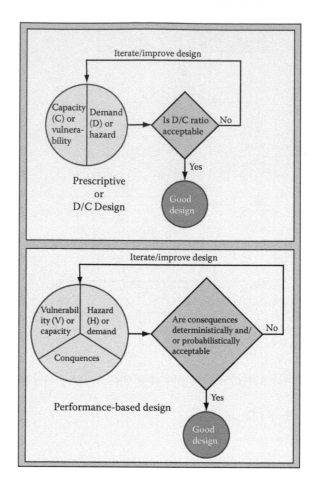

FIGURE 2.16 Prescriptive versus performance-based paradigms.

TABLE 2.1
Differences in Steps between Prescriptive Design and PBD

Design Considerations	Prescriptive	Performance
Seismic hazard level	Determined by codes	Chosen during the process
Structure damage level (acceptance criteria)	Specific	Chosen during the process
Consequences	NA	Computed as part of the process (qualitatively for first generation PBD, quantitatively for second generation PBD)
Uncertainties (performance probabilities)	NA	Computed as part of the process (second generation PBD)

immediate occupancy after events that can be of importance for sensitive facilities, and a clear quantitative picture on how the facility will perform during a designed event, and what the consequences of such performance would be with no surprises to the stakeholders. As a rule, PBD is based on concepts of risk, where CD design is based on concepts of reliability. The main difference is that PBD accounts for consequences (or outcomes) whereas CD design does not. More in-depth analysis of risk and reliability are provided in Chapter 8.

2.3.3.3   Modern versus Traditional PBD

Seismic PBD was introduced in FEMA 273/274, which were then reissued, with minor modifications as FEMA 356. It is generally accepted that those efforts constitute the first generation of seismic PBD. An ASCE standard (ASCE 2007) has been issued on the basis of that first generation PBD effort. There has been an effort by FEMA to generate a second generation of seismic PBD since 2002. That effort is continuing as of the writing of this article. Several differences exist between the first and second generation of the PBD effort. Among them are the continuous quantitative performance measures compared to the discrete qualitative measures offered by the first generation. Uncertainties are also addressed quantitatively in the second generation seismic effort. In addition, the second generation utilizes component and system fragilities and presents more details of analytical and design techniques.

2.3.3.4   Elements of PBD

2.3.3.4.1   Overview

Performance-based design roughly comprises three steps that are usually performed consecutively: (a) estimation of hazard, (b) evaluation of vulnerability, and (c) computation of consequences. Figure 2.17 shows schematically these steps, which will be discussed next. Seismic hazard and building structures are used as an example in this section due to the use of PBD in this field for several years, as noted in the previous section.

2.3.3.4.2   Hazard Level

The choice of hazard level includes choosing seismic input characteristics for the project. Seismic input characteristics can be simple, such as the choice of hazard level and the shape of the design spectra. It can be more involved, such as generating an ensemble of seismic acceleration time histories. It also can be deterministic-based, or probabilistic-based. In most situations, the designer needs to address issues such as return period (the length of time one earthquake of a given level is expected to occur), maximum earthquake ground acceleration, and so on. In the second generation seismic PBD effort, the probability of the chosen seismic hazard is an integral part of the design input needs. This is needed to compute the expected consequences of the design, as shown in Figure 2.18. Another feature of second generation seismic PBD is that it can be either based on a single scenario, such as a unique earthquake level, or be based on multiple earthquake levels, with varied return periods. This latter approach is obviously more time consuming, since designs must be performed for each of the scenarios. The advantage of the multiple scenario approach is that it gives a more complete picture of the total life cycle of the building. Table 2.2 shows the seismic hazard evaluation differences between prescriptive and performance-based seismic designs.

2.3.3.4.3   Damage Level (Vulnerability)

After the seismic input is defined, the building design process starts. In general, the process is fairly similar to the conventional prescriptive design process. There are, however, some important differences, as shown in Table 2.3.

Computing types, levels, and probabilities of earthquake structural or nonstructural damages in buildings, or any other type of infrastructure, are not an easy task. An advanced analysis techniques, as well as an experienced analyst are needed. This is one area that is currently undergoing extensive research and development. An emerging technique for relating earthquake damage to uncertain inputs, and computing the needed damage uncertainties is the fragility technique. Figure 2.18 shows how fragilities are used in a PBD context. We note that component seismic fragilities have been under development for some time. Efficient, practical, and general methods for system level fragility, on the other hand, are just starting to be considered by the community.

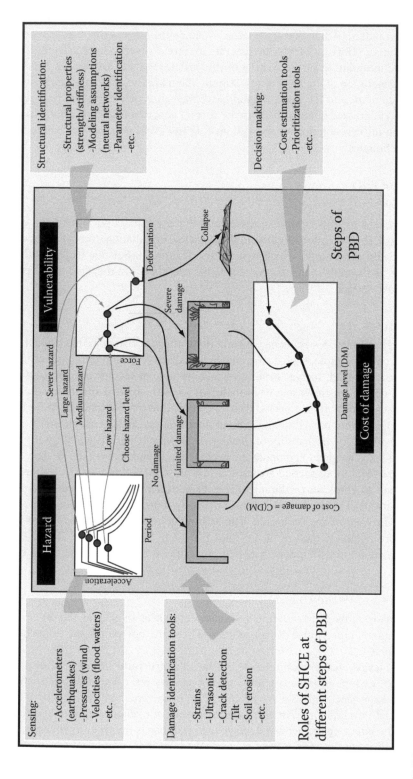

FIGURE 2.17 Steps of PBD.

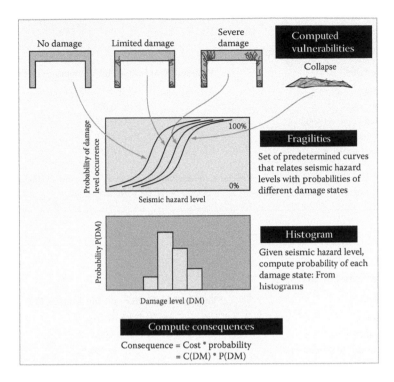

FIGURE 2.18   Computations of risk.

TABLE 2.2
Seismic Hazard Evaluation

Issue	Prescriptive	Performance Based
Earthquake amplitude and wave form (time domain), or design spectra	Prescribed	Varies, depending on desired performance
Probabilities of occurrence	Usually not an issue	Needed to compute consequence

TABLE 2.3
Design (Vulnerability) Evaluation

Issue	Prescriptive	Performance Based
Acceptance criteria	Prescribed, mainly to ensure life safety	Varies, depending on desired level of earthquake damage to the building
Analysis techniques	Mostly linear	Can be linear. However, nonlinear analysis is preferred
Analysis objectives	Limited. Only check to see if specific acceptance limits are met	Wide range. Check for acceptance limits. Also computes damage types and levels, which are needed for consequence analysis
Probabilities of damage/ response occurrence	Usually not an issue	Needed to compute consequence

TABLE 2.4
Failure Rates of Buildings

Building Type	Probability of Failure during Average Life Span
Apartment	0.3 in 10,000,000
Mixed housing	5 in 10,000
Controlled domestic housing	1 in 100,000
Mixed housing	1 in 1,000
Engineered structures	1 in 10,000

Source: Based on Melchers, R. E.: *Structural Reliability Analysis and Prediction.* 2002. Copyright Wiley-VCH Verlag GmbH & Co. KGaA. Reproduced with permission.

TABLE 2.5
Failure Rates of Bridges

Bridge Type	Probability of Failure during Average Life Span
Steel railway	1 in 1000
Large Suspension	3 in 1000
Cantilever and suspended span	1.5 in 1000
General	1 in 100

Source: Based on Melchers, R. E.: *Structural Reliability Analysis and Prediction.* 2002. Copyright Wiley-VCH Verlag GmbH & Co. KGaA. Reproduced with permission.

Melchers (2002) gave probability of failure for buildings that vary from one in 10,000 for engineered structures in Canada to 0.3 per million in apartment buildings in Denmark during their average life span. Melchers (2002) also gave typical failure rates for bridges between 1900 and 1960 that vary from 1 in 100 for general bridges based on an Australian source to 3 in 1000 for large suspension bridges in the world. It is interesting to note that the probability of failure is generally an order higher for the bridges when compared to buildings even though the average life spans are very similar.

Table 2.6 gives the hazard related deaths in United States during a period of 25 years from 1970 to 2004. Similar information including risk of death in various activities can be found in the literature (e.g., Melchers 2002). Melchers (2002) reported that the risk of death per year due to structural failures is about 0.1×10^{-6}, that is 10,000 times more than that of cigarette smoking and 1200 times more than that of boating. Table 2.7 gives generally accepted risk tolerance from infrastructure, based on values reported in the literature (see, e.g., Melchers 2002). This general criterion assumes that the risk due to a hazard should not exceed the risk faced by an individual in everyday life, that is, approximately one in 1,000 chance of annual risk of death.

2.3.3.4.4 Consequences

Computing consequences of seismic event on design of buildings is perhaps the most important difference between prescriptive design and PBD. In the context of PBD, consequences generally relate to the owner, or user. The consequences to the neighborhood or other regional effects are beyond the scope of current PBD efforts. In the current efforts of FEMA, two types of consequences are considered: monetary and casualty. Figure 2.18 illustrates how consequences are estimated as the final step of the PBD effort. The probabilities of different types of damages, as estimated by fragility curves, for example, are combined with predetermined damage level damage costs relationships. The estimated cost of the earthquake event can then be computed as shown in Figure 2.18. This

TABLE 2.6
Hazard Related Deaths in the United States during a 25-year period (1970–2004)

Natural Hazard	Deaths (%)
Volcano	0.3
Wildfire	0.4
Avalanche and landslides	0.8
Fog	0.8
Earthquakes	1.2
Hurricanes/tropical storms	1.5
Coastal	2.3
Wind	8
Severe storms/thunder storms	10
Lightning	11.3
Tornado	11.6
Flooding	14
Winter weather	18.1
Heat	19.6

Source: Borden, K. A., Natural hazards mortality in the United States, Ph. D. thesis, University of South Carolina, SC., 2008.

TABLE 2.7
Risk Tolerance

Category	Annual Risk of Death
Broad acceptance (no efforts required in risk mitigation)	Less than one in a million
Tolerable (not negligible and mitigation suggested)	Between one in 10,000 to one in a million
Unacceptable (mitigation required)	Greater than one in 10,000

method of computing cost based on uncertainties is one of the many definitions of risk (Chapter 8). Thus, we can see that PBD is a risk-based paradigm, as stated earlier.

After the cost (risk) of the seismic event, given the chosen performance levels, is determined, the stakeholders (owner, architect, engineer, users, insurance companies, etc.) must decide if it is an acceptable cost (risk). If it is acceptable, the design should go forward as is. If the costs proved to be too high the performance levels are then changed, and the whole procedure is repeated until an acceptable cost (risk) is reached (Figure 2.16).

2.3.3.4.5 Uncertainties

One of the main advances that second generation seismic PBD paradigm offer is that it acknowledges the uncertainty present during the process of seismic, or any other hazard, design of buildings, or any other infrastructure. The uncertainties in defining the seismic hazard, the design process (vulnerabilities), and estimating consequences are all included within the PBD paradigm. This is in sharp contrast with prescriptive designs. Admittedly, uncertainties are also accommodated to a certain extent in prescriptive designs: allowable stress design (ASD) utilizes factors of safety and load and resistance factor design (LRFD) accounts for load factors and strength reduction factors, as the name implies. Yet PBD allows for far more freedom in prescribing desired degrees of exceedance levels and probabilistic levels for the building and events on hand. For example, a particular building stakeholder might decide that a nonexceedance probability of 95% is needed for the performance of the building in a case of a seismic event. Another stakeholder, for another building, might decide that an 85% nonexceedance probability is more adequate for the building. Such a freedom of uncertainty level choice is one of the major advantages of PBD.

2.3.3.5 Future Trends of PBD

Performance-based design for seismic and other hazards is gaining interest from the professional community, for the reasons stated earlier. There are several other emerging paradigms that will make PBD of more importance to the infrastructure design community. Some of these paradigms are briefly discussed below.

2.3.3.5.1 Multihazards Considerations

The rising costs of different hazards, seismic, wind, floods, bomb blasts, progressive collapse, and so on have prompted interest in optimal designs that consider all of those hazards so as to increase safety and reduce costs, Ettouney et al. (2005). For a given building design, two hazards, seismic and blast, for example, can both exhibit higher performance levels as the magnitudes of *both* hazards increase (Figure 2.19). In other building configuration, the performance levels as the building responds to each of the hazards might decrease (Figure 2.19). The former situation can be costly, while the latter situation can be unsafe. Careful multihazards considerations are needed as a companion to PBD approach.

2.3.3.5.2 Life Cycle Analysis

Life cycle analysis is a byproduct of PBD. The knowledge of life cycle behavior is of immense importance to asset managers in their decision making efforts: inspection, prioritizing, budgeting, maintenance, and so on. Current prescriptive design paradigm does not offer such knowledge.

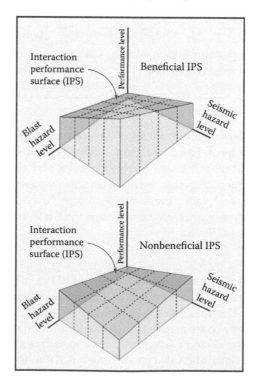

FIGURE 2.19 PBD and multihazards considerations.

Performance-based design offer numerous advantages as discussed above. It also presents several challenges: need for learning, smooth multidisciplinary integration, and added expertise of professionals. The advantages of PBD make meeting these challenges a worthwhile goal.

2.3.3.6 Role of SHCE and PBD

2.3.3.6.1 Overview

Structural health monitoring is also emerging as an essential paradigm for preserving the health of infrastructure. There is a natural symbiosis between PBD and SHM. On one hand, SHM produces the needed plethora of information by PBD. Conversely, PBD techniques provide to SHM professionals some valuable information about expected damages in the structure that can aid them in placing sensors and identification of structures and damages. Figure 2.20 shows how the three basic components of PBD paradigm interact with SHM/SHCE methods and components.

2.3.3.6.2 Fragilities

We discussed earlier the use of fragility concepts in PBD. Since fragilities offer simple way of describing probabilities of different limit states, their use in infrastructure design and assessment is gaining popularity. For example Reinhorn et al. (2001) used fragilities in seismic assessment of structures. FEMA-445 (2006) used fragilities extensively in developing PBD in seismic mitigation of nonstructural components in buildings. A fragility relationship as used in Figure 2.21 is defined as

$$P(D \geq D_i) = f(H) \tag{2.2}$$

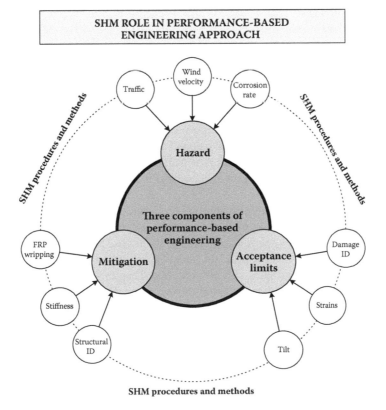

FIGURE 2.20   SHM methods as related to performance-based engineering.

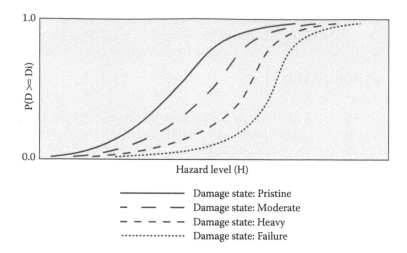

FIGURE 2.21   Typical fragility curves.

where $P(D \geq D_i)$ is the probability that the damage D of the system, or component, of interest is at damage state that is equal or more severe than D_i due to a hazard level of H. The function $0.0 \leq f(H)$ ≤ 1.0 is the fragility function. Numerous authors have studied and produced fragility relationships. See, for example, Mostafa and Grigoriu (1999) and Dumova-Jovanoska (2000).

When fragility relationships are known, it can help officials in assessing competence of the system. Also, it is a invaluable step in PBD, as shown in Figure 2.18. The main problem is: how can fragilities be developed accurately? Fragility is usually developed by analytical techniques. Common steps are as follows:

1. For a given component, estimate statistical parameters of uncertain parameters. For example: if a bridge girder is under consideration, properties such as the moment of inertia, modulus of elasticity, and so on can be assumed to be uncertain. For each of these uncertain properties, the mean and variance should be estimated.
2. Choose a damage state D_i. The damage can be any type of degradation in the system of interest. For example damage can be deformation level, cracking of any type, settlement of foundation, or state of paint or corrosion. The damage states can vary from pristine to complete failure. Note that for a total of damage states of N, $i = 1, 2, ..., N$. Since the system parameters of # 1 are random, it is obvious that the damage measures, for example deformation levels, are random.
3. For the hazard of interest, choose an appropriate hazard level H_j with $j = 1$.
4. Using appropriate probabilistic method, estimate the statistics or the histogram of the damage level D when the system is subjected to the hazard level H_j. Monte Carlo simulation method is used for this purpose for complex systems. For simple systems, analytical or semianalytical techniques can be used. See Chapter 8 for more details. From the histogram, the probabilities that the system is in a given damage state D_i due to H_j can be computed.
5. Change hazard level to H_{j+1}, and repeat #4 given above.
6. Repeat steps 3 through 5 until all reasonable levels of hazard are considered.
7. The N fragility relationships $P(D \geq D_i) = f(H)$, with $i = 1, 2, ..., N$ are formed. A set of N fragility curves can be plotted as shown in Figure 2.21.

The above-mentioned analytical method in computing fragilities is fairly straightforward. However, it relies in its accuracy on several assumptions, such as the knowledge of the statistical parameters of the system properties. Another implicit, yet very important, assumption is that we assumed that we *know* the exact relationship between the hazard, and the resulting damage during the analysis step # 4. Such an assumption does not produce accurate results in many situations.

Structural health monitoring techniques can be used to reduce the effects of some of these knowledge gaps. For example, on-site sensing can produce enough data points to form a reasonable data set that can produce accurate statistical estimates of the desired system properties. The hazard–damage relationship can be accurately computed using many of the SI/damage identification methods of Chapters 6 and 7. Next, we present two useful uses of fragilities in bridge management that utilize SHM techniques.

2.3.3.6.2.1 Fragilities and Deterioration Transition Matrices Let us consider next bridge deterioration as a damage measure and pose the question: can we use SHM techniques to produce bridge, or bridge component, deterioration fragilities? We offer the following method to produce such curves. We start by using the well-known bridge deterioration data that is generated from inspection reports; see Agrawal et al. (2008), or DeLisle et al. (2002), for example. We formalize these curves as

$$D = D(T) \tag{2.3}$$

where D is an estimate of the deterioration and T is the time. A typical deterioration curve is shown in Figure 2.22. A measure of deterioration can be bridge rating, which is one of the outcomes of a manual bridge inspection. Obviously, at anytime, the estimation of D is uncertain. As such D can be considered a random variable. DeLisle et al. (2002) studied this subject and developed a method to

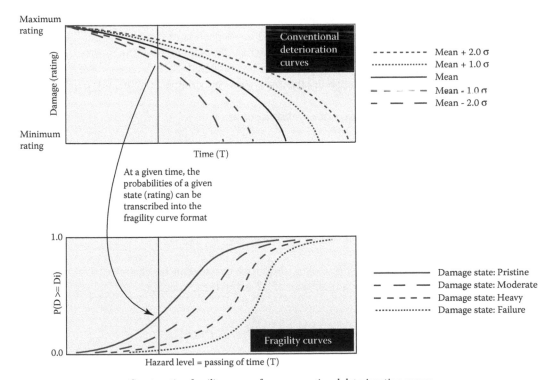

Constructing fragility curves from conventional deterioration curves

FIGURE 2.22 Relationship of deterioration and fragility curves.

estimate the uncertainties of D. Confidence intervals can be produced using that method as shown in Figure 2.22. Close inspection of Figure 2.22 reveals that the damage, as measured by bridge deterioration, is an outcome of the passing of time T. Thus we propose that the time T is the hazard that produces deterioration. On the basis of this we observe that Figure 2.22 actually includes the three basic ingredients of fragilities:

1. It includes an increased hazard (passing of time)
2. It includes damage states (deterioration levels as represented by bridge, or component, rating)
3. It includes probabilities of occurrence as represented by confidence intervals

Armed with these observations, we can transpose conventional deterioration curves into fragility curves using the conceptual method of Figure 2.22.

Bridge, or bridge component, deterioration fragility curves enable the decision maker with a powerful tool to use PBD techniques to reach accurate decisions. This can be accomplished by improving the estimates of the probability transition matrices $[P]$ that are commonly used in Markov processes in bridge management systems (see Chapter 8). The matrix $[P]$ is a square matrix with size of n that represents the number of possible damage states. When considering the bridge deterioration as damaged states, it is customary to consider n as the number of bridge rating states, see Agrawal et al. (2008b). As such, in NYSDOT practice, $n = 7$. The components of the matrix p_{ij} indicates the probability that the damage state of the system, or the component, will change from damage state D_i to damage state D_j in a given time interval ΔT. The values of p_{ij} are computed using historical records and measurements. They are usually assumed to be stationary, that is, time-independent. Such an assumption is not an accurate assumption in many situations. We recognize these probabilities as the prior estimates of probabilities; we redefine them as $p_{ij}|_{PRIOR}$.

Fragility deterioration curves, as described above, can be used to improve on as well as provide time dependency of the traditional estimates of p_{ij}. Close inspection of the fragility curves (note that there is a curve for each damage, or rating state) show that the probabilities of the system, or component, in a given damage state, at a particular time can be computed from the curves. We define these sets of probabilities $p(T,D_i)$ as the probability that the system, or component, is at a given damage state D_i at time T.

We use Bayes theorem to improve prior probabilities as

$$p_{ij}\Big|_{POSTERIOR} = \frac{p_{ij}\big|_{PRIOR} \cdot p(T,D_i)}{\sum_{j=1}^{j=n} p_{ij}\big|_{PRIOR} \cdot p(T,D_j)} \tag{2.4}$$

Note that the expressions of Equations 2.4 are time dependent; they are also improved estimates of the prior transition probabilities.

2.3.3.6.2.2 Fragilities and Life Cycle Costs   Fragility curves can be easily used in estimating deterioration life cycle costs during a time span of ΔT using the expression

$$LCC(\Delta T) = \int_{T=T_0}^{T=T_0+\Delta T} \left[\sum_{i=1}^{i=n} p(T,D_i) \cdot C(D_i) \right] dT \tag{2.5}$$

The initial time is T_0. The cost of damage state D_i is $C(D)$. Equation 2.5 is useful in estimating costs during shorter time spans. For longer time spans, a dynamic programming, or Monte Carlo technique should be used (see Chapter 9 of Ettouney and Alampalli 2012).

2.3.3.6.3 *Hazards, Vulnerabilities, and Consequences*

SHCE can play an even bigger role in enhancing PBD methods. Table 2.8 shows some examples of how SHCE techniques can aid in accurate estimations of hazards, vulnerabilities, and consequences for different situations. Judicial use of the techniques can improve safety at reasonable costs.

TABLE 2.8
SHM Role in Estimating Hazards, Vulnerabilities, and Consequences

Class	Application	Hazard	Vulnerability	Consequence
Deterioration	Paint	Paint systems deteriorate as a direct response to environment. Monitoring different environmental factors, such as temperature, freeze/thaw, moisture, and so on can give an accurate estimate on how the paint would survive the environmental onslaught	When paint deteriorates, irregularities on the surface would start forming. At first, these irregularities can't be detected by visual inspection. A laser-based surface inspection can help in detecting such small size irregularities. Of course, as the size of paint irregularities increases, it becomes possible to detect them using visual inspection	One of the major consequences of loss of paint protection is the formation of rust. Rust mitigation can be costly. Ultimately, severe loss of structural resistance due to rusting can lead to local, or more extensive levels of failure. Decision making techniques can help in forming optimal safe and cost effective decisions regarding rust mitigation. See Chapters 3 and 4 in Ettouney and Alampalli (2012) for examples of decision making processes regarding rust and corrosion
	Corrosion	See Chapters 3 and 4 in Ettouney and Alampalli (2012)		
	Fatigue	See Chapter 5 in Ettouney and Alampalli (2012)		
	Settlement	Displacement and tilt sensors, (Chapter 5 of this volume), can be used to monitor potential settlement	Strain monitors, located at optimal locations, are most suited to estimate vulnerability to settlement. Note that STRID methods can help in identifying accurately potential damages from settlement	Consequences of ductile settlement (gradual and noticeable settlement) are usually manageable, if detected early on. Consequences of large and sudden settlements can be large, including large-scale collapse. Scour is the leading cause for such sudden settlements. For detailed discussions on SHM and scour, see Chapter 1 in Ettouney and Alampalli (2012)
Hazard	Earthquakes	See Chapter 2 in Ettouney and Alampalli (2012)		
	Scour	See Chapter 1 in Ettouney and Alampalli (2012)		
	Live loads (Truck Weights)	Monitor truck loads, frequency and dynamics. Also, if possible, monitor the dynamic bridge-truck interaction	Monitor deflections, strains and dynamic amplifications. Ensure that the locations of sensors are in optimal positions, see Chapter 5 for more on optimal sensor locations	Consequences of overloads vary from shorter fatigue life (Chapter 5 in Ettouney and Alampalli [2012]) to brittle failure (I-35, MN collapse as described in Chapter 3 of this volume)

2.4   NDT VERSUS SHM

This section explores the interrelationship and differences between methods and techniques of nondestructive testing (NDT) and SHCE/SHM. NDT offer wealth of methods and techniques that can compliment and enhances traditional visual inspection methods (see Figure 2.23). These methods and techniques offer opportunities for the SHM community. We argue that the use of NDT methods has been successful in SHCE/SHM when hazard-centric needs of SHCE/SHM and physical demands of these hazards are recognized.

2.4.1   Overview

Nondestructive Testing methods have been in use and in continuous development for more than a century. Their main use has been to provide better understanding of material and system properties. There are numerous publications on this subject including several handbooks such as *NDT Methods Handbook of the American Society* for NDT. SHCE and its subcomponent SHM are emerging fields in the infrastructure arena and are intended to provide more information about structural condition (health) to make appropriate decisions. Even though there are a vast array of NDT methods and techniques that are available to the SHCE/SHM professionals, the use of NDT methods in the SHCE/SHM field has been relatively slow.

2.4.2   NDT Tools for Different SHM/SHCE Components

As noted earlier, SHM can be subdivided into three components: sensing/measurements, STRID, and damage identification. The subject of SHCE contains SHM and decision making. Tables 2.9 through 2.12 show how sensing, SI, damage identification, and decision making relate to both SHCE and NDT fields, respectively.

2.4.3   NDT, SHCE/SHM Tools and Bridge Hazards

The previous section showed some of the differences between NDT and SHCE/SHM concepts. In this section, beneficial uses of NDT tools in the field of SHCE/SHM are explored. We investigate this subject by recognizing that the field of SHCE/SHM aims mostly to counter different hazards

FIGURE 2.23   Visual inspection of bridges. (Courtesy Barton Newton.)

TABLE 2.9
General Comparison of NDT and SHCE/SHM for Sensing/Measurements

NDT	SHCE/SHM
Localized in nature	Generally covers larger area
Needs good idea as to the location of damage	Location of damage is not known before
Mostly a localized event in time	Time spans can be fairly long: almost as long as the lifespan of the target system. Can be continuous monitoring or long-term monitoring
Sensing technology is very advanced	Sensing technology is developing (wireless sensing, remote sensing, etc.)
Mostly manual	Mostly automatic or semiautomatic
Mostly single type of sensor	Mostly multiple types of sensors

TABLE 2.10
General Comparison of NDT and SHCE/SHM for STRID

NDT	SHCE/SHM
STRID methods use limited subset of NDT methods, since STRID deals with global structural behavior rather than localized structural regions	STRID is an integral part of most SHM projects
Need to develop STRID methods that make better utilization of NDT methods	STRID methods are well developed
Most NDT projects do not utilize/need STRID methods	Utilize/need STRID methods in many projects

TABLE 2.11
General Comparison of NDT and SHCE/SHM for Damage Identification

NDT	SHCE/SHM
Localized identification of damage sources	Efficient solutions are still being developed
If type and approximate location of damage is known, would give accurate results	Damage location can be in a wider area; and cannot be detected easily. Still under investigation
Due to localized nature, global structural information is not needed.	Requires more information regarding structure (geometry, properties, etc.)
Simple computational techniques	Need extensive analytical methods
Higher reliability. No need for extensive statistical and reliability analyses	Reliability unknown yet. Need extensive statistical and reliability analyses

TABLE 2.12
General Comparison of NDT and SHCE/SHM for Decision Making

NDT	SHCE/SHM
Many tools are available for NDE	More involved decision making tools and processes are needed due to the global nature (both in space and time) of SHM
Consequences of global failure are not usually addressed	Consequences of global failure must be considered
Vulnerability of global systems is not an issue due to the local nature of the NDT	Global system vulnerability is important
Due to short time scope of NDT, stochastic modeling is limited	Time scope can be long (as long as the useful life of the target system) and hence, stochastic modeling is important
Traditionally, risk as a tool, is not used. Only reliability (probability of failure) is used extensively	Risk as a tool is used in addition to reliability (or probability of failure) issues

that might affect bridges. Hence, we discuss some NDT technologies as they relate to four of the common hazards—scour, earthquakes, corrosion, and fatigue. Figure 2.24 shows an experiment to detect remaining fatigue life in a steel girder bridge. These four hazards affect the bridge in several manners and thus, the use and efficiency of different NDT methods can vary considerably.

The following commonly used NDT methods are considered to illustrate the above.

Ultrasound/Sonar: Ultrasound methods are based on generating stress waves in the target system at frequencies in the range of 0.10–35 MHz. The measured response of the system will reveal information about the defects. Sonar waves also are used for similar purposes. In a recent NYSDOT project, underwater sonar was used in an SHM setup that calls for continued monitoring of soil level around bridge piers as shown in Figure 2.25 (Scour 2006). When the soil level falls below a critical scour level, it is detected by the sonar system and the bridge officials can be warned.

FIGURE 2.24   Remaining fatigue life experiment. (Courtesy Keith Ramsey.)

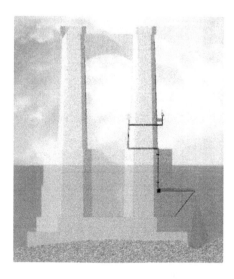

FIGURE 2.25   Sonar setup for a scour project, elevation view. (Courtesy of New York State Department of Transportation.)

Ultrasound Methods: Other ultrasound methods are used extensively in bridge health projects. Figure 2.26 shows an ultrasound detection process for bridge pin. The accuracy of the methods is needed due to the importance of bridge pins to the overall well being of the bridge. Impact echo is a popular and simple ultrasound technique that is also used successfully in bridge monitoring.

Acoustic Emission: The damage process produces stress waves that propagate away from the damage locations, that is, acoustic emissions. The frequencies of these waves are generally in the range of 20 KHz to ~ 1.2 MHz. Recording and analyzing these stress waves can assist in damage detection.

Thermography: Infrared thermography is used widely as an NDT technique to detect near-surface defects. It is based on stimulating the material by heating/cooling it and observing its surface temperature response to this stimulus. Knowing thermal properties of the material, material defects can be detected. Possible use of infrared thermography to inspect bridge column wrapped with fiber reinforced polymer materials is shown in Figure 2.27. Figure 2.28 shows results of infrared thermography when detecting possible damage in a bridge pavement. If the method is well developed, a base line image can be taken after construction and can be used to monitor the deck periodically during its life.

FIGURE 2.26   Ultrasound monitoring of bridge pins. (Courtesy Barton Newton.)

FIGURE 2.27   Infrared thermography: bridge column. (Courtesy of New York State Department of Transportation.)

FIGURE 2.28   Infrared thermography: bridge pavement. (Courtesy Keith Ramsey.)

FIGURE 2.29   Monitoring of cable stayed bridge. (Courtesy Keith Ramsey.)

Penetrating Radiation: When high frequency waves, such as X-rays or gamma-rays penetrate solids, their back-scattered or penetrated traces can reveal information about the interior of the solid, including any defects.

Vibration Monitoring: This technique is also used extensively in STRID and DMID of bridges. Figure 2.29 shows an SHM project for monitoring cable stayed bridge behavior.

Other Methods: There are several other NDT methods that are used to monitor bridge health. Some of these methods include magnetic-particle methods, magnetic flux leakage (Figure 2.30), dye penetrant (Figure 2.31) and ground penetrating radar.

Table 2.13 shows a sample of the applicability of these NDT methods for use in SHCE/SHM projects as they pertain to different hazards.

FIGURE 2.30 Monitoring for cracks using magnetic flux NDT method. (Courtesy Barton Newton.)

FIGURE 2.31 Dye penetrant capability in detecting surface cracks in bridge structures. (Courtesy Barton Newton.)

2.4.4 CLOSING REMARKS

This section explored the interrelationships between NDT and SHCE/SHM and showed some of the differences between the two fields. It also illustrates the manner that the well-established NDT field can help the emerging SHCE/SHM field to achieve their goals of safer infrastructure at reduced cost. In conclusion, NDT has immense potential in advancing the SHCE/SHM field but should address the specific and unique needs of SHCE/SHM.

We close this section by observing that for NDT methods to become more valuable in the SHCE/SHM field, they should be

- Reliable
- Easily understood
- Justified in terms of value (benefit) versus costs
- Easy to use by routine inspectors

TABLE 2.13
Sample Applicability of NDT for SHCE/SHM in the Bridge Hazard Field

Bridge Hazard	NDT Method			
	Ultrasound/Sonic	Acoustic Emission	Thermography	Ground Penetrating Radar
Scour	Sound waves in water (Sonar) can identify critical scour depth of bridge foundations See Chapter 1 of Ettouney and Alampalli (2012)	No clear direct use of the technologies. An indirect use would be to spot-check structural damages due to scour settlement or failures		
Earthquake	Identifies material properties and condition assessment for structural identification before and after seismic events	Identifies damage after seismic events. Useful for most construction materials	Condition assessment before and after seismic events. Mainly for FRP-based construction	Has been is used extensively for condition assessment of bridge decks. Can be used for condition assessment of connections after seismic events (Housner and Masri 1996)
Corrosion	Detects corrosion presence indirectly, through detecting cracks or changes in material properties	Was shown to be capable of detecting emissions from the rusting process. May be used to continuously monitor corrosion activities	Can detect damages in FRP-wrapping material due to degradation in underlying concrete material	Can be used for condition monitoring of bridge decks
Fatigue	Detects cracks due to fatigue	Has potential to detect active fatigue cracks	Can detect damages due to fatigue in FRP-based construction	Can be used for fatigue-related condition monitoring of bridge decks

- Easy to implement at network level
- Durable and maintenance free (or close)
- Validated theoretically in the early stages of development

2.5   VALUE OF INFORMATION

2.5.1   INTRODUCTION

SHCE is all about gaining information about the structure on hand. Executing any SHCE project would entail cost expenditure. Costs can be direct financial, labor, social, and so on. In numerous situations most of the SHCE costs are required upfront or at early stages of the project. In return, mostly information about the structure is obtained. This brings one of the most important questions of the subject of this volume:

Are the information obtained through a particular SHCE project worth the costs involved?

Clearly, if the answer to this question is "no," then the particular project under consideration is not worth pursuing. If the answer is "yes," then that project worth pursuing.

To find an accurate answer to the above question, we need to evaluate both the cost of the project C as well as the benefit of the project B. If

$$C > B \qquad (2.6)$$

then the project is not worth pursuing, and if

$$C < B \qquad (2.7)$$

then the project is worth pursuing.

We can simplify the above equations even further by introducing the concept of value V where

$$V = \frac{B}{C} \qquad (2.8)$$

So the evaluation of an SHCE project is reduced simply to evaluating the value of the project: if

$$V > 1 \qquad (2.9)$$

then pursue the project, and if

$$V < 1 \qquad (2.10)$$

then the project is not worth pursuing.

Since C, B, and V are paramount to the subject of SHCE, we will be addressing those issues in most of the subjects of this volume, and will be trying to illustrate different methods and techniques that would enable the professionals and decision makers to make accurate decisions regarding specific SHCE projects. Of particular importance is that we need to define both C and B using same units, to be able to find a numerical expression for V. It is easy to express C in monetary terms. Thus, it is desired to express B also in monetary terms. Since both C and B are functions of time their monetary estimates will also be the function of discount rate I throughout the desired life span T.

The remainder of this section will be devoted to the more specific application of the above question. We can restate the question as: can there be a value to a given type and quality of information? We will offer a general approach to addressing the issue next.

2.5.2 GENERAL APPROACH TO EVALUATE VALUE OF INFORMATION

Cost Estimation: Generally speaking, estimating C of any SHCE project is much easier than estimating B. Elements of costs, as was mentioned earlier, include labor, equipments, analysis, design, operations, and so on. The cost of the project C can be expressed as

$$C = \sum_{i=1}^{i=NC} C_i(I) p_i \qquad (2.11)$$

where $C_i(I)$ is the present value of the cost of the ith cost component. The probability of occurrence of the ith cost component is p_i. The total number of cost components is NC.

Benefit (Direct) Estimation: Elements of benefit are much more difficult to enumerate, especially in monetary terms. For example, some possible benefits of an SHCE project are direct, such as increasing the life span of the structure, increasing the safety of the users, or reducing the costs of operations. The direct benefits of the project B_D can be expressed as

$$B_D = \sum_{i=1}^{i=NC} B_{Di}(I) \ p_{BDi} \tag{2.12}$$

where $B_{Di}(I)$ is the present value of the direct benefit of the ith direct benefit component. The probability of occurrence of the ith direct benefit component is p_{BDi}.

Benefit (indirect) Estimation: Some indirect benefits of SHCE are gained knowledge about particular monitoring technique, evaluation of specific type of sensors, applications of structural or damage identification methods on an experimental basis. Can we put a monetary estimate to these indirect benefits?

Indirect benefits are those benefits that might be achieved after several steps have been accomplished. For example: if an experiment in the validity of ultrasonic laser in damage identification (Chapter 7) is under consideration. Since it is an experimental project, there are no direct benefits. Of course, there are indirect benefits. To estimate those indirect benefits, we establish that as a direct result of the experiment, there are two possible outcomes: success and failure. Each of those outcomes has a particular probability of occurrence. In case of success, an experimental, system-wide project will be established. Such a system-wide experiment will have a potential of success or failure, each with its own occurrence probability. This will continue until the final effects of the experiments will reach a self sustaining beneficial use. In this case, the potential use of ultrasonic laser in identifying damage. It is at this moment that an actual monetary estimate of the final benefit of the project can be made. The indirect direct benefits of the project B_I can then be expressed as

$$B_I = \sum_{i=1}^{i=NI} B_{Ii}(I) \ \prod_{j=1}^{j=NS} p_{ij} \tag{2.13}$$

Where $B_{Ii}(I)$ is the present value of the indirect benefit of the ith indirect benefit component. The probability of success of the ith indirect benefit component at the jth step is p_{ij}.

The above method of estimating indirect benefit is a simple form of sequential decision tree analysis. See Hamburg (1977) and/or Benjamin and Cornel (1970) for more details. In the next section we present a method of estimating value of information with uncertainties.

Note that B_I can be considered as a monetary measure of the benefits of information. We note also that Equation 2.13 is a special form of Equation 2.12, with $NC = 1$.

On the basis of the above, the value of the information gained from an SHCE project is

$$V = \frac{C}{B_I + B_D} \tag{2.14}$$

We can then conclude that as a result of any SHCE project there always exists a finite value V for the information that result from the project. It behooves the professional to ensure that the value of such information is well understood and documented.

REFERENCES

Agarwal, A., Kawaguchi, A., and Qian, G., (2008). "Bridge Element Deterioration Rates: Phase I, Report," TIRC/NYSDOT Project # C-01–51, NYSDOT, Albany, NY.

Alampalli, S., (1998) "Effects of Testing, Analysis, Damage, and Environment on Modal Parameters." Proceedings of Modal Analysis & Testing. NATO-Advanced Study Institute, Sesimbra, Portugal, pp. 427–443.

Alampalli, S. and Ettouney, M., (2004) "Observations, Recommendations and Items of Interest to Health Monitoring Community," Proceedings of 2nd International Workshop on Structural Health Monitoring of Innovative Civil Engineering Structures, ISIS Canada Corporation, Manitoba, Canada.

ASCE., (2007) *Seismic Rehabilitation of Existing Buildings*, ASCE / SEI Standard No 41-06, American Society of Civil Engineers, Reston, VA.

Benjamin, J., and Cornell, A, (1970). *Probability, Statistics, and Decision for Civil Engineers*, McGraw-Hill, New York, NY.

Borden, K. A. (2008) "Natural hazards Mortality in the United States," Ph.D. thesis, University of South Carolina, SC.

Cobb, R. G., (1996) "Structural damage identification from limited measurement data," Ph. D. dissertation, School of Engineering, Airforce Institute of Technology, Wright Patterson AFB, OH.

DeLisle, R., Sullo, P., and Grivas, D., (2002) "A network-level pavement performance model that incorporate censored data," TRP preprint Number 03–4281.

Dumova-Jovanoska, E., (2000) "Fragility curves for reinforced concrete structures in Skopje (Macedonia) region," *Soil Dynamics and Earthquake Engineering, Elsevier,* 19, 455–466.

Eghbali, G.H. and Elsayed, E.A., (1997) "Reliability Estimation Based on Degradation Data," Working Paper No. 97-117, Department of Industrial Engineering, Rutgers University, Piscataway, NJ.

Elsayed, E. A., (1996) *Reliability Engineering*, Addison-Wesley, Reading, MA.

Enright, M. P. and Frangopol, D. M., (1998) "Probabilistic Analysis of Resistance Degradation of Reinforced Concrete Bridge Beams Under Corrosion," *Engineering Structures*, 20(11), 960–971.

Ettouney, M. and Alampalli, S., (2000) "Engineering Structural Health" Proceedings of the Structural Engineers World Congress, Philadelphia, PA.

Ettouney, M., and Alampalli, S., (2008) "Virtual Sensing in Structural Health Monitoring," XXVI International Modal Analysis Conference (IMAC), Orlando, FL, February 2008.

Ettouney, M. and Alampalli, S., (2012) *Infrastructure Health in Civil Engineering: Applications and Management*, CRC Press, Boca Raton, FL.

Ettouney, M., Alampalli, S., and Agrawal, A., (2005) "Theory of multihazards for bridge structures," *Journal of Bridge Engineering,* IOS Press, 1(3), 281–291, Amsterdam, Netherlands.

Ettouney, M. and Elsayed, E. A., (1999) "Reliability Estimation of Degraded Structural Components Subject to Corrosion," Fifth ISSAT International Conference, Las Vegas, NV.

Ettouney M., Daddazio, R. and Hapij, A., (1998) "Health Monitoring of Complex Structures" SPIE Smart Structures and Materials Conference, San Diego, CA.

Ettouney, M., Daddazio, R., and Hapij, A., (1999) "Optimal sensor locations for structures with multiple loading conditions" SPIE Smart structures and Materials Conference, San Diego, CA.

Farrar, C. Duffy, T., Doebling, S., and Nix, D., (2000) "A statistical Pattern Recognition Paradigm for Vibration-Based Structural Health Monitoring," Proceedings of 2nd International Workshop on Structural Health Monitoring, Stanford University, Stanford, CA.

FEMA-445., (2006) Next-Generation Performance-Based Seismic Design Guidelines, FEMA Report No 445, Washington, DC.

Ganji, V., Tabrizi, K. and Vittilo, N., (March 2000) "Project Level Application of Portable Seismic Pavement Analyzer," Proceedings of the SMT/NDT Conference, NYS-DOT, NJ-DOT and FHA, Atlantic city, NJ.

Hamburg, M., (1977) *Statistical Analysis for Decision Making*, Harcourt Brace Jovanovich, Inc., New York, NY.

Helmicki, A., Hunt, V. J., and Lenett, M., (2002) "Non-Destructive Testing and Evaluation of the Ironton-Russell Bridge," Proceedings of NDE Conference on Civil Engineering, ASNT, Cincinnati, OH.

Housner, G. and Masri, S., (1996) "Structural Control Issues Arising from the Northridge and Kobe Earthquakes," 11th World Conference on Earthquake Engineering, Mexico, Paper No. 2009.

IBC., (2006) *2006 International Building Code*, International Building Council, Washington, DC.

Melchers, R. E., (1987), *Structural Reliability*, Ellis Horwood Limited.

Melchers, R. E., (2002) *Structural Reliability Analysis and Prediction*, John Wiley and Sons, New York, NY.

Mostafa, E. and Grigoriu, M., (1999) "Fragility curves for non-structural systems," Report No. 99-2, School of Civil and Environmental Engineering, Cornell University, Ithaca, NY.

Oswald, G. F. and Schuëller, G. I., (1983) "On the Reliability of Deteriorating Structures," in *Reliability Theory and Its Application in Structural and Soil Mechanics*, Editors: Thoft-Christensen, P. and Nijhoff, M. The Hague, Netherlands.

Pugsley, A., (1962) *Safety of Structures*, Edward Arnold, London.

Reinhorn, A., Barron-Corvera, R., and Ayala, A., (2001) "Spectral Evaluation of Seismic Fragility of Structures," in *Structural Safety and Reliability (ICOSSAR 2001)*, Editors: Corotic et al., Swets & Zeitlinger, Balkema.

"Scour Monitoring Program Manual," Hardesty & Hanover, prepared for New York State Department of Transportation, July 2006.

UBC-97., (1997) *Uniform Building Code*, International Conference of Building Officials, Whittier, CA.

3 Lives and Deaths of Infrastructure

3.1 OVERVIEW

We explore in more detail the concept of bridges (or any other civil infrastructure) as an organism that is born, lives and dies (fails). We note that, similar to humans, there are three distinct phases of bridge life cycle: (a) its birth, (b) its normal existence (life), and (c) its death (failure). We propose that ensuring health of bridges applies to all phases of their life cycles. We immediately define the main attributes of healthy bridge during all phases as follows:

- A healthy bridge meets or exceeds stated performance goals
- At reasonable costs

Among performance goals we can state further that the bridge should experience no unwarranted service interruptions due to distress, and partial or total failure. A healthy bridge should also perform its stated functions, such as load-carrying specifications, in a safe and secure fashion. Near the end of the life cycle of the bridge, it is essential to control the decommissioning process, as opposed to abrupt failure/distress.

Overall, we aim at discussing categories that would enable decision makers to quantify different phases of life cycle of the bridge; we also explore different roles structural health monitoring (SHM)/structural health in civil engineering (SHCE) during life cycle of the bridge, and how those roles change as the bridge life span change. Understanding these details can enhance the optimal health of bridges during all the phases of its existence.

3.1.1 ENABLING VERSUS TRIGGERING CAUSES

To simplify examinations of health concepts of bridges, we categorize causes that affect bridge (or any other type of structure) into enabling and triggering causes. The categorization was used by Wardhana and Hadipriono (2003) to explore bridge failures. We extend the definition into other phases of bridge life span (construction, service life, and ultimate failure). Table 3.1 shows a general comparison between the two types of causes.

3.1.2 THIS CHAPTER

This chapter explores the health of a bridge at different phases of its life cycle as shown in Figure 3.3. First the health of the bridge during the birth phase is described. The different issues regarding health during the life span of the bridge is then surveyed. The third phase, bridge failure, or decommissioning, is discussed next. In all life cycle phases the triggering and enabling causes of structural health are defined and explored. The many roles SHM methodologies and techniques play at different phases are discussed in details. Whenever possible, cost–benefit issues of different SHM roles are sited. The chapter ends with an in-depth survey of important bridge failure examples in the United States in the past one hundred years.

TABLE 3.1
Enabling versus Triggering Causes

Causes	Enabling	Triggering
Temporal behavior	Occurs over a long period of time (see Figure 3.1)	Occurs suddenly (see Figure 3. 2)
Severity	Both can have severe effects on bridges	
NDT role	Can be useful	Effective only before and after event
SHM role	Long-term monitoring; intermittent monitoring	Before, during and after event monitoring might require different strategies

FIGURE 3.1 Example of enabling cause: corrosion of bridge column. (Courtesy of New York State Department of Transportation.)

FIGURE 3.2 Example of triggering cause: overload on a bridge. (Courtesy of New York State Department of Transportation.)

3.2 BIRTH OF BRIDGES

3.2.1 Ensuring Healthy Life

Healthy life of structures starts at "birth." Several phases of birth include (a) preliminary planning, (b) analysis, design, and detailing, (c) construction, and (d) quality assurance/quality control (QA/ QC). SHM/SHCE can play many roles during all of those phases. These roles are described next.

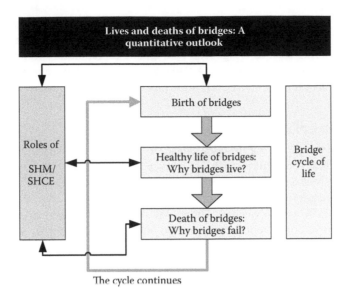

FIGURE 3.3 Organization of Chapter 3.

3.2.1.1 Preliminary Planning

Preliminary planning of a bridge includes numerous roles of SHM techniques and methods. Perhaps the most obvious are different decision making (DM) processes. Activities such as cost–benefit analysis, finance and economics implications, and social studies must precede any bridge construction phase. Also an exhaustive risk and life cycle studies must be done. Risk analysis would include different uncertainties that the bridge might encounter during its life span. Life cycle analysis of the bridge is performed. Implications of different decisions regarding the bridge should be analyzed. Decisions on optimal attributes of the bridge can then be made. Those DM methodologies are discussed in detail in Chapter 8.

Other SHM components also have roles in the preliminary planning phase. For example, soil explorations can affect decisions regarding the bridge, including the very viability of constructing it. STRID activities that make use of available data, such as soil exploration results can help in ensuring healthy life of the bridge, after it is built.

3.2.1.2 Design Phase

Analysis phase of new bridges can include STRID methods for optimal designs. Since STRID methods (Chapter 6) need actual test results as an input, simulated test results might be used. Simulated results can be generated using similar previous tests or any other appropriate technique. Another important measure can be to prepare base analytic models for future STRID/DMID activities. Such base analytic models can be for the global bridge structure, or for important local details. As future actual datasets start accumulating, those analytic models can be updated and improved to help in different future decisions and activities such as DMID, inspection, maintenance, and rehabilitation.

Design phase can utilize available test results (such as soil test results) to aid design paradigms, whether it is the capacity/demand or performance-based designs. Also, the design of future maintenance, rehabilitation, retrofits, and decommissioning activities has to be compatible with the overall bridge design. The designer should ensure that all such designs are optimal. A comprehensive life cycle analysis plan should be intertwined with the initial designs of the bridge. Such life cycle analysis plans should be scalable, flexible, and easy to be adjusted to future events.

Details of the bridge SHM projects must be included with the initial structural details. Such SHM details should be for (a) long-term monitoring, (b) intermittent monitoring, (c) one-time specialized in situ monitoring such as bridge load testing, and (d) laboratory testing, such as different NDT techniques. The bridge designer should also try to accommodate conventional visual inspection by using simple detailing as well as easy-to-inspect detailing such that other expensive options can be minimized.

3.2.1.3 Construction Phase

Innovative uses of SHM during construction phase of large civil infrastructure can produce many dividends. An example of such innovations was reported by Ni et al. (2008). The subject of the project is the Guangzhou New Television Tower, which is located in Guangzhou China (see Figure 3.4). It is a 610-m high structure that is constructed of a mix of reinforced concrete and concrete-filled steel tubes. The Hong Kong Polytechnic University designed the SHM project with two major goals: (a) to monitor in-construction performance, and (b) to monitor in-service performance. Both goals are done in real time. In addition, the SHM project has other innovative features such as (a) a modular design of the SHM system, (b) efficient wireless-sensing array, (c) advanced fiber optics sensing that is based on Bragg grating technology, and (d) structural identification processing using static and dynamic monitoring data. The project also has a sophisticated system for data validation and benchmark case studies.

What interests us in this section is the implementation of a during-construction SHM system by Ni et al. (2008) (see Figure 3.5). They used a total of 527 sensors that were placed at 12 cross-sections of the tower to monitor performance during construction. The choices of the sensor locations were obtained by analyzing the structure during construction process and placing them at sections with higher stress levels. Most of the sensors were strain sensors (total of 416) to monitor strains, creep, and shrinkage (vibrating wire strain gauges). In addition, large number of temperature sensors was used (total of 96). Some additional weather, wind speed, inclination, leveling of floors and tower, and displacement sensing were utilized. The use of these SHM sensors enabled the project construction engineers to ensure higher quality and safe construction process of the tower.

FIGURE 3.4 Guangzhou New Television Tower. (With permission from John Wiley & Sons Ltd.)

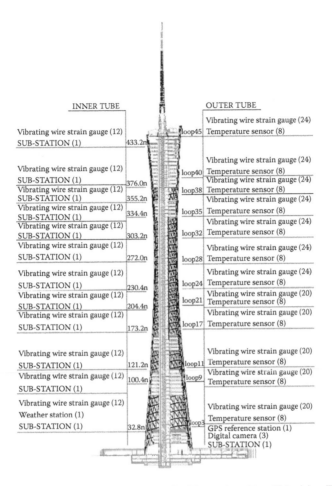

INNER TUBE OUTER TUBE

Vibrating wire strain gauge (24)
Temperature sensor (8)

Vibrating wire strain gauge (12)
SUB-STATION (1)   433.2m   loop45

Vibrating wire strain gauge (24)
Temperature sensor (8)

Vibrating wire strain gauge (12)
SUB-STATION (1)   376.0m   loop40
Vibrating wire strain gauge (24)
Temperature sensor (8)
Vibrating wire strain gauge (12)
SUB-STATION (1)   355.2m   loop38
Vibrating wire strain gauge (24)
Temperature sensor (8)
Vibrating wire strain gauge (12)
SUB-STATION (1)   334.4m   loop35
Vibrating wire strain gauge (24)
Temperature sensor (8)
Vibrating wire strain gauge (12)
SUB-STATION (1)   303.2m   loop32
Vibrating wire strain gauge (24)
Temperature sensor (8)
Vibrating wire strain gauge (12)
SUB-STATION (1)   272.0m   loop28
Vibrating wire strain gauge (24)
Temperature sensor (8)

Vibrating wire strain gauge (12)
SUB-STATION (1)   230.4m   loop24
Vibrating wire strain gauge (24)
Temperature sensor (8)
Vibrating wire strain gauge (12)
SUB-STATION (1)   204.4m   loop21
Vibrating wire strain gauge (20)
Temperature sensor (8)
Vibrating wire strain gauge (12)
SUB-STATION (1)   173.2m   loop17
Vibrating wire strain gauge (20)
Temperature sensor (8)

Vibrating wire strain gauge (12)
SUB-STATION (1)   121.2m   loop11
Vibrating wire strain gauge (20)
Temperature sensor (8)
Vibrating wire strain gauge (12)
SUB-STATION (1)   100.4m   loop9
Vibrating wire strain gauge (20)
Temperature sensor (8)

Vibrating wire strain gauge (12)
Weather station (1)
SUB-STATION (1)   32.8m   loop3
Vibrating wire strain gauge (20)
Temperature sensor (8)
GPS reference station (1)
Digital camera (3)
SUB-STATION (1)

FIGURE 3.5 In-construction SHM sensor placement for Guangzhou New Television Tower. (With permission from John Wiley & Sons Ltd.)

We should also note that one of the innovations of the SHM project was to integrate the in-construction monitoring with a long-term in-service monitoring. In addition, 280 sensors were used for the in-service SHM task. Of these, arrays of fiber optics Bragg grating sensor network were installed in permanent steel conduits. A total of 120 Bragg grating strain and temperature sensors were utilized for this task.

3.2.1.4 Quality Control/Quality Assurance (QA/QC)

Bridges, as well as all other major civil infrastructure, involve complex and lengthy construction process. The importance of performing construction tasks in accordance with the design and detailing specifications is exemplified in a usually rigorous and detailed QA and QC strategies. SHM methods are emerging as an accurate and inexpensive construction phase QA/QC device. Simple monitoring of different attributes such as concrete temperature, humidity, ambient temperature, and/or member alignment, can help in QA/QC processes. Consider, for example, the SHM experiment by Hansen and Surlaker (2006). They aimed at coupling the maturity principle of concrete during the curing phase and real time SHM activities in providing real time and accurate QA/QC information that can save costs, and ensure life cycle safety of the system. The maturity principle simply relates the curing time and the curing temperature to the concrete strength. Thus, by knowing the elapsed time, and concrete temperature, it was shown that an accurate estimate of concrete strength is possible. Two important reasons as to why accurate estimation of cured concrete strength is important: (a) it permits the removal of temporary forms when the concrete reaches postulated minimum strength, and (b) it signals the optimal time for post-tensioning in post-tensioned systems. The authors used a set of embedded radio frequency ID (RFID) sensors in two pilot structures: a parking garage, and a concrete highway pavement. They reported the following major benefits for their technique.

- In situ estimation of concrete strength eliminates the need for using laboratory cylinder tests, thus saving time and cost, while ensuring an even more accurate strength estimation.
- When used for a post-tensioned system, an online estimation of concrete strength would reveal the exact time when the concrete strength would permit the post-tensioning process. This optimizes the time utilization during construction, while ensuring safe post-tensioning operations.
- By placing RFID sensors across the depth of the pavement, it was possible to monitor the temperature gradient across the depth. This information can help predict any curling of the pavement; such curling can lead to a premature surface fatigue cracking and reduction of the service life of pavements.

Several other benefits of the experiment were reported by Hansen and Surlaker (2006). The authors provided estimates of cost savings that the use of their technique can provide the owners during the construction phase. We observe that there are additional savings that can be had during the life span of the system. Those additional savings result from factors such as improvements to fatigue strength limits, and reduced maintenance costs (by reducing or eliminating early tensile pavement cracking). It is clear that such innovative, yet simple and inexpensive use of SHM technologies can help in QA/QC of bridges during their construction phase.

One of the applications where NDT is very useful is quality control and assurance of bridges and pavements during construction. Garg et al. (2008) described a portable seismic property analyzer (PSPA) used to determine the modulus of concrete pavement slabs at FAA's National Airport Pavement Test Facility (Figure 3.6).

PSPA is a portable device (see Figure 3.7) and consists of a receiver transducers and a source transducer; and is based on generating and detecting stress waves in a layered medium. The data collected is processed by spectral analysis to determine the modulus of the layer. Advantages of using PSPA include nondestructive in nature, rapid to perform, immediate availability of results, repeatable, in situ testing of pavement in its natural state, and easy and convenient to operate.

FIGURE 3.6   FAA National Airport Pavement Test Facility. (Reprinted from ASNT Publication.)

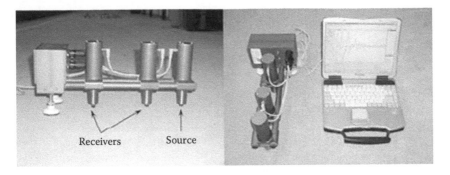

FIGURE 3.7   Portable seismic property analyzer (PSPA). (Reprinted from ASNT Publication.)

PSPA used estimate modulus values from tests on three concrete slabs under three different support conditions at different times after the concrete pours. The results compared well with the modulus values measured by free-free resonance tests performed on the concrete beam and cylindrical specimens prepared at the same time as the concrete slabs. The same PSPA device was also used to determine the modulus of hot mix asphalt (HMA) layer during CC2-OL (HMA overlay over rubblized concrete) project except that the active length of the seismic path was shortened. These two studies demonstrated the application potential of a NDT device such as PSPA for assessing the early-age strength of concrete and the modulus of HMA layer.

3.2.2   Role of SHCE

Structural health monitoring/structural health in civil engineering tools and methods can play major role during early phases of bridge life span. Table 3.2 shows some examples of such a role.

3.3   WHY BRIDGES LIVE?

3.3.1   Definitions of Healthy Life

Healthy life of bridges can be expressed in three facets: (a) meeting functional goals, (b) adequate performance, and (c) at reasonable costs. Functional goals include traffic, pedestrian, and commercial needs. Performance metrics include ensuring safety and security of users. This can be accomplished, in technical terms, by meeting adequate engineering criteria such as deflection, vibration, and stress/strain limits. Reasonable costs, or adequate cost benefits, are a major issue that is discussed in numerous parts of this volume.

This section addresses healthy life of bridges, its causes, and how to quantify a healthy bridge life. We offer some specific examples as to how SHM/SHCE field can improve healthy bridge life.

3.3.2   Classification of Healthy Life Causes

3.3.2.1   Enabling Causes of Healthy Life

By definition, enabling causes are continuous: they occur nearly all the time and their effects are felt over the long term. There are many enabling causes that would ensure healthy life of a bridge. To continue the framework of quantification of healthy life, we assign formal variables to these causes as discussed in Table 3.3.

TABLE 3.2
Roles of SHM in Early Phases in Bridge Life Span

Ensuring Healthy Life for New Bridges	Sensing/Measurements	STRID	DMID	Decision Making
Preliminary planning	Soil conditions need to be measured. In some situations, special sensing might be needed, such as propensity for liquefaction in areas with high seismic activities	Conventional structural analysis methods can be used for preliminary planning	Literature searches for damages and methods of DMID for bridges and sites that are comparable to the bridge and the site that are under considerations. The results of these searches can have two benefits: • Affects decisions for the current project • Affects decisions on any DMID method to be used in an SHM setup for the current project, if any	• Cost–benefit analysis • Risk analysis • Life cycle analysis
Design/detailing	Placement of future sensors. Types, numbers, and locations of the sensors Optimal NDT methods	Base analytical STRID models for updating at different stages of the bridge life cycle	Optimal integration of SHM activities and structural design of the bridge	Optimal decisions that accommodate • Cost–benefit • Life cycle analysis
Construction	Safe construction process can be monitored by strain, displacement or vibration sensors that are placed in optimal locations. Those locations are predetermined using structural analysis tools			
QA/QC	Adequate curing of concrete can be monitored using embedded RFID temperature sensors Quality of welds can be inspected using different NDT techniques			

TABLE 3.3
Enabling Causes for Healthy Life

Enabling Cause	Objective Rating
Sound management	$0 \leq EL_1 \leq 10$
Design	$0 \leq EL_2 \leq 10$
Detailing	$0 \leq EL_3 \leq 10$
Construction	$0 \leq EL_4 \leq 10$
Scheduled maintenance	$0 \leq EL_5 \leq 10$
Inspection/monitoring	$0 \leq EL_6 \leq 10$
Other	To be added as needed

We thus establish the enabling measure of a bridge healthy life as

$$HLTH_E = W_E \sum_{i=1}^{i=N_E} \left(W_{Ei} \cdot EL_i \right)$$

(3.1)

where

W_{Ei} = Appropriate weight for the ith parameter
W_E = Appropriate weight for the whole enabling causes
N_E = Total number of enabling causes

Note that in Table 3.3 $N_E = 6$. The number of enabling causes can change, depending on the situation on hand. The different weights in Equation 3.1 are assigned based on experience and research results. The measure $HLTH_E$ is thus a relative measure that can be used in DM situations, such as allocation of budgeting, and so on.

3.3.2.2 Triggering Causes of Healthy Life

Triggering healthy life causes are relatively sudden: they occur suddenly on the time scale of the bridge life span; their effects are felt suddenly in the form of improved bridge health. There are many triggering causes that would ensure healthy life of a bridge. Similar to the enabling causes, we assign formal variables to triggering causes as discussed in Table 3.4.

We thus establish the triggering measure of a bridge healthy life as

$$HLTH_T = W_T \sum_{i=1}^{i=N_T} \left(W_{Ti} \cdot TL_i \right)$$

(3.2)

where

W_{Ti} = Appropriate weight for the ith parameter
W_T = Appropriate weight for the whole triggering causes
N_T = Total number of triggering causes

Note that in Table 3.4 $N_T = 4$. The number of triggering causes can change, depending on the situation on hand. The different weights in Equation 3.2 are assigned based on experience and research results. The measure $HLTH_T$ is thus a relative measure that can be used in DM situations, such as allocation of budgeting, and so on.

3.3.3 QUANTIFICATION OF HEALTHY LIFE

3.3.3.1 Conventional Metrics

There are numerous methods to quantify the health of a bridge; see, for example, NYSDOT (2008). Among some of those methods:

Bridge Condition Rating: A simple rating system from 1 (totally deteriorated condition) to N (new condition). The value of N vary. For example, New York State uses $N = 7$. The value of the condition rating indicates the health condition of the bridge. Table 3.5 shows qualitative description of the condition ratings as used in New York State.

TABLE 3.4
Triggering Causes for Healthy Life

Triggering Cause	Objective Rating
Retrofits	$0 \le TL_1 \le 10$
Rehabilitation	$0 \le TL_2 \le 10$
Special maintenance	$0 \le TL_3 \le 10$
Replacement	$0 \le TL_4 \le 10$
Other	To be added as needed

TABLE 3.5
Definitions of Bridge Condition Ratings

Rating	Definition
1	Totally deteriorated or failed condition
2	Used for condition between 1 and 3
3	Serious deterioration or not functioning as originally designed
4	Used for conditions between 3 and 5
5	Minor deterioration and is functioning as originally designed
6	Used for conditions between 5 and 7
7	New condition

TABLE 3.6
Categories of Bridge Deficiencies

Condition Rating	Deficiency Level	Description	Potential Remedial Action
<3	Severe	Comprehensive serious deterioration of the bridge structural elements	Priority to remediate and repair
3.0 and ≤ 3.999	Moderate	Serious deterioration to some of the bridge main structural elements	Comprehensive structural work is likely. Rehabilitation and replacement options are generally available
4.0 and ≤ 4.999	Marginal	Moderate structural deterioration. Minor deterioration to primary support elements Efficient structural performance is rarely compromised	Minor rehabilitation/major maintenance activities might be needed

One common way to quantify the bridge condition rating is described in detail in NYSDOT (2008). The procedure is summarized as follows: during each general inspection of the bridge, various components, or elements of each bridge span are rated by the inspector according to the extent of deterioration and the ability of the component to function structurally, relative to when it was newly designed and constructed. These element rating values are then combined using a weighted average formula to compute an overall bridge condition rating value for each bridge. This formula assigns greater weights to the ratings of the bridge elements having the greatest structural importance and uses lesser weights for minor structural and nonstructural elements. If a bridge has multiple spans, each element common to multiple spans is rated on a span-by-span basis; the lowest individual span element rating is used in the overall condition rating formula.

Deficiency: A deficient bridge can be defined using its condition rating value. For example, NYSDOT defines deficient bridges as those with a condition rating less than 5 (on a scale from 1 to 7). A deficient condition rating indicates the presence of sufficient deterioration and/or loss of original function to require corrective maintenance or rehabilitation to restore the bridge to its fully functional, nondeficient condition. It does not mean that the bridge is unsafe. In most cases, bridges have enough excess or reserve structural capacity to accommodate some deterioration or degradation of structural function as indicated by a deficient condition rating. Figure 3.13 shows the number of deficient bridges in New York State between 1992 and 2006. The trend of the deficient bridge numbers is decreasing. To correlate state of deficiency with DM, NYSDOT (2001) in a report on bridge overloading identified further different states of bridge deficiency, their condition rating and potential remedial actions in Table 3.6.

Posting: When the rating of a bridge is reduced even further to a rating of 3 or less, some protective measures might be needed. One example is the *posting* of weight limits that are allowed on the bridge. Also, the bridge might be *closed* until it can be repaired, rehabilitated, replaced or permanently closed. Another qualitative indication of bridge health is *red flags*. These flags identify potentially or imminently unsafe structural conditions and require the owner to take prompt, certified corrective, or protective actions to resolve the flag. These include repair, posting, or closure. Figure 3.14 shows the number of red flags issued in New York State between 1989 and 2006.

Alampalli and McCowan (2008) described in detail the flagging procedure, which is a practical measure to describe state of bridge health, in New York State. It is part of the comprehensive bridge inspection program implemented in New York State to identify serious deficiencies, both structural and nonstructural, affecting public safety so that owners can take appropriate action in a timely fashion. The procedure is very robust, safety oriented, and establishes requirements for certifying that appropriate corrective or protective measures are taken within an appropriate time frame. The critical inspection findings (flags) can be either structural or safety related. The structural flags are further subdivided into two categories.

Red structural flags are used to report the failure of a critical primary structural component or a failure that is likely before the next scheduled inspection (see Figures 3.8 and 3.9).

Yellow structural flags are used to report a potentially hazardous condition which, if left unattended beyond the next anticipated inspection, would likely become a clear and present danger. This flag is also used to report the actual or imminent failure of a noncritical structural component, where such failure may reduce the reserve capacity or redundancy of the bridge, but would not result in a structural collapse by the time of the next scheduled inspection interval (see Figures 3.10 and 3.11).

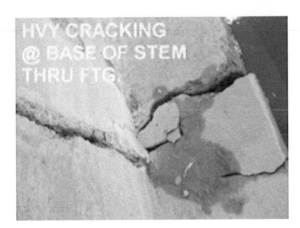

FIGURE 3.8 Red flag due to serious cracking in abutment stem. (Courtesy of New York State Department of Transportation.)

FIGURE 3.9 Red flag due to overextended bearing. (Courtesy of New York State Department of Transportation.)

FIGURE 3.10 Yellow flag due to a crack in the primary redundant member. (Courtesy of New York State Department of Transportation.)

FIGURE 3.11 Yellow flag due to loss of bearing area in a pedestal. (Courtesy of New York State Department of Transportation.)

Nonstructural conditions are reported using a "Safety Flag." The safety flag is used to report a condition presenting a clear and present danger to vehicle or pedestrian traffic, but is in no danger of structural failure or collapse. Safety flags can also be issued on closed bridges whose condition presents a threat to vehicular or pedestrian traffic underneath the bridge (see Figure 3.12).

The notification and response procedures vary depending on the flag type.

US Federal Rating System: The federal ratings result from an overall condition assessment of each bridge's three or four major components and do not require the multielement evaluations mandated by other states, such as NYSDOT inspection program. The federal ratings are used to identify bridges that do not meet contemporary Federal Highway Administration (FHWA) standards. Those bridges are classified as either "structurally deficient" or "functionally obsolete." Bridges are considered "structurally deficient," according to the FHWA, if significant load-carrying elements are found to be in poor or worse condition due to deterioration and/or damage, the bridge has inadequate load capacity or repeated bridge flooding causes traffic delays. The fact that a bridge is "structurally deficient" does not imply that it is unsafe or is likely to collapse.

A "structurally deficient" bridge, when left open to traffic, typically requires significant maintenance and repair to remain in service and eventual rehabilitation or replacement to address the

FIGURE 3.12 Safety flag due to exposed wiring. (Courtesy of New York State Department of Transportation.)

TABLE 3.7
Count of Bridges by Construction Material

Count of Bridges	Concrete	Steel	Prestressed Concrete	Wood	Others[a]
Total stock	248,739	188,551	132,033	26,682	3802
Structurally deficient	18,506	38,419	5036	9855	710
Functionally obsolete	28,187	34,678	12,773	3482	685

[a] Includes: Masonry, aluminum, wrought iron, or cast iron, and other.

deficiencies. To remain in service, structurally deficient bridges are often posted with weight limits. "Functionally obsolete" refers to a bridge's inability to meet current standards for managing the volume of traffic it carries, not its structural integrity. For example, a bridge may be functionally obsolete if it has narrow lanes, no shoulders, or low clearances.

The federal bridge rating scale addresses both structural condition and functional adequacy, using different criteria than the above mentioned NYSDOT inspection condition rating scale. Table 3.7 and Figure 3.13 show structurally deficient and functionally obsolete highway bridges in the United States as categorized by construction material. Table 3.8 and Figure 3.14 show structurally deficient and functionally obsolete bridges in the United States as categorized by bridge type.

We end this section by observing that a good state inspection practice should include both Federal ratings as well as the state's own ratings, if available (Figures 3.15 and 3.16).

3.3.3.2 Healthy Life Causes as Metrics
We can quantify the health of a bridge as

$$HLTH = HLTH_E + HLTH_T \tag{3.3}$$

Note that the health of the bridge measure $HLTH$ is a qualitative measure that should be used for DM situations. It should be used as a complementary measure to other bridge performance measures such as bridge ratings. The main difference between $HLTH$ and bridge rating is that bridge rating evaluated the field condition of bridge. On the other hand, $HLTH$ evaluates other factors that *lead* to the current bridge condition. Both measures account for different set of parameters, thus the consideration of both can give a more complete picture of the bridge health.

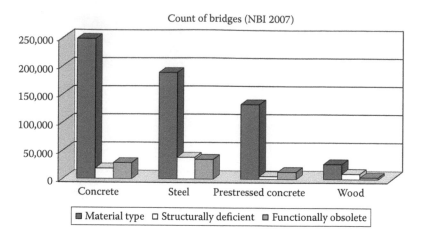

FIGURE 3.13 Count of bridges by construction material. (From FHWA, National Bridge Inventory, NBI, Federal Highway Administration, Washington, DC, 2007.)

TABLE 3.8
Count of Bridges by Structure Type

Count of Bridges	Slab	Stringer/ Multi Beam or Girder	Girder and Floor Beam System	TEE Beam	Box Beam or Girders (Single or Spread/Multiple)	Truss (Deck and thru)	Suspension/ Stayed Girder	Others[a] (Culvert)
Total stock	79,879	249,238	7432	36,444	55,330	12,608	133	158,726 (126,401)
Structurally deficient	6600	39,911	2594	4770	3189	7161	36	8262 (2944)
Functionally obsolete	9612	39,819	1792	7807	6660	2455	44	11,609 (5543)

[a] Includes: Frame; orthotropic; arch-deck and arch-thru; movable (lift, bascule, and swing); tunnel; mixed types; segmental box girder; channel beam; other; and culverts.

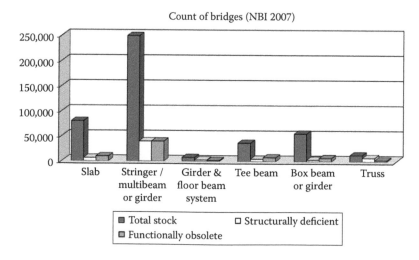

FIGURE 3.14 Count of bridges by structure type. (From FHWA, National Bridge Inventory, NBI, Federal Highway Administration, Washington, DC, 2007.)

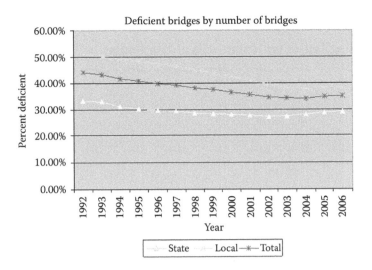

FIGURE 3.15 Percent of deficient bridges. (Courtesy of New York State Department of Transportation, NYSDOT 2008.)

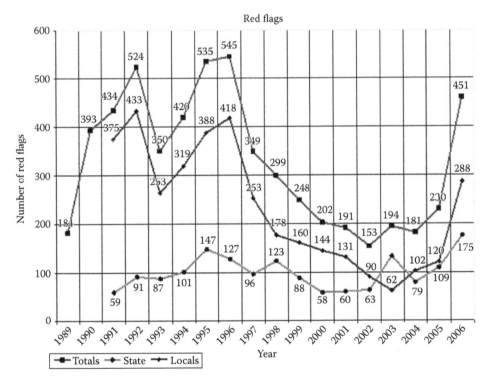

FIGURE 3.16 Number of red flagged bridges. (Courtesy of New York State Department of Transportation, NYSDOT 2008.)

To give an example of the use of *HLTH* let us consider an example of two bridges. The official needs to compare the health of the bridges. By making an in-depth study of all enabling and triggering causes, the different estimated factors were computed (qualitatively) as shown in Tables 3.9 and 3.10.

Notice that the relative weights in both Tables 3.9 and 3.10 are assigned by the official. More studies are needed for the nature and the values of these weights. The final estimates for *HLTH* are 695 and 763 for bridges A and B, respectively. This outcome indicates that bridge B should be healthier than bridge A.

TABLE 3.9
Example of Enabling Causes for Healthy Life of Two Bridges

Enabling Cause	Weight	Objective Rating of Bridge A	Objective Rating of Bridge B
Sound management	10	7	9
Design	8	6	6
Detailing	10	4	7
Construction	10	7	5
Scheduled maintenance	12	8	6
Inspection/monitoring	12	8	9
Total		420	438

TABLE 3.10
Example of Triggering Causes for Healthy Life of Two Bridges

Triggering Cause	Weight	Objective Rating of Bridge A	Objective Rating of Bridge B
Retrofits	10	5	5
Rehabilitation	10	6	5
Special maintenance	15	8	8
Replacement	15	3	7
Total		275	325

3.3.4 ROLE OF SHM

Structural health monitoring/structural health in civil engineering can play a major role in almost all enabling and triggering bridge healthy life. Some specific examples are shown in Table 3.11.

The project that was described by Kosnik and Hopwood (2008) provides an example of healthy bridge life that was aided by HSM monitoring. The John F. Kennedy Memorial Bridge, a large cantilever through truss opened in 1963, carries Interstate 65 across the Ohio River between Louisville, KY, and Jeffersonville, IN. An overall view of the bridge is shown in Figure 3.17. The bridge is restrained by pairs of bearing assemblies on each bank of the river. The bearings resist considerable designed-in uplift forces, particularly on the Indiana side. In 2006, one of the anchor bolts on the northwest bearing was found to have been fractured (see Figure 3.18). The washer on the bolt opposite the failed bolt could be spun in place by hand, indicating that there was very little tension in that bolt. The bearing assembly with the fractured bolt also moved visibly under live traffic.

Live strain, displacement, and acceleration data were collected on uplift bearing anchor bolts for a total of approximately 17 hours over several weeks under a variety of weather and traffic conditions and both before and after replacement of a fractured anchor bolt (see Figure 3.19). On the basis of initial data it was determined that the compromised North West, NW bearing assembly was subject to large live strains in two of the three remaining anchor bolts. The fractured North West-North East (NW-NE) anchor bolt was replaced with a threaded rod instrumented with strain gauge arrays, and strain data on the three original anchor bolts and the replacement bolt were recorded before and after installation and tightening of the replacement. The live strains in the original anchor bolts decreased after tightening of the replacement bolt, and no further indications of bending of the anchor bolts were observed.

TABLE 3.11
SHM Role in Some Bridge Healthy Life

No.	Healthy Life Issues	SHM Applications
1	Rehabilitation/repair choice, closing/opening/usage-restriction of the structure, evacuation for safety	Monitoring and evaluating condition state as is. STRID and DMID projects
2	Evaluation/implementation of new designs/construction methodology at local or global level	New and existing engineering paradigms
3	Quality of life (improving mobility, etc. —opening appropriate lanes or redirecting traffic in-time)	Bridge testing and life cycle analysis.
4	Maintenance decisions (routine to using just-in-time concept—say sending a salt truck to a bridge only when you see that such conditions warrant that will help structural durability, saves money, and improves safety)	Bridge management tools and methods Applying reliability and risk methods for maintenance decisions
5	Safety (users and structural safety)	Engineering design and analysis paradigms
6	Evaluation (durability, design/analysis option evaluation, etc.)	Deterioration monitoring and decision making tools
7	Research, development, and technology transfer a. Improving the state-of-the-art b. Improving the state-of-the-practice c. Bridging the gap between the state-of-the-art and state-of-the practice	SHM/SHCE in civil infrastructure is in its infancy (as of writing of this volume) more research is needed in all components of the fields
8	Security (relatively new in bridge field, but probably common for security applications you deal with)	There are many common needs and applications for bridge security and SHM (see Chapter 11 of Ettouney and Alampalli 2012)

FIGURE 3.17   Overall view of the I-65 John F. Kennedy Memorial Bridge. (Reprinted from ASNT Publication.)

3.4   WHY BRIDGES FAIL/DIE?

3.4.1   Definitions of Failure

There are several definitions of bridge failure. Most of those definitions are qualitative. A general failure definition was provided by NYSDOT (2001), which categorized failure into three groups as follows

Catastrophic: The structure is vulnerable to a sudden and complete collapse of a superstructure span or spans. This failure maybe the result of partial or total failure of either the superstructure or the substructure. A failure of this type would endanger the lives of those on or under the structure.

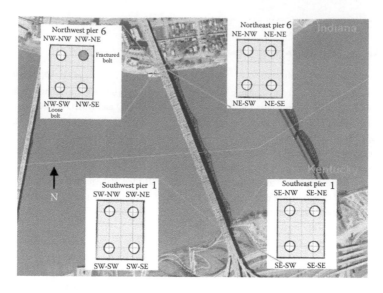

FIGURE 3.18   Anchor bolt identification. (Reprinted from ASNT Publication.)

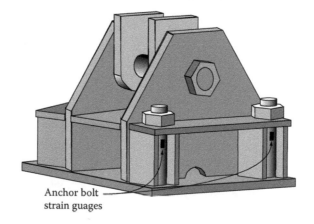

FIGURE 3.19   Anchor bolt strain gauge locations. (Reprinted from ASNT Publication.)

Partial Collapse: The structure is vulnerable to major deformation or discontinuities of a span. (This would result in loss of service to traffic on or under the bridge.) This failure may be the result of tipping or tilting of the substructure causing deformation in the superstructure. A failure of this type may endanger the lives of some of those crossing or those under the structure.

Structural Damage: The structure is vulnerable to localized failures. This failure may be the result of excessive deformation or cracking in the primary superstructure or substructure members of the bridge. A failure of this type maybe unnoticed by the travelling public but would require repair once it is discovered. This type of damage would also make a bridge more susceptible to overload failures.

Wardhana and Hadipriono (2003) showed the relationship between classes of failure that are similar to the above classification of NYSDOT and the phasing (construction vs. in-service phases). The number of failures (up to 2003) according to failure phasing is shown in Table 3.12.

TABLE 3.12
Number of Bridge Failures According to Failure Categories

Types of Failures	Construction	Service	Unknown
Distress	0	17	0
Partial collapse	3	80	13
Total collapse	5	12	21
Unknown	0	277	75
Total	8	386	109

The distress category is fairly similar to the NYSDOT Structural Damage category.

Note that each of the above failure classifications/groups has two attributes: (a) The scale and extent of the damage, and (b) The consequences of such failure. This makes those definitions to be closer to a risk-based definition that concerns itself with vulnerabilities, threats, *and* consequences. Such general and encompassing definition makes it possible to quantify failure. A general risk-based approach would be

$$F_R = T \cdot V \cdot C \tag{3.4}$$

where

F_R = Failure severity estimate
T = Threat/hazard level estimate
V = Vulnerability of structure to failure due to T
C = Consequences of failure

We recognize Equation 3.4 as the classical risk equation FEMA 452 (2010) and FEMA (2009). We now try to make Equation (3.4) more specific for quantification of failure risk by defining

$$S_{TV} = T \cdot V \tag{3.5}$$

where
S_{TV} = Propensity to failure

Thus, a suitable failure severity estimate can be expressed as

$$F_R = S_{TV} \cdot C \tag{3.6}$$

Assuming that the limits of S_{TV} and C to be

$$1 \le S_{TV} \le 10 \tag{3.7}$$

$$1 \le C \le 10 \tag{3.8}$$

The failure estimate should have limits as

$$1 \le F_R \le 100 \tag{3.9}$$

Assuming that the propensity to failure of a given bridge "A" is $S_{TV} = 4.5$, while the consequence of this failure is $S_{TV} = 7.5$ the failure severity estimate is $F_R = 33.75$. For a different bridge "B," if the propensity to failure is $S_{TV} = 7.3$, while the consequence of this failure is $S_{TV} = 3.0$ the failure severity estimate is $F_R = 21.9$. So, the severity of failure of bridge "A" is much more than the severity of failure of bridge "B," even though bridge "B" is more vulnerable to the postulated threats than bridge "A." Equations 3.4 through (3.9) can be used to quantify severity of failure across a number of bridge networks.

3.4.2 CLASSIFICATION OF FAILURE CAUSES

Failure causes have been classified by Wardhana and Hadipriono (2003) into enabling and triggering causes. The differences between the two types of causes were discussed earlier; in general, enabling causes occur over longer period of time, while triggering causes occur suddenly. We discuss different enabling and triggering failure causes first. Next we try to associate those failure causes with Equations 3.4 through 3.9 for estimating f F_R. We finally discuss the role of SHM in evaluating failure causes.

3.4.2.1 Triggering Causes

Triggering causes of failure include hydraulic/flood (scour, debris, drift, etc.), collision/impact, overload (Figures 3.20 through 3.22), fire, ice, earthquakes (including tsunamis), wind (hurricanes/tornadoes, etc.), and soil failures. Note that all of these causes occur suddenly, and last relatively short period of time. Figures 3.23 and 3.24 show percentages of bridge failure rates according to different triggering causes in the United States and New York State, respectively. Note that this database is not a comprehensive database of all the failures but only the failures recorded by the NYSDOT based on data collected from periodic survey of other states.

FIGURE 3.20   Impact failure.

FIGURE 3.21   Overload failure. (Courtesy of International Association of Structural Movers.)

FIGURE 3.22 Traffic overload. (Courtesy of International Association of Structural Movers.)

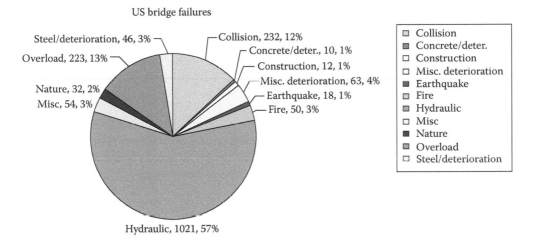

FIGURE 3.23 Failure rates (US). (Courtesy of New York State Department of Transportation.)

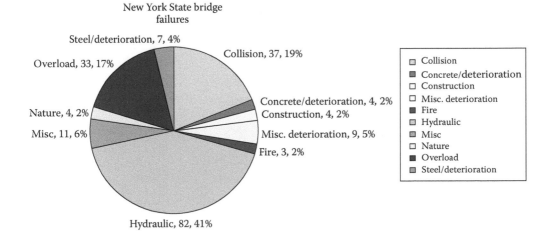

FIGURE 3.24 Failure rates (New York State). (Courtesy of New York State Department of Transportation.)

3.4.2.2 Enabling Causes

Wardhana and Hadipriono (2003) subdivided enabling causes (they called it principal causes) as design, detailing (Figures 3.25 and 3.26), construction, maintenance, material related (including deterioration, normal wear and tear, fatigue, and/or corrosion). Of these, maintenance and material related causes were the largest causes for collapse or distress. Note that all of these causes occur over relatively long period of time. Figure 3.27 shows effects of construction material type on bridge failures. Figure 3.28 shows the effects of bridge construction dates and Figure 3.29 shows bridge type on failure rates.

3.4.2.3 Estimating Bridge Failure Propensity to Causes

We just presented a plethora of bridge failure causes. For a given bridge, many of those causes might not be applicable; yet several others might have an effect on the bridge. We present a method for evaluating total failure severity estimate when there are several potential causes of failure that we define as F_{R_TOTAL}. A generalized form of Equation 3.4 can be offered as

$$F_{R\_TOTAL} = \alpha \left(\sum_{i=1}^{i=N_{CAUSE}} \left(F_{Ri} \right)^n \right)^{1/n} \tag{3.10}$$

where

F_{Ri} = Failure severity estimate for i^{th} failure cause
N_{CAUSE} = Number of pertinent failure causes
n = Suitable power number
α = Scaling factor

The form of Equation 3.10 was used in FEMA (2009) to sum the individual risks of uncorrelated threats; it is reasonable to use to sum the severity of failures due to uncorrelated failure causes in our current endeavor. Note that $n = 2$ reduces Equation 3.10 to the popular square root of sum of squares

FIGURE 3.25 Failure due to concrete details (a) overview of failure site and (b) failed box girder. (Courtesy of New York State Department of Transportation.)

FIGURE 3.26   Failure due to steel details. (Courtesy of New York State Department of Transportation.)

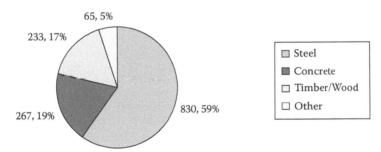

FIGURE 3.27   Failure of bridges according to material of construction. (Courtesy of New York State Department of Transportation.)

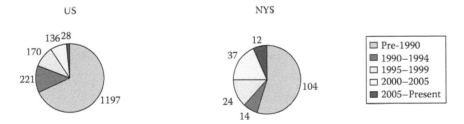

FIGURE 3.28   Failure of bridges according to construction date. (Courtesy of New York State Department of Transportation.)

(SRSS) method. However, we propose to use a much higher value for n, say $n = 10$, which would produce more realistic results, see FEMA (2009) for more discussion of this approach. Proceeding as before

$$F_{Ri} = T_i \cdot V_i \cdot C_i \tag{3.11}$$

where

   T_i = Threat/hazard level estimate ith failure cause
   V_i = Vulnerability of structure to failure due to T ith failure cause
   C_i = Consequences of failure of the ith failure cause

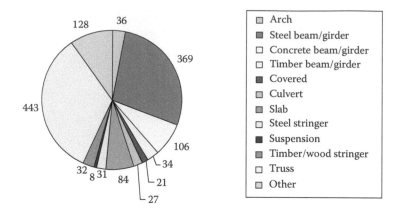

FIGURE 3.29 Failure of bridges according to bridge type. (Courtesy of New York State Department of Transportation.)

Also, defining

$$S_{TVi} = T_i \cdot V_i \tag{3.12}$$

where

S_{TVi} = Propensity to failure due to T_i

Thus, a suitable failure severity estimate can be expressed as

$$F_{Ri} = S_{TVi} \cdot C_i \tag{3.13}$$

Assuming that the limits of S_{TV} and C to be

$$1 \leq S_{TVi} \leq 10 \tag{3.14}$$
$$1 \leq C_i \leq 10 \tag{3.15}$$

By adjusting the value of α, the total failure severity estimate would have limits as

$$1 \leq F_{R\_TOTAL} \leq 100 \tag{3.16}$$

The limits in Equations 3.14 through 3.16 are arbitrary. Careful studies should be conducted to evaluate the effects of those limits, and provide realistic limits, for practical bridge situations.

As an example of estimating failure severity estimate for multiple causes, let us consider two bridges, "A," and "B," with different estimated failure causes, propensity of failures, and consequences of failure and shown in Table 3.13. In this example, the power number is assumed to be $n = 10$. The number of pertinent hazards are $N_{CAUSE} = 6$ and $N_{CAUSE} = 5$ for bridges "A," and "B," respectively. Bridge "A" does not have hydraulics as a realistic failure cause, since it does not cross a water body. Bridge "B" is not located in a seismically prone area; it does not have wind as a potential cause of failure. The failure severity estimates of bridges "A," and "B," are 30.15 and 25.65, respectively. Studying the numbers in Table 3.13, it is clear that the consequences of bridge "A" failures caused these results even though bridge "A" seem to have lesser threats and vulnerabilities than bridge "B."

Equations 3.10 through 3.16 can be used to quantify severity of failure due to any number of failure cause combinations across a number of bridge networks.

TABLE 3.13
Failure Severity Estimates for Two Bridges

Cause	Bridge "A"			Bridge "B"		
	STVi	Ci	Fri	STVi	Ci	Fri
Construction	2	9	18	4	4	16
Hydraulic	NA	NA	NA	6	5	30
Seismic	4	9	36	NA	NA	NA
Wind	2	9	18	NA	NA	NA
Collision/impact	2	7	14	6	3	18
Fatigue	6	4	24	6	3	18
Overload	3	5	15	7	3	21
N_{CAUSE}		6			5	
α		0.84			0.85	
Scaled totals		30.15			25.65	

TABLE 3.14
SHM Roles in Bridge Failure—General

Causes of Failure		Sensing/Measurements	STRID	DMID
Enabling	Design Detailing	Use sensing to verify design procedures for unusual geometries that are not within the bounds of conventional codes such as bridges with large skew angles	Modal or parameter identification methods can be used to enhance analytical bridge models, especially for existing bridges	Using in-field damage information, such as remaining fatigue life, provide for an efficient and safe retrofit bridge designs
	Construction	See Table 3.15		
	Maintenance	NDT methods can compliment conventional visual inspection practices	Different STRID methods produce more accurate analysis results for maintenance projects	Detecting damage in an accurate and timely manner would focus maintenance efforts in safer directions
	Material related	See Chapters 3 and 4(for concrete deterioration), Chapter 5 (for fatigue) and Chapters 6 and 7 (for fiber reinforced polymers) in Ettouney and Alampalli (2012)		
Triggering	External events	See Table 3.15		

3.4.2.4 SHM/SHCE Role

Structural health monitoring plays a major role in all aspects of bridge failure. Some examples of the different roles each of the three components of SHM can play in enabling and triggering causes of bridge failure are shown in Table 3.14 through 3.16.

3.4.3 ATTRIBUTES OF FAILURE

The failure event of a bridge can be used as an information source to help understand failure mechanisms, causes, attributes, and so on. When bridge fails, it might be instructive to collect information regarding the event as follows:

Bridge Type: The structural type, including superstructure, substructure, foundation, and soil
Bridge Attributes: Span length, support conditions (before and after the failure)

TABLE 3.15
SHM Roles in Bridge Failure—Triggering Causes

Hazard	Sensing/Measurements	STRID	DMID
Collision/accidents (Figure 3.30)	On bridge vibration monitoring can reveal occurrence and extent of collision events	Before event and after event STRID efforts can give estimates of extent of global, and perhaps local, damage	Global or local DMID methods can aid in producing safe and cost-effective retrofit efforts
Concrete deterioration	See Chapters 3 and 4 in Ettouney and Alampalli (2012)		
Construction	Sensing behavior of systems during construction projects would provide an invaluable QA/QC insight that helps in improving safety during construction and potential clarifications of future behavior	Modal identification methods can show abnormal dynamic behavior during construction. Thus providing additional safety measures to the construction process	DMID techniques can show extent of damages after construction accidents
General deterioration	SHM procedures for general deterioration are similar to those procedures in Chapters 3 and 4 in Ettouney and Alampalli (2012)		
Fire (Figures 3.31 and 3.32)	After fire events, sensing would produce valuable information regarding extent of fire damage (deflections, tilt, etc.)	Modal identification methods can reveal potential changes in global bridge behavior after fire	NDT methods such as acoustic emission (local damage) or thermography (regional damage) can provide damage information
Hydraulic/scour/flood	See Chapter 1 in Ettouney and Alampalli (2012)		
Nature	See Chapters 1 through 5 in Ettouney and Alampalli (2012)		
Overload	Capacity and demands can be correlated in real time by measuring bridge response (capacity: strains, displacements, etc.) and live load demands (truck loads, volume and frequency)	STRID methods offer a unique opportunity for improving modeling technique of overloaded systems	Global or local DMID methods can aid in producing safe and cost-effective retrofit efforts when a system is overloaded
Steel deterioration	See Chapters 3, 4, and 5 in Ettouney and Alampalli (2012)		
Seismic (Figure 3.33)	See Chapter 2 in Ettouney and Alampalli (2012)		
Fatigue	See Chapter 5 in Ettouney and Alampalli (2012)		

Material: Record the material of construction of different bridge components

Abutments and Piers (before and after the failure): Type, height, size, bearing types, foundations, seat widths, and so on.

Age: Year Built, year failed, and if there were any indications of sudden deterioration or loss of function

Failure Type: Describe the type of failure (as described throughout this section)

Failure Cause: The cause (or causes) of failure (as described throughout this section)

Consequences: This includes direct consequences such as number of Fatalities and/or number of injuries. Additional consequences might be of interest such as economic, social or other local/regional consequences

Multimedia: Recording failure aftermath with photographs and/or video can provide valuable database. This is particularly important since failure sites are usually reconstructed soon after the failure event

TABLE 3.16
SHM Role in Bridge Failure—Age of Bridge

Bridge Age	Sensing/Measurements	STRID	DMID
Older	Monitoring systems on older bridges can be costlier than their counterparts on new systems	Can validate rehabilitation and existing analytical models	Different DMID procedures can help in detecting damages of older bridges, thus extending their life span
Newer	Newer bridges offer opportunities of placing monitoring systems that are more interconnected and consistent with the bridge system than older bridges	Can validate rehabilitation and existing analytical models. They can help in validating and improving accuracy of construction-related analytical models	Newer bridges are less exposed to aging-type damages, such as deterioration or increased traffic demands. However, they can be vulnerable to abnormal hazards that exceed their design limits, such as an abnormally high flood or abnormally strong earthquake

FIGURE 3.30 Failure due to accidents. (Courtesy of New York State Department of Transportation.)

FIGURE 3.31 Fire failures: steel girders have deformed extensively. (Courtesy of New York State Department of Transportation.)

FIGURE 3.32   Fire failure: deck damage. (Courtesy of New York State Department of Transportation.)

FIGURE 3.33   Seismic failure. (Courtesy of New York State Department of Transportation.)

3.4.4 STRUCTURAL MODES OF FAILURE

Bridges, as well as any other infrastructure can fail only in one of two temporal structural modes: either brittle or ductile. Spatially, the structures can either fail progressively, or globally. We explore those structural modes of failure next. At the end of this section, we examine the role and limitations of SHM in each of those modes. Note that we differentiate between damage and failure in this discussion. Damage would be a local event in spatial terms, while failure is more extensive in spatial terms.

3.4.4.1 Ductile Failure

Ductile structural failure occurs over long period of time. Signs of distress are usually observable. As such most conventional SHM techniques can be used to monitor the damages and alert for any impeding failure. Examples of ductile failure are failures due to foundation settlements, or excessive wind-induced deflections and vibrations. The failures of Tacoma Narrows Bridge (Figure 3.45) and the Tay Rail Bridge (Figure 3.42) are ductile-type failures. The Tacoma Narrows collapse due to excessive wind vibrations lasted long enough to prevent any human causality. Unfortunately, the Tay Rail Bridge collapse occurred suddenly, thus causing great loss in life. We consider the Tay Rail Bridge failure as a ductile failure since the signs of poor performance were observed long before the collapse, such as rattling of steel connections, and excessive vibrations when trains crossed the bridge.

3.4.4.2 Brittle Failure

Brittle structural failure occurs suddenly. Signs of distress are usually subtle, and difficult to observe. As such care is needed when applying conventional SHM techniques. It is important to clearly identify the brittle modes of failure that the SHM project aims to uncover, and ensure that the techniques used are capable of identifying those brittle failure modes. Examples of brittle failure are failures due to corrosion (Figure 3.34) or fatigue (Figure 3.35) since both types of failure occur suddenly, even though the damage occurred over a long time period. The failure of the I-35 Bridge in Minnesota (Figure 3.59) is clearly a brittle failure; it occurred suddenly and caused major catastrophic consequences.

FIGURE 3.34 Corrosion failure. (Courtesy of New York State Department of Transportation.)

FIGURE 3.35 Fatigue failure. (Courtesy of New York State Department of Transportation.)

3.4.4.3   Global Failure

Brittle or ductile failure can lead to global bridge failure. Global failure occurs when the whole structural system becomes unstable. Global instability is perhaps the most severe failure mode and the most difficult to detect or monitor. When the bridge is newly constructed, its global stability condition is usually investigated and ensured. As the bridge ages, this global stability condition changes: the global stability safety factor starts decreasing. After a suddenly applied abnormal loading condition, such as an earthquake event, the structural stability safety factor decreases even further, due to the nonlinear changes in the system. Such nonlinearities would decrease global stability resilience. Ettouney et al. (2006) explored the effects of such nonlinearities on the global stability of structural systems. For highly redundant structural systems reductions of global stability safety factors are slow. For low redundant systems, investigating global structural stability should be performed often; for such investigations to be beneficial, they should be done using actual bridge properties, not theoretical properties. An example of the use of SHM in observing stability condition of a steel truss is given in Chapter 8.

3.4.4.4   Progressive Failure (Collapse)

Progressive collapse condition occurs when the failure starts locally, then progress further away from the initial location of the failure. Ettouney et al. (2004) presented a progressive collapse theory that stated that progressive collapse event can lead to one of two outcomes: (a) if the failure front is arrested, say by reaching a resilient support, or (b) if large portion of the structure have failed such that the remaining structure becomes globally unstable. The latter scenario was discussed earlier. We discuss now the first outcome, that is, the progression of failure front. Extent of progressive collapse in bridges can be described in terms of four scales, as follows.

Limited Progression: The collapse progression is limited in scale. Examples are fatigue cracks, local buckling, and limited corrosion
Local Progression: Collapse progressed to a single and or continuous spans; superstructure only
Regional Progression: Collapse progressed to a single and or continuous spans; superstructure, as well as substructure
Global Progression: The collapse would propagate across many spans and over many piers and supports

Figure 3.36 shows schematics of the failure scales. Clearly, SHM techniques can be of value in monitoring and assessing progressive collapse at different collapse scales, as shown in Table 3.17.

3.4.4.5   Mitigation Strategies and Failure Modes

Obviously, *optimal* failure mitigation strategies should depend on the mode of failure. An optimal strategy should aim at delaying or eliminating the failure under consideration at reasonable costs. Quantitatively, this means that we should strive to reduce probability of failure. Since in this section we are concerned with bridge (or system) failure, we recall immediately that an optimal failure mitigation strategy should aim at increasing the bridge system reliability. Note that for optimal strategy, reduction of threat (corrosion prevention measures, for example) might offer a more cost-effective solution than reducing vulnerability. Thus an optimal mitigation strategy might be defined as the one that would reduce risk, provide for higher benefit to cost ratio, or produce the optimal life cycle analysis (costs, benefits and life spans). Chapters 8, 9, and 10 in Ettouney and Alampalli (2012) discuss those issues in greater detail. In what follows, we explore specific optimal mitigation needs for different structural failure modes.

Ductile Failure: The goal of ductile failure mitigation should be to eliminate or reduce the chance of such a failure. Inspection (Figure 3.38) or monitoring can help in such a situation. Regular maintenance

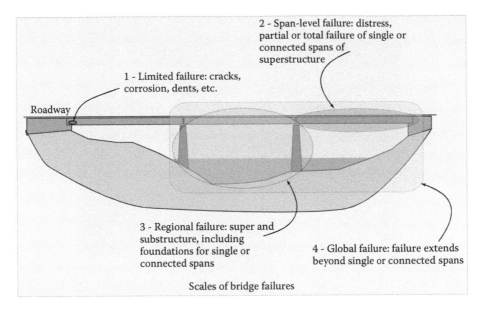

2 - Span-level failure: distress, partial or total failure of single or connected spans of superstructure

1 - Limited failure: cracks, corrosion, dents, etc.

Roadway

3 - Regional failure: super and substructure, including foundations for single or connected spans

4 - Global failure: failure extends beyond single or connected spans

Scales of bridge failures

FIGURE 3.36   Scales of bridge failures.

TABLE 3.17
SHM Role in Different Progressive Collapse Scales

Failure Scale	Example	SHM Role		
		Sensing	STRID	DMID
Limited	Fatigue crack	See Chapter 5 in Ettouney and Alampalli (2012)		
Local	Barge collision	**Before event**: proximity sensors **During event**: vibration signatures (time histories of accelerations, velocities, or strains) **After event**: different NDT procedures to detect damages	Before and after event STRID model evaluation can help in damage assessment, accurate retrofit design, and decision-making processes	Global DMID methods, e.g., thermography or more localized DMID methods, e.g., ultrasonic or acoustic emission can aid in detecting presence, type, location and extent of damage
Regional	Scour failure of footing or underlying soils (Figure 3.37)	See Chapter 1 in Ettouney and Alampalli (2012)		
Global	Failure of single span will propagate to other spans by dynamics, or catenary action	Placing vibration sensors over the whole length of the bridge (in an optimal manner as in Ettouney and Alampalli [2012]) can help in accurate understanding of the progressive collapse event	Simulating progressive collapse of any structural system in an accurate fashion is not an easy task. Using different STRID methods, coupled with optimal sensing data, would result in an improved accuracy of progressive collapse modeling	NDT methods can help in forensic analysis of failed structure

FIGURE 3.37   Scour failure. (Courtesy of New York State Department of Transportation.)

FIGURE 3.38   Bridge inspection.

and rehabilitation efforts should be adequate in mitigating ductile failure. The bridge in Figure 3.39 is being rehabilitated after a general deterioration in its conditions was observed over the course of several years. Note that the rehab efforts are striving to preserve the historic nature of the bridge.

Brittle Failure: Brittle failure is much more difficult to mitigate. This is due to the difficulty in continuously assessing the actual global stability condition. We immediately notice two contributing factors:

- *Governing Redundancy Factor*: It is essential to identify all pertinent potential failure modes and the redundancies that control each of these modes. Traditionally, theory of structures provided some help in assessing some structural redundancies. For example, configurations of members in trusses (Figure 3.40) or the number of supports (Figure 3.41) can indicate the degree of redundancy in a given system. Noting that such general assessment rely on the implicit assumption that all structural components are in pristine condition. Sudden damages or normal wear and tear can alter this situation, reducing what might

FIGURE 3.39   Rehabilitation to avoid ductile failure. (Courtesy of New York State Department of Transportation.)

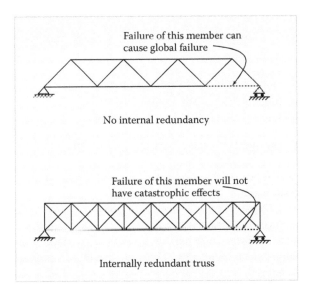

FIGURE 3.40   Internal redundancy.

have been a highly redundant system into a system with low redundancy that is susceptible to brittle failures.

- *Time Factor:* As indicated earlier, passing of time can reduce the governing redundancy of a bridge system. An optimal brittle failure mitigation strategy should account for the as-built condition, not the designed condition.

An effective mitigation strategy should follow some or all of the following steps

1. Identify all governing redundancies in the system. This should be done using a well-designed STRID effort
2. Identify and monitor actual conditions of items that control governing redundancies of #1. A well integrated system of inspection and SHM is needed
3. Use the different SHM principles in this volume: duality, scaling and serendipity to help identifying low redundancy that might not have been detected in #1 or #2. For example, a subtle change in local resonance might indicate a reduction in a governing redundancy

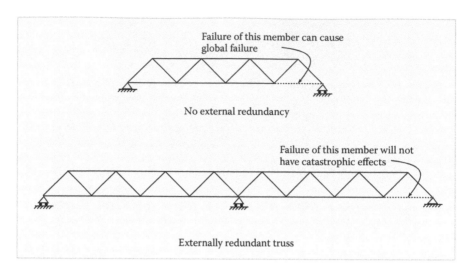

FIGURE 3.41   External redundancy.

When embarking on brittle failure mitigation efforts, the decision maker needs to ask the following questions:

- Can visual inspection alone detect brittle failure potential?
- Can sensing strains *alone* detect brittle failure potential?
- What kind of sensing can detect brittle failure potential?

The detailed answers to these questions can ensure an *effective* and *optimal* brittle failure mitigation strategy.

Global Failure: Global failure mitigation strategies share same attributes to both ductile and brittle failure mitigation strategies.

Progressive Failure (Collapse): First, we need to define progressive collapse mitigation strategy (PCMS). We propose two types of strategies. An active PCMS would try to prevent local failures that might cause progressive failures from happening in the first place. A passive PCMS would try to prevent the scale of failure from increasing. Thus, passive PCMS implies accepting the fact that initial failure might/can occur.

Passive PCMS include the use of two concepts: (a) improving redundancy of the structural system, and/or (b) hardening different components of the structural system. Detailed and careful studies are needed to ensure that

- The PCMS is adequate to meeting the project objectives
- The PCMS does not have side effects that can reduce resiliency during other types of threats (multihazards effects)

The active PCMS aims at reducing or eliminating local failures that can cause progressive collapse. Different SHM techniques, such as NDT methods (Chapter 6), bridge security methods (see Chapter 11 in Ettouney and Alampalli 2012) or bridge management methods (see Chapter 9 in Ettouney and Alampalli 2012) can be used for optimal active PCMS.

Decision Making Role: Mitigation of bridge failure is complex and overreaching subject, as evident from the above overview. There are numerous problems to resolve, and there are even more choices for mitigation approaches. Issues such as hardening versus improving redundancy, governing redundancies, passive PCMS versus active PCMS provide overwhelming choices to the decision maker. Pairings of problems and solutions are not easy to find. We recommend a utilization of different quantitative DM techniques in finding such pairings. Not only DM offer potential of using optimal mitigation solution, it also can be automated (due to its quantitative nature) to any SHM, or management practice. The cost saving and consistency of decisions are two obvious benefits of using DM techniques in this field.

3.5 EXAMPLES OF BRIDGE FAILURES

3.5.1 INTRODUCTION

Competent personnel using well-established material and structural specifications design bridges. In recent years, most bridge designs do employ quality control and assurance measures. Similarly, they are also inspected routinely to identify issues that affect their structural safety to carry the in-service loads as well as routine overloads imposed on them. Appropriate actions are taken based on inspection results depending on the urgency, effect of the critical findings on structural safety and durability, and available resources. But, most of these are reactive in nature and applies to slow form of deterioration such as corrosion and age-related issues. But, bridges do face several hazards and are designed to some of these hazards based on a certain return interval while not designed for other hazards that the structure has a very low probability with low consequence of encountering in its service life. Due to several uncertainties and due to the limitations associated with specifications, constructability, design/analysis difficulties and time, variability of materials used, and other factors, designs are generally very conservative in nature.

Despite all the conservatism, failures do happen often and in some cases with loss of life and significant economic impacts, due to several factors. These include design errors, unanticipated level of loads or hazards, material issues, failure of quality control or assurance, limitations associated with inspection, communication issues between different groups involved with bridge service life, neglect. But, very few failures can be attributed to hazard level exceeding the designed hazard level. In most cases, bridges do not fail due to a single factor but a combination/convergence of a number of factors. Even though some of these can be detected through inspection process several such as design errors, items concealed from general inspections, and severe problems that can develop rapidly before the next cycle are hard to detect once bridge is in service. This section discusses some of the historic failures and documented reasons to show that the bridge failures in general are due to convergence of several factors progressing for several years than a simple, sudden, single reason.

3.5.2 TAY RAIL BRIDGE, UK

The 3.5-km-long Tay Rail Bridge spanned the Firth of Tay in Scotland, between the city of Dundee and the suburb of Wormit in Fife, replacing an early train ferry. On December 28, 1879, the centre section known as the "High Girders" (thru truss) collapsed killing 75 people when a train on it went down with the bridge.

The Tay Rail Bridge was a lattice-grid (truss) design, combining cast and wrought iron and was opened on June 1, 1878. The bridge had several issues early in the construction that was attributed to original bedrock surveys. During the construction, it was found that the bedrock near the banks was deeper than predicted earlier. This led to the redesign of piers and reduced number of piers making superstructure spans longer that originally planned. The bridge was a combination of deck truss ("Low Girders") and thru truss sections.

The official forensic investigations noted that the bridge was "badly designed, badly built and badly maintained, and that its downfall was due to inherent defects in the structure, which must sooner or later have brought it down" (Court of Inquiry 1879). Thus several factors contributed to the failure including the following: (a) Allowance for wind load had been made by the designer Bouch (Seim 2008) as he was advised that this was unnecessary for girders shorter than 200 ft. The designer apparently did not make any allowances when the revised design involved longer girders.

The middle section of the bridge was a thru truss to allow a higher clearance for the passage of ships underneath and was potentially top heavy and very vulnerable to high winds; (b) The cylindrical cast iron columns supporting the 13 longest spans of the bridge, each 75 m long, were of poor quality. Many had been cast horizontally, with the result that the walls were not of even thickness, and there was some evidence that imperfect castings were disguised from the quality control inspections. In particular, some of the lugs used as attachment points for the wrought iron bracing bars had been "burnt on" rather than cast with the columns. However, no evidence of the burnt-on lugs has survived, and the normal lugs were very weak. They were tested for the *Inquiry* by David Kirkaldy and were proved to break at only about 20 t rather than the expected load of 61 t. These lugs failed and destabilized the entire centre of the bridge during the storm; (c) There was clear evidence that the central structure had been deteriorating for months before the final accident. A few months after the bridge was opened, the bridge inspector noticed that several joints had loosened making many of the tie-bars useless for bracing the cast-iron piers; and (d) Recent research also indicates that the cast iron used to join the columns of the bridge together might have become brittle under great strain and might have contributed to the failure.

This bridge failure shows the value of preliminary site data collection, proper design, selection of structural material, proper quality control (during the design, fabrication, construction, and inspection), and proper maintenance when deficiencies are noted during the bridge inspection. It is interesting to that some of the same errors contributed to the failure of the recent collapse in Minnesota that failed more than100 years after the failure of the Tay Rail Bridge (Figures 3.42 and 3.43).

3.5.3 QUEBEC BRIDGE FAILURE

One of the oldest failures of major bridges, killing 75 of the 86 people on the bridge, during the construction can be the First Quebec Bridge in August of 1907. Failure of this bridge shows the value of checking the design calculations after any significant change in structures design, quality control in the design process, involvement of the design engineer during construction through field visits, and supervision of a competent reengineer during the construction phase.

During 1907, the Quebec Bridge was supposed to be the bridge with the longest central span as well as longest cantilever span once built. But, the southern half of the bridge that was nearing completion collapsed suddenly. Theodore Cooper, then famous engineer, directed the project. The engineer soon after his appointment recommended increasing the span from 1600 ft to 1800 ft to reduce the vulnerability of the piers to scour by moving them to shallow water and to reduce the project time by making it easy to construct. To offset the increase steel costs, he also modified specifications that would allow for higher unit stresses. Because the St. Lawrence was a shipping lane, the 2800-ft bridge needed almost 150 ft clearance above the water to allow the ocean-going vessels to pass. Further, the bridge was to be multifunctional and was required to be 67-ft-wide to accommodate two railway tracks, two streetcar tracks, and two roadways.

FIGURE 3.42   Tay Rail Bridge after construction.

FIGURE 3.43   Tay Rail Bridge after collapse.

In the rush to provide shop drawings for steel fabrication preliminary weight calculations pre-pared in early stages were never properly checked when the design was finalized using revised specification. The real bridge weight exceeded significantly than the preliminary estimated weights and thus the bridge dead load capacity. Due to the inability of the project director to visit the field, an engineer with less experience for the major bridge construction was in charge of the field site. By the time, a material inspector reported the excess weights, south anchor arm, tower, and two panels of the south cantilever arm were fabricated and six panels of the anchor arm were already in place. The project director concluded that the increased stresses are within allowable limits and proceeded with the construction of the bridge. When the bridge was nearing completion during the 1907 summer, the effect of increased dead loads were revealed in the form of distortions of lower chord compression members that are key structural members. Signs of buckling were observed in them and also in the splices between some lower chord members. The deflection of one of the chords of the south anchor arm grew from 0.75 in to 2.25 in. But the work continued as it was felt that there is no immediate danger and some attributed the bends presence before the installation. By the time, project director was informed, a decision was made and conveyed to field personnel to stop the construction, two compression chords of the anchor arm buckled and the subsequently bridge collapsed in matter of seconds. For detailed discussions about the Quebec Bridge collapse, see Lienhard (2008), Smith (2008), and Ricketts (2008) (Figure 3.44).

3.5.4   TACOMA NARROWS, WA

In 1937, the Washington State legislature created the Washington State Toll Bridge Authority and appropriated $5,000 to study the request by Tacoma and Pierce County for a bridge over the Narrows. The first Tacoma Narrows Bridge, a cable-supported bridge, was opened to traffic on July 1, (See Figures 3.45 and 3.46). It collapsed 4 months later on November 7, 1940, at 11:00 AM (Pacific time) due to a physical phenomenon known as aeroelastic flutter caused by a 67 km per hour wind in a moderate windstorm (Ketchum 2011c).

The bridge was the third longest suspension bridge when built with a center span of 2800 ft, 29-ft-wide, and had the greatest ratio of length to width ever built. The false bottom type of cais-sons were used in sinking of cellular piers in water 120-ft-deep. Total height of west and east piers was 198, and 247 ft, respectively. Shore anchorages for suspension cables each contained 2500 cubic yards of concrete weighing 52, 500 tons. Two steel towers, 425 ft in height and each weighing

The 640-foot central span of the mighty Quebec bridge in place: Undismayed by tragic failures in 1907 an 1910, the builders have at last achieved one of the greatest triumphs in the history of bridge building

The central span, weighing over 5000 tons, being raised by practically the same method as was employed last year, with certain added precautions

Ready for the hoist of 150 feet: The span is about to be lifted from its pontoons by means of hydraulic jacks and chains, composed of massive bars of steel. The raising occupied practically four days

FIGURE 3.44 Collapsed Quebec Bridge: sequence of events that led to collapse.

FIGURE 3.45   Completed bridge, before opening. (Courtesy of University of Washington Libraries, Specail Collections.)

FIGURE 3.46   Opening of bridge for traffic. (Courtesy of University of Washington Libraries, Specail Collections.)

1927 tons, surmount the piers and support the suspension cables. The two suspension cables, 17.5 in diameter and more than 1 mile long, each consists of 6308 #6 parallel wires formed into 19 individual strands of 332 wires each. This forms a total of 14,000 miles of wire weighing 3817 tons (Washington University 2008a and 2008b).

The Tacoma experience taught engineers that wind causes not only static loads on the bridge, but also significant dynamic actions. It is commonly presented as an example of failure due to resonance when fundamental frequency of the bridge coincides with the external periodic frequency of the load. A cable-supported bridge is subject to wind-induced drag (the static component), flutter (the instability that occurred at Tacoma Narrows), and buffeting (where gusts "shake" the bridge). Adequate aerodynamic performance is required with respect to each of these effects. For modest span bridges, drag generally controls the strength required to resist wind. Flutter becomes critical when the wind acting on the structure reaches a critical velocity that triggers a self-excited unstable condition. The task in design is to assure that the critical wind velocity is high enough so that it has a very low probability of occurrence. This can be achieved by providing a stiff structure and/ or an aerodynamically streamlined superstructure shape. Buffeting influences fatigue of the bridge materials as well as users' comfort. The magnitude of buffeting response under higher probability wind conditions must be controlled. Addressing these issues in an engineering context requires the use of wind tunnel models. Current practice is converging on use of such models for the aerodynamic properties of the bridge shape only. The mechanical properties of the bridge, and the final wind evaluation, are performed using computer models that incorporate the wind tunnel results (Ketchum 2008). Figure 3.47 illustrates the sequence of events that led to the bridge collapse.

(a)

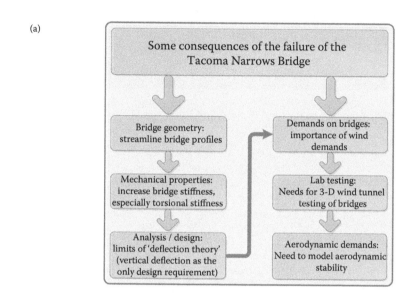

(b)

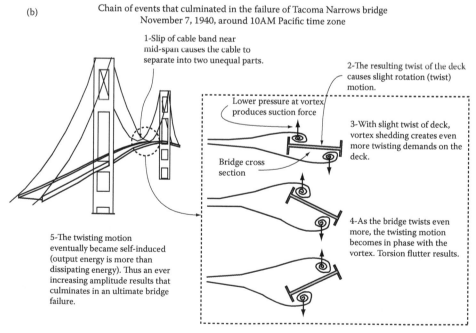

FIGURE 3.47 Sequence of bridge collapse (a) interrelationship between geometry and increased wind demands and (b) wind flutter.

3.5.5 Silver Bridge, OH

The Silver-Point Pleasant Bridge (known in history as Silver Bridge) collapsed on December 15, 1967 evening killing 46 people and injuring 9 when 31 of the 37 vehicles on the bridge plunged into water. Since the bridge on the U.S. Highway 35 connects Point Pleasant, WV and Kanauga, OH over the Ohio River, a major transportation route connecting both the States was destroyed causing a major disruption to many and caused enormous sensation across the nation on the safety of bridges.

The Silver Bridge got its name as it was the country's first aluminum painted bridge. The 2235-ft-long Silver Bridge was a two-lane eye-bar suspension type bridge designed according to American Society of Civil Engineers specifications and constructed in 1928 by the General Corporation and the American Bridge Company. It was designed with a 22-ft roadway and one 5-ft sidewalk with "High Tension" eye-bar chains, a unique anchorage system, and rocker towers. At the time of its construction, the maximum permitted truck gross weight was about 20,000 lb compared to the large truck limit of 60,000 lb or more in 1967. See Figure 3.48.

The Silver Bridge was the first eye-bar suspension bridge of its type to be constructed in the United States. The bridge's eye bars were linked together in pairs like a chain. A huge pin passed through the eye and linked each piece to the next. Each chain link consisted of a pair of 2 × 12 in bars and was connected by an 11 in pin. The length of each chain varied depending upon its location on the bridge. Such bridges had usually been constructed from redundant bar links, using rows of four to six bars, sometimes using several such chains in parallel. The eye bars in the Silver Bridge were not redundant as links were composed of only two bars each of high strength steel (more than twice as strong as common mild steel), rather than a thick stack of thinner bars of modest material strength "combed' together as is usual for redundancy. With only two bars, the failure of one could impose excessive loading on the second, causing total failure—unlikely if more bars are used. While a low-redundancy chain can be engineered to the design requirements, the safety is completely dependent upon correct, high quality manufacturing, and assembly.

"Rocker' towers were used that was also a unique feature of the Silver Bridge. About 131-ft-high towers allowed the bridge to move due to shifting loads and changes in the chain lengths due to temperature variations. A curved fitting was placed next to a flat one at the bottom of the piers. The rocker was then fitted with dowel rods to keep the structure from shifting horizontally. With this type of connection, the piers were not fixed to the bases. These allow the bridge to respond to various live loads by a slight tipping of the supporting towers that were parted at the deck level, rather than passing the suspension chain over a lubricated or tipping saddle or by stressing the towers in bending. Thus the towers required the chain on both sides for their support, so failure of any one link on either side, in any of the three chain spans would result in the complete failure of the entire bridge.

The forensic investigations concluded that the heat-treated carbon steel eye-bar (#330, on the north of the Ohio subsidiary chain, the first link below the top of the Ohio tower) broke due to a small crack that formed through fretting wear at the bearing. Over the years, partly due to increased

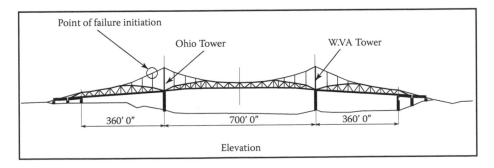

FIGURE 3.48 Silver Bridge collapse—before collapse.

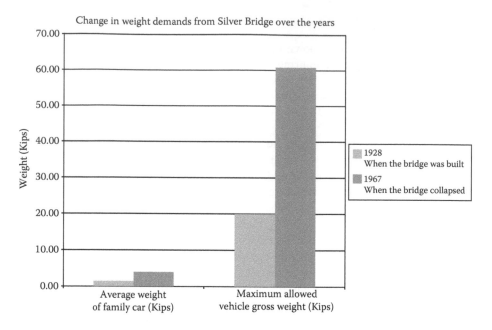

FIGURE 3.49   Changes in demands for Silver Bridge.

loads (Figure 3.49) when compared to design loads, stress corrosion, and corrosion fatigue allowed the crack to grow, and when became critical, the member broke in a brittle fashion. This placed undue stress on other members of the bridge. The remaining steel frame buckled and fell due to the newly concentrated stresses. The entire structure collapsed in the span of a minute.

The visual inspections performed cannot detect such cracks formed in the eye bar without disassembling the eye bar. These defects are still hard to detect even in present days. This failure prompted President Johnson to initiate a taskforce charged to determine procedures available to preclude future disasters and implement changes, if needed. This led to the establishment of National Bridge Inspection Standards by the Federal Government and current routine bridge inspections with qualified inspection personnel. Details of the bridge and its failure can be found at National Transportation Safety Board (NTSB) (1971) and LeRose (2001). Figure 3.50 illustrate some of the factors that led to collapse. Figure 3.51 shows the aftermath of the collapse.

3.5.6   Mianus River Bridge Collapse

Mianus River Bridge carried Interstate 95 over the Mianus River near Greenwich, CT. A 100-ft suspended eastbound span between piers 20 and 21 collapsed on June 28, 1983, killing three and seriously injuring three others when their vehicles fell into the river with the bridge. The collapse was attributed to the failure of the pin and hanger connection that was commonly used decades ago as this offered easier analysis and low construction costs. For more details, see Nationmaster.com (2011d).

The suspended span that collapsed was attached to the bridge structure at each of its four corners. A pin and hanger assembly attached each corner to the girders of the cantilever arm of an adjacent anchor span to support the weight of the northeast and southeast corners of the suspended span, The pin and hanger assembly includes an upper pin attached through the 2 1/2-in-thick web of the girder of the cantilever arm and a lower pin attached through the 2 1/2-in-thick web of the girder of the suspended span. One and one half-inch-thick steel hangers connect the upper and lower pins-one on the inside and one on the outside of the web (NTSB 1984). Figure 3.52 shows a pin and hanger system that is similar to the one used in Mianus River Bridge.

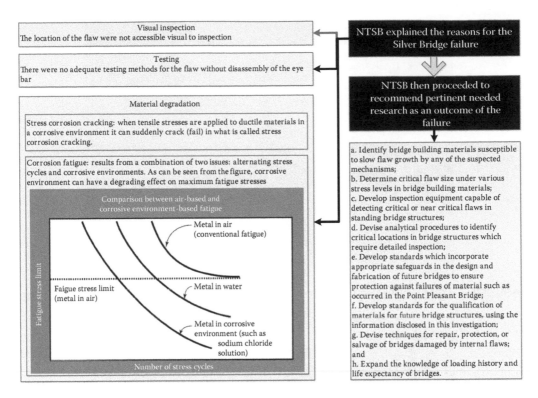

Visual inspection
The location of the flaw were not accessible visual to inspection

Testing
There were no adequate testing methods for the flaw without disassembly of the eye bar

Material degradation

Stress corrosion cracking: when tensile stresses are applied to ductile materials in a corrosive environment it can suddenly crack (fail) in what is called stress corrosion cracking.

Corrosion fatigue: results from a combination of two issues: alternating stress cycles and corrosive environments. As can be seen from the figure, corrosive environment can have a degrading effect on maximum fatigue stresses

Comparison between air-based and corrosive environment-based fatigue

Metal in air (conventional fatigue)

Faigue stress limit (metal in air)

Metal in water

Metal in corrosive environment (such as sodium chloride solution)

Fatigue stress limit

Number of stress cycles

NTSB explained the reasons for the Silver Bridge failure

NTSB then proceeded to recommend pertinent needed research as an outcome of the failure

a. Identify bridge building materials susceptible to slow flaw growth by any of the suspected mechanisms;
b. Determine critical flaw size under various stress levels in bridge building materials;
c. Develop inspection equipment capable of detecting critical or near critical flaws in standing bridge structures;
d. Devise analytical procedures to identify critical locations in bridge structures which require detailed inspection;
e. Develop standards which incorporate appropriate safeguards in the design and fabrication of future bridges to ensure protection against failures of material such as occurred in the Point Pleasant Bridge;
f. Develop standards for the qualification of materials for future bridge structures, using the information disclosed in this investigation;
g. Devise techniques for repair, protection, or salvage of bridges damaged by internal flaws; and
h. Expand the knowledge of loading history and life expectancy of bridges.

FIGURE 3.50 Silver Bridge collapse—reasons for collapse.

FIGURE 3.51 Silver Bridge collapse—after collapse. (With permission from National Highway Institute.)

Sometime before the collapse of the suspended span, the inside hanger in the southeast corner of the span came off of the inside end of the lower pin. This action shifted the entire weight of the southeast corner of the span onto the outside hanger. The outside hanger gradually worked its way farther outward on the pin, and over a period of time, a fatigue crack developed in the top outside end of the upper pin. The shoulder of the pin fractured off, the pin and hanger assembly failed, and the span collapsed into the river. The NTSB determined that the probable cause of the collapse of the Mianus River Bridge span was the undetected lateral displacement of the hangers of the pin and

FIGURE 3.52   Typical pin and hanger construction.

FIGURE 3.53   Failed span. (With permission from National Highway Institute.)

hanger suspension assembly in the southeast corner of the span by corrosion-induced forces due to deficiencies in the State of Connecticut's bridge safety inspection and bridge maintenance program (NTSB 1984). Figure 3.53 shows the failed bridge. Figure 3.54 illustrates the sequence of failure.

The investigations cited corrosion from water build up due to inadequate drainage as a cause. The highway drains had been deliberately blocked during road mending some 10 years before and water leaked down through the pin bearings, causing them to rust. The outer bearings were safety-critical and nonredundant, a design flaw of this particular type of structure. The bearings were difficult to inspect close up, although traces of rust could be seen near the affected bearings.

This highlights the need for thorough inspection and maintenance program, use of nondestructive procedures when needed, and the value of redundancy.

3.5.7   SCHOHARIE BRIDGE COLLAPSE

On April 5, 1987, two spans of the five-span Schoharie Creek Bridge collapsed causing five vehicles falling into the river with 10 fatalities. A third span collapsed 90 minutes later. The bridge was

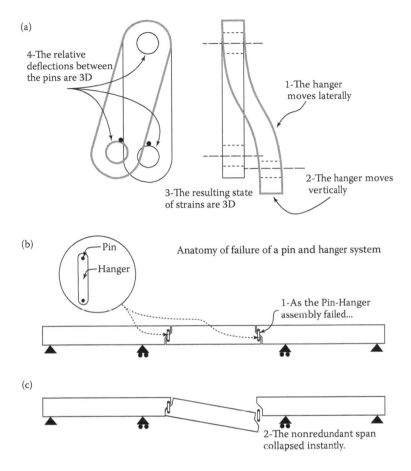

(a)

4-The relative deflections between the pins are 3D

1-The hanger moves laterally

3-The resulting state of strains are 3D

2-The hanger moves vertically

(b)

Pin

Hanger

Anatomy of failure of a pin and hanger system

1-As the Pin-Hanger assembly failed...

(c)

2-The nonredundant span collapsed instantly.

FIGURE 3.54 Failure sequence of a pin and hanger system (a) lateral and longitudinal expansion of hangers, (b) as-designed pin-hanger system, and (c) failure of bridge due to lack of system redundancy.

built in 1954 and owned by the New York State Thruway Authority. It was located in Mohawk valley northwest of Albany, NY. The 540-ft-long and 112-ft-wide bridge carried four lanes of traffic, median area, and shoulders over Schoharie creek. The average height of the bridge above the creek was about 80 ft. The bridge was designed to 1949 edition of AASHTO Standard Specifications for Highway Bridges. It was a five-span steel superstructure with reinforced concrete deck. The piers consists of two slightly tapered reinforced concrete columns connected by a tie beam near column tops and supported on a common pedestal and footing. Two of the piers were in the Creek with other two piers on the creek banks to support the structure along with two end abutments.

The bridge collapsed on April 5, 1987 during the spring flood caused by a rainfall totaling 150 mm combined with snowmelt to produce an estimated 50 year flood (WJE Associates 1987). The collapse was initiated by the toppling of pier three, which caused the progressive collapse of spans three and four into the flooded creek. Pier two and span two fell 90 minutes after span three dropped, and pier one and span one shifted 2 hours after that (Thornton-Tomasetti 1987). The NTSB suggested that pier two collapsed because the wreckage of pier three and the two spans partially blocked the river, redirecting the water to pier two and increasing the stream velocity (NTSB 1988).

According to investigations, scouring under piers began shortly after the bridge was built in 1955 as the bridge footings experienced floodwater flows unanticipated in the design of the bridge, a 100-year flood. It is believed that the majority of the scouring energy was dissipated into moving the original riprap layer from around the footings. Once the backfill had been exposed, the years of peak flows removed the backfill material, and the backfill material in turn was replaced by sediment settling into the scoured. Furthermore, the riprap placed at construction had probably been washed away during the 1955 flood, and had not been replaced (WJE Associates 1987).

The investigations concluded that the collapse of the Schoharie Creek Bridge was due to the extensive scour under pier three that was affected by four important factors (Thornton-Tomasetti 1987): (a) the depth of shallow footings, bearing on soil, was not enough to take them below the probable limit of scour; (b) the foundation of pier three was bearing on erodable soil that allowed high velocity floodwaters to penetrate the bearing stratum; (c) the as-built footing excavations and backfill could not resist scour; and (d) riprap protection, inspection and maintenance were inadequate.

Besides scour issues, several factors including two common practices when the bridge was designed in the 1950s contributed to the severity of the collapse (Thornton-Tomasetti 1987). These were the bridge bearings allowed the spans to lift or slide off of the concrete piers and the simple spans were not redundant. Thus, the use of continuous spans, rather than simple spans, would have provided redundancy once pier three failed, and perhaps allowed for the redistribution of forces between the spans. The failure also shows that it is important for bridge owners to identify the critical features that can lead to the collapse of a bridge and to ensure that those critical features are inspected frequently and adequately (NTSB 1988). Although the Schoharie Creek Bridge had been inspected annually or biennially since 1968, an underwater inspection of the piers footings had never been performed. The bridge was scheduled for an underwater inspection in 1987, but the bridge collapsed before the inspection took place (NTSB 1988). The failures also emphasize how important it is to design footings deep enough to avoid loss of support capacity due to scour (Shephard and Frost 1995) (Figures 3.55 and 3.56).

3.5.8 I-35 Bridge, Minneapolis, MN

On August 1, 2007, evening the eight lane, 1907-ft-long I-35W highway bridge over the Mississippi River in Minneapolis, Minnesota, experienced a catastrophic failure in the main span of the deck truss killing 13 and injuring 145 persons. One thousand feet of the deck truss collapsed, with about 456 ft of the main span falling 108 ft into the 15-ft-deep river.

The I-35W Bridge was located about 1 mile northeast of the junction of I-35W with Interstate 94. In addition to spanning the Mississippi River, the bridge also extended across Minnesota Commercial Railway railroad tracks and three roadways: West River Parkway, 2nd Street, and the access road to the lock and dam. The bridge was opened to traffic in 1967 with 141,000 ADT in

FIGURE 3.55 Collapsed bridge, from http://ny.water.usgs.gov/projects/scour/text.html

FIGURE 3.56   Scour hole, from http://ny.water.usgs.gov/projects/scour/fig5.html

FIGURE 3.57   I-35 Bridge spanning the Mississippi river. (NTSB 2008.)

2004. The 14-span structure used welded built-up steel beams for girders and truss members, with riveted and bolted connections. The bridge was 1907-ft-long and carried eight lanes of traffic, four northbound and four southbound. The 1064-ft-long deck truss portion of the bridge encompassed a portion of span 5; all of spans 6, 7, and 8; and a portion of span 9. The deck truss was supported by four piers (see Figure 3.57).

Riveted steel gusset plates at each of the 112 nodes (connection points) of the two main trusses tied the ends of the truss members to one another and to the rest of the structure. The gusset plates were riveted to the side plates of the box members and to the flanges of the H members. All nodes had at least two gusset plates, one on either side of the connection point. A typical I-35W main truss node, with gusset plates, is shown in Figure 3.58.

The NTSB report (NTSB 2008) attributed the probable cause of the collapse of the bridge (Figure 3.59) to the inadequate load capacity due to a design error of the gusset plates at the U10 nodes (see Figure 3.60), which failed under a combination of substantial increases in the weight of the bridge, which resulted from previous bridge modifications, and the traffic and concentrated construction loads on the bridge on the day of the collapse. Contributing to the design error was the failure of quality control procedures by the design company to ensure that the appropriate main truss gusset plate calculations were performed for the I-35W Bridge and the inadequate design review by Federal and State transportation officials. Contributing to the accident was the generally accepted practice among Federal and State transportation officials of giving inadequate attention to

FIGURE 3.58 Typical five-member node (two upper chord members, one vertical member, and two diagonal members) on I-35W Bridge. (NTSB 2008).

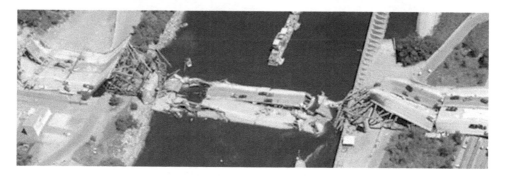

FIGURE 3.59 Collapsed bridge. (NTSB 2008).

FIGURE 3.60 Locations of components. (NTSB 2008.)

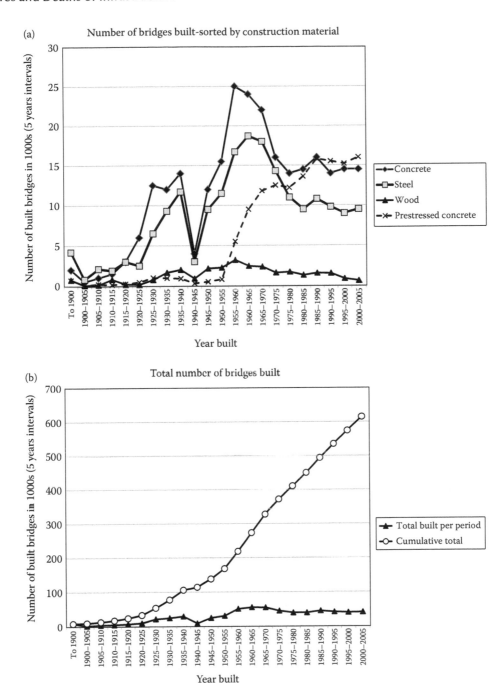

FIGURE 3.61 Count of bridges in the United States: (a) Number of bridges built - sorted by construction material, and (b) total number of bridges built.

TABLE 3.18
Count of Bridge Inventory in the United States

Year Built	Concrete	Steel	Prestressed Concrete	Wood
Before 1900	1738	4196	642	731
1901–1905	283	825	126	34
1906–1910	882	2129	222	69
1911–1915	1556	1597	161	47
1916–1920	3248	2899	381	182
1921–1925	5564	2395	403	152
1926–1930	12,121	6529	852	716
1931–1935	11,874	9255	752	1650
1936–1940	13,883	11,566	657	1994
1941–1945	4463	2949	198	898
1946–1950	11,824	9509	418	2121
1951–1955	15,185	11,268	728	2255
1956–1960	24,584	16,684	5278	3090
1961–1965	24,345	18,621	9176	2453
1966–1970	22,017	18,002	11,709	2309
1971–1975	15,333	14,304	12,513	1631
1976–1980	13,801	10,990	12,195	1701
1981–1985	14,186	9321	13,510	1286
1986–1990	15,409	10,742	15,685	1589
1991–1995	14,265	9611	15,580	1534
1996–2000	14,500	8845	15,178	870
2001–2007	14,545	9275	15,977	568

gusset plates during inspections for conditions of distortion, such as bowing, and of excluding gusset plates in load rating analyses.

3.6   APPENDIX I: COUNT OF BRIDGES IN THE UNITED STATES BY CONSTRUCTION MATERIAL

The number of highway bridges in the United States is a massive number indeed as shown in Figure 3.61 and Table 3.18. The primary materials used in bridge construction are concrete, pre-stressed concrete, steel, and wood, as shown in Figure 3.61a. The utilization of wood as a bridge construction material peaked in the mid-twentieth century. Around the same time, the utilization of pre-stressed concrete increased. At present, the use of steel, concrete, and pre-stressed concrete dominate bridge construction. Figure 3.61b indicates the total bridges built in 5-year intervals. As of the writing of this volume, the total number of bridges in the United States is over 600,000. Note that the numbers in Figure 3.61b does not account for the decommissioned, failed, or demolished bridges.

REFERENCES

Alampalli, S. and McCowan, P., (2008) "Addressing Critical Findings from Highway Bridge Inspection: New York State Procedures," *NDE/NDT for Highways and Bridges: Structural Materials Technology (SMT)*, ASNT, Oakland, CA.

Court of Inquiry, (1879), "Tay Bridge Disaster," Report Of The Court of Inquiry, and Report Of Mr. Rothery, "Upon the Circumstances Attending the Fall of a Portion of the Tay Bridge on the 28th December 1879" place this as the second reference". *The Railways Archive*. Retrieved on 2007-02-05.

Ettouney, M., and Alampalli, S., (2012) *Infrastructure Health in Civil Engineering: Applications and Management*, CRC Press, Boca Raton, FL.

Ettouney, M., Smilowtz, R., Tang, M., and Hapij, A., (2006) "Global System Considerations for Progressive Collapse with Extensions to Other Natural and Man-Made Hazard," *Journal of Performance of Constructed Facilities*, 20(1), 403–417.

FHWA., (2007) "National Bridge Inventory, NBI," Federal Highway Administration, Washington, DC.

FEMA., (2009) *Handbook for Rapid Visual Screening of Buildings to Evaluate Terrorism Risk*, Federal Emergency Management Agency, FEMA Publication, Washington, DC.

FEMA., (2010) *A How-To Guide to Mitigate Potential Terrorist Attacks*, Federal Emergency Management Agency, FEMA Publication, Washington, DC.

Garg, N., Larkin, A., and Ricalde, L., (2008) "Evaluation of the Application Potential of PSPA in determination of Modulus of Concrete HMA pavements at the FAA's National Airport Pavement Test Facility," *NDE/NDT for Highways and Bridges: Structural Materials Technology (SMT)*, ASNT, Oakland, CA.

Hansen, W. and Surlaker, S., (2006) *Embedded Wireless Temperature Monitoring Systems For Concrete Quality Control*, University of Michigan, An Arbor, MI.

Ketchum, M., (2011c) "Tacoma Narrows" http://www.ketchum.org/wind.html [accessed February 20, 2011].

Kosnik, D. and Hopwood, T., (2008) "Monitoring of In-Situ Strains in Bearing Assembly Anchor Bolts on a Large Through - Truss Bridge," *NDE/NDT for Highways and Bridges: Structural Materials Technology (SMT)*, ASNT, Oakland, CA.

LeRose, C., (2001) *The Collapse of the Silver Bridge*, West Virginia Historical Society Quarterly Vol. XV, No. 4.

Lienhard, J., (2008) "Quebec Bridge Collapse," http://www.uh.edu/engines/epi220.htm [accessed February 20, 2011].

National Transportation Safety Board (NTSB)., (1988) "Collapse of New York Thruway (1–90) Bridge over the Schoharie Creek, near Amsterdam, New York, April 5, 1987." Highway Accident Report: NTSB/HAR-88/02, Washington, D.C.

Nationmaster.com (2011d) "Mianus River Bridge," http://www.nationmaster.com/encyclopedia/Mianus-River-Bridge [accessed February 20, 2011].

Ni, Y., Xia, Y, Liao, W., and Ko, J., (2008) "Technology Innovation in Developing the Structural Health Monitoring System for Guangzhou New TV Tower," *International Symposium on Structural Control and Health Monitoring*, Taichung, Taiwan.

NTSB., (1971) "Mianus River Bridge Collapse" http://www.ntsb.gov/Publictn/1971/HAR7101.htm, accessed on September 2, 2008. [accessed February 20, 2011].

NTSB., (1984) "Silver Bridge Collapse" report date 1984 accessed on 5 – 15 – 2008, http://www.ntsb.gov/publictn/1984/HAR8403.htm [accessed February 20, 2011].

NTSB., (2008) "Highway Accident Report, Collapse of I-35W Highway Bridge Minneapolis, Minnesota, August 1, 2007," National Transportation Safety Board, Washington, D.C., NTSB/HAR-08/03, PB2008–916203, Notation 7975C.

NYSDOT., (2008) *SFY 2006–07 Annual Report of Bridge Management and Inspection Programs*, New York State Department of Transportation, Albany, NY.

Ricketts, B., (2008) "The Collapse of the Quebec City Bridge," accessed from the web on 2008-02-05, http://www.mysteriesofcanada.com/Quebec/quebec_bridge_collapse.htm [accessed February 20, 2011].

Seim, C., (2008) "Why Bridge Have Failed Throughout History," *Civil Engineering Vol 78* (5): 64–71, 84–87, http://www.ASCE.org [accessed February 20, 2011].

Shepherd, R. and Frost, J. D., (1995) *Failures in Civil Engineering: Structural, Foundation and Geoenvironmental Case Studies.* American Society of Civil Engineers, New York.

Smith, C., (2008) "A Disaster in the Making," accessed from the web on 2008-02-05, http://www.civeng.carleton.ca/ECL/reports/ECL270/Prelude.html [accessed February 20, 2011].

Thornton-Tomasetti, P. C., (1987) "Overview Report Investigation of the New York State Thruway Schoharie Creek Bridge Collapse." Prepared for New York State Disaster Preparedness Commission, December.

Wardhana, K. and Hadipriono, F., (2003) "Analysis of Recent Bridge Failures in the United States," *ASCE J. of Performance of Constructed Facilities, Vol 17*, No. 3.

Washington University., (2008a) Data sheet on the 1940 Tacoma Narrows Bridge, 1940, http://content.lib.washington.edu/cdm4/item_viewer.php?CISOROOT=/farquharson&CISOPTR=108&CISOBOX=1&REC=1

WJE, Associates, Inc., and Mueser Rutledge Consulting Engineers., (1987) "Collapse of Thruway Bridge at Schoharie Creek," Final Report, Prepared for: New York State Thruway Authority, November.

4 Theories and Principles of IHCE

4.1 INTRODUCTION

Conventional experiments in bridge engineering have been mostly of the nondestructive testing (NDT) type. NDT experiments are mostly local, both in time and space. They are very adapting in revealing accurate picture of the condition of a local part of the bridge at specific time. Due to such limited scope of NDT, the value (or benefit-to-cost) of NDT experiments were not a major issue, since the objectives of the NDT experiments are usually well defined, and the tools of executing such an experiment are well suited for obtaining the objectives of the experiments. The structural health monitoring (SHM) paradigm, which purport a bridge monitoring experimentation that is wider in both locale and time than the conventional NDT experimentation, have been gaining interest lately. The overall system of SHM/SHCE/IHCE is shown schematically in Figure 4.1. (Note that SHM is used in this chapter [and in most of this volume] in lieu of SHCE/IHCE without loss of generality.) SHM projects and experiments are demanding, as shown in Figure 4.2. Due to the larger space and time range of SHM experimentation, the optimum experiment parameters, such as number of sensors, type of sensor mix, type of data acquisition, labor costs, and the value of additional information from the experiment, needs to be carefully studied. Currently, there are no quantifying methods that help in designing an SHM experiment that studies and optimizes the overall value of such an experiment.

This chapter introduces several theoretical SHM concepts. We discuss the concepts of cost and value in SHM as metrics for optimizing different projects. We then introduce a general theory of experimentation (GTE). The general theory addresses the optimum value of SHM experiment. The theory lays the groundwork for quantifying the value any SHM experiment, thus providing the decision maker a subjective tool to justify the SHM project *a priori*, with clearly quantifiable value of the results. A special theory of experimentation (STE) is also introduced that concentrates on the narrower subject of sensors within the SHM field. This special theory addresses the type, number, and location of sensors in an overall SHM project. It also includes the cost of material, labor, and the overall expected value of using the sensor topology. The purpose is to optimize such a value, given the direct and indirect costs. Finally, a theory of SHM triangulation is offered. The theory addresses the all important concept of sensor, or detection, types. It lays down the theoretical basis for using different detection schemes during SHM projects. Some applications in the SHM field will be demonstrated.

The chapter then presents several principles of SHM, namely: the duality, scaling, and serendipity principles. The three principles can help in achieving efficient and cost-effective executions of SHM projects.

It is believed that the theories and the principles of this chapter will help in providing qualitative means to enumerate the values of SHM projects. This should, in turn, help in promoting SHM projects, thus fulfilling the SHM goals of improving safety while reducing costs.

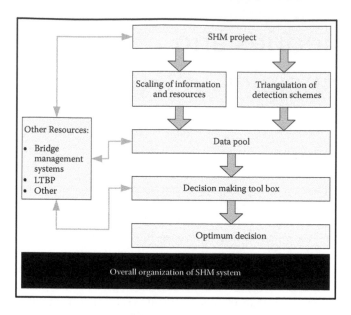

FIGURE 4.1 Organization of SHM/SHCE System.

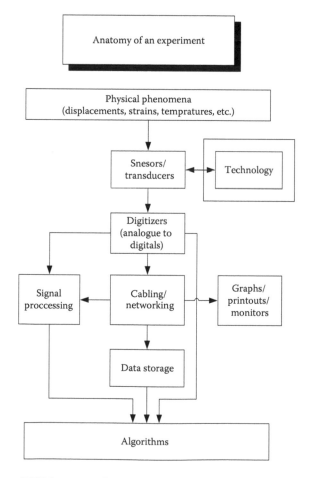

FIGURE 4.2 Anatomy of SHM, or general, experiments.

4.2 GENERAL THEORY OF EXPERIMENTATION (GTE)

4.2.1 INTRODUCTION

We discussed the importance of SHM and the SHCE in general for preserving and enhancing structural health in the previous chapter. In embarking on an SHM project, one of the most important concerns that the decision makers face is: how valuable such a project is? We already have seen that the value of any activity related to costs of such an activity and the benefits that such an activity would bring. We try in this section to formalize the value versus cost of experiments. Figures 4.3 and 4.4 show the different parameters that might be needed for quantifying cost and value of experiments. We introduce a general theory for experimentation that would help in such formalization. We then provide two practical examples to show how the theory can be used to ensure optimum value for the experiment on hand.

Before introducing the theory, we would like to discuss its governing parameters first. In several other locations of this volume, we identified value as a relationship between costs and benefits. For simplicity, we opted to relate the value directly with costs in the theory below. By doing this, we couple benefits and value in a single expression. There are several advantages for doing that: (a) it simplifies the theory and its applications, (b) it allows for a qualitative treatment of the concept of value in a nondimensional manner, (c) it avoid the intricacies and complications of relating value to benefits and costs, and (d) it allows for incorporating non-quantifiable benefits of experiments in estimating the value (such benefits include, for example, added redundancies by using additional sensors, or educational benefits). Of course, the concept of benefit can be incorporated to the application of the theory if desired, without any loss of generality or accuracy.

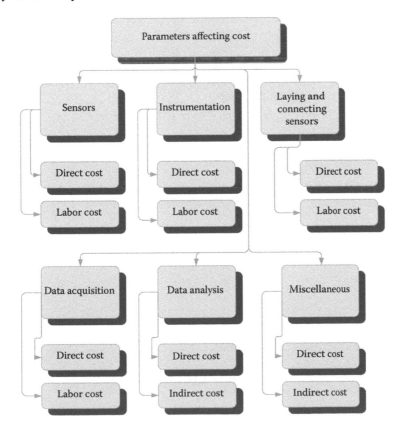

FIGURE 4.3 Parameters affecting cost and value

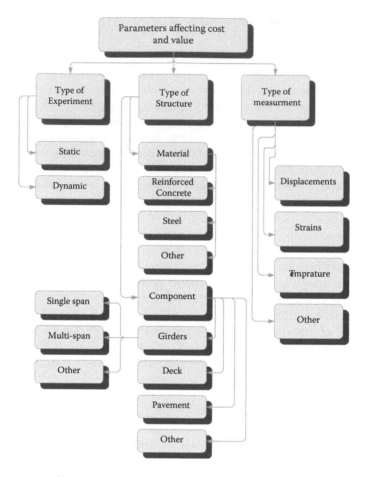

FIGURE 4.4   Parameters affecting cost and value.

4.2.2 Overview of GTE

The GTE can be stated as follows:

There exists an intrinsic value for every experiment. The value, as a function of cost, posses a critical point; beyond this point, any additional expenditure is not cost-effective.

we discuss several aspects of the theory next:
Analytically, the theory can be enumerated by first defining a value for the SHE as

$$V = V(C) \tag{4.1}$$

where
C = Cost of the experiment

Define the critical value point, V_{cr} such that

$$V_{cr} = V(C_{cr}) \tag{4.2}$$

where
V_{cr} = Critical value of the experiment
C_{cr} = Critical cost of the experiment

According to the theory, when the cost of the experiment $C > C_{cr}$ it would be logical not to expend such additional cost since any additional cost beyond and above the critical cost would produce either a diminishing, or no additional, value.

In addition to the critical point, (C_{cr}, V_{cr}), in the $C - V$ space, it is also beneficial to define an ultimate point, (C_{ult}, V_{ult}), such that

V_{ult} = Ultimate value of the experiment

C_{ult} = Ultimate cost of the experiment

The ultimate value of an experiment V_{ult} is the value beyond which there can be no added value to an experiment, in spite of any additional cost. The ultimate cost of an experiment C_{ult} is the cost that corresponds to V_{ult}, such that

$$V_{ult} = V\left(C_{ult}\right) \tag{4.3}$$

Formally

$$\left.\frac{dV}{dC}\right|_{C=C_{ult}} = 0 \tag{4.4}$$

4.2.3 PROOF OF GTE

4.2.3.1 General

For the function $V = V(C)$, a small incremental added value of the experiment ΔV can be expressed as

$$\Delta V = \frac{dV(C)}{dC}\,\Delta C \tag{4.5}$$

The term $dV(C)/dC$ is the rate of value versus cost, it can be defined as

$$R_V = \frac{dV(C)}{dC} \tag{4.6}$$

Consider now the parameters that affect V as

$$a_1, a_2, a_3, \cdots a_{NE}$$

where

$a_i = i^{th}$ parameter of the experiment

NE = Total number of parameters that affect the experiment

Moreover, the cost C can be subdivided into two parts:

$$C = C_V + C_F \tag{4.7}$$

C_V = Costs that are dependent on the parameters a_i, as such, they are the costs that directly affect the value

C_F = Costs that are independent on the parameters a_i, as such, they are the costs that do not affect the value

In general, C_F represents the fixed costs of the experiment. These costs are the fixed costs that do not change much for a given experiment. As such, these costs are considered fixed costs. Fixed costs of an experiment can be data acquisition systems, labor for traffic protection, software for data collection and analysis, test vehicles, and so on. In most cases, changing these costs does not result in any meaningful changes in the resultant value of the experiment.

On the other hand, C_V represents variable costs, that is, the costs that can affect the value of the experiment. Examples of these variable costs are costs of sensors, instrumentation, and data acquisition and analysis.

As such, variable costs can be expressed as

$$C_V = C_V\left(a_1, a_2, a_3, \cdots a_{NE}\right) \tag{4.8}$$

The parameters a_i represent any/all of the important discrete parameters that affect the value of a particular experiment. For example, these parameters can represent individual sensors, components of instrumentation system, data acquisition system(s), and so on. Equation 4.8 can be simplified, without any loss of generality, to

$$C_V = \sum_{i=1}^{i=NE} C_{Vi}\left(a_i\right) \tag{4.9}$$

Similarly, the value can be expressed as

$$V = \sum_{i=1}^{i=NE} V_i\left(a_i\right) \tag{4.10}$$

From Equations 4.6, 4.9, and 4.10, the rate of value can be expressed as

$$R_V = \sum_{i=1}^{i=NE} \left[\frac{\partial V_i\left(a_i\right)/\partial a_i}{\partial C_i\left(a_i\right)/\partial a_i}\right] \tag{4.11}$$

The terms $\partial V_i(a_i)/\partial a_i$ and $\partial C_i(a_i)/\partial a_i$ represents the rates of change of value V_i and cost C_i as the parameter a_i changes, respectively. It is reasonable to conclude that as a_i increases, the magnitude of $\partial V_i(a_i)/\partial a_i$ will eventually decreases. Ultimately, there exists a point beyond which there would be no added value to the experiment, no matter how much a_i is increased. This can be expressed as

$$\left.\frac{\partial V_i\left(a_i\right)}{\partial a_i}\right|_{a_i \to \infty} = 0 \tag{4.12}$$

On the other hand, it is reasonable to assume that as a_i increases, the cost $C_{Vi}(a_i)$ will always increase: it will never reach a value of 0. As such

$$\left.\frac{\partial C_{Vi}\left(a_i\right)}{\partial a_i}\right|_{a_i \to \infty} > 0 \tag{4.13}$$

From Equations 4.11, 4.12, and 4.13 it can be stated that

$$R_V\big| \to 0, \tag{4.14}$$

as the cost $C \approx C_{ult}$. In other words, at a given level of cost expenditures C_{ult} there can be no additional value to the experiment.

At the lower end of the magnitude of the parameters a_i, it is reasonable to assume that

$$R_V \gg 0 \tag{4.15}$$

This means that at a lower level of cost, there is a finite magnitude of R_V; there is an appreciable added value to the experiment for additional costs.

Equations 4.14 and 4.15 provide the limits of R_V. It indicates that the maximum rate of value of any given experiment is between very low and high cost levels: there is an ultimate cost point beyond which there is no additional added value.

We have proved so far the existence of an ultimate value and cost point for every experiment. How about the critical value and its associated critical cost?

We need to recall that the critical cost of experiment is the cost point beyond which any additional cost will result in a diminishing, or no, additional value. We can then establish the following relationships

$$V_{cr} \leq V_{ult} \tag{4.16}$$

and

$$C_{cr} \leq C_{ult} \tag{4.17}$$

In addition, we propose the following relationship

$$V_{cr} = \alpha_{GTE} V_{ult} \tag{4.18}$$

From Equations 4.16 and 4.18 it is clear that $\alpha_{GTE} \leq 1.0$. Also, it is logical that $0 < \alpha_{GTE}$.

Since the ultimate value of an experiment exists, based on Equation 4.18, we can conclude that the critical value and its associated costs, do exist. This completes the proof of GTE.

4.2.3.2   Corollary # 1: Types of Experiments and α_{GTE}

We differentiated between V_{cr} and V_{ult} for a very practical reason. Such a reason becomes evident by categorizing experiments into two types.

1. Type I experiments, where a finite $C_{ult} < \infty$ and $V_{ult} < \infty$ exist, or
2. Type II experiments, where only an infinite $C_{ult} = \infty$ and $V_{ult} = \infty$ exist. In this type of experiments there is always an added value for additional costs, albeit with an increasingly diminishing rate of return.

Figures 4.5 through 4.7 show the two different types of experiments. Note that for type I:

$$C_{cr} = C_{ult} \tag{4.19}$$
$$V_{cr} = V_{ult} \tag{4.20}$$

Thus, for type I

$$\alpha_{GTE} = 1.0 \tag{4.21}$$

For Type II experiments, since both C_{ult} and V_{ult} exist only at ∞, the critical point (C_{cr}, V_{cr}) is subjectively located on the $C - V$ based on the shape of the curve, usually at a point when there is a sharp downturn of the magnitude of R_V as shown in Figure 4.7. For practical reasons, V_{cr} is always less than V_{ult}.

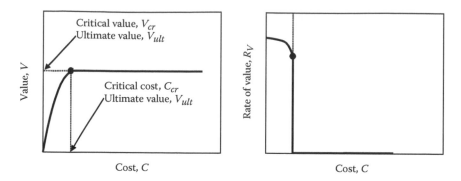

FIGURE 4.5   Type I experiment.

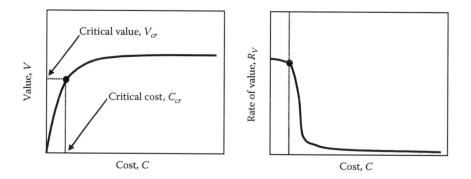

FIGURE 4.6   Type I experiment.

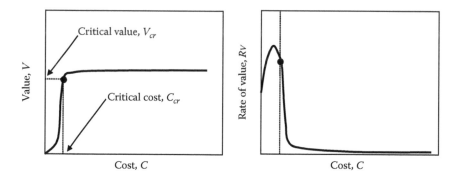

FIGURE 4.7   Type II experiment.

This leaves a final question: what is the magnitude of α_{GTE} for type II experiments? Unfortunately, there is no possible quantitative answer for this question, beyond the knowledge of the limits $0 < \alpha_{GTE} \leq 1.0$. Nor there is any body of knowledge as of the writing of this volume that can help in answering such a question. We propose that a detailed study of value-cost (V-C) relationship of experiments are performed, and a reasonable choice of α_{GTE}, and (C_{cr}, V_{cr}) be done based on such a study.

4.2.3.3 Corollary # 2: Initial Conditions

In the $C - V$ space of an experiment, there can be situations where the experiment needs an initial outlay of funding, before any value is realized. This can be expressed as

$$V(C)\big|_{0 \leq C \leq C_{min}} = 0 \tag{4.22}$$

Here, C_{min} is the minimum outlay of funding (cost) that needs to be spent before any value can be realized. Figure 4.8 shows the $C - V$ space for this situation. Generally speaking, most experiments would have a finite magnitude of C_{min}.

In practice, a reasonable estimate of C_{min} must be established before embarking on any experiment. The decision maker must then compare C_{min} with the available funding F_A and decide if the executing the experiment is still viable option. Two examples of this process are shown in Figure 4.8.

4.2.4 Practical Applications

4.2.4.1 General

The existence of an optimal configuration of experiments was discussed in an indirect manner by Hyland and Fry (2000). They presented an outline of a scheme to optimize the performance of SHM systems. Their approach constituted two parts: local and global parts. The local part deals with sensor performance in detecting damage. They recommended an Artificial Neural Network (ANN) to train the sensors to detect damages. The global part of the scheme is a genetic algorithm for continued evolution of the overall SHM system. This includes continued evaluation of sensor

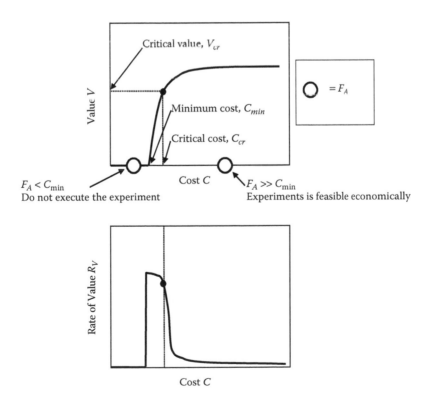

FIGURE 4.8 Effects of initial costs on experiment.

performance, and making decisions as to ways to improve such performance. The type, number, and placement of sensors are always evaluated and improved by the genetic algorithm.

The recognition of the existence of optimal sensor types, number, and placement by Hyland and Fry is, indirectly, supportive of our general and special theories of experimentations. The special theory deals specifically with the optimal number of sensors for each experiment. This optimal condition can be found, as the theory imply, by any suitable method, including, but not limited to, ANN. Also, our GTE indicates that there is an optimal setting for each experiment. Such optimality is controlled by the objective(s) of the SHM project. Again, the methods of studying the optimal conditions should be done on a case by case basis. Genetic technique is one of the many approaches that can be used to study and achieve such optimality. In addition, our GTE includes both the technical performance of the experiment as well as the cost of such performance (value of experiment). The genetic approach of Hyland and Fry (2000) stresses mostly on technical performance of experiments. In realistic settings, cost implications must be an integral part of any optimal solutions of an SHM project.

4.2.4.2   Identifying Goals

In identifying goals of experiment several issues must be considered. The technical goal is foremost of those goals to be identified. For example: is the purpose to monitor corrosion progression, fatigue effects, or load capacity? Is the goal to diagnose a problem that was observed? (For example: an excessive vibration). Is the goal to validate a particular design or particular analytical techniques? The length of the experiment (short term or long term) must be decided next, depending on the goal of the experiment. Length of experiments can be a major factor in evaluating V_{cr} and C_{cr}.

4.2.4.3   Identifying Algorithms

After identifying the goals of the experiment, the algorithms that would be used to achieve these goals are chosen next. This document includes numerous examples on how to choose appropriate algorithms to achieve specific goals. Some examples are given in Table 4.1.

4.2.4.4   Needed Type of Measurements (TOM)

Selection of appropriate algorithms is a logical step before choosing the measurements. This will ensure efficient operations. It is the algorithm that drives the measurements, not the other way around. Table 4.2 shows some examples of needed TOMs.

4.2.4.5   Values of Parameters

After assigning all pertinent parameters to the experiment, a value should be assigned to each parameter. The values can be either monetary, or dimensionless. The dimensionless values can have an arbitrary scale, say from 1 to 10, or 1 to 100. The only obvious condition in choosing the values is that they need to be consistent for all the parameters. Relative magnitudes of values of

TABLE 4.1
Algorithm Examples

Goal	Algorithm
Fatigue at a given point	Rain flow algorithm remaining life, etc. (see Chapter 9 of Ettouney and Alampalli 2012)
Model verification	Finite elements code, optimization schemes and tolerances, neural networks (see Chapter 6 of this volume)
Modal testing and excessive vibrations	Modal identification method based on type of excitation source and pertinent analytical method (see Chapter 6 of this volume)
Load rating	Diagnostic testing, influence lines (see Chapter 8 of Ettouney and Alampalli 2012, etc. and analysis type)

TABLE 4.2
Needed Measurements

Algorithm	Needed Measurements
Remaining life (Fatigue)	Strains as a function of time
Model verification and modal testing	Time histories of displacements or velocities at given points
Influence lines	Displacements. For dynamic influence lines, need displacements as a function of time
Load rating	Depending on the type of load rating analysis: displacements, strains, weights, etc.

each parameter should reflect the relative importance of that parameter to achieving the objectives of the experiment. The values of the two examples given later in this section are dimensionless, with a scale from 0.0 to 1.0. Of course, any null value indicates that the parameter has no useful role in the experiment.

4.2.4.6 Parameter Cost Evaluation

Costs of each parameter should also be assigned next. As discussed earlier, there are two types of costs: fixed and variable costs. Fixed costs are those costs that are needed for each experiment, yet they have no effect on the value of the experiment. They play no role in determining V_{cr} or V_{ult}. Variable costs need to be estimated for each parameter.

Similar to values, costs can be either monetary or dimensionless. The dimensionless costs can have arbitrary scales; say from 1 to 10, or 1 to 100. Again, the only condition in choosing parameter costs is that they need to be consistent for all the parameters. Relative magnitudes of costs of each parameter should reflect the relative cost of that parameter. The costs of the two examples given later in this section are dimensionless. Of course, any null cost indicates that the parameter has no effect on V_{cr} or V_{ult}; no-cost parameters should be considered as fixed costs.

4.2.4.7 Value–Cost (V–C) Diagram

After assigning values and costs for each parameter two vectors, $\{V\}$ and $\{C\}$, containing the values and costs, respectively, can be populated. The size of the vectors is *NE*. Before they can be used logically, they should be sorted in an ascending order using the value as the controlling index. After sorting, the cumulative cost and value vectors should be prepared. Using these cumulative results, the cumulative value-cumulative cost curve can be charted, and this will produce the required $C - V$ space.

4.2.4.8 Critical Magnitudes: V_{cr} and C_{cr}

The $C - V$ space, either type I or type II, as shown in Figures 4.5, 4.6, or 4.7, can be used to determine V_{cr}, C_{cr}. If the $C - V$ space is of type I, it should be a straightforward determination. If the $C - V$ space is of type II, then a logical decision regarding the value of α_{GTE} is needed, as described above.

4.2.4.9 Summary of the Steps

The theory can be applied to almost any SHM/SHCE situation. The benefits of applying this theory are to

1. Get the optimum value for any experiment
2. Gain more understanding to the different issues pertaining to any SHM/SHCE situation
3. For complex/large SHM/SHCE project, applying the theory can shed light to shortcomings of the proposed project

The practical steps of applying the theory are as follows.

1. Identify the goal of the SHM
2. Identify the algorithms to be used in reaching the above goal
3. Identify the needed measurements for the algorithm
4. From 2 and 3, identify all parameters a_i
5. Identify all costs for measuring a_i. These costs include both variable and fixed costs
6. Assign relative values for all a_i in a descending order, starting from the most essential parameter down to the least essential parameter
7. Assign costs to all variable costs (for each parameter a_i) and fixed costs
8. Evaluate the nondimensional costs for each of the parameters a_i
9. From 6 and 8, construct the (V-C) diagram
10. From 9, evaluate V_{cr} and C_{cr}

4.2.5 EXAMPLES OF GENERAL THEORY OF SHM

4.2.5.1 Frequency Analysis

4.2.5.1.1 The Problem

Let us consider a monitoring system for a simple-span bridge/beam that aims at evaluating seismic vulnerability of the bridge. These natural frequencies might be utilized in a later evaluation of possible damages in the bridge, or in validating a finite element model for the bridge.

Since only the natural frequencies of the simple-span beam are of interest, a possible algorithm for the experiment can be one of the several modal identification algorithms of Chapter 6. Any of those algorithms requires a time history recording of the vertical displacement of the bridge $U(t)$. It is implied that only vertical modes of vibration are of interest in this example. On the basis of this it is reasonable to place the sensor at the center of the bridge. Theoretically, only one sensor is needed to achieve the goal of the experiment. It should be noted that the measured center displacements would only generate symmetric modes of vibration. Generally speaking, for such a system, symmetric modes may be the most important modes in evaluating the behavior of the bridge. Of lesser importance are the asymmetric modes. If these modes are needed, then placing a second sensor at quarter span is needed. In addition, sensors might be added for redundancy purposes.

4.2.5.1.2 The Test Setup

The experimental setup is to utilize NE number of vibration detection sensors. These sensors will be mounted on the center line of the beam. A list of the parameters a_i that control the value of this experiment is shown in Table 4.3. A normalized (dimensionless) costs as well as values are used for the example. Note that the maximum value is obtained from placing the first sensor. The additional sensors would add lesser value to the experiment. Also note that the cost of cabling and data acquisition is independent of the number of sensors. For simplicity, we only included the variable costs of the project in this example.

4.2.5.1.3 Discussion of Results

We use the information of Table 4.3 to produce a $C - V$ plot, as shown in Table 4.4 and Figure 4.9. Note that there is no value to the experiment until the cost of the first sensor is spent ($C = 5$). After the value of the experiment jumps from 0 to 3, the additional costs of additional sensors produce lower value.

It is clear that this experiment is type II experiment. The question is what is V_{cr} and the associated C_{cr}. The answer is mostly qualitative. It depends on available funding, quality of sensors, and experience of personnel. It appears that the most prudent cost-value choice is 7 and 3.6, respectively. Any additional cost beyond 7 seems to offer negligible additional value.

TABLE 4.3
Variable Costs and Value of Experiment

Parameters			Dimensionless Value and Cost	
Type	Parameter # i	Description	$C_{Vi}(a_i)$	$V_i(a_i)$
Variable cost	1	Sensor # 1, midspan	1	1
parameters	1	Cabling	2	1
	1	Data acquisition system	2	1
	2	Sensor # 2, quarter span left	1	0.3
	3	Sensor # 3, quarter span right	1	0.3
	4	Sensor # 4, midspan	0.9	.1
	5	Sensor # 5, quarter span left	.9	.05
	6	Sensor # 6, quarter span right	.9	.05

TABLE 4.4
Cost-Value Relationship of
Frequency Testing

I	Dimensionless Cost	Dimensionless Value
0	0	0
1	5	3
2	6	3.3
3	7	3.6
4	7.9	3.7
5	8.8	3.75
6	9.7	3.8

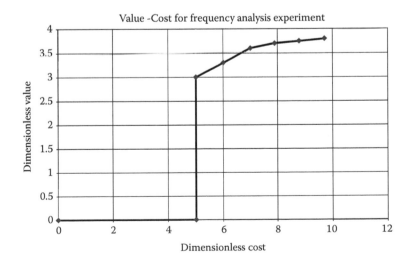

FIGURE 4.9   Value–cost relationship for frequency analysis experiment.

4.2.5.1.4 Concluding Remarks

This simple example can be expanded and utilized for the popular SHM vibration testing of bridges. First, it encourages the testers to study the objectives of the test, i.e., how many resonant frequencies are needed and what frequency range. This example also shows that problem definition is probably the most important step, both for the test itself as well as for evaluating the value of the test as a function of costs (applying GTE). If the objective of the test is just the evaluation of the first mode, one sensor (such as a strain gage or accelerometer reacting fast enough) is probably enough, as long as it is kept anywhere on the beam except at supports. At the same time, if the sensor is kept at the center, there is a better chance of getting good data due to better signal to noise ratio as modal magnitude should be highest there. If the first two modes are needed, still one sensor may be enough, although it cannot be placed at the mid-span of the beam. So, number of sensors will become a combination of needed number of frequencies, frequency range, etc. If mode shapes are needed, then there is a need for more sensors, measurement locations, or a combination of two to achieve the goal of the test depending on how many nodal points one will encounter. In other words, the aforementioned simple example problem can be used to resolve many of the decisions that are needed during vibration testing, and ensure a cost-effective test.

4.2.5.2 Corrosion FRP Mitigation Efficiency

Consider now the Congress Street Bridge experiment (New York State Department of Transportation [NYSDOT]) that is described in Chapter 7 of Ettouney and Alampalli (2012). The experiment was to monitor the long-term effects of corrosion mitigation of columns by utilizing an fiber reinforced polymers FRP wrapping techniques. Let us explore a similar project. Figure 4.10 shows schematics of the bridge geometry. Let us apply the theory of experimentation of this chapter to identify the critical cost-value point.

The goal of the experiment is to investigate the efficiency of using FRP layer to control and slow corrosion rate for the bridge columns. To accomplish this, all fixed costs of the experiment are estimated. This includes agency costs, management costs, as well as data acquisition and analysis costs. Due to the small number of sensors in this experiment, it is assumed that data acquisition costs are independent of number of sensors; hence they were assumed to be fixed costs. As was mentioned earlier, fixed costs of an experiment are spent without any gained value to the experiment in the C-V charts. For our example, all costs and values will be dimensionless, as was described earlier.

In designing the experiment, each column station is assumed to have three locations on three sides of the column. Each location will have three types of sensors: corrosion rate sensor, temperature sensor and humidity sensor. Another sensor will be added at the FRP layer to make sure appropriate bond exists between the FRP and the concrete it is attached to. All the sensors in a given column station will be assigned same value. It is essential for this experiment to assess the correlation between corrosion rate, temperature and humidity (humidity will indicate if moisture is present for corrosion process). It should be noted that such a correlation might not be essential in other types of corrosion experiments; in such a case, different values for each type of sensor might be assumed.

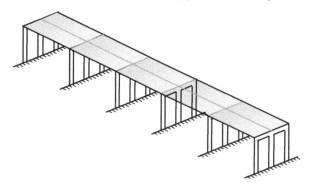

FIGURE 4.10   Illustration of Congress Street Bridge.

Generally speaking, corrosion monitoring situations for columns, no more than three column stations are needed: near top, near center, and near bottom. For this experiment, the observed damage of corrosion along the height of the column was judged to be near uniform. As such it was decided not to monitor the center of the columns. The top station monitoring has more value than the bottom station since the top of the column is near the source of leaking joints. However, the cost of using the column top as monitoring station is several times more than the cost of monitoring of the bottom stations. The sensor installation costs as well as the monitoring costs for such elevated station contribute to such higher costs.

The bridge as shown in Figure 4.10 is built on a series of frames. Each frame has three columns (North, Center, and South). The relative values of sensors at different columns were assigned based on the following assumptions

- Assign higher value for the north column
- Next is the value for the center column. The center columns, by contrast, can add a baseline measure
- Since the south columns are similar to the north columns, they were assigned the lowest value. It is assumed that the south column information is fairly redundant when the north column information is available

Given the framing configuration of Figure 4.10, two types of frames can be identified: end frames and middle frames. It is noted that the columns of the end frames are more susceptible to corrosion damage, since they are adjacent to the joints, which are the primary source of leaking that would lead to corrosion. As such, it is decided that the value of monitoring the columns in an end frame is higher than the value of monitoring the columns in an intermediate frame. Costs of monitoring intermediate frames are slightly less than the cost involved with monitoring end frames. It is assumed that the management will start by installing sensors at end frames (due to its higher value for the experiment). Adding sensors to an intermediate frame should cost less; some of the cost can be shared with the cost of installing sensors of the end frame.

On the basis of the above discussions of relative values and estimated relative costs, Table 4.5 was developed. It shows the parameters of the experiment and the aggregated dimensionless costs and values assigned to each parameter. Note that there can be several components that can have identical parameter number. For such components, it is assumed that they are all needed to work together, that is, they must be treated as a *parameter group*. As such there were fifty parameters that were identified for this situation. These parameters were consolidated to only twelve parameter groups.

Table 4.6 shows the consolidations of Table 4.5. Finally, Figure 4.11 shows the corresponding $C - V$ curve and $C - (dV/dC)$ curves. The curves show clearly that the critical point (C_{cr}, V_{cr}) of this experiment is achieved with instrumenting only one station at the bottom three columns at the end frame, total dimensionless cost of 43.1 and dimensionless value of 18.8. All these data are for illustration purposes only. Value analysis is one of the tools in a decision-maker's tool chest and should be used wisely.

4.3 SPECIAL THEORY OF EXPERIMENTATION (STE)

4.3.1 INTRODUCTION

GTE showed a process of optimizing the value of any experiment. It explored the overall parameters that affect an experiment. Many of the arrangements of sensors were presumed to be *fixed* during the execution of GTE. In this section, we explore the optimality of sensor arrangements that includes location, number, and type of sensors. The problem is well known in the NDT community. For example, Annis (2005) posed the important question: *what is the optimum number of specimens in*

TABLE 4.5
Variable Costs and Values of Experiment

Column ID	Parameter # i	Parameter group # i	Description	Cost	Value	Total Cost	Total Value
	1	1	Management, OH, etc.	10	0	19	4
	2		Data acquisition System	4	0		
N1 - Bottom	3		CP, HP, and TP	1	1		
	4		CP, HP, and TP	1	1		
	5		CP, HP, and TP	1	1		
	6		FRP layer	2	1		
C1 - Bottom	7	2	CP, HP, and TP	1	0.95	24	7.8
	8		CP, HP, and TP	1	0.95		
	9		CP, HP, and TP	1	0.95		
	10		FRP layer	2	0.95		
S1 - Bottom	11	3	CP, HP, and TP	1	0.9	29	11.4
	12		CP, HP, and TP	1	0.9		
	13		CP, HP, and TP	1	0.9		
	14		FRP layer	2	0.9		
N2 - Bottom	15	4	CP, HP, and TP	0.9	0.7	33.7	14.2
	16		CP, HP, and TP	0.9	0.7		
	17		CP, HP, and TP	0.9	0.7		
	18		FRP layer	2	0.7		
C2 - Bottom	19	5	CP, HP, and TP	0.9	0.6	38.4	16.6
	20		CP, HP, and TP	0.9	0.6		
	21		CP, HP, and TP	0.9	0.6		
	22		FRP layer	2	0.6		
S2 - Bottom	23	6	CP, HP, and TP	0.9	0.55	43.1	18.8
	24		CP, HP, and TP	0.9	0.55		
	25		CP, HP, and TP	0.9	0.55		
	26		FRP layer	2	0.55		
N1 - Top	27	7	CP, HP, and TP	4	1	61.1	22.8
	28		CP, HP, and TP	4	1		
	29		CP, HP, and TP	4	1		
	30		FRP layer	6	1		
C1 - Top	31	8	CP, HP, and TP	4	0.95	79.1	26.6
	32		CP, HP, and TP	4	0.95		
	33		CP, HP, and TP	4	0.95		
	34		FRP layer	6	0.95		
S1 - Top	35	9	CP, HP, and TP	4	0.9	97.1	30.2
	36		CP, HP, and TP	4	0.9		
	37		CP, HP, and TP	4	0.9		
	38		FRP layer	6	0.9		
N2 - Top	39	10	CP, HP, and TP	4	0.75	115.1	33.2
	40		CP, HP, and TP	4	0.75		
	41		CP, HP, and TP	4	0.75		
	42		FRP layer	6	0.75		

TABLE 4.5 (continued)
Variable Costs and Values of Experiment

Column ID	Parameter # i	Parameter group # i	Description	Cost	Value	Total Cost	Total Value
C2 - Top	43	11	CP, HP, and TP	4	0.7	133.1	36
	44		CP, HP, and TP	4	0.7		
	45		CP, HP, and TP	4	0.7		
	46		FRP layer	6	0.7		
S2 - Top	47	12	CP, HP, and TP	4	0.65	151.1	38.6
	48		CP, HP, and TP	4	0.65		
	49		CP, HP, and TP	4	0.65		
	50		FRP layer	6	0.65		

CP = corrosion rate sensor, HP = humidity sensor, TP = temperature sensor, and FRP Layer = sensor at FRP layer. N = North, S = South, and C = Center.

TABLE 4.6
Cost-Value Relationship of Corrosion Monitoring

Parameter	Cumulative Cost	Cumulative Value	Rate of Value
0	0	0	0
0	14	0	0
0	14	0	0.210526316
1	19	4	0.78
2	24	7.8	0.74
3	29	11.4	0.659793814
4	33.7	14.2	0.553191489
5	38.4	16.6	0.489361702
6	43.1	18.8	0.273127753
7	61.1	22.8	0.216666667
8	79.1	26.6	0.205555556
9	97.1	30.2	0.183333333
10	115.1	33.2	0.161111111
11	133.1	36	0.15
12	151.1	38.6	0.144444444

an NDT project? He showed that the answer to such question is to use as many samples as needed to reduce the uncertainty of the result to an acceptable level. We will now generalize the question in two ways: (a) From the local confines of NDT to the more global applications of SHM, and (b) We will consider, in addition to sensor numbers, both sensor locations and sensor types.

So, we ask the question: *what is the optimum arrangement (number, location, and type) of sensors in an SHM project*? To find a simple answer to such a complex question, we introduce a STE. The existence of an optimal sensor arrangement of any experiment (NDT and/or SHM) would encourage professionals to attempt finding such an optimal arrangement in an objective fashion (see Figure 4.12).

We note that in addition to answering such an important question, this volume offers methods and steps for obtaining such an optimal sensor arrangement (see Chapter 5).

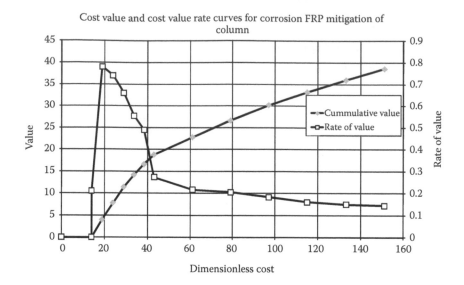

FIGURE 4.11 Cost-value and rate of value of corrosion monitoring.

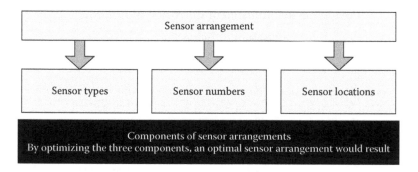

FIGURE 4.12 Components of sensor arrangements.

4.3.2 OVERVIEW OF STE

A STE can be stated as follows:

There exists an optimum arrangement of sensors for any experiment. The sensor arrangement includes type, number, and location of sensors.

The STE states that the optimal arrangement of sensors in any experiment depends on three factors: sensor types, number, and locations of sensors. We first need to define a cost function that is to be optimized. Similar to the GTE, the sensor arrangement need to minimize costs, and maximize benefits; that is, the maximize value. It seems reasonable then to express STE formally as

Given the function $f(\ )$

$$V = f\left(N_S, L_S, T_S\right) \tag{4.23}$$

V = The value of the sensor arrangement
N_S = Measure for number of sensors
L_S = Measure for location of sensors
T_S = Measure for type of detection schemes (type of sensors)

There is an optimal magnitude for V that depends on the measures N_S, L_S, and T_S.

We can prove the theory by first making the reasonable simplifying assumptions

- Depending on the goals of the experiment, there is an optimal N_S
- For a given N_S, there is an optimal L_S (optimal sensor location, OSL), and
- There is an optimal arrangement of detection schemes T_S

On the basis of the above assumptions, then it is reasonable to state

$$V_{OPT} = f\left(N_S\big|_{OPT}, L_S\big|_{OPT}, T_S\big|_{OPT}\right) \tag{4.24}$$

V_{OPT} = Optimal value of the sensor arrangement

$N_S\big|_{OPT}$ = Optimal measure for number of sensors

$L_S\big|_{OPT}$ = Optimal measure for location of sensors

$T_S\big|_{OPT}$ = Optimal measure for type of detection schemes (type of sensors)

Equation 4.24 proves of the existence of optimal V. But, this requires proof of the above mentioned three assumptions. Next section discusses these in more detail.

4.3.3 PROOF OF STE

4.3.3.1 Optimality of Sensor Type

Using different detection schemes optimally can result in an V_{OPT}. This can be done by triangulating the detection schemes. A theory of triangulation that proves this concept is introduced later in this chapter.

4.3.3.2 Optimality of Sensor Number

We prove the optimality concept of N_S, that is, the existence of $N_S\big|_{OPT}$ by introducing a sensor demand and sensor capacity regions. We then optimize the interrelationship of these two concepts and show that such an optimization would produce $N_S\big|_{OPT}$.

Demand Region: Let us define the demand region DR of an experiment as that region of the structure that the professional need to obtain information about. For example, for an FRP bridge deck, the professional might need to know the delamination extent in the entire upper surface of the deck. For a post-tensioned reinforced concrete beam, the corrosion and/or hydrogen embrittlement extent along the tendons might be of interests. Another example for DR might be the displacement of a single-span bridge during a load testing experiment. Note that the DR can be 3D (volume), 2D (area), 1D (length), or a single point. The shape and/or extent of the DR are dependent on the type and objectives of the experiment. Table 4.7 shows some examples of DR.

We can safely state that there is a DR for every SHM, as well as every NDT experiment. The extent of DR is one of the several distinguishing factors between SHM and NDT. Generally speaking,

TABLE 4.7
Demand Regions for Different SHM Experiments

Type of *DR*	Example
Point	Maximum strain at center of single-span bridge
Linear	Extent of damage (corrosion, hydrogen embrittlement) of post-tensioned tendons
Surface/area	Delaminations of composite (FRP) plates in an FRP bridge deck
Volume	Extent of damage in reinforced concrete bent after seismic event

SHM would be involved with larger DR than NDT. By definition, the DR must be well defined by the project official (it is advisable to define the DR such that it is consistent with the project goals, and with an eye on the potential decision making processes that might result as the SHM experiment progresses).

Capacity Region: We define next the capacity region of a sensor CR as the coverage region of the sensor of interest. For example, local fiber optics, for example, Fabry-Perot, or Fiber Bragg Grating, has a point CR. Similarly, most mechanical or thin film strain sensors are point CR. Linear CR can be guided wave ultrasonic transducers that are used for detecting pipeline damages (see Chapter 7). Similarly, distributed Brillouine Fiber Optic Sensors are linear CR (see Chapter 5). Area CR sensors are those that have surface applications. For example piezoelectric transducers that pulses in a composite plate so as to identify damage location (using time of flight (TOF) technique, in conjunction with other sensors) would have a surface type CR. Another application for surface CR is using acoustic emission sensors to detect damage within a plate, again in conjunction of a set of acoustic emission sensors using TOF technique. Similar examples can be used for volume CR.

Units Mismatch: For simplicity, we will assume that CR and DR have compatible units, that is, point CR sensors will be used for sensing DR, and so on. Sometimes this is not the case. For example, Fiber Bragg Grating sensors (which are point CR sensor) might be used for predicting damage in a surface DR such as FRP bridge decks. Such a situation is more complex and will not be considered here. It should be mentioned that the added complexity will not detract from the generalization of the STE.

Optimum Number of Sensors: We can now express the minimum number of sensors for a give experiment as

$$n_{DCR} = \frac{DR}{CR} \tag{4.25}$$

The fraction in Equation 4.25 is then approximated upward to the next higher integer number N_{DCR}. Note that for the trivial case (point DR and CR),

$$CR = DR = 0 \tag{4.26}$$

$$N_{DCR} = n_{DCR} \rightarrow 1 \tag{4.27}$$

Uncertainties: Equation 4.25 computes n_{DCR} using the deterministic values of DR and CR. Generally speaking, the value of DR should be deterministic, with little or no uncertainties. Its value is determined by the goals of the SHM experiment. As such, it is dependent of the needs for cost savings, or increased safety. In other words, the value of DR should be determined based on economic and/or social reasons.

Most of the uncertainties in Equation 4.25 are found in estimating CR. We note that an upper bound of CR might be found in the specifications of the sensor of interest. Using such an upper bound of CR will result in the minimum number of sensors n_{DCR}. There are several potential uncertainties that can affect CR such as material properties, installation errors, quality of manufacturing (tolerances), environmental conditions, potential interactions between different sensors and the material they are embedded in/on, and so on. Current practice is to use a reduced value of CR to evaluate the number of sensors using Equation 4.25. This is due to the fact that the statistical properties of such uncertainties are impossible to estimate generically. Is it possible to develop a quantitative method to estimate the required value of CR?

Let us assume that the uncertain capacity region cr is uniformly distributed. A truncated normal distribution might be more accurate, however, the uniform distribution is simpler to express. If the readers desire they can develop similar expressions for the truncated uniform distribution case. The maximum value of CR is CR_{max}, as discussed earlier. We need to define the minimum value of CR as CR_{min}. The estimation of CR_{min} will depend on the experiment itself, as well as the expertise of the professionals who are involved in the project. Defining X as the probability that a value of CR will exceed certain value CR_X, we can evaluate CR_X as

$$CR_X = CR_{min} + (1 - X)(CR_{max} - CR_{min}) \qquad (4.28)$$

The optimum number of sensors becomes

$$n_{DCR} = \frac{DR}{CR_X} \qquad (4.29)$$

As an example, let us assume that in a particular TOF piezoelectric experiment, the total area of the plate to be monitored is $DR = 3000$ in². Each sensor has a CR_{max} range of 30 in². The professional estimated that, based on previous experience, $CR_{min} = 10$ in². Moreover, the professional needed a 90% nonexceedance probability for that experiment. Thus $X = 0.90$. Applying Equation 4.28 we find $CR_X = 12$ in². The optimum number of sensors that accommodates uncertainties of this situation is 250 sensors (utilizing Equation 4.29). By comparison, applying Equation 4.25 that does not account for uncertainties would have required only 100 sensors. Utilizing such low number of sensors might have resulted in an inaccurate SHM results.

4.3.3.3 Optimality of Sensor Location

Chapter 5 shows a detailed method of computing optimum sensor locations (OSL) for measuring structural responses to gravity, seismic, and wind loads. Similar approaches can be used to compute OSL for any other type of sensor arrangement.

4.4 THEORY OF SHM TRIANGULATION

4.4.1 THEORY STATEMENT

Suppose that a particular structural state needs to be detected by the use of some monitoring activities. Such a structural state can be the overall stability condition of the structure, current scour condition under the bridge foundation, or the extent of damage in post-tensioned tendons. Each of these conditions can be detected by one or more of SHM techniques. Equally, the detection of each of these conditions is not an easy task; even with good planning, there is always a finite probability that successful detection would not be achieved. We immediately recognize that this situation describes a general situation, which can be formalized as follows:

1. There is an unknown structural state S_{DT}.
2. There is a particular monitoring resource scheme SR_i. The subscript i is a counter that refers to the ith monitoring scheme. We assume that there are N monitoring schemes that might be used to achieve the task.
3. There is an uncertainty of SR_i being able to detect S_{DT}. Such uncertainty can be assigned a variance V_i.

Obviously, the probability that S_{DT} can be detected, $P(S_{DT})$ is the same as the probability that SR_i can detect S_{DT}, $P(SR_i)$, if only single detection scheme i is used. Suppose that and additional detection scheme SR_j, with $i \neq j$, for a total of two schemes, is used. Would that improve the probability of overall detection?

The problem can be illustrated qualitatively in Figure 4.13. The figure shows that each detection scheme has a detection range. Whenever the detection ranges intersect, the probability of detection increases. The qualitative illustration in Figure 4.13 shows three detection schemes: the region they intersect at has a higher potential of being detected than surrounding ranges.

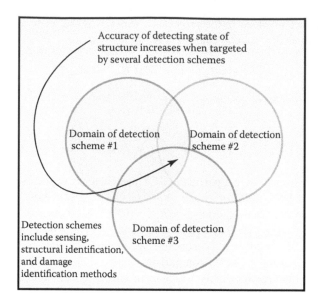

FIGURE 4.13   Triangulation scheme in SHM.

We recognize the proposal in Figure 4.13 to be a form of triangulation. It is also intuitive that the answer to the above posed question is "yes." On the basis of this, we propose the following SHM triangulation theory: *Efficiency of detecting the structural state is improved by using a mix of detection schemes.*

4.4.2   PROOF OF THEORY

Even though the theory as stated above is fairly intuitive, there exists a simple proof by using probability theory. Let us consider two potential detection schemes SR_i and SR_j, with $i \neq j$. The probability of detecting the structural state S_{DT} if each of the schemes is used separately is $P(SR_i)$ and $P(SR_j)$, respectively. If the two schemes are used together, the probability that S_{DT} can be detected, $P(S_{DT})$ can be shown to be

$$P\left(S_{DT}\right) = P\left(SR_i \cap SR_j\right)$$ (4.30)

which implies that the desired detection can be achieved by either SR_i, or SR_j. From elementary probability theory

$$P\left(SR_i \cap SR_j\right) = P\left(SR_i\right) + P\left(SR_j\right) - P\left(SR_i\right)P\left(SR_j\right)$$ (4.31)

Note that, by definition, $0.0 \leq P(SR_i) \leq 1.0$ and $0.0 \leq P(SR_j) \leq 1.0$. Let us disregard the trivial situations when $P(SR_i) = 0.0$ or $P(SR_j) = 0.0$, which would revert to the single detection scheme. We can easily then, by inspection, reach the conclusions that

$$P\left(S_{DT}\right) > P\left(SR_i\right)$$ (4.32)

and

$$P\left(S_{DT}\right) > P\left(SR_j\right)$$ (4.33)

This proves the theory, without any reservation.

The above proof makes the realistic assumption that the two detection schemes are mutually exclusive, that is, the use of one scheme does not have an effect on the other scheme. Such an assumption is reasonable in SHM situations. Furthermore, we note that the above proof can be easily extended to situations where more than two detection schemes are used.

As an example, let us assume that it is desired to detect extent of damage in a small bridge. The professional is considering two detection schemes: strain measurements and natural frequency (vibrations) measurements. Previous experience indicates that the probabilities of successful detection for each of the two schemes, when used separately, are 75% and 60%, respectively. When the two schemes are used together, by applying Equation 4.31, the probability of detection improves to 90%. The improvement in detection potential is certainly obvious. Such improvements can be even greater in situations where detection probabilities of individual schemes are low to start with. For example, if the probabilities in the previous example are 50% and 35% instead, the detection probability of the combined detection scheme, according to Equation 4.31 is 67.5%. The improvement in detection probability is noteworthy indeed. Figure 4.14 shows overall triangulation effects for two detection schemes.

4.4.3 DISCUSSION AND APPLICATIONS

4.4.3.1 General Discussion

According to the theory, for example, if it is desired to monitor potential of brittle failure, it would be more efficient to use a mix of accelerometers and stain gauges, than to use only one ToM, say strain gauges. The explanation is simple: strains are direct measure of deformation; they will offer insight to stiffness, but will not offer a clear insight to an imminent loss of stability. Velocities offer a direct measure of inertia; careful monitoring would reveal a relationship between stiffness and effective inertia (Timoshenko 1955). Effective inertia is a direct measure of imminent loss of stability. Thus, a careful mix of different ToM can offer more insight to the structural behavior. By careful use of different ToM more information can be revealed. A step by step application of triangulation is as follows.

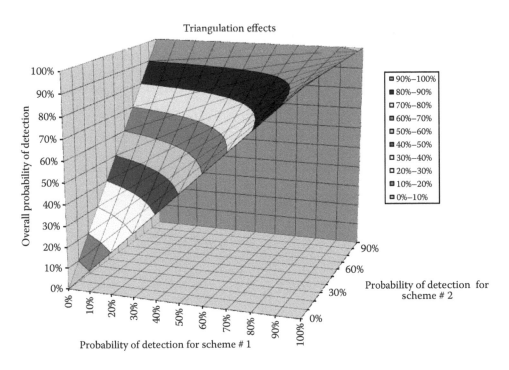

FIGURE 4.14   Triangulation probabilities for two-detection schemes situation.

State goals of Monitoring, for example, detecting changes in boundary conditions, or connection stiffness

1. Identify parameters that cause, or result from, such changes. In the above example, a frozen bearing can cause boundary condition change, while a large strain can result from boundary condition change
2. Place appropriate tilt and strain gauges to monitor the boundary condition changes
3. Using a mix of predetermined acceptance rules, if observed changes reach a threshold, the facility owner is informed

In order to generalize the triangulation concept even further, we can qualitatively define structural/damage identification by a qualitative measure *DETEC* (for detectability). The detectability of the structural state as a function of a specific sensor type *ST* and number *SN* can be described formally as

$$DETEC_i = f\left(ST_i, SN_i\right) \tag{4.34}$$

The subscript i is used to identify the specific experiment. It there are two different sensing types each will result in a monotonically increasing *DETEC* as shown in Figure 4.15. The needed threshold detection might be reached at a specific combination ST_1 and SN_1 for the first type or ST_2 and SN_2 for the second type. Figure 4.15 shows a situation where $DETEC_1$ reaches the required threshold of detection, while $DETEC_2$ failed to detect the desired structural state. By using a combination of ST_1, SN_1, ST_2, and SN_2 the detectability rate and potential increases, for similar sensing resources, as shown in Figure 4.15 ($DETEC_3$). The qualitative relationship shown in Figure 4.15 will be studied and quantified during the course of the project.

The above examples and discussions can be generalized so as to cover most generic SHM situations. This can be accomplished by following the above steps. Note that in the above example, we used a mix of only two ToMs. There is no theoretical limit to the number of ToM mix that can be used to triangulate and identify the desired behavior. For example, temperature, accelerations, acoustic emission, ultrasound or even radiation waves can be used in the mix. There would be, of course an optimum number and location of each type of sensors that would satisfy the SHM objectives in an efficient and accurate manner. Table 4.8 shows the four generic steps in the triangulation method, and the issues that need to be considered for each step.

Let us have a more detailed look at some of the practical situations that were mentioned earlier, and see how triangulation might be applied to them. Table 4.9 shows practical situations and the different possible SHM detection schemes that can triangulate the detection process.

4.4.3.2 Case Study

Maser (2008) offered a practical example on the utility of integrating two technologies and how such an integration can offer increased efficiency and accuracy. In the experiment, ground penetrating radar (GPR) and infrared thermography (IR) have been applied for bridge deck condition

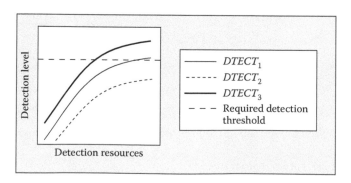

FIGURE 4.15 Detection versus resources relationships.

TABLE 4.8
Triangulation Issues in SHM

Triangulation Steps	Issues
SHM goals	Diagnostics, performance, safety, hazard (earthquake, scour, fatigue, etc.) effects, etc.
Parameters	Strains, displacements, modal parameters, tilting, corrosion, normal wear and tear, bridge rating, etc.
Appropriate sensors	Strains, accelerometers, pressure, etc.
Acceptance rules	Confidence levels, acceptable risk, reliability (probability of failure), etc.

TABLE 4.9
Triangulation Examples

Problem	Triangulation scheme
Structural stability	For stability (especially elastic stability) evaluation, both capacity and demand need to be sensed. For an accurate triangulation effort, use a combination of vibration, strain, and load sensors.
Scour condition	See Chapter 1 by Ettouney and Alampalli (2012)
Damage in post-tensioned tendons	Use impact-echo process (Section 7.4.2.3). The triangulation is achieved by applying the process at uniform grid on the surface of the post-tensions system. The direction of the detection scheme is perpendicular to the direction of the tendons. This can be combined with methods such as ground penetrating radar.

assessment. GPR is equally effective on bare concrete decks as well as on decks with concrete and bituminous overlays. GPR data requires office processing to reveal deteriorated areas. While producing good quantity estimates and general locations, it is not strong at precisely locating delaminated areas. While GPR is generally effective for deck slabs on girders, it has limitations with one-way slab type of decks. IR produces a visual delaminations image immediately in the field that can be checked on the spot using sounding and cores, and thus provides good location accuracy. However, as a surface temperature method, its detection capability is depth limited. While it is very effective for non-overlaid decks, its effectiveness can be reduced in the presence of overlays. IR data collection takes place at 5 mph and requires a moving closure, and solar radiation conditions must be adequate to produce the required temperature differentials. This project combined the data from the two methods to improve the data quality by minimizing limitations of each method.

Forty five bridge decks in Wisconsin were tested using both techniques along with under deck visual surveys, and selective core sampling (see Figure 4.16). GPR rebar depth data was used for each deck to determine the depth limitations of IR detection. Where the rebar was too deep, IR was used to identify overlay debonding, while GPR was used to identify rebar depth deterioration. For decks that were not well suited for GPR, IR provided necessary deterioration information. The author reports that combination of IR and GPR maximized the capabilities of each method and compensates for the limitations (see Table 4.10).

4.4.4 EVALUATING DETECTABILITY OF SHM SCHEMES

In the previous sections of this chapter, we have made extensive use of the probability detection $P(SR_i)$ of a given SHM scheme SR_i. In this section, we offer methods of evaluating this probability. Specifically, we offer a quantitative and a qualitative method. The qualitative method is based on evaluating previous experiences in a numerical fashion, while the qualitative method depends on the experiences of the analyst. Obviously, quantitative method is preferable, but it requires more effort.

(b)

(a)

FIGURE 4.16 GPR and infrared equipment. (a) back of the equipment vehicle and (b) front of the equipment vehicle. (Reprinted from ASNT Publication.)

TABLE 4.10
Summary of Core Results

Detected by	Total (#)	Debonded (#)	Delaminated (#)	Correct (%)	False alarms (%)	Misses #	– (%)
IR &/or GPR	105	25	61	83	17	5	5
IR only	64	20	35	86	14	22	26
GPR only	59	9	33	71	29	18	17

TABLE 4.11
Consistent Categories for Histogram

Category	Comment
Type of SHM experiment	Vibration measurement, remaining fatigue life, scour effects, corrosion extent, etc.
Type of bridge	Simple span, multiple spans, construction material, structural system type and other pertinent bridge attributes
Sensors	Relatively similar types of sensors. Number and interconnectivity of sensors should be consistent. If wireless sensing is used, the wireless network and topology should be consistent
Instrumentation	Instrumentation hardware should be consistent with sensors, bridge type, and experiment objectives.

4.4.4.1   Numerical Method for Evaluating Detectability

The steps for such an evaluation are as follows:

Identify Suitable SHM Database: This step requires identifying and enumerating all previous SHM schemes that are consistent with the SHM scheme of interest. Some examples of schemes of interest can be to detect modes of vibration, remaining fatigue life or scour performance. Table 4.11 shows other consistent categories that need to be considered for proper inclusion in the histogram database.

Identify Adequate Detectability Measurement Scale: A scale should be established for measuring detectability SR_i, for the remainder of this section we drop the subscript, without any loss of accuracy. One possible scale can range from 0 to 10. As shown in Table 4.12. Admittedly, the descriptions of each value of SR are qualitative. This approach is not uncommon in many engineering application and thus, we will be adopting it for our current purpose.

TABLE 4.12
Scale of *SR*

$SR_{SCALE} =$ Scale of *SR*	Description
0	Complete failure of experiment. No beneficial information were gathered
1	About 30% of the objectives were realized
2	About 30% of the objectives were realized. Some additional information was realized
3	About 50% of the objectives were realized
4	About 50% of the objectives were realized. Some additional information was realized
5	About 75% of the objectives were realized
6	About 75% of the objectives were realized. Some additional information was realized
7	About 90% of the objectives were realized
8	Complete success for all objectives. No additional objectives were realized
9	Complete success for all objectives. Some additional objectives were realized
10	Complete success for all objectives. Many additional objectives were realized

Note that there is an added value in the estimation of SR to realizing additional information (Serendipity principle).

TABLE 4.13
***DRL* Scale**

DRL Scale	Description
0	Trivial case: no resources for the SHM scheme
1	Very limited resources
2	Medium-low resources
3	Medium resources
4	Medium-high resources
5	High level of resources

Using the database of consistent SHM schemes and evaluating the results of those schemes, a numeric value of *SR* can now be estimated using tables similar to Table 4.12.

Identify measure of detection resource and detection resource level: Since SHM schemes vary greatly in regards to the resources used for the experiment, it is logical to assume that detectability of a given SHM scheme *SR* is a function of the SHM detection resource level *DRL*. The *DRL* is the horizontal scale in Figure 4.15. The *DRL* can be described on a sliding scale, say from 0 to 5, with each level qualitatively defined as in Table 4.13.

The database that is collected in previous step would be categorized further according to their estimated *DRL*.

Prepare Histogram: For all the records in the database within a specific *DRL*, a set of the established values of *SR* form the basis of a histogram. The size of the histogram is N_{RECORD}: the number of consistent SHM schemes in the database. See Chapter 8 for a description of this process.

Compute Desired Probability: Using the histogram, the probability of detection *P(SR)* can be evaluated as the cumulative probability

$$P(SR) = P\left(X \geq SR_{SCALE}\big|_i\right) \tag{4.35}$$

The choice of i depends on the desired standards of management. For example $i = 10$ is too demand-ing, while a choice of $i = 7$ is more realistic and less demanding.

Note that the method as described above can be performed by sending questionnaires to bridge managers or experts for reporting their experiences with different SHM schemes. By archiving the results, a probability database can be established in a central location. Such probability database can be used by the SHM community for future SHM projects.

4.4.4.2 Subjective Method for Evaluating Detectability

When the information needed for the numerical method is not available, the users can rely on their per-sonal experiences and judgments. The desired probabilities $P(SR)$ can be estimated by the users. Such an approach is judgmental and is less accurate than the numerical approach. It is simple and less costly.

4.4.5 Triangulation Cost–Benefit Analysis

4.4.5.1 Analytical Considerations

We have been focusing on the methods and benefits of triangulations so far. Of course, there is a cost for achieving such benefits. Whenever an additional detection scheme is used, there will be additional costs. The balance of cost–benefit needs to be considered carefully before embarking on a triangulation technique. The remainder of this cost–benefit section will consider the two-scheme situation. More schemes can be analyzed in similar manner.

Cost of Triangulation: Cost of triangulation is simple to compute. It is the total cost of adding another SHM scheme. For example, if the cost of the first scheme is C_i, and the cost of adding another scheme is C_j, then the total cost of triangulation is

$$C_{Triangulation} = C_i + C_j \tag{4.36}$$

For ease of computations, we use monetary measures for all costs and benefits.

Benefit of Triangulation: As usual, estimating monetary value of benefits is not an easy task. We try to devise a technique to avoid this difficulty, without any loss of generality. We assume first that the monetary benefit of a 100% successful SHM project is B_{SHM}. For the ith and jth SHM schemes, the probability of detection are $P(SR_i)$ and $P(SR_j)$, respectively. The probability detec-tion when using both the ith and jth SHM schemes is $P(S_{DT})$. The benefit when using ith scheme is

$$B_i = P(SR_i)B_{SHM} \tag{4.37}$$

Similarly, the benefits of the jth and the triangulated SHM schemes are

$$B_j = P(SR_j)B_{SHM} \tag{4.38}$$

$$B_{Triangulated} = P(DT)B_{SHM} \tag{4.39}$$

$P(DT)$ is the probability of detection with triangulated scheme. Note that, unlike the cost, the tri-angulated benefits are not the simple sum of the benefits of the two schemes. Having identified all costs and benefits, we can move to the benefit cost analysis phase.

Benefit-Cost Analysis: What concern us here is the changes in benefit-cost ratios from the base-line case (ith SHM scheme only) to the triangulated situation, where both ith and jth SHM schemes are used. The baseline benefit-cost ratio is

$$BC_i = \frac{B_i}{C_i} \tag{4.40}$$

The triangulated benefit-cost is

$$BC_{Triangulated} = \frac{B_{Triangulated}}{C_{Triangulated}} \tag{4.41}$$

Expressing the benefit costs in terms of B_{SHM}

$$BC_i = \frac{P(SR_i)B_{SHM}}{C_i} \tag{4.42}$$

The triangulated benefit cost is

$$BC_{Triangulated} = \frac{P(DT)B_{SHM}}{C_i + C_j} \tag{4.43}$$

The ratio of the benefit cost of triangulated and baseline schemes is

$$\frac{BC_{Triangulated}}{BC_i} = \frac{P(DT)}{P(SR_i)} \frac{1}{1 + \frac{C_j}{C_i}} \tag{4.44}$$

The condition the a triangulated scheme would produce a better benefit cost is

$$\frac{BC_{Triangulated}}{BC_i} > 1.0 \tag{4.45}$$

This condition is satisfied when

$$\frac{P(DT)}{P(SR_i)} > 1 + \frac{C_j}{C_i} \tag{4.46}$$

This equality is an interesting result: even though the probability of detection in a triangulated SHM setup is always higher than an individual SHM scheme, the benefit cost of triangulation is not always higher than a single scheme.

Let us continue the previous example of a baseline detection probability and triangulated probability of 75% and 90%. If the costs of the baseline and additional SHM schemes are $1,000, and $600, the left-hand side of the equality is 1.2, while the right-hand side is 1.6: the equality is not satisfied, and triangulation is not benefit-cost effective. If the costs of the two schemes are $1,000, and $150, the left-hand side of the equality is still 1.2, while the right-hand side is 1.15: the equality is satisfied, and triangulation is benefit-cost effective.

4.4.5.2 Case Studies

Gray and Gray (2005) discussed the cost benefits of the use of multitudes of nondestructive evaluation (NDE) techniques in a given situation. Since different NDE techniques provide differing data (information), the combination of those differing data was defined as the data fusion. They argued that all available NDE data must be considered in the data fusion paradigm. Of course, they raised important question: is the additional available date worth the additional cost? We can extend the logic of the data fusion into the SHM realm. The data fusion in SHM is the combination of data that result from varying SHM projects, or techniques. For example, in an SHM corrosion evaluation project for bridge piers, appropriate data fusion approach would be periodic measurements of pH levels in concrete, ambient temperature as well as periodic visual inspection. Note that each of these parameters can result in an estimate of corrosion in concrete individually. However, correlating the data (fusing them) can result in a more reliable and accurate corrosion estimates. Even though the use of data fusion in any SHM project would increase reliability and accuracy of the results, the SHM professional should ask the same question suggest by Gray and Gray (2005): is the cost for the additional data source worth the effort?

Obviously, the answer to such a question is a case-dependent. We can offer the following guidelines for efficiently using additional data sources to incorporate a data fusion process in an SHM project:

1. Utilize a multihazards process, if appropriate
2. Utilize optimum sensors location techniques whenever possible
3. Utilize different analytical techniques for evaluating value and cost–benefit
4. Utilize different decision making tools

4.4.6 CONCLUDING REMARKS

This section presented the concept and theory of triangulation as applied to SHM. We proved that by using a mutually exclusive detection schemes, the odds of successful detection increases. Few examples showed that there are several other potential applications to the SHM triangulation theory. Further explorations to these potential applications can be performed by future researchers. We provided benefit-cost analysis of triangulation, and showed that triangulation can be cost–benefit effective in certain conditions only. For now, we summarize some benefits of SHM Triangulation as (a) improved efficiency (both cost and accuracy), (b) improved expandability of SHM efforts, (c) can aid in objective evaluation of overall SHM schemes, and (c) makes it easier to link SHM results to other efforts, such as bridge management systems.

4.5 DUALITY PRINCIPLE IN SHM

Structural health monitoring can be considered as field experimentation application: it strives to monitor the structural behavior through a combination of automatic sensing and visual observations. The GTE and STE mentioned earlier address the efficiency of experiments by discussing the value, in a qualitative manner, of both the experiment itself and/or the sensor arrangement. For us to present the duality principle, we need to define the value in a more quantitative fashion as

$$V_E = \frac{B_E}{C_E} \tag{4.47}$$

The cost, in today's dollars, of the SHM experiment is C_E. The benefit of the experiment is B_E. The cost C_E is relatively easy to define, while benefit B_E is more difficult to quantify. For our immediate purposes, let us define a qualitative benefit scale, as shown in Table 4.14. (Note that the benefits B_E in Table 4.14 are only for demonstration. Magnitudes of B_E should be assigned by the users to suit individual circumstances. Note that the scale and range of these magnitudes should be consistent with cost scale and range.) The designer of the SHM experiment would assign the appropriate B_E in Equation 4.47, to determine V_E. Since we are interested in the relative merits of the experiment SHM field, this simple definition of B_E would suffice.

Having laid down the quantitative foundations for evaluating the value of an SHM experiment, we turn our attention to the duality issue. We will limit the scope of Equation 4.47 to only those SHM experiments that handle specific hazard (earthquakes, wind, corrosion, fatigue, scour, excessive

---

TABLE 4.14

Categories of SHM Experimental Benefit, B_E

Description of SHM Experiment	Benefit B_E
Experimental results would not lead to direct benefit (system behavior, safety concerns, economic concerns)	1
Somewhat beneficial behavioral information. No immediate, or long term, direct result to safety or economic concerns	2
Considerable behavioral information. Some long-term results to either safety or economic concerns	3
Considerable behavioral information. Considerable long-term results to either safety or economic concerns	4
Considerable behavioral information. Considerable immediate and long term direct results to either safety or economic concerns	5

---

vibrations, etc.). Let us assume that a particular SHM experiment is to be performed for hazard i. The value of such an experiment is

$$V_{Ei} = \frac{B_{Ei}}{C_{Ei}} \tag{4.48}$$

In a multihazards environment, the value Equation 4.48 can be generalized to

$$V_{Ei} = \sum_{j=1}^{j=NH} V_{Eij} \tag{4.49}$$

$$V_{Eij} = \frac{B_{Eij}}{C_{Eij}} \tag{4.50}$$

The value of SHM experiment for the considerations of the primary hazard ($i = j$) is V_{Eii}. The cost of the primary experiment is C_{Eii}, while the benefit of the primary experiment is B_{Eii}. The $i \neq j$ terms in Equation 4.50 can be defined as the multihazards terms. Thus $V_{EAi} = \sum_{j=1}^{j=NH} V_{Eij}|_{i \neq j}$ represents the added value to the primary experiment if multihazard considerations are considered. The added cost to the SHM experiment with the ith primary hazard to consider aspects of the jth hazard is $C_{Eij}|_{i \neq j}$. The added benefit to the SHM experiment with the ith primary hazard to consider aspects of the jth hazard is $B_{Eij}|_{i \neq j}$.

The multihazard SHM value equation can now be represented as

$$V_{EI} = V_{Eii} + V_{EAi} \tag{4.51}$$

Equation 4.51 represents another manifestation of the multihazards theory; see Ettouney et al. (2005). An implicit assumption that was made in developing the equation is that the structure would behave in a way that manifests multiple hazards at once; we call this physical behavior the *duality principle*. It behooves the SHM designer to take advantage of that behavior to maximize the value of the SHM experiment. For example, accelerations resulting from the impact damages of Figures 4.17 and 4.18 might have been detected if there was an earthquake monitoring system on those systems. Such monitoring could have alerted the officials of the accident in real time. Accelerations, wave forms and amplitudes could have given a better indication of severity of damage. Figure 4.19 shows the concept of the principle.

FIGURE 4.17   Impact damage. (Courtesy of New York State Department of Transportation.)

FIGURE 4.18   Truck impact damage. (Courtesy of New York State Department of Transportation.)

4.6   SCALING PRINCIPLE IN SHM

4.6.1   OVERVIEW

The categorization of SHM into global and local is well known in the field. The feasibility of integration of global and local SHM activities is one of the main objectives of this proposal. We argue that one of the most promising way to integrate global and local SHM efforts is by approaching it as a multiscale problem, and use both qualitative and quantitative techniques to scale parameters so as to achieve the desired SHM integration. The scaling process involves different techniques at all levels and for all components of SHM.

4.6.2   ONE-DIMENSIONAL SCALES

4.6.2.1   Frequency Spectrum

Frequency scale, which is shown in the electromagnetic spectrum, as explained in Chapter 7, is widely utilized in NDT applications. The unit of the scale is in Hz. Sometimes wave length is

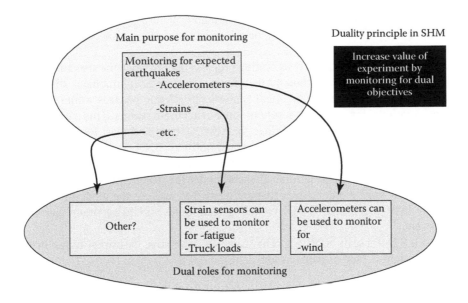

FIGURE 4.19 Duality principle in SHM.

also used as a companion scale. Frequency scale spans most of NDT methods and applications. It typifies the very concept of scaling: different methods and techniques are applicable at different frequency scales. When designing an SHM projects, it makes sense to try to utilize different ranges in the frequency scale so as to achieve the goals of the project. For example, it might be beneficial to utilize acoustic emission at higher frequency range to detect smaller fatigue cracks, while using conventional accelerometers at much lower frequency range to detect the global effects of larger size damages.

4.6.2.2 Sensor Density

Sensor density scale ρ_{SENSOR} can be defined as the average number of sensors N in a given area A. Formally,

$$\rho_{SENSOR} = \frac{N}{A} \tag{4.52}$$

As ρ_{SENSOR} increases the capability of detection of the ToM increases. There is an optimal value of ρ_{SENSOR} that depends on sensor sensitivity, costs, and SHM project objectives. It is prudent to try to estimate this optimal vale of ρ_{SENSOR} before embarking on the experiment. Before the emergence of wireless sensing as a primary tool of SHM, ρ_{SENSOR} was fairly small, because of the limitations imposed by wires. The increasingly wide use of wireless sensing enabled SHM practitioners to use higher values of ρ_{SENSOR}.

4.6.2.3 Scaling in Structural Identification

Structural scaling problem have been well known in other field, such as acoustics and vibration fields (Ettouney et al. 1997). Simply stated: low frequency range (0 ~ 100 Hz), medium frequency range (~ 100 ~ 1000 Hz), and high frequency range (higher than ~ 1000 Hz) all require different sets of structural analysis techniques. In problems such as NDT/SHM, immense frequency range do exists. Some NDT methods such as ultrasound or acoustic emission operate in the KHz or MHz ranges, while conventional modal identification methods operate in the low frequency range. There is a need to interrelate information of such a wide frequency range. Scaling techniques offer potential of such interrelationships. By having a continuous picture of the entire range of the system in a single data bank (visual, or numerical), decisions as to the state of the structural health can be easily and accurately analyzed.

Another important structural analysis scaling problem occurs when the structural model is built to model different hazards. Even though, the physical system accommodates different hazards, including their interdependencies, numerical analysis of different hazards would require separate analysis considerations. Ignoring this would produce inaccurate results. This can be simply exemplified through multihazard design of bridge piers for seismic and blast load cases. For seismic loading, the design and analysis model can be a simplified SDOF model. On the other hand, a higher precision finite element model may be required for analysis and design for blast load cases. Figure 4.20 shows a high resolution finite element model for studying blast load on bridges. Note the extremely small finite element sizes that are needed to capture the modeling demands of the high frequency blast pressures. STRID modeling techniques should accommodate such scaling principles closely.

We introduce STRID scale STR_{SCALE}. The scale is a qualitative descriptor of the structural analytical model resolution. One possible formal definition of STR_{SCALE} is

$$STR_{SCALE} = \frac{Min\left(S_{ELEMENT}\right)}{\lambda_{MIN}}$$

(4.53)

$S_{ELEMENT}$ = Descriptor of the size of elements in a finite element model
λ_{MIN} = Minimum wave length of interest in the SHM experiment

The minimum element size $Min(S_{ELEMENT})$ should adequately transmit the minimum wave in the model whose wave length is λ_{MIN}.

4.6.2.4 Scaling in Damage Identification

There are numerous methods for detecting damages in structural systems. Each of these methods is designed to detect specific type of damages. Currently there are no techniques to scale all of these methods and damage types in a coherent and general fashion. Alampalli and Ettouney (2007) presented a chart for categorizing and scaling damage in a comprehensive fashion. Such damage scaling organization can be easily used to categorize varying NDT/SHM results (which seems at first

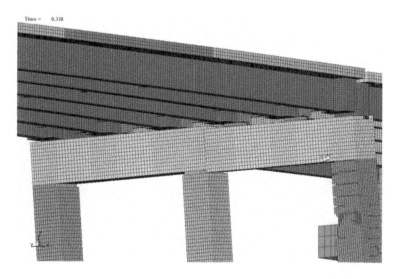

FIGURE 4.20   High resolution numerical finite element model. (Courtesy of Dr. Anil Agrawal.)

glance to be completely unrelated) and input them into a decision making tool box, thus utilizing all pieces of information that is available from any NDT/SHM effort.

Similar to STRID, we introduce a DMID scale DM_{SCALE}. The scale is a qualitative descriptor of the damage size as related to a specific size of the system. One possible formal definition of DM_{SCALE} is

$$DM_{SCALE} = \frac{Min(DM_{SIZE})}{S_{SYSTEM}} \tag{4.54}$$

DM_{SIZE} = Descriptor of the size of monitored damage in the SHM project
S_{SYSTEM} = Adequate descriptor of size of system

Different NDT or SHM methods are adequate for different ranges of damage sizes. Thus, the chosen method of monitoring must scale the required range of DM_{SCALE}. If the method can't scale the required range, more than one method may be needed (see theory of triangulation in this chapter). Figure 4.21 illustrates the importance of damage scale in SHM.

4.6.2.5 Hazard (Demand) Scale

We concern ourselves with spatial scale of hazards. Hazards of any type can be categorized by its spatial attribute. Figure 4.25 shows qualitatively spatial values of different hazards as they affect structures. Hazards with local spatial range, such as fatigue or corrosion (Figure 4.22), will result in local effects on structure. Hazards with larger spatial range, such as overload (Figure 4.23), wind, flood (Figure 4.24), or earthquakes will affect the structure both locally and globally. The concept is shown in Figure 4.25.

Table 4.15 summarizes some scaling issues, for illustration purposes, for different components of SHM field that this proposal will consider.

4.6.3 Two-Dimensional Scale

We observe that the linear scales described so far do interact together; thus generating what might be described as two-dimensional scale. The importance of interactions between the different SHM scales varies in importance. Table 4.16 shows a qualitative estimate of such importance for illustration purposes. Only the upper side of Table 4.16 is filled, since the table is symmetric.

4.6.4 Scales of Higher Dimensions: Role of Decision Making

We identified several scales in SHM. Some or all these scales need to be covered, either partially or fully, during SHM projects. We also showed that in most situations those scales are related,

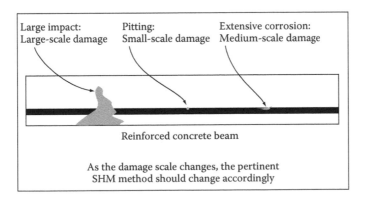

FIGURE 4.21 Damage scale concept in SHM.

FIGURE 4.22   Corrosion hazard: localized. (Courtesy of New York State Department of Transportation.)

FIGURE 4.23   Overload hazard: global. (Courtesy of New York State Department of Transportation.)

together, forming 2D or higher dimensions. Clearly, there is no single technology or algorithms that can handle all of those multiple scales simultaneously. A good management scheme is needed. We offer the following steps for multiple, or single, scaling situations:

1. Identify pertinent scales
2. Identify technologies that are needed to cover required range of scales for the project on hand
3. Identify algorithms that would span different scale ranges, first on each linear scale, then across multiple scales
4. Place adequate decision making rules

FIGURE 4.24   Flood hazard: global. (Courtesy of New York State Department of Transportation.)

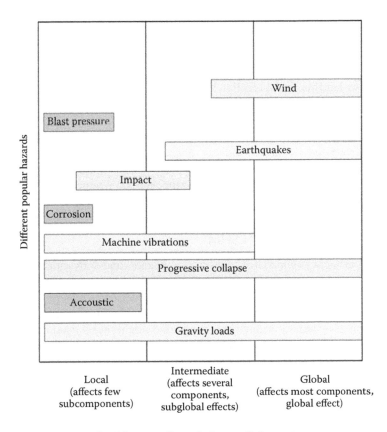

Spatial extent of hazard when applied to system

FIGURE 4.25   Spatial scale and multihazards considerations.

TABLE 4.15
Important Scaling Issues in SHM

SHM Component	Scaling Issues
Sensing	Type of measurement (ToM), space/distance, time (real, intermittent), sampling rates
Structural identification	Frequency range, number of modes, mode shapes, standing versus propagating waves, wave numbers/lengths, element type and size
Damage identification	Size, extent, source of damages, multihazards issues

TABLE 4.16
Qualitative Importance of SHM Scale Interactions

Scale	Frequency	Sensor Density	STRID	Damage Scale	Hazard Scale
Frequency	NA	M	H	H	L
Sensor density		NA	M	H	M
STRID			NA	H	H
Damage scale				NA	
Hazard scale					NA

NA = not applicable, L = low, M = medium, H = high

4.6.5 BENEFITS OF SCALE CONSIDERATIONS IN SHM

There are immense benefits and potential for having successful scaling of SHM methods as described above. For example, by scaling sensor arrays, it is possible to measure and detect higher order mode shapes (thus being able to make a judgment for the reason of a frequency shift as to whether a bearing is frozen, or loss of composite action). Scaling structural identification methods would enable the observations of effects of large range of wave numbers (or wave lengths). Another benefit of scaling SHM techniques would be the smooth integration of current SHM applications which are geared in general to global behavior with the well-known techniques of NDT, which are geared to localized behavior.

4.7 SERENDIPITY PRINCIPLE IN SHM

When an SHM experiment is conducted to monitor a hazard, and if by an accident or by chance, some beneficial information were obtained regarding another hazard, we call this event the serendipity principle in SHM. Serendipity principle, as defined, is definitely a by-product of the multihazards theory as introduced earlier. It is natural to expect that while investigating a particular hazard, an un-anticipated bit of information about another hazard might reveal itself due to the interdependencies that the multihazard theory describes. Figure 4.26 illustrates the serendipity principle in SHM.

We observe that the serendipity principle does not have a rigorous objective approach to it. As such, it might be difficult to defend its inclusion in the context of this chapter. However, there are measures that the owner might follow to maximize the utility of the serendipity principle, and the chances that unintended, yet beneficial, information might result from an SHM experiment. Some of these measures are as follows:

- Increase awareness of the system under consideration. This includes construction material, structural system, load paths, redundancies, support conditions, etc.

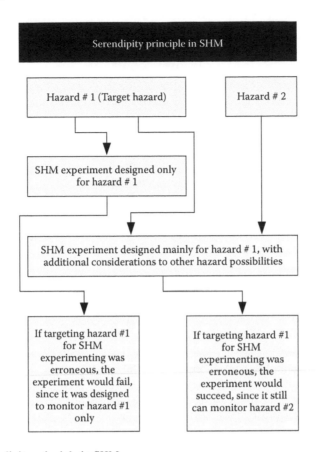

FIGURE 4.26 Serendipity principle in SHM.

- Perform careful design of sensors, instrumentation and data collection, and analysis algorithms
- Ensure the compatibility of the SHM experiment to the intended goal of the experiment
- Keep an open mind of any environments that have not been included in the SHM experiment design phase
- Pay attention to all the observed trends, especially those trends which might not have been expected; do not be hasty in discounting unexpected trends as "just noise"

The above are just some common-sense rules that would increase chances of success of a SHM experiment through the serendipity principle.

As an example to the serendipity principle, let us consider the Newburgh-Beacon SHM experiment (Shenton et al. 2002). Before the experiment, field tests have indicated large magnitude stress spikes that occurred in the bridge hangers. The level of the stress spikes was high enough to prompt the launching of an SHM experiment in search for the causes of those spikes. In planning for the experiment, the authors arranged for thirteen active strain gauges. In addition, sensors for measuring wind speed and direction, vertical accelerations, relative bearing displacement, and temperature were included in the overall instruments. The authors also provided two "dummy" strain gages. One of those dummy strain gauges (named "zero dummy gauge") had no lead wires and was used for control purpose (it was not attached to the structure). The other "dummy" strain gage was not attached to the structure either, however, it had a 100-ft lead wire that was draped over the catwalk (hence it was called "100 ft. dummy gauge"). Neither of the "dummy" sensors was expected to register any strain spikes during the experimentation.

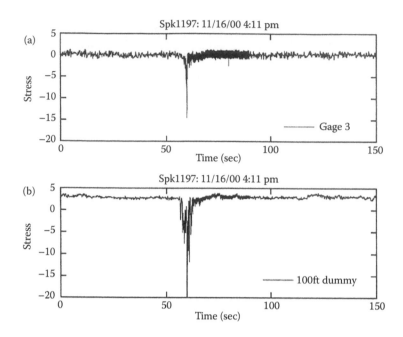

FIGURE 4.27 Stress measurements at actual and dummy gauges (a) measurements during weekend, and (b) dummy gage measurements. (Reprinted from ASNT Publication.)

The researchers made three observations during the experimentation; (a) The active strain gages recorded spikes only during work days, but had almost no recorded spikes during weekends, see top graphic in Figure 4.27; (b) All the truss members that intersect in an instrumented joint recorded compressive spikes; and (c) While the zero dummy gage recorded no spikes, as expected, the 100-ft dummy gage did record spikes with large strain amplitudes, see bottom graphic in Figure 4.27. The first observation pointed to the potential that the spikes did not result from human activities, and the second observation pointed to the fact that the spikes can't be due to a physical behavior of the bridge (not *all* members in a single joint can be compressive at the same time). Finally, the recordings of the 100-ft dummy gage pointed the attention to what the researchers suspected: the reason for the spikes is noise/interferences, not a real stress. In addition, experimentation confirmed this otherwise unexpected finding.

This experiment shows that careful planning and foresight, and a little extra effort (by adding the dummy gauges) yielded a successful experiment, which is a good example for the serendipity principle: what started as an investigation for sources of overstress spikes in bridge components, ended in finding completely different source for the spikes: a harmless noise/interference.

REFERENCES

Alampalli, S. and Ettouney, M., (April 2007) "Structural Identification, Damage Identification and Structural Health Monitoring," Nondestructive Characterization for Composite Materials, Aerospace Engineering, Civil Infrastructure, and Homeland Security 2007, SPIE Volume 6531, San Diego, CA.

Annis, C., (2005) "How Many Specimens is Enough?" Proceedings of the 2005 ASNT Fall Conference, Columbus, OH.

Ettouney, M. and Alampalli, S., (2012), *Infrastructure Health in Civil Engineering: Applications and Management*, CRC Press, Boca Raton, FL.

Ettouney, M., Alampalli, S., and Agrawal, A., (2005) "Theory of Multihazards for Bridge Structures," Bridge Structures, 1(3): 281–291.

Ettouney, M., Daddazio, R., and Abboud, N., (1997) "Some Practical Applications of the use of Scale Independent Elements for Dynamic Analysis of Vibrating Systems," Computers and Structures, 65(3): 423–432.

Gray, I. and Gray, J., (2005) "Assessing UT, EC, and RT Image Processing and Information Fusion with NDE Computer Simulation," Proceedings of the 2005 ASNT Fall Conference, Columbus, OH.

Hyland, D. and Fry, G., (2000) "A Neural-Genetic Hybrid Approach for Optimizing Structural Health Monitoring Systems," Proceedings of 2nd International Workshop on Structural Health Monitoring, Stanford University, Stanford, CA.

Maser, K., (2008) "Integration of Ground Penetrating Radar and Infrared Thermography for Bridge Deck Condition Evaluation," *NDE/NDT for Highwaya and Bridges: Structural Materials Technology (SMT)*, ASNT, Oakland, CA.

Shenton, H., Chajes, M., Finch, W., and Sivakumar, B., (2002) "Long-Term Monitoring of the Newburgh-Beacon Bridge," Proceedings, Structural Materials Technology V: An NDT Conference, NYSDOT, FHWA and ASNT, Cincinnati, OH.

Timoshenko, S., (1955) Vibration Problems in Engineering, Van Nostrand Reinhold Company, New York, NY.

5 Sensors and Infrastructure Health

5.1 INTRODUCTION

5.1.1 Sensing and Infrastructure Health

In a general overview, sensing can be defined as the process of knowing specific information about the system of interest. With this general definition, we can include human inspection as a sensing activity. A more technical definition of sensing is that it is a process where sensors interrogate (interact) the system, and produce certain information about the state of the system. We have thus identified several components in the sensing process: (a) the sensors, (b) the information, (c) the system. The sensors are those vehicles that will detect the state of the system and produce the appropriate information to the user. The information (defined later as type of measurements [ToM]) is transmitted to end user by the sensors: they describe specific states of the system. The system is the infrastructure on hand, or any component within it. Figure 5.1 shows this relationship.

The well-being of any infrastructure from its very inception to the last day of its decommissioning is a continuous exercise of this system → sensor → information process. In addition, we note that as an outcome to the process, there will be some kind of a decision making that may affect the system (a null decision is also a type of decision making). Throughout history such a process existed, and will always exist as long as there is infrastructure. What is changeable are the (a) type, quantity, and quality of information; (b) type, shape, and accuracy of sensors; (c) type, size, and function of infrastructure and (d) type of resulting decisions. As we mentioned earlier, humans can, and should be, considered as a type of sensor. Human sensing abilities are remarkable. Such abilities can be extremely accurate at times. Unfortunately, there are limits to human sensing abilities. In such situations, the need for more accurate sensing devices arises.

The interrelationship between the triumvirate system → sensor → information, and their interaction with the decision making processes, is fairly complex. There are immeasurable combinations of systems, sensors, and information. Because of this, the choice of sensing method, or sensor, should be done carefully. The situation can be described as an optimization problem: given all the problem parameters, it is decided to find the optimal sensor and sensing method that would result in the most efficient resolution. The efficiency metric in this case can be cost effectiveness, accuracy of results, time savings, or any other suitable metric.

We thus reach an important conclusion: sensors (including humans) are an integral and invaluable component in the well-being of infrastructure health. Not only the sensors as separate entities are important, but also their interrelationships with all other aspects of infrastructure are important. This chapter will attempt to address sensors both as separate units and as a component within the complex infrastructure health system.

5.1.2 This Chapter

This chapter will explore categories of sensors that are used in the infrastructure first. We then review briefly to mechanisms and theories that govern the behavior of major sensing mechanisms. We believe that some understanding of the basics of sensors is important for optimal utilization of sensors in any experiment or project. After that we turn our attention to the utilization of sensors

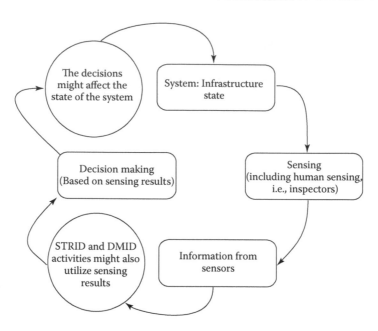

FIGURE 5.1   Sensing paradigm.

in structural health monitoring (SHM). We will balance the theory of sensors with the practical applications of those sensors. The following two sections will be devoted to two emerging sensing techniques: fiber optics sensors (FOR) and wireless sensing. These will be followed by a section discussing smart structures (SS). It is obvious that structures are only smart because they can sense the surrounding environment. Two practical sections will follow: a section describing parameters that can affect choices of sensors for a particular project and optimum sensor locations (OSL). We then offer a guide to help practitioners to choose appropriate sensors. The final section of this chapter is devoted to another emerging sensing field: remote sensing (RS). We immediately note that RS is not the same as wireless sensing. Wireless sensing places sensors in contact with the system and transmit the measurements (information) wirelessly. RS performs the measurements remotely: with no contact with the system. The measurements of remote system can then be transmitted either in a wired (tethered) mode, or in a wireless mode. Figure 5.2 shows the composition of this chapter.

5.2 SENSOR CATEGORIZATIONS

Sensing is the first and most utilized component in SHM and infrastructure health in civil engineering (IHCE). There has been countless number of manuscript, research, and development regarding sensors, measurements, and their use in SHM as well as civil infrastructure. For our purposes in this volume, we start exploring the subject of sensing by considering logical sensor categorizations. This is essential step if we desire to quantify such concerns as optimum sensor selection (OSS), OSL, or any other class of decision making processes that involve using sensors. We propose, to categorize sensors into four categories in cascading order as follows (see Figure 5.3):

Scope: This is a fairly qualitative category that we deemed necessary to differentiate between non-destructive testing (NDT) and SHM fields. Such differentiation is not well defined, however, it does exist. We propose to subdivide sensing within the SHM field into two categories: a general sensing and an NDT sensing categories. General sensing categories are used to sense conventional behavioral metrics such as temperature, displacements, strains, humidity, and so on. They also have the important

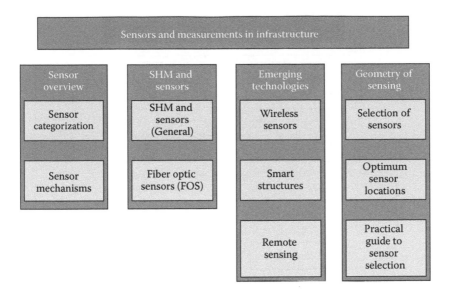

FIGURE 5.2 Composition of sensors chapter.

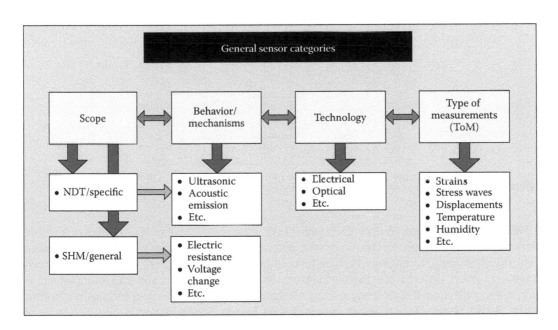

FIGURE 5.3 Sensor categories.

distinction in that they usually do not aim to identify damage in the system. As such, they are general in scope, and fairly limited in objectives. NDT category offer the opposite properties: they are more limited in scope, but they are more general in objectives. For example: acoustic emission (AE), as an NDT technology has a very well-defined scope: measure stress wave emissions that emanate from damages in systems to locate the damage. The objective of AE goes far beyond just measuring stress emissions: its objective it to identify damage attributes in the system. We note that in SHM field, a mixture of NDT and general purpose sensing usually offer the most efficient project solutions.

Technology: This level of categorization addresses the technological basis of sensors. As can be seen from Figure 5.4, there are several technologies that are used for sensing.

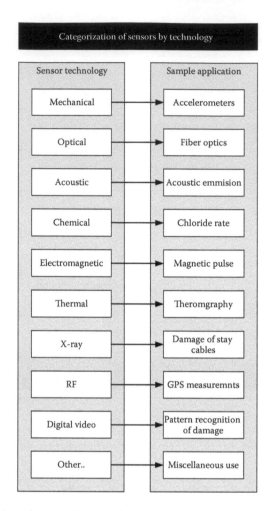

FIGURE 5.4   Categorization of sensors by technology.

Behavior/Mechanism: Within each technology, there exists several ways to sense a physical phenomenon. For example, electric resistance, and electric voltage change are different mechanisms within electric circuit technology that can be used for sensing. In the NDT field, electromagnetism technology offers different sensing mechanisms such as eddy current and magnetic particle. For wide-ranging discussion of different sensing mechanisms for general purpose sensing, see Northrop (2005). ASNT (1996) offers in-depth review of NDT-based mechanisms.

Type of Measurement: Another important sensing category is the ToM category. ToMs can bay any physical metric that is of interest to an SHM project. They can be strains, velocities, displacements, temperature, AE, wave length, or corrosion rates. Figure 5.5 shows some examples of ToMs.

5.3   BASICS OF SENSOR BEHAVIOR

Sensing mechanisms can be categorized into several classes. Among the popular classes are electrical, optical, and chemical, etc. Within each class of sensing mechanisms there are usually several subclasses. In this section, we discuss electrical-based sensing mechanisms and optical-based sensing mechanisms. Those mechanisms are traditionally not categorized within the NDT field. We conclude the section by discussing the distinction between the classes of sensing mechanisms that are presented here and the NDT mechanisms that are explored in Chapter 8.

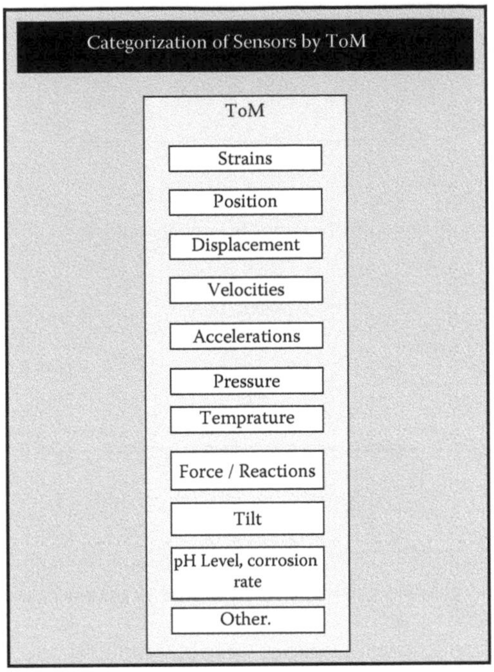

```
┌─────────────────────────────────────────┐
│        Categorization of Sensors by ToM  │
│                                           │
│        ┌───────────────────────────┐     │
│        │           ToM             │     │
│        │   ┌───────────────────┐   │     │
│        │   │     Strains       │   │     │
│        │   └───────────────────┘   │     │
│        │   ┌───────────────────┐   │     │
│        │   │     Position      │   │     │
│        │   └───────────────────┘   │     │
│        │   ┌───────────────────┐   │     │
│        │   │   Displacement    │   │     │
│        │   └───────────────────┘   │     │
│        │   ┌───────────────────┐   │     │
│        │   │    Velocities     │   │     │
│        │   └───────────────────┘   │     │
│        │   ┌───────────────────┐   │     │
│        │   │   Accelerations   │   │     │
│        │   └───────────────────┘   │     │
│        │   ┌───────────────────┐   │     │
│        │   │     Pressure      │   │     │
│        │   └───────────────────┘   │     │
│        │   ┌───────────────────┐   │     │
│        │   │    Temprature     │   │     │
│        │   └───────────────────┘   │     │
│        │   ┌───────────────────┐   │     │
│        │   │  Force / Reactions│   │     │
│        │   └───────────────────┘   │     │
│        │   ┌───────────────────┐   │     │
│        │   │       Tilt        │   │     │
│        │   └───────────────────┘   │     │
│        │   ┌───────────────────┐   │     │
│        │   │ pH Level, corrosion│  │     │
│        │   │       rate        │   │     │
│        │   └───────────────────┘   │     │
│        │   ┌───────────────────┐   │     │
│        │   │      Other.       │   │     │
│        │   └───────────────────┘   │     │
│        └───────────────────────────┘     │
└─────────────────────────────────────────┘
```

FIGURE 5.5   Categorization of sensors by type of measurement (ToM).

5.3.1   VOLTAGE-GENERATING SENSORS

In some situations, when a physical environment changes, an electric magnetic field (EMF) is generated. Perhaps the most popular example is piezoelectric materials that generate EMF when subjected to certain changes. By relating the generated voltages to the properties of these materials, the physical parameters that caused the change can be measured. We discuss the basic principles of piezoelectric sensors first, then summarize other types of voltage-generating sensing techniques.

5.3.1.1   Piezoelectric

Figure 5.6 shows the basic concept of piezoelectric strain transducer. When a transducer is in its baseline stage, its thickness is h. The voltage across the transducer is V_0. When a force is applied to the transducer, it changes its thickness by Δh. Such a change will generate a change in the voltage across the transducer to V_1. The relationship between the change of thickness of the transducer and the voltage can be shown to be as follows:

$$V_1 = g\, E\, \Delta h \tag{5.1}$$

Where g and E are the voltage sensitivity and the modulus of elasticity of the material, respectively. From Equation 5.1, by measuring V_1, the strains in the transducer can be evaluated as

$$\varepsilon = \frac{\Delta h}{h} \tag{5.2}$$

Finally, the strain measurement as a function of voltage is

$$\varepsilon = \left(\frac{1}{g\, E\, h}\right) V_1 \tag{5.3}$$

(a)

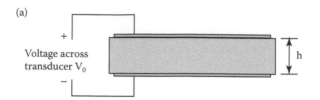

Piezoelectric transducer: not strained

(b)

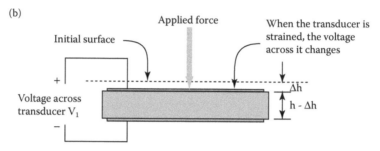

Piezoelectric transducer: after applying a normal force

FIGURE 5.6 Principles of piezoelectric transducers (a) before straining, and (b) after straining.

There are many other uses of piezoelectric materials in SHM. This includes foil strain gauges and accelerometers. Both are discussed later in this chapter.

5.3.1.2 Other

Among other systems that relate EMF voltages and physical parameters are thermocouples that generate EMF when subjected to changes in temperatures. Certain accelerometers also generate EMF voltages when subjected to accelerations. These accelerations can be recorded by observing the voltage in the circuit. Velocities can also be measured using generated voltages in linear velocity sensors (LVS).

5.3.2 RESISTIVE SENSORS

Resistive type sensors rely on changes in electric resistance of a given circuit for measuring the desired physical parameter. If the changes in the electric resistance ΔR, is small compared to a baseline resistance R_0 that is, $\Delta R/R_0 \gg 1$, then the resistance changes are linearly proportional to the voltage change ΔV in the circuit. The basic relationship is

$$\frac{\Delta R}{\Delta V} = \frac{R_0}{V_0} \tag{5.4}$$

In Equation 5.4, V_0 is a baseline voltage. By measuring the voltage changes it is possible to infer the changes in the desired physical parameter. For example, it is known that electric resistance R is related to temperature T such that

$$R = R(T) \tag{5.5}$$

A change in temperature ΔT will generate a change, in electric resistance of ΔR. Such a change will produce a change in voltage ΔV. If the inverse of relationship Equation 5.5 is known as

$$T = T(R) \tag{5.6}$$

then the change of temperature can be measured as

$$\Delta T = T(R) = T(\Delta V, R_0) \tag{5.7}$$

Resistive strain sensors operate with similar principles. Since strains are related to changes in length, which can be related to electric resistance, it is possible to develop a relationship between strains and changes in voltages that can be measured in electric circuits. In addition, resistive sensors are also used to measure changes in humidity, linear position, and angular position.

Several electric circuit topologies are used for this purpose, among them are Wheatstone and Kelvin bridge circuits.

5.3.2.1 Wheatstone Bridge

Wheatstone bridges are a type of direct current (DC) electrical circuit that is used to make precise measurements of sensitive metrics, such as strains in structural components. The basic operational concept of the Wheatstone bridge is the null concept. In a null concept, the input and output voltages are balanced so as to produce a null voltage across the bridge. Figure 5.7 shows the basic operational concept of a null DC instrument.

Utilizing the null DC concept, the Wheatstone bridge (Figure 5.8), is a popular technique that is used for SHM measurements, for example, strain foil transducers. In a Wheatstone bridge circuit, the input and output voltages V_{INP} and V_{OUT}, respectively, are balanced so as to produce a null voltage across the circuit such that

$$V_0 = V_{INP} - V_{OUT} = 0 \tag{5.8}$$

From Figure 5.8, the resistance ratio of the bridge is

$$\frac{R_1}{R_4} = \frac{R_2}{R_3} \tag{5.9}$$

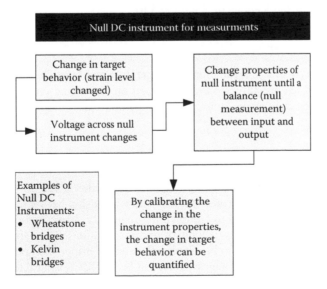

FIGURE 5.7 Operational concept of a null DC instrument.

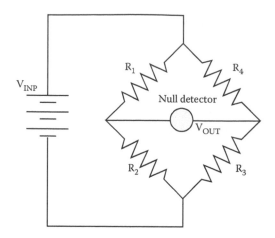

FIGURE 5.8   Wheatstone Bridge configuration.

The relationship between V_{INP} and V_{OUT} can be shown to be

$$V_{OUT} = V_{INP} \frac{R_1 R_3}{(R_1 + R_2)(R_3 + R_4)} \left(-\frac{\Delta R_1}{R_1} + \frac{\Delta R_2}{R_2} - \frac{\Delta R_3}{R_3} + \frac{\Delta R_4}{R_4} \right) \qquad (5.10)$$

Where R_1, R_2, R_3, and R_4 are the resistance magnitudes in the Wheatstone bridge components, and ΔR_1, ΔR_2, ΔR_3, and ΔR_4 are the incremental changes in the resistance. If one of the components, say R_1, represents a sensor that undergoes a change ΔR_1 that is proportional to a strain (see foil strain sensors section later this chapter), while no other components change their resistances, that is,

$$\Delta R_2 = \Delta R_3 = \Delta R_4 = 0 \qquad (5.11)$$

Equation 5.10 can be used to obtain the desired strain level. We note that the Wheatstone bridge can be used in more complex sensing arrangements, such as calibrating temperature drift effects for some sensors, by careful applications of Equation 5.10.

5.3.3   CHANGE OF CAPACITANCE

Capacitance is a property that is a function of temperature and humidity. Because of this, several types of humidity and temperature sensors are based on measuring changes in capacitance and detecting corresponding changes in humidity or temperature accordingly, (see Northrop 2005 for more details).

5.3.4   LINEAR VARIABLE DIFFERENTIAL TRANSFORMERS (LVDT)

This class of sensor is based on measuring magnetic fields that result from relative movements between coils that are arranged in certain manner. The relationship between the magnetic field and the relative motion can then be used to compute the required relative motions. This class of sensors is used to measure displacements, velocities, and accelerations. They are widely used in SHM projects. For example, they are used to measure deflections of bridge members by converting measured displacements to equivalent voltages which can be recorded using data acquisition systems. Creation of a stable support to serve as a datum is a major problem when measuring deflections on bridges. Field measurement of deflections is not easy, even though they can be related to member stiffness more directly than other parameters such as stresses, strains, and inclinations. However, in field testing of bridges this is not always practical and more than often not possible. Some examples are given in the next section.

5.3.5   OPTICS

Optics-based sensors can be subdivided into two general categories: FOS, and mechano-optics sensor. Fiber optic sensors (FOS), as the name implies, are based on the passage of light waves through fibers, and through the interaction of the light with the surrounding media, the desired measurements can be obtained. Due to their importance and emerging wide use in SHM field, FOS will be described in detail later in this chapter.

The mechano-optics sensors are based on transmitting light onto the object of interest. The reflection, or transmission, of the light from or through the object will be affected by the motion of the object. By processing the reflected light, some information regarding the motion of the object can be inferred. This type of sensing is explored further in the remote sensing section of this chapter.

Another type of mechano-optics sensors is based on transmitting a light beam (laser) on the object. The light will interact with the object; such an interaction will generate ultrasound waves within the object. These waves can be body waves, surface waves, or plate waves (in case of thin plate systems). The waves are then detected by a receiver that would infer properties of the object, such as presence of defects in the path of the ultrasound waves. This type of sensing is also explored further in the remote sensing section of this chapter.

5.3.6   OTHER

There are numerous other types of sensing mechanisms. Some of those are based on chemical or radiating wave physics. We discuss the chemical sensing for monitoring corrosion process later this section. Radiating wave sensing is discussed in detail in Chapter 7.

5.3.7   NDT-SPECIFIC

There is another large class of sensing mechanisms that are used in SHM projects. These sensing mechanisms are used in the NDT field. Among these mechanisms are ultrasonic sensing, electromagnetic sensing, thermal sensing, radiating wave sensing, and AE. What differentiate these sensing mechanisms, and the whole of NDT field from the type of sensing mechanisms that is discussed in this section is that the NDT mechanisms is more integrated. Most of the sensor mechanisms of this section concern themselves only with the ToM they are designed for. For example, an LVDT accelerometer will measure a given acceleration wave form only. The analysis of the wave form is left to the tester to perform depending on the objectives of the sensing project. Meanwhile, NDT-type mechanism measure a particular ToM (such as stress wave form) and tries to analyze it and detect damages, if any: all in one integrated method.

Recognizing this subtle, but important difference, we decided to discuss many of the important NDT sensing and evaluation methods in Chapter 7, on damage identification, instead of this chapter.

5.4   SENSOR MEASUREMENTS IN SHM

5.4.1   OVERVIEW

Previous section presented different types of sensing mechanisms. This section explores sensors from ToM view point. Important ToMs for civil infrastructure are shown in Figure 5.5. We explore all of them in this section, including types of sensors used and the underlying physical behavior of them. We also point to the fact that there are several different ways that each ToM can be measured. This is important point, since it can, and should, be used when choosing sensors for specific project, as will be explained in Section 5.8.

5.4.2 Strain Sensors

5.4.2.1 General

Strain gages are widely used to measure strains in such structural components as girders, rebars, and concrete decks. Measured data can then be employed to calculate physical parameters, evaluate bridge load capacity, and monitor structural safety. Environmental conditions may play important roles in bridge testing. For example, self-temperature compensating type gages are generally preferred when wide temperature variations are expected. Gages are welded or attached with adhesives to the structure. Two or four arm gages are also favored as they increase the signal to noise ratios, and several gage combinations are sometimes adopted to measure different components of strain (such as shear, bending, and axial). Normally, strain gages require DC input in the range 0–15 volts, and give outputs in the same range. Measured voltages are then converted to strains/stresses using appropriate engineering relations.

5.4.2.2 Foil Strain Gages

Strain gages are the sensors which are very widely used in civil engineering testing and research to measure the structural behavior under loads. Figures 5.9, shows schematics of foil strain gage, Figure 5.10 shows schematics of piezoelectric foil gages Figure 5.11 shows a typical strain gage in the field. Figure 5.12 shows a strain gauge in an SHM experiment. Since strain is directly related to stress up to elastic limit and most structures are designed to operate within elastic limits, stress is very critical to failure of civil structures. Bonded strain gages became very commonly used sensors in the world. Bonded strain gages become integral to the structure they are bonded to and, if applied properly, will undergo same strains as the parent structure and gives direct indication of the strains experienced by the parent structure. Ideal strain gage characteristics depend on the type of application and nature of the strain field experienced during the testing. For more on foil strain gauges see Perry and Lissner (1962) and Hannah and Reed (1991).

The change in electrical resistance in the strain gage is related to strain; and this principle using simple Wheatstone bridge circuit forms the basis for strain gage operations. Strain gages primarily form one, two, three, or all legs of the Wheatstone bridge as described in the previous section. In general, in a balanced bridge, the change in strain generally translates to change in current through the gage or the change in voltage across the gage resulting in an unbalanced Wheatstone bridge.

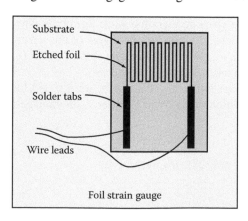

FIGURE 5.9 Concepts of foil strain gauges.

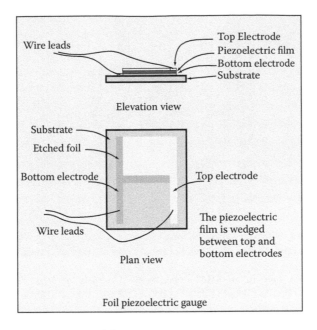

FIGURE 5.10 Concepts of piezoelectric foil gauges.

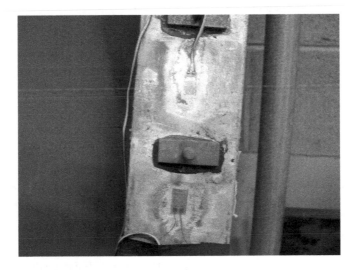

FIGURE 5.11 Typical foil strain gage.

This unbalance can be translated into appropriate strain using the electrical principles and strain gage characteristics following the relation

$$\frac{\Delta R}{R} = G\varepsilon \tag{5.12}$$

Where ε is the desired strain, R is the baseline resistance, and ΔR is the measured change in resistance that produce a null balance in the Wheatstone Bridge. The gauge factor G is a constant that depends on the sensor configurations.

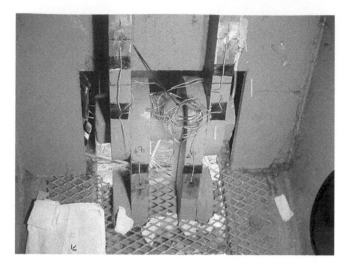

FIGURE 5.12 Foil strain gage in an SHM experiment. (Courtesy of New York State Department of Transportation)

Advantages

- Foil gages are easy to bond or weld
- Available in small to big sizes. Hence, ideal for measuring stress concentrations as well as in applications where an average strain is needed (in Reinforced concrete applications). Can be grouped easily to get tri-axial strains
- Become integral part of the structure and direct indication of the strain in parent structure
- Can give both static and dynamic strains
- Not expensive
- Low mass

Disadvantages

- Needs good understanding of the strain gage principles for good interpretation of data
- Where environmental conditions, such as temperature, rapidly vary, comprehensive understanding of strain age principles and the effect of environment on structural behavior are needed to get good data
- Instrumentation required could be significantly expensive
- Not very suitable for long-term continual use. Drift can be a big issue due to several factors that cause unbalance of the Wheatstone bridge circuit
- Not reusable
- Needs constant power supply
- Needs expertise for good and reliable installation of gages
- Needs protection from environment for long-term installations
- Needs extensive surface cleaning and preparation for good bonding

5.4.2.3 Demountable Strain Gages/Strain Transducers

Demountable strain gages in general employ number of bonded strain gages, typically with a full Wheatstone bridge to eliminate dummy gages, into a fixed enclosure such as a ring or a plate. This makes the gage reusable (see Figures 5.13 and 5.14 for typical BDI transducers, a demountable strain gage) and hence the name demountable strain gage. These gauges are attached to the structure at

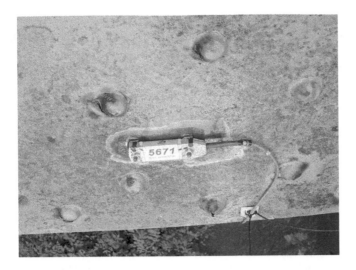

FIGURE 5.13 Demountable strain gauges.

FIGURE 5.14 Close up of a Demountable strain gauges.

two ends, with a known distance between them, and the change in gage length is related to strain in the structure.

Advantages

- Eliminates the need for extensive surface preparation
- Easy to fix and use
- Reusable
- Needs less expertise for installation and operation
- Eliminates the need for dummy gages and balancing circuits; and hence instrumentation can be cheaper

Disadvantages

- Since these use strain gages in their assembly, some of the inherent disadvantages with strain gages are not eliminated

- Reliable calibration is required
- In general, needs big gage lengths and cannot be easily installed at tight locations and closure to connections, where stress concentrations can be important
- Since, they are not integral to the structure, the structural deformations can create errors in response, if not calibrated and understood well
- As these are relatively big, it is not easy to install a group of gages together to measure stresses in several directions at same location
- Use is generally limited to shorter durations
- Mismatch of structural material and gage materials can create issues with temperature and other environmental factors. Hence, good understanding of structural behavior and sensor behavior are essential

5.4.2.4   Vibrating Wire Gages

Vibrating wire strain gages use the principle that a wire vibrates when held in tension and a force is applied. Any relative movement of the fixity at ends (supports) will effect a change in measured frequency of vibration, and can be used to measure the movement. This principle is used in strain gages, load cell piezometers, and crack meters, but predominantly used for measuring strains in infrastructure area.

The natural frequency of a vibrating wire f can be related to strain (ε) as

$$\varepsilon = K f^2 \tag{5.13}$$

where K is a constant factor for the sensor and Gage-Technique (2011) depends on (a) wire density, (b) gravity constant, (c) wire length, and (d) wire modulus of elasticity. Thus, any relative change in strain can be measured knowing the frequency at both initial and current states using the following equation:

$$\Delta\varepsilon = K\left(f_1^2 - f_2^2\right) \tag{5.14}$$

Many of the vibrating wire gages are designed to be fixed to the surface of the structure requiring strains using epoxy or welds or embedded inside. These normally have two end blocks with a wire between them. The end blocks are attached to the surface of interest. When the surface deforms or is strained, the end blocks move relative to each other. This causes change in the length of the wire and thus the natural frequency of the wire. The data loggers measure the frequency and thus associated strain. These gauges are used for long-term static applications due to the advantages they offer. More details on demountable strain gauges can be found at Perry and Lissner (1962), Hannah and Reed (1991), Hag-Elsafi et al. (2003) and Gage-Technique (2011) (Figures 5.15 and 5.16).

Advantages

- The response from these gages is analogue . Hence, the resolution is infinite and is limited by the devise used to read and display the data
- Immune to electrical noise
- Tolerate wet environment
- Can transmit data to long distance without loss of accuracy
- Stable for long-term applications as there is no drift
- Can compensate for temperatures with a built-in temperature measurement
- Relatively rugged and durable
- Easy to install

FIGURE 5.15   Vibrating wire strain gauges (surface mounted).

FIGURE 5.16   Vibrating wire strain gauges.

Disadvantages

- Not suitable for dynamic measurements
- Restively bigger than foil gages
- Data analysis could be tricky, in long-term applications, if structural environment in not understood well

5.4.3   POSITION

Linear variable differential transformer gives position of an object they are mechanically attached to. This converts position into DC voltage and can be read using an appropriate reader; and does not require physical connection to the extension like a potentiometer. The extension valve shaft (or control rod) in an LVDT moves between primary and secondary windings of a transformer causing inductance between the two windings to vary. This varies the output voltage proportional to the position of the valve extension. The theoretical principles of the LVDT are shown in Figure 5.17.

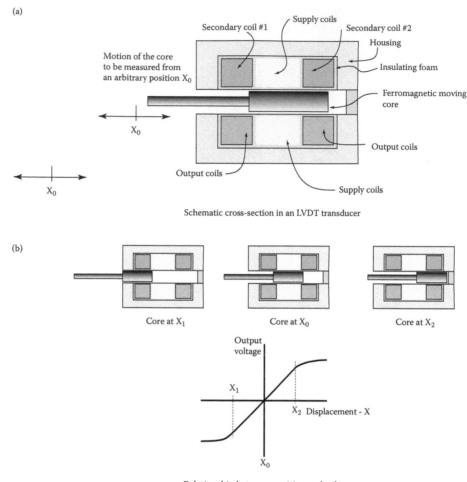

FIGURE 5.17 Principles of LVDT transducers (a) schematics, and (b) basic principles.

Figure 5.18 shows a typical LVDT unit. Figure 5.19 shows an LVDT unit that is installed to measure displacements during an SHM project. Several online references are available such as Sensorland (2011), Engineersedge (2011), Singer (2011), and Macrosensors (2011).

Advantages

- Very reliable
- Rare failures
- Can measure very small movements (as small as 0.01 microns)
- Rugged

Disadvantages

- High variation in ambient temperatures can affect quality and hence long-term measurements
- Setup on field structures could be difficult
- Operating ranges could be fixed and limited
- Limited sensitivity

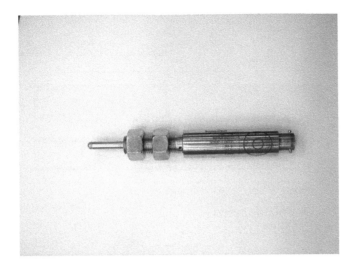

FIGURE 5.18 Typical LVDT unit.

FIGURE 5.19 Field application of LVDT unit for measuring displacement. (Courtesy of New York State Department of Transportation.)

5.4.4 ACCELEROMETERS (ANGULAR AND LINEAR)

Accelerometers are sensors that utilize different concepts for measuring accelerations. We explore two different types of accelerometers: LVDT and piezoelectric types

5.4.4.1 LVDT-Based Accelerometers

Linear variable differential transformers based accelerometers, (see Figure 5.20), are based on the following equation of motion:

$$M \ddot{x}_1 + C \, \Delta \dot{x} + K \, \Delta x = 0 \qquad (5.15)$$

with

$$\Delta x = x_1 - x_0 \qquad (5.16)$$

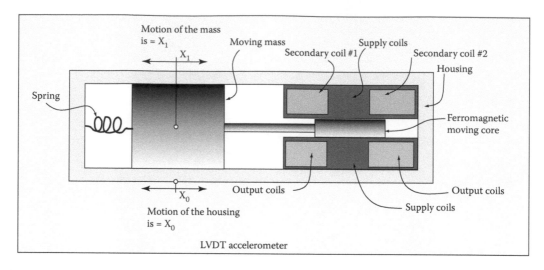

FIGURE 5.20   LVDT-based accelerometer basics.

The mass, damping and stiffness parameters of Equation (5.15) are M, C, and K, respectively. They are sensor properties that are assumed to be known. The relative motion between the mass and the base of the sensor Δx is proportional to the measured LVDT voltage V as

$$V = K\,\Delta x \qquad\qquad (5.17)$$

In Equation (5.17), K is a known LVDT factor. Thus, by measuring the output voltage, the acceleration can be obtained from Equations (5.15), (5.16), and (5.17).

5.4.4.2   Piezoelectric Sensor Accelerometers

Piezoelectric elements-based accelerometers are designed to produce electric signal proportional to forces induced due to vibration of the structures they are attached to. Quartz is a popular piezoelectric element in many commonly used accelerometers due to its long-term stability, and temperature stability, but requires voltage amplifiers due to the high voltage sensitivities they exhibit. Voltage amplifiers limit the signal to noise ratios but allow high level of vibration measurements. Polarized lead zirconate titinate (PZT) is another alternative material used in accelerometers, PCB (2011) but are limited to low level vibrations. These use quieter microelectronic charge amplifiers and also require frequent calibration. The selection of an accelerometer depends on the amplitude range, frequency range, environment or in-service conditions, and duration of use. Figure 5.21 shows the basic concepts of piezoelectric accelerometer. Figures 5.22 and 5.23 show the use of piezoelectric accelerometers in lab testing and *in-situ* testing, respectively. See Northrop (2005) and PCB (2011) for more on piezoelectric accelerometers.

Advantages

- Easy to use
- Easy to install
- Wide variety
- Relatively low cost

Disadvantages

- Requires thorough understanding of structural vibration theories
- Do not directly correlate to structural health
- Turn-key systems are uncommon for civil applications

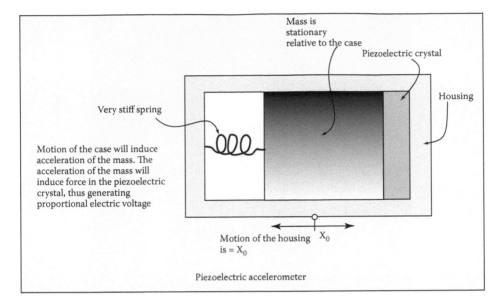

Mass is
stationary
relative to the case

Piezoelectric crystal

Housing

Very stiff spring

Motion of the case will induce
acceleration of the mass. The
acceleration of the mass will
induce force in the piezoelectric
crystal, thus generating
proportional electric voltage

Motion of the housing X_0
is = X_0

Piezoelectric accelerometer

FIGURE 5.21   Concepts on piezoelectric accelerometers.

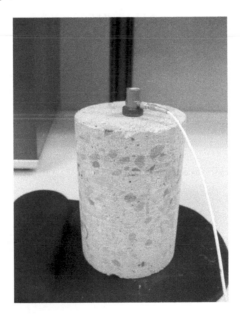

FIGURE 5.22   Piezoelectric sensor on concrete specimen. (Courtesy of New York State Department of Transportation.)

- Needs experienced personnel for good quality results
- Long-term installations are hard to maintain
- Can be damaged relatively easily
- Accompanying equipment is relatively expensive

5.4.5   VELOCITIES (ANGULAR AND LINEAR)

Velocities can be measured indirectly by using an LVDT accelerometer as in Figure 5.20. The accelerations are computed using Equations 5.15, 5.16, and 5.17, then the velocity can be estimated

FIGURE 5.23   Piezoelectric sensor under steel girder.

by integrating the measured accelerations as

$$\dot{x}_1 = \left(\int_{t_0}^{t_1} \ddot{x}_1 \, dt \right) + \dot{x}_1 \Big|_{t_0} \tag{5.18}$$

with the initial velocity as $\dot{x}_1 \big|_{t_0}$.

A direct linear velocity transducer/sensor (LVT/LVS) method of computing velocity is shown in Figure 5.24. It is based on slightly different form of Equation 5.17 such that is directly relates the voltage to the relative velocity between the coil and the magnet $\Delta \dot{x}$. The basic equation becomes

$$V = K_{VEL} \Delta \dot{x} \tag{5.19}$$

where K_{VEL} is a known LVT/LVS factor and

$$\Delta \dot{x} = \dot{x}_1 - \dot{x}_0 \tag{5.20}$$

Thus, by measuring the output voltage, the relative velocity can be obtained from Equations 5.19 and 5.20.

5.4.6   Displacements (Angular and Linear)

Linear variable differential transformers concepts can be easily used to measure displacements. The relative displacement between a baseline x_0 and a displaced object x_1 as in Equation 5.16 can be computed by measuring the LVDT generated voltage as in Equation 5.17.

5.4.7   Force

Load cells are mechanical devices used to measure naturally or mechanically induced loads, the structure is subjected to (see Figures 5.25 and 5.26). Load cells convert the load or force into electrical output. There are several types of load cells and are characterized by type of output signal generated (pneumatic, hydraulic, or electric) and the weight detection mechanism (bending, shear, tension, etc.). These include mechanical load cells (such as hydraulic or pneumatic load cells), strain gage based load cells (such as shear beam, ring and pancake, bending beam load cells), and other

(a)

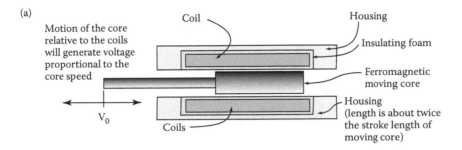

Schematic cross-section in an LVT /LVS transducer

(b)

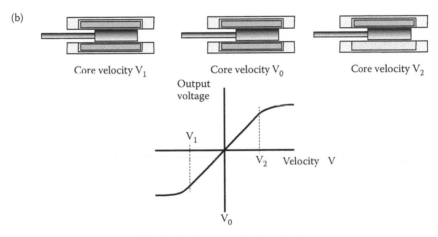

Relationship between velocity and voltage

FIGURE 5.24 Concepts of LVT/LVS instruments. (a) schematics, and (b) basic principles.

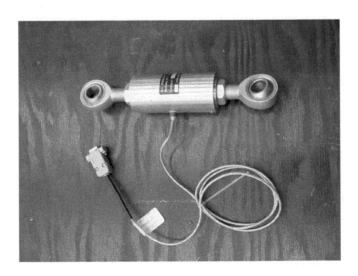

FIGURE 5.25 Load cell.

FIGURE 5.26   Load cell cable.

load cells (such as fiber optic or piezo-resistive load cells). For more on load cell sensing, see Perry and Lissner (1962), Hannah and Reed (1991), Hag-Elsafi et al. (2003), and Omega (2011)

But, in recent days, most commonly used load cells are transducers built based on strain gages and principles associated with them. For example, if a strain gage is attached to an axial member (see Figure 5.26), by knowing the modulus of elasticity, cross-section, and the strain the member is subjected to, the load can be calculated as follows:

$$P = \varepsilon\left(E A\right) \qquad (5.21)$$

Note that P is the desired load. The strain, modulus of elasticity, and area of cross-section are ε, E, and A, respectively. Thus, as long as the load applied keeps the axial member in the linear rage of modulus of elasticity, the load cell can be used to measure the load by multiplying the strain with a constant calibration factor. Many load cells also use a full bridge (i.e., four gauges, usually two in tension and two in compression), to avoid the use of dummy gages to complete the balancing circuit, to avoid temperature compensation, and to obtain maximum sensitivity. These can also be used such a way that the electrical output to load can be increased Omega (2011).

In bridges, load cells are normally used in bridge testing to measure naturally or mechanically induced loads for which structures are being subjected. Figure 5.27 shows a high capacity, hydraulic load cell that utilizes the change in fluid pressure in estimating the applied load level. Again, conversion of output voltages to loads is direct and is accomplished in a similar manner to that for inclinometers.

Advantages

- Relatively small in size, as small as a button
- Good linearity within the elastic range
- Can measure tension and compression
- Can be made weather resistant
- Low profile and can be integrated into restricted areas
- Connections can be custom made
- Offer good long-term stability
- Low cost
- Low thermal effects

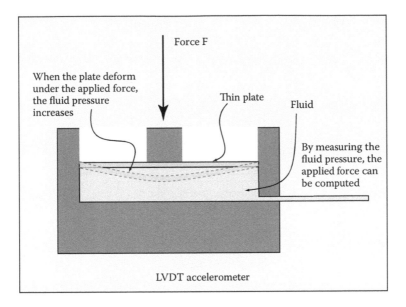

FIGURE 5.27   Hydraulic load cell.

Disadvantages

- Issues inherent to strain gages still exists, that is, drift and so on
- Need to be water tight to avoid moisture and humidity as they can impact the bonding between the strain gages and the material they are affixed to
- Low electrical output associated with strain gages may necessitate external amplification of the signal (between the gage and the instrumentation)

5.4.8   INCLINOMETERS

Inclinometers are used to measure inclination of various structural elements due to distresses in the system (see Figure 5.29). They are mainly used to evaluate fixity of bridge girders at supports and to monitor long-term movements of piers, abutments, and girders for general assessment of structural integrity. Normally, inclinometers require DC input in the range 5–15 volts and give outputs in the same range. Measured voltages are then converted to inclinations using their sensitivity. Furthermore, this class of sensors are used to measure inclination or slope or elevation of elements they are attached to with respect to gravity. These are also known as tilt sensors or slope sensors or grade indicators. There are several types of inclinometers. Simple, hydrostatically based inclinometers are based on hydrostatic leveling of water, as shown in Figure 5.28. The angle of incline, θ is measured as

$$\theta = \frac{H_1 - H_2}{L} \tag{5.22}$$

Advantages

- Gives absolute slope with respect to gravity or reference
- No need of interpolation
- Easy to attach to concrete or metal structures
- Very sensitive

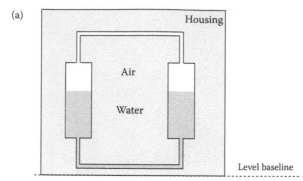

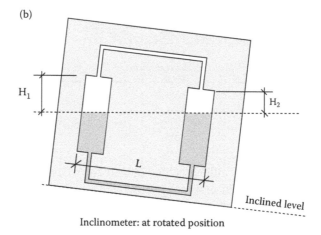

FIGURE 5.28   Basis of hydrostatic inclinometers (a) level position, and (b) rotated position.

FIGURE 5.29   Typical electrically based inclinometer.

Disadvantage

- Very sensitive in small range, making it hard to install
- Do not react very fast. Hence, not possible to make dynamic measurements
- Installation method is very important and can be difficult

5.4.9 CORROSION

Corrosion sensors, (see Corrosionsource 2011) are normally used to measure the rate of corrosion in steel or concrete. These are relatively new to infrastructure arena and are used for long-term measurements, in most cases to monitor anticorrosive performance of new materials. Corrosion monitoring systems/sensors vary in complexity and size. But, potential cost savings from the implementation of these in the future could be significant to infrastructure arena. The technologies vary from use of polarization resistance to time domain reflectometry. Size varies from the size of a pebble to a small box depending on the technology. Many are coupled with temperature and other measurements.

A popular corrosion measuring method is the half cell approach, ASTM 1987. Half cell sensing is used to measure corrosion levels as well as possibility of corrosion development in reinforcing steel in concrete structures. It is well known that the passive oxide film with high pH level that cover steel rebar is meant to protect the rebar from corrosion process. However, when concrete structures are subjected to environments with high chlorine levels, such as salt water, these chlorine ions penetrate reinforced concrete to breakdown the protective oxide film and initiate the corrosion process (rust forming) of the now unprotected steel.

The half cell sensing method measures the voltage drop between a known reference point and any other point at the reinforced concrete surface. Such drop in electric potential can be mapped as contours. The concentrations of the contour lines indicate areas of high probability of corrosion.

Some known disadvantages of the method are

- Can't be used for post-tensioned construction, since the tendons do not come into contact with the concrete itself, rendering the half cell potential readings not indicative of corrosion condition of the tendons themselves
- Since the half cell method results are affected with humidity levels, its results can be inaccurate in dry environment
- Compensation for concrete cover might be needed
- The sensors give localized corrosion rate. Hence, may not represent the entire structure and care should be given in choosing appropriate locations
- Temperature and humidity can influence the corrosion rates
- Installation is not very easy for some of these
- No standards or formal guidelines for field application
- Many are intrusive

The advantages of the half cell potential are

- Simplicity and ease of implementation
- Can be embedded inside, or mounted on the surface of the structure (see, respectively, Figures 5.30 and 5.31).
- Can be monitored over long time periods; this feature makes the method very suitable for SHM projects.
- Gives rate of corrosion directly
- Versatile and reliable
- Continuous monitoring can be accomplished
- Many are passive
- Makes remote corrosion monitoring possible

FIGURE 5.30 Embedded half cell potential sensor. (Courtesy of New York State Department of Transportation.)

FIGURE 5.31 Half cell potential sensor after installation.

5.4.10 Pressure

There are several methods to measure pressure. Many of the concepts that are discussed earlier in this section can be used to measure pressure. For example:

- *Piezoelectric pressure sensors* use the generated voltage that results from the strain ε of a piezoelectric material to relate the pressure $p = P/A$ to (see Equation 5.21) to the generated voltage from the material strain.

- The LVDT principle can be used to measure pressure in *LVDT pressure sensors*. When a coil is displaced relative to a magnet, the resulting voltage can be measured. The displacement can be related to the pressure of interest.
- *Vibrating wire pressure* sensors are used for highly transient pressures, such as pressures from explosives (see Equations 5.13 and 5.14). When a pressure is applied to the sensor, the natural frequency of the sensor changes, by measuring the change, similar to Equation 5.14, the magnitude of the pressure can be estimated.

For detailed pressure sensing discussion, see Norton (1989).

5.4.11 TEMPERATURE

Temperature measurements are often needed in SHM projects. Corrosion, fatigue, FRP as well as general thermal expansion of structures are of interest to decision makers. There are numerous methods of measuring temperature effects. For example:

- A *biomaterial temperature sensor* that uses two different materials. When the temperature changes from the baseline temperature, sensor deforms according to the differences of the thermal expansion coefficient and the length of each material within the sensor. The deformation can be used to detect the temperature changes.
- *Electrical resistance thermometer* uses the relationship between the temperature and the electric resistance of materials to detect temperature changes. It is used in applications such as underwater submerged temperature measurements as well as pipelines, tanks, and air.
- *Thermocouples thermometers* use electrical properties of two dissimilar materials to accurately measure temperature. Thermocouples thermostats are widely used in many forms.
- *Pyroelectric thermometers* are based on using pyroelectric materials that generate electric charge when subjected to heat.

Other forms of thermometers are fiber optic temperature sensors and infrared (IR) thermometers. For extensive discussion of thermometers, see Webester (1999).

5.4.12 OTHER SENSORS IN SHM

There are several other types of sensors that are used in SHM projects. Some of these sensors and their use in SHM are shown in Table 5.1.

The reader is referred to more comprehensive sensing references such as Northrop (2005), Norton (1989), and Webester (1999).

5.4.13 ADVANCED SENSING TECHNOLOGIES

5.4.13.1 Microelectromechanical systems (MEMS) Technology

MEMS is an emerging sensing technology that holds great promise for SHM in civil infrastructure. An overview of MEMS technology is offered by El-Sheimy et al. (2004).

TABLE 5.1
Use of Other Types of ToM in SHM

ToM	SHM Use
Humidity	Corrosion detection and mitigation
UV exposure	FRP behavior
Radiation	Location of hidden defects in concrete structures
Fluid flow	Scour investigation

5.4.13.2   Friction-Based Strain Sensing

Utilization of this new type of strain sensors is described by Ojio et al. (2004). The authors used the new sensor in monitoring bridge behavior in Japan.

5.4.13.3   Scour Sensing

The scour hazard has been and is still a major concern for bridge owners and it contributes to more failures than other hazards. Measuring scour depth is still hard and is important where conventional analyses are not available in several cases (tidal scour is an example). Mercado and Rao (2006) describe the development of a pneumatic scour detection system (PSDS) designed to monitor scour depth on an as-needed or real time basis during either normal or extreme flood conditions (see Figure 5.32). It is based on differential resistance to air (or liquid) flow through an array of stainless steel porous filters. This system is intended to eliminate common deficiencies found with other conventional systems, that is, debris accumulation around piers masking sound signals, and strong turbulence and impact of large debris during floods.

Pressure versus time data is recorded at filters situated at various depths of the probe driven into a river bed. Scour depth is determined by measuring the resistance to air flow through the filters as a function of the nature of the material against which the filter is embedded. Low resistance indicates water, where as high resistance indicates that the filter is sealed against competent soil (see Figure 5.33).

The system was tested in the laboratory and then was demonstrated in the field with an excavated pit. The laboratory and simulated field test results showed that system is consistent and produces clear results.

5.4.13.4   Innovative ETDR Sensing

Electric time domain reflectometry (ETDR) sensing technology is used by Chen (2004) to develop crack detection sensors. The method is based on the propagation of electromagnetic waves in an electric cable or a transmission line, which functions both as a signal carrier and sensor. Discontinuity causes a reflected signal due to an electric property change. Arrival time of the reflected signal and its amplitude gives an indication of location and degree of discontinuity, respectively, (see Figure 5.34). A cable sensor embedded in concrete can thus detect both the location and width of the crack.

Two innovative cable sensors (see Figure 5.35) were designed based on the change in topology, or electric structure, to utilize them as crack sensor. The exact dimensions were finalized based on available materials in the market and sensitivity analysis.

The operation is illustrated in Figures 5.36 and 5.37. The presence of a partial or complete separation between adjacent spirals, which act as the outer conductor of the cable sensor, will force the return current on the transmission line outer shield to change its flow path.

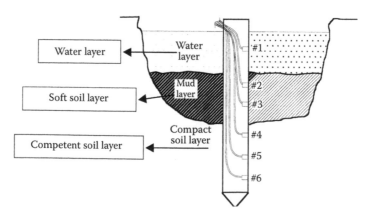

FIGURE 5.32   Sketch of the pneumatic scour detection probe. (Reprinted from ASNT Publication.)

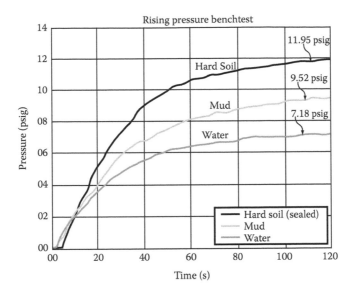

FIGURE 5.33 Rising pressure tests in the laboratory. (Reprinted from ASNT Publication.)

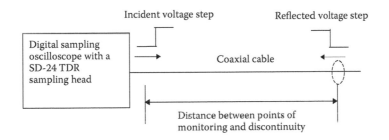

FIGURE 5.34 Principle of cable sensor measurement. (Reprinted from ASNT Publication.)

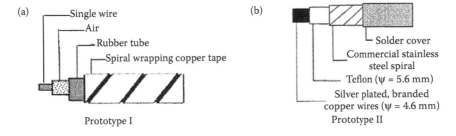

FIGURE 5.35 Schematic view of two crack sensors (a) Prototype I, and (b) Prototype II. (Reprinted from ASNT Publication.)

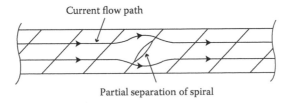

FIGURE 5.36 Change of current flow path. (Reprinted from ASNT Publication.)

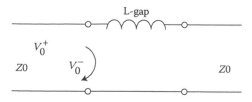

FIGURE 5.37 Equivalent transmission line model. (Reprinted from ASNT Publication.)

Static tests and dynamic loading tests were conducted to investigate the sensitivity, signal loss, and partial resolution. Both types of sensors were found to be sensitive to cracks of various sizes, and can identify crack locations.

5.5 EMERGING TECHNOLOGY: FIBER OPTICS SENSORS

5.5.1 INTRODUCTION

5.5.1.1 General

Fiber optic sensors use light as both a transduction and signal transmission mechanism. Optical effects that are useful for transduction include intensity modulation, modal domain interference, Bragg grating wavelength selective transmission and reflection, noncoherent white light interference, and Brillouin backscatter. Intensity-based FOSs use mechanical effects to modulate transmitted or reflected light intensity. Light intensity can be detected easily with phototransistors, photoresistors, and photodiodes, and can be used to make absolute intensity sensors that are useful in SHM. Optical interference is a useful transduction mechanism for FOSs. When a beam of light is split into two parts, each of which is sent on a separate path, and then rejoined, interference patterns may form. A variation in the length of one of the paths causes the interference pattern to shift. Measuring interference pattern shifts can give very precise and accurate information about changes in physical parameters that affect effective light path lengths. Both coherent (usually laser light) and incoherent (white or light-emitting diode [LED] light) interferometry are useful for sensing applications. FOS can be sued for measuring strain, temperature, inclination, and acceleration among others. Ansari (2002) provided an overview of different FOS usage in SHM, theory, and different types. The author also described advantages and disadvantages of FOS. Tennyson et al. (2004) presented theoretical study about validation of FOS in pipelines. A practical application of using FOS in bridge monitoring was offered by Benmokrane and El-Salkawy (2002). Park et al. (2004) discussed the theoretical basis and advantages of using long gauge FOS. A landmark paper that introduced the expression "CIVIONICS" was presented by Rivera et al. (2004). The authors observed that the integration of automatic monitoring of civil infrastructure, which relies on sensing, is analogous to automatic sensing in aviations community that utilizes the expression "AVIONICS." They also introduced specifications of FOS systems, which aimed at standardizing the use of FOS in civil infrastructure applications.

A typical construction of FOS is shown in Figure 5.38. The fiber core that transmits light signals is made of materials such as plastics or glass. The fibers are protected by a cladding tube that helps in guiding light signals along the fiber. An environmental cover that protects the system is on the outside.

Fiber optics sensing can be categorized in several manners (see Figure 5.39). The mode of sensing can be either intrinsic or extrinsic. There are several sensing methods: light wave phase sensing (Figure 5.40), time of flight (TOF) sensing, polarity, and intensity are among those methods. The geometry of sensing can be either local (at discrete points), or distributed (continuously along the cable). Finally the method FOS is attached to the test article can be either embedded in the interior, or attached to the exterior. Table 5.2 summarizes some of the classifications of FOS. This section will explore many of the FOS modes, especially as they are applied to civil infrastructure. We then discuss cost–benefit of FOS. Several practical applications are offered at the conclusion of the chapter.

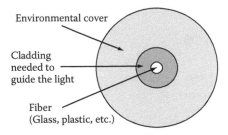

Basic cross-section of fiber optic sensor

FIGURE 5.38   Basic cross-section of fiber optic sensor (FOS).

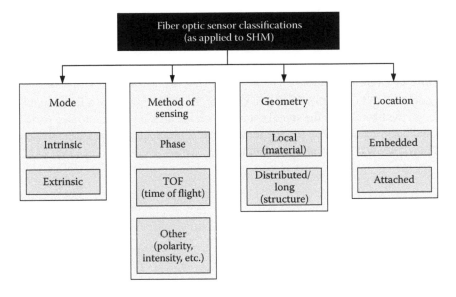

FIGURE 5.39   Classifications of fiber optic sensors.

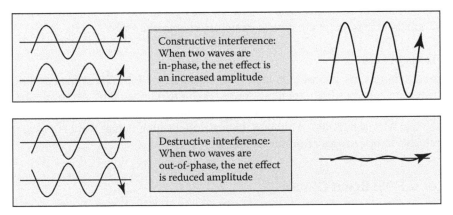

Effects of phase: destructive and constructive interferences

FIGURE 5.40   Phase measurement concept in FOS.

TABLE 5.2
Classifications of Some Popular FOS

Classification	Type	Fabry-Perot	Bragg Grating	Optical Time Domain Reflectometry	Brillouin Optical Time Domain Analysis
Mode	Intrinsic	Yes (only some applications)	Yes	Yes	
	Extrinsic	Yes			Yes
Method of sensing	Phase		Yes		Yes
	TOF	Yes		Yes	Yes
	Other?				
Geometry	Local	Yes	Yes		
	Distributed			Yes (incrementally)	Yes (continuous)
Location	Embedded	Yes	Yes	Yes	Yes
	Attached	Yes	Yes	Yes	Yes

5.5.1.2   Time of Flight and FOS

One of the concepts in FOS is the TOF, (see Chapter 7). In TOF sensing, a light signal is sent at a time T_0 through the fiber. When the signal encounters a scaterer, it is split into two parts in the form of forward and back scatter. When the scattered signal is observed at time T_1, the distance between the observation point and the scaterer is computed as

$$D = \frac{V}{T_1 - T_0} \tag{5.23}$$

The speed of light in the fiber is V.

5.5.1.3   Different Modes of FOS

There are two modes of sensing using Fiber Optics (FO): intrinsic and extrinsic. Intrinsic FOS is when the sensing occurs totally within the fiber. In this case, the light does not leave the fiber (see Figure 5.41). Time domain reflectometry (TDR), and Bragg grating are some intrinsic FOS. The extrinsic FOS occurs by some interaction between the light in the fiber and the exterior environment. Brillouin and Fabry-Perot are typical examples of extrinsic FOS.

5.5.1.4   ToMs of FOS

In civil infrastructure applications, FOS can sense most needed ToM. These include strains, accelerations, displacements (position), rotation, forces, pressure, temperature, pH level, and humidity. An overview of different uses of FOS to detect temperature, strains, and crack growth was offered by Tennyson and Morison (2004). Although the main application in that work is pipes, the lessons can be applicable to other types of infrastructure.

5.5.2   LOCAL FOS: BRAGG GRATING

Bragg grating is a popular FOS, see, for example, Chuang et al. (2004). It is an intrinsic type-sensing. The basic operations of the sensor are based on defining n as the refractive index

$$n = \frac{\lambda_{GLASS}}{\lambda_{AIR}} \tag{5.24}$$

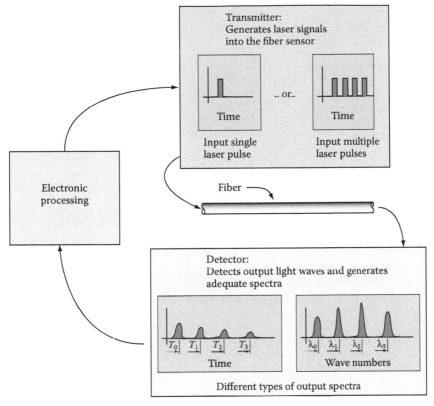

Basic fiber optic sensing system

FIGURE 5.41 Intrinsic fiber optic sensing basics.

λ is the wave length of light, which is frequency dependent. The Bragg grating governing equation is

$$\lambda_{REFLECTED} = 2\,n\,D \tag{5.25}$$

With D as the distance between the gratings, and $\lambda_{REFLECTED}$ is measured in air.

For a strained fiber grating, the distance between the gratings, the reflected wave length, and the refracted index will all change as $\Delta\lambda$, ΔD, and Δn. They are governed by

$$\frac{\Delta\lambda}{\lambda} = \frac{\Delta D}{D} + \frac{\Delta n}{n} \tag{5.26}$$

The initial and strained wave lengths λ_0 and $\lambda_1 = \lambda_0 + \Delta\lambda$ are observed during the sensing. The old and strained refractive indices n and $n + \Delta n$ can also be observed by using an external light source. From the above, the required strain can then be evaluated as

$$\varepsilon = \frac{\Delta D}{D} = \frac{\Delta\lambda}{\lambda} - \frac{\Delta n}{n} \tag{5.27}$$

Figure 5.42 shows the basic operating principles of Bragg grating FOS.

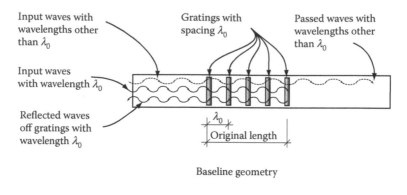

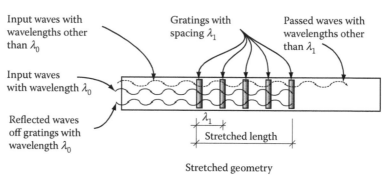

FIGURE 5.42 Fiber Bragg gratings (FBG) optic strain sensor.

5.5.2.1 Applications of FBG

Demountable Sensors: Metje et al. (2004) explained how a demountable sensor is developed, calibrated, and tested before put in operation. A sensor is developed using fiber rods along with multiple fiber optic Bragg grating sensors for monitoring movements in tunnels for safety, maintenance operations, and long-term monitoring. It is meant to measure settlement, rotation, and distortion. Several connected fiber rods will be introduced into tunnel lining for monitoring purposes. The new sensor is based on the principle that strain will be measured at discrete points; and by knowing the geometry of these fixed positions strains can be converted into displacements, (see Figure 5.43).

At first, the authors tested the ring with regular strain gages to understand its behavior, fibers were tested for understanding their mechanical properties then tested with Bragg grating sensors incorporated in to the rods, and then testing them under simple conditions that can be interpreted with relative ease.

Soil Settlement: The settlement of soil under concrete slab road will cause road damage; and a gap is formed between the slab and the soil with air is trapped in between, (see Figure 5.44). Since thermal properties of the air and the soil are different, thermal disparity between them was used by Cai et al. (2008) to address the issue of soil settlement under the concrete slab using fiber Bragg grating (FBG) sensor for temperature monitoring, (see Figure 5.45). It was concluded that by measuring the daily temperature variation under the bridge approach slab, the settlement and erosion of the soil beneath can be conceptually monitored and predicted.

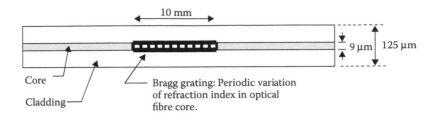

FIGURE 5.43 Schematics of Bragg grating and smart rod. (Reprinted from ASNT Publication.)

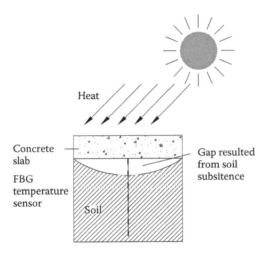

FIGURE 5.44 Soil settlement monitoring method. (Reprinted from ASNT Publication.)

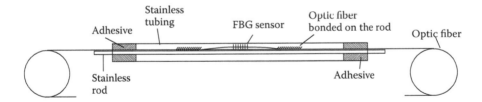

FIGURE 5.45 Fiber Bragg grating (FBG) temperature sensor design. (Reprinted from ASNT Publication.)

5.5.3 LOCAL FOS: FABRY-PEROT

The reflected wave from the near and far edges of the Fabry-Perot gap will interfere either constructively or destructively, depending on the phase between the different waves. For wave numbers with phase of multiple of 2π, the interferences will be constructive. The length of the Fabry-Perot gap can then be expressed as

$$L_1 = \frac{(\varphi_2 - \phi_1)}{4\pi} \frac{\lambda_1 \lambda_2}{(\lambda_2 - \lambda_1)}$$ (5.28)

where λ_1 and λ_2 are any constructive reflected wave numbers and $(\phi_2 - \phi_1)$ is the phase difference between these two waves. For two consecutive waves, such a phase difference is 2π. The above relationship reduces to

$$L_1 = \frac{1}{2} \frac{\lambda_1 \lambda_2}{(\lambda_2 - \lambda_1)}$$
(5.29)

For the strained sensor, the relationship becomes

$$L_2 = \frac{1}{2} \frac{\overline{\lambda}_1 \overline{\lambda}_2}{(\overline{\lambda}_2 - \overline{\lambda}_1)}$$
(5.30)

where $\overline{\lambda}_1$ and $\overline{\lambda}_2$ are the strained constructive reflected wave numbers. The required strain of the sensor is

$$\varepsilon = \frac{(L_2 - L_1)}{L_1}$$
(5.31)

Figure 5.46 shows the basic operating principles of Fabry-Perot FOS.

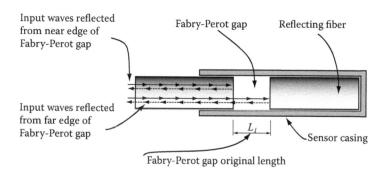

Baseline geometry

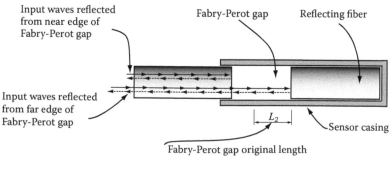

Stretched geometry

FIGURE 5.46   Fabry-Perot FOS.

5.5.4 Distributed (Long) FOS: Optical Time Domain Reflectometry (OTDR)

Monitoring the times of arrival of reflected light pulses, the following relationship can be established

$$T_{ij} = T_i - T_j = \frac{L_i - L_j}{V_{PULSE}} \tag{5.32}$$

where T_i and T_j are the times of arrival of the ith and j^{th} pulses, respectively. The distances L_i and L_j are the distances of the ith and j^{th} partial reflectors, while V_{PULSE} is the speed of laser pulse in the fiber. For the strained condition, the above relationship changes to

$$\overline{T}_{ij} = \overline{T}_i - \overline{T}_j = \frac{\overline{L}_i - \overline{L}_j}{V_{PULSE}} \tag{5.33}$$

Where $\overline{T}_i$ and $\overline{T}_j$ are the new times of arrival of the ith and j^{th} pulses, respectively. The distances $\overline{L}_i$ and $\overline{L}_j$ are the distances of the ith and j^{th} partial reflectors in the strained configuration. The average axial strain between the ith and j^{th} station is

$$\varepsilon = \frac{\overline{T}_{ij} - T_{ij}}{T_{ij}} \tag{5.34}$$

OTDR sensing is appropriate for long uniform constructs such as wires, cables, pipelines, and so on. Figure 5.47 shows the basic operating principles of OTDR.

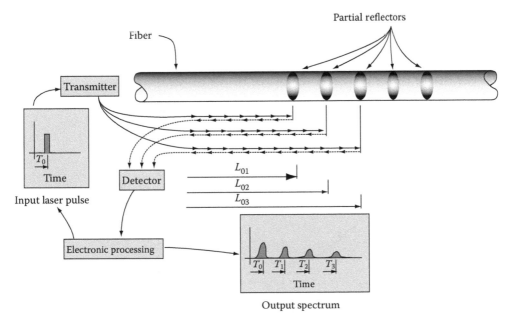

Fiber optic sensor: optical time domain reflectometry (OTDR)

FIGURE 5.47 OTDR basic operating principles.

5.5.4.1 BOTDR in SHM

Wu and Xu (2002) found that distributed BOTDR can be effective. The bonding length, if short, can affect the strain accuracy. They recommended some work-around for this problem in the conclusions of their work. Yamaura et al. (2000) showed the applicability of using Brillouin Optical Time Domain Reflectometer (BOTDR) to measure strains along steel plats and reinforced concrete beams. Both experiments show good accuracy of the BOTDR when compared to conventional strain gauges, up to the point of crack development. The accuracy of the BOTDR dropped slightly after the formation of the crack in the beam. As the applied load increased, the conventional strain gauge did break, while the BOTDR kept on functioning. We can conclude that the BOTDR can develop an accurate strain measurements along the length of the specimen of interest up to development of cracks, beyond which the accuracy of the BOTDR might drop; the overall measured strain trends, are still of use up to the point of ultimate strains.

Another comparison of the adequacy of BOTDR of sensing strains along a distance is given by Oka et al. (2000). Figure 5.48 shows the predicted axial strains when the actual FOS itself is subjected to an axial tensile force. The predicted strains agreed well with expected strains. The authors reported that the sensor was able to predict strains as high as 1%. In a more realistic testing configuration, they subjected a 0.3-m beam to two points loading. The deflections of the beam were measured by integrating measured strains from the BOTDR sensors, conventional strain gauges, and from a displacement meter. Figure 5.49 shows a comparison of the three measurements. The BOTDR-based displacements agreed well with the displacement meter results.

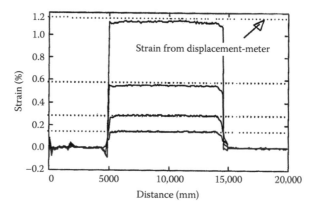

FIGURE 5.48   Axial strain distribution. (With permission from Dr. Fu-Kuo Chang.)

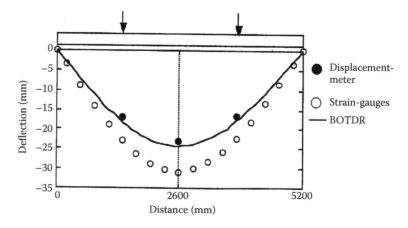

FIGURE 5.49   Displacement comparison. (With permission from Dr. Fu-Kuo Chang.)

5.5.5 Distributed (Long) FOS: Brillouin Optical Time Domain Analysis (BOTDA)

Brillouin frequency shift v_B is directly proportional to acoustic wave velocity of the material such that (see Texier and Pamukcu 2003, and Cho and Lee 2004)

$$v_B = \frac{2nV_a}{\lambda_{PUMP}} \tag{5.35}$$

where n is the refractive index, λ_{PUMP} is the wave length of the incident laser wave (pump), and V_a is the local acoustic wave velocity. The refractive index n and the incident wave number λ_{PUMP} are controlled during the experiment. Thus by measuring the Brillouin frequency shift v_B, the local acoustic wave velocity can be measured. When the material undergoes compressive, or tensile, strains the changes in local V_a can be measured. Such changes are directly related to the local normal strains in the material. By combining a series of pulses, and observing times of flights, the strains at different lengths along the fiber can be computed. Equation 5.35 shown above can be simplified as

$$v_B = A_\varepsilon \, \varepsilon \tag{5.36}$$

where A_ε and ε are a factor of strain proportionality and normal local strains, respectively. It should be mentioned that the Brillouin shift can also be directly related to temperature, see Hecht (2006) and OZOptics (2006). The Brillouin shift is generalized to

$$v_B = A_\varepsilon \, \varepsilon + A_T \, T \tag{5.37}$$

where A_T and T are a factor of temperature proportionality and local temperature, respectively. Equation 5.37 shows the power of Brillouin sensing: it can sense both temperature and strains along the whole length of the fiber. In addition, BOTDA is capable of distributed sensing at high spatial resolution. This makes it suitable for infrastructure applications such as prestressed, post-tensioned tendons (corrosion), and measuring hydrogen embrittlement for high strength wires, complex reinforced concrete or connection behavior. The locations of potential damage in these applications are not known *a priori*, which makes BOTDA a suitable sensing technique. Figures 5.50 and 5.51 show the basic operating principles of BOTDA.

Many authors studied the use of BOTDA in infrastructure applications. For example, Zeng et al. (2002) described using distributed Brillouin scattering fiber optics in measuring temperatures. They argued for the use of reference temperature measurements to ensure accurate results. Zou et al. (2004) showed the interdependence between strains and temperature measurements.

5.5.5.1 Smart BOTDA

Bastianinin et al. (2006) explored a technique using "smart" FRP that smears the measured strains along a crack, thus avoiding the step functions effect of conventional FOS. The technique uses Brillouin FOS, which features a distributed strain sensing, that is, capability to simultaneously measure the strain level and locate the strained point along the sensor. To mitigate the issues commonly observed with traditional optical fiber cables (easy breakage or insulation rendering it not useful as a strain sensor), the following Brillouin FOS-specific solutions have been developed (see Figure 5.52):

1. Sensor featuring an inner FRP core hosting the optical fiber and an outer molded coating that carries helical grooves intended to enhance adhesion to concrete
2. A "Smart" composite obtained by adding one or more optical fibers to the glass or carbon or aramid weft of a woven textile
3. A thermoplastic ribbon featuring multiple optical fiber for strain for strain sensing and thermal compensation

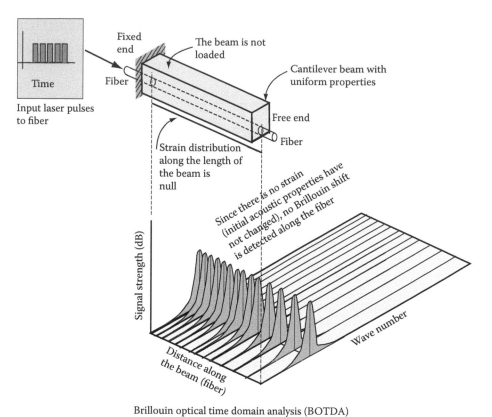

Brillouin optical time domain analysis (BOTDA)
non-strained beam

FIGURE 5.50   BOTDA: no-strain state.

The "Smart" FRP sensor has an advantage over traditional sensor in the sense that it smears the measured strains along a crack, thus avoiding the step functions effect of conventional FOS. Thus, these can maintain sufficient accuracy even in the case of cracked substrates, and hence may be a good solution for reinforce concrete (RC) structures susceptible for cracking.

These sensors were used on a RC bridge and a large steel bridge. The load test data form RC bridge showed that the peak strains expected by analysis matched closely with the mart sensor, where as the traditional gages and traditional FOS showed low values. This was attributed to the inability of these sensors in the presence of cracks and crack openings were observed during the load tests. The data from steel bridge showed good comparison to those obtained using analytical tools. Figure 5.53 shows measured strain profiles along two main girders of the test bridge at two load passes (bold lines) compared with theoretical strains from finite element analysis (FEA) and LRFD design provisions (AASHTO).

5.5.6 COMPARISON BETWEEN LOCAL AND DISTRIBUTED FOS

A comparison between local and distributed FOS is shown Table 5.3; for more details, see ISIS Canada (2006), and OZOptics (2006)

5.5.7 FOS USE IN NDT VERSUS SHM

When it comes to the use of FOS in civil infrastructure, the distinctions between NDT and SHM is reduced to a question of time and space. For temporary tests/observations, the use of FOS would

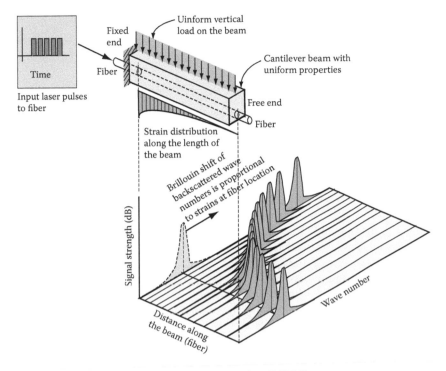

Brillouin optical time domain analysis (BOTDA)
strained beam

FIGURE 5.51   BOTDA: strained state. (Reprinted from ASNT Publication.)

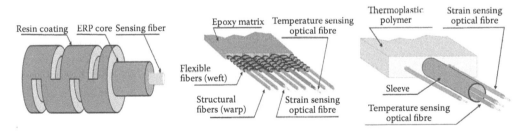

FIGURE 5.52   Brillouin SHM sensor concepts. (Reprinted from ASNT Publication.)

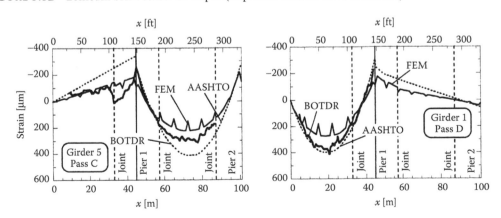

FIGURE 5.53   Measured strain profiles. (Reprinted from ASNT Publication.)

TABLE 5.3
Local and distributed FOS

Type of FOS	Local	Distributed
Optimum use	Material behavior	General structural behavior
Measuring resolution	High	Medium
Application	Single point	Very long distances
Fiber configuration	Multiple fibers	Can be either single or loop-type fibers configuration

resemble a classical NDT. While for a permanent, or semipermanent, installation of FOS, the project would be an SHM project. Spatially, we observe that local FOS can be considered either an NDT or an SHM tool. For distributed FOS, it is perhaps more appropriate SHM tool, since it covers larger range, without loss of accuracy. Finally, it is safe to state that an embedded FOS should be considered as an SHM application, due to its permanent setting. Attached FOS, on the other hand, can be either NDT or SHM application, depending on other factors, as discussed above.

5.5.8 Value (Cost–Benefit) of FOS

This section we present a qualitative summary of FOS in civil infrastructure monitoring. We follow it by introducing an objective method to evaluate costs-benefits and value of using FOS.

5.5.8.1 Advantages and Disadvantages of FOS

Disadvantages of FOS include the following:

Cost: Relatively speaking, FOS can cost more than conventional sensors. As newer manufacturing techniques emerge, the relative cost of FOS is reduced.

Experience: FOS is relatively new when compared with other conventional sensors. Body of experience is thus not as large as conventional sensors.

Durability: Some FOS might not be adequately rugged, specially when used in demanding civil infrastructure environment (heat, humidity, abnormal loads and responses, etc.).

Size: Some FOS can be bulky (conduits, jackets, etc.).

Advantages of FOS include the following:

Miniaturization: Some FOS can be miniaturized, see Newton (1999).

Embedment: Capability of FOS for embedment in solids is attractive when measuring internal responses and environments, such as strains of reinforcing steel rebars in concrete.

Sensitivity and range: FOS can sense wide range of ToMs, see, for example, Kao and Taylor (1996).

Nonelectrical: FOS operations depend on light transmission. Lack of electrical or electromagnetic use or need can be attractive for certain material applications, as well as when used in certain environmental conditions. In addition, there is no need for electrical power since the light can travel through the fiber for long distances.

Can be inexpensive: Careful planning and efficient utilization can produce adequate cost–benefit and life cycle utilization.

Distributed sensing: Some FOS is capable of continuous (distributed) sensing such as Brillouin FOS. This capability offers potential sensing of whole fields of strains, for example.

Geometrically efficient: Signal can be transmitted to long distances along the fibers. This provides for efficiency, and potential cost savings.

Multifunctional: In nondistributed FOS applications, several measurement points along one cable is possible. Bragg grating FOS is a typical example of this usage.

Other advantages: Other potential advantages of FOS are as follows (a) it can provide both relative and absolute measurements, (b) provides ease of installations: they can be bonded or embedded easily, (c) capable of both static and dynamic sensing, and (d) can be used for long-term applications.

5.5.8.2 Objective Cost–Benefit Analysis of FOS

We are now faced with the problem of quantifying the value of FOS for a particular SHM project. We use a utility-type approach (see Chapter 8) in quantifying the costs and the benefits. The value can then be established by associating the costs and the benefits. Since the approach is based on qualitative estimates, it can be used mainly for comparison purposes, such as comparing FOS with other sensors.

Estimate of Benefits: Potential benefits of using FOS can be estimated qualitatively by the professional as B_{FOS_i} subjected to

$$1 \leq B_{FOS\_i} \leq 10 \qquad (5.38)$$

With $i = 1,2\ldots9$. The descriptions of B_{FOS_i} are shown in Table 5.4. For a given use of FOS, the magnitudes of B_{FOS_i} are estimated according to the particular conditions and demands of the experiment.

The total benefits of using FOS is

$$B_{FOS} = \sum_{i=1}^{i=9} W_{Bi}\, B_{FOS\_i} \qquad (5.39)$$

The weights W_{Bi} should be assigned by the professional to reflect relative importance of different benefits of Table 5.4.

Estimate of Costs: Potential costs of FOS can be estimated qualitatively by the professional as C_{FOS_i} subjected to

$$1 \leq C_{FOS\_i} \leq 10 \qquad (5.40)$$

TABLE 5.4
Benefit Factors

i	B_{FOS_i}
1	Importance of miniaturization
2	Embedment need/potential
3	Sensor sensitivity
4	Sensor measurement range
5	Importance of nonelectrical properties of sensor
6	Importance of electrical power cable and topology
7	Importance of distributed sensing
8	Advantage of using cables, rather than discrete conventional sensors; multifunctionality
9	Other factors

TABLE 5.5
Cost Factors

i	C_{FOS_i}
1	Out of pocket cost
2	Lack of experience
3	Potential durability damages
4	Potential size disadvantages
5	Other cost/disadvantage factors

with $i = 1, 2, \ldots 5$. The descriptions of C_{FOS_i} are shown in Table 5.5. For a given use of FOS the magnitudes of C_{FOS_i} are estimated according to the particular conditions and demands of the experiment.

The total cost of using FOS is

$$C_{FOS} = \sum_{i=1}^{i=5} W_{Ci} \, C_{FOS\_i} \tag{5.41}$$

The weights W_{Ci} should be assigned by the professional to reflect relative importance of different costs/disadvantages of Table 5.5. It is important to assign the cost weights W_{Ci} so as to be consistent with the benefit weights W_{Bi}. Such consistency is needed for ensuring accurate estimates of the FOS value.

Value of FOS: The value of using FOS can now be computed as

$$V_{FOS} = \frac{B_{FOS}}{C_{FOS}} \tag{5.41}$$

Comparison with other types of sensors: Obviously, as benefits outweigh the costs, the value of using FOS is assured. In some situations, other sensing technologies might be available, and a comparison between values of competing solutions might be needed. In such a situation, and to ensure a reasonable comparison, The costs, benefits, and value of the other technology can be evaluated using similar equations from Equation 5.38 through Equation 5.41 such that

$$B_{OTHER} = \sum_{i=1}^{i=9} W_{Bi} \, B_{OTHER\_i} \tag{5.42}$$

$$C_{OTHER} = \sum_{i=1}^{i=5} W_{Ci} \, C_{OTHER\_i} \tag{5.43}$$

$$V_{OTHER} = \frac{B_{OTHER}}{C_{OTHER}} \tag{5.44}$$

The benefits and costs for the other technology have assigned relative weights for different parameters as B_{OTHER_i} and C_{OTHER_i}. These assignments are consistent with the definitions of Table 5.4 and Table 5.5.

The relative values of FOS and any other competing technology can now be compared using Equations 5.41 and 5.44.

Example: Let us illustrate the methodology by considering the situation where an SHM project manager considered the use of FOS and a competing technology. The manager assigned relative weights for the situation as in Table 5.6. Clearly, the cost parameter is a major consideration in the decision. Also, the benefits from using FOS were not in need for that project. The resulting two values for FOS and the competing technology reflects those issues. The value of the competing technology is higher than FOS.

Another situation, where benefits of FOS are judged to be important to the project, and the cost is relatively similar between the competing technologies, is shown in Table 5.7. The value of using FOS in this case is superior to the competing technology.

5.5.9 Case Studies—Applications

5.5.9.1 Overall SHM Experiment

Example of using Fiber Optics in an overall SHM experiment was offered by Neubauer et al. (2004). They described a monitoring system for a bridge in Iowa with high performance steel and integral abutments. It had some typical and atypical fatigue sensitive details such as web gaps (can lead to distortion induced cracking), stiffeners welded to tension flanges, and weld concentrations at web and flange splices. Thus the long-term bridge performance was being evaluated using monitoring system. All sensors are fiber Bragg grating sensors to accommodate long-term monitoring without bias, drift, and so on. Figure 5.54 shows the schematics of the SHM system.

The three main objectives of the monitoring and evaluation portion of this project are as follows:

- Evaluate bridge performance continuously
- Monitor for deterioration of the bridge over time and develop a baseline record for identifying structural performance changes
- Conduct a detailed fatigue evaluation

Both global performance (distribution factors, neutral axis, and restraint ratios) and local performance were considered in designing these systems and sensor locations. Local effects of traffic,

TABLE 5.6
Cost, Benefit, and Value of FOS and Competing Technology, Case 1

Item	i	Parameter	Parameter Wt.	Sensor Wt. FOS	Sensor Wt. Competing Tech.
Benefits	1	Importance of miniaturization	1	5	4
	2	Embedment need/potential	1	7	4
	3	Sensor sensitivity	7	6	6
	4	Sensor measurement range	8	7	7
	5	Importance of nonelectrical properties of sensor	2	8	8
	6	Importance of electrical power cable and topology	2	8	6
	7	Importance of distributed sensing	3	9	1
	8	Advantage of using cables, rather than discrete conventional sensors; multifunctionality	1	9	4
Costs	1	Out of pocket cost	10	8	3
	2	Lack of experience	7	4	2
	3	Potential durability damages	5	6	6
	4	Potential size disadvantages	5	5	6
Total Benefits				**178**	**141**
Total Costs				**163**	**104**
Value				**1.09**	**1.36**

TABLE 5.7
Cost, Benefit, and Value of FOS and Competing Technology, Case 2

Item	i	Parameter	Parameter Wt.	Sensor Wt. FOS	Sensor Wt. Competing Tech.
Benefits	1	Importance of miniaturization	1	5	4
	2	Embedment need/potential	7	7	2
	3	Sensor sensitivity	7	6	6
	4	Sensor measurement range	8	7	7
	5	Importance of nonelectrical properties of sensor	9	8	3
	6	Importance of electrical power cable and topology	9	8	3
	7	Importance of distributed sensing	9	9	1
	8	Advantage of using cables, rather than discrete conventional sensors; multifunctionality	5	9	4
Costs	1	Out of pocket cost	10	8	3
	2	Lack of experience	7	4	2
	3	Potential durability damages	5	6	6
	4	Potential size disadvantages	5	5	6
Total Benefits				**422**	**199**
Total Costs				**163**	**104**
Value				**2.59**	**1.91**

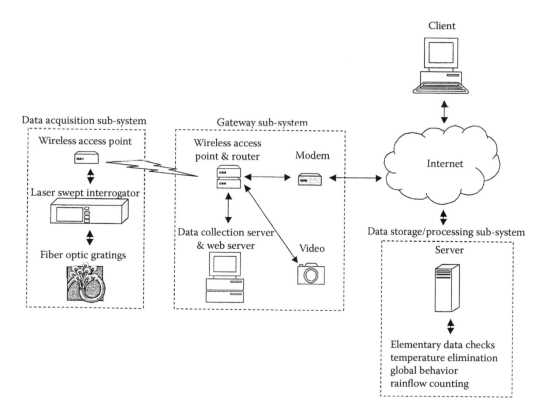

FIGURE 5.54   Block diagram of SHM system. (Reprinted from ASNT Publication.)

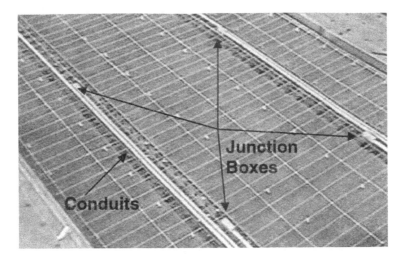

FIGURE 5.55   Fiber optic sensor use for sensing composite rebars—overview. (Courtesy ISIS Canada.)

web gap regions, and stress concentration areas were instrumented along with temperature sensors. Some gages were not attached to the structure and hence they give the effects of temperature and can be used for compensating for the temperature. Rainflow algorithm is used to estimate the remaining life of the details and the susceptibility of details.

The system allows the client to view real-time video of traffic crossing the bridge and the corresponding real-time strain data at various bridge locations.

5.5.9.2 Monitoring Composite Rebars

The use of composite rebars in reinforced concrete systems has been gaining interest since they are not as susceptible to corrosion problems as steel rebars. Because of their recent introduction to civil infrastructure systems, continuous or intermittent monitoring their behavior is of importance. ISIS Canada (2006) reported a monitoring project that utilizes the attributes of FOS to monitor composite rebars, (Figures 5.55 and 5.56).

5.6 WIRELESS SENSORS

5.6.1 Overview

This section explores briefly the wireless sensor units and components. Finally, the utility of wireless sensors in SHM field are presented.

5.6.1.1 Composition of Wireless Sensor Units

Wireless sensors can be divided into two general categories: passive and active, as shown in Figure 5.57. In passive sensing, as the name implies, the sensor will measure a particular ToM (displacement, acceleration, pressure, etc.) by responding passively to the conditions of the sensor environment. Passive wireless sensors are defined, in a generic manner, by four issues: (i) the data acquisition specifications, (ii) embedded computing specifications, (iii) wireless channel specifications, and (iv) the physical attributes of the sensor itself. In active sensing, the sensor will generate signals in a controlled manner, and then sense the response of the system to these signals. An additional unit is included in active sensors that would generate the signal is included in active sensors. A comprehensive discussion of wireless sensor unit was offered by Lynch and Loh (2006).

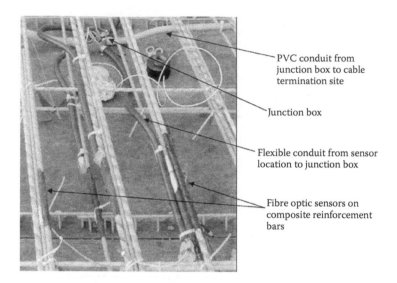

PVC conduit from junction box to cable termination site

Junction box

Flexible conduit from sensor location to junction box

Fibre optic sensors on composite reinforcement bars

FIGURE 5.56   Fiber optic sensor use for sensing composite rebars—detail. (Courtesy ISIS Canada.)

Components of wireless sensor unit

Active wireless unit

Passive wireless unit

Data acquisition specifications	Embedded computing specifications	Wireless channel specifications	Physical attributes	Actuation specifications
• Number of channels • Sample rate • A/D resolution • Number of digital inputs	• Type of processor • Bus size • Clock speed • Program memory • Data memory	• Type of radio • Frequency band • Wireless standard • Spread spectrum used? • Outdoor range • Indoor range • Data rate	• Physical dimensions • Power • Power source	• Number of D/A channels • Sample rate • D/A resolution • Voltage outputs

FIGURE 5.57   Composition of wireless sensor unit. (Based on Lynch, J. and Loh, K., *Shock and Vibration Digest*, 38(2), 2006. With permission.)

5.6.1.2   Composition of Wireless Sensor Network

What differentiate wireless sensing from conventional sensing is exemplified when studying the components of wireless sensors, as shown in Figure 5.58. First is the sensing unit itself, which can be active or passive. The wireless radio protocol that offer communications between the sensors as well as the project data processing center is another component. The local data processing and

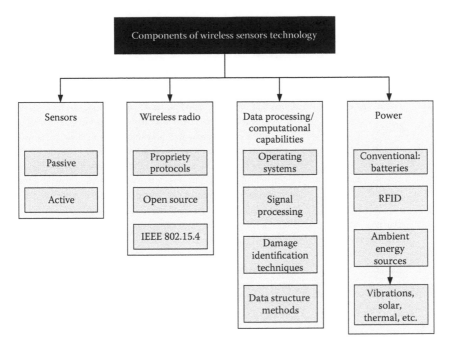

FIGURE 5.58 Wireless sensor components. (Based on Lynch, J. and Loh, K., *Shock and Vibration Digest*, 38(2), 2006. With permission.)

TABLE 5.8
Importance of Components of Wireless Sensors to SHM

Wireless Components	Applicability to SHM
Data acquisition specs.	Frequency limits would affect detection frequency ranges. Sensitivity of sensors would affect measured amplitudes
Embedded computing specs.	Can be extremely efficient in producing real time or near real time results.
Wireless channel specs.	Can affect amount of transmitted data
Physical attributes	Size and ruggedness of the sensors can affect longevity of the SHM project
Actuation specs. (for active sensors)	Higher sensitivity can produce more accurate experiment

computational capabilities can play a major role in the efficiency of the whole monitoring system. By processing the data locally (e.g., by generating frequency spectra from the measured temporal data) and reaching preliminary, or final, conclusions the computational demands on the central data processing resources are reduced. Another important component of wireless units is power source. Several power sources can be used by wireless sensing systems. This includes conventional batteries, radio-frequency identification (RFID), ambient energy sources, solar, vibration, thermal, and so on. The different components of wireless sensors are offered in detail by Lynch and Loh (2006)

5.6.1.3 Wireless Sensors and SHM

This section discusses the different modes of interrelationships between wireless sensing and SHM. As discussed above, wireless sensing contains several components. The importance of these components in an SHM project is discussed in Table 5.8. The utility of wireless sensors in SHM project is discussed in Table 5.9. Table 5.10 compare tethered and wireless sensors from SHM view point. Finally, Table 5.11 shows the applicability of wireless sensing and different NDT/SHM technologies. Clearly, the utilization of wireless sensing enables the use of NDT methods in the more general and demanding SHM projects.

TABLE 5.9
Use of Wireless Sensors in SHM Projects

Component	Issue	SHM Applicability
Data processing/ computational capabilities	Operating systems	
	Signal processing	Applications to wavelet and Fourier Transform methods are well proven (see Pendat and Piersol 2010). Need to develop wireless-sensing-specific methods that take advantage of potential of large number of sensor network (see discussion of this subject later in this section)
	Structural identification	Applications to conventional structural identification methods are well proven (see Lynch and Loh 2006). Need to develop wireless-sensing-specific methods that take advantage of potential of large number of sensor network (see discussion of this subject later in this section)
	Damage identification	Need to develop wireless-sensing-specific methods that take advantage of potential of large number of sensor network (see discussion of this subject later in this section)
Power	Batteries	Use of conventional batteries might be adequate for short term and small number of sensors in an SHM project. For long term and/or large number of sensors, a new generation of longer lasting batteries is needed
	RFID	A promising technology for utilization in SHM projects Deployments in realistic and large scale SHM projects, such as bridges, have not been demonstrated yet
	Ambient energy sources	Promising technologies that might have a great impact on the wider usage of wireless sensors should be further explored

TABLE 5.10
Comparison between Wireless and Tethered Sensors

Issue	Tethered	Wireless
Connecting hardware	Extensive demands of wiring, safety, and expense	Limited demands
Power	Power is supplied via wires	Power is provided via batteries, or via innovative methods such as solar, or ambient vibrations
Sensors (Type)	Most sensors can be used in either mode	
Sensors (Number)	Number of sensors can be limited due to tethering demands	Higher number of sensors can be used wirelessly
Labor	Maintaining the tethering equipments requires higher labor effort	
Overall cost	Cost per sensor is lower for wireless sensing	
Accuracy	Tethered sensors provide accurate sensing results	Wireless sensing accuracy has been improving

5.6.2 CASE STUDIES

5.6.2.1 Trout Brook Bridge

Gangone et al. (2006) showed in a real-life experiment some of the important attributes of wireless sensing. The experimental bridge is located in St. Lawrence County in New York State. It is a 56-ft-span integral abutment bridge. The reinforced concrete deck is supported by four steel beams that are cast into the abutment. Ten accelerometers and ten strain gauges were mounted wirelessly on the steel beams, as shown in Figure 5.59. Figure 5.60 shows the actual layout of the sensors. Figure 5.61 shows the wireless arrangement of both strain and acceleration sensors. The measured accelerations were processed and STRID analyses were performed to find the experimental natural

TABLE 5.11
Applicability of Wireless Sensors to Different NDT/SHM Technologies

NDT/SHM Technology	Applicability of Wireless Sensors to NDT/SHM Technology
Ultrasound	Fairly suited to wireless sensing
Acoustic emission	
Corrosion	
Microwave	Due to security need, wireless sensing might not be suitable for use with NDT microwave
Radiography	processes
Fiber optics	A combination of wireless and fiber optic sensing might not be feasible. However, such combination might not be very efficient
Vibrations	Fairly suited to wireless sensing
Thermography	A combination of wireless and thermography might be feasible
Eddy current	Eddy currents require manual expertise, as such it might not be suited to wireless sensing

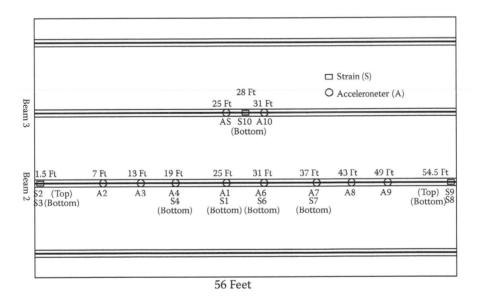

FIGURE 5.59   Sensor locations. (Courtesy Dr. Jerome Lynch.)

frequencies and mode shapes. These frequencies and mode shapes were compared with analytical (FEA) results as shown in Figure 5.62. A good comparison shows the adequacy of wireless sensing in this SHM project.

5.6.2.2   Geumdang Bridge

Some of the objectives of this project by Lynch et al. (2006) were (a) to investigate different aspects of cost-effective wireless accelerometer sensors, and (b) to compare the wireless sensor results to conventional tethered sensors. The sensors utilized in the project are shown in Figure 5.63. Figure 5.64 shows the assembled unit. The test bridge is the Geumdang Bridge in Korea. It is a single box girder concrete bridge (Figure 5.65). The wireless and tethered sensor arrangements are shown in Figure 5.66. A typical installation of a wireless sensor is shown in Figure 5.67. The collected accelerations from both wireless and tethered sensors were processed, and a follow-up STRID produced the frequencies and mode shapes of Figures 5.68 and 5.69 for the wireless and tethered

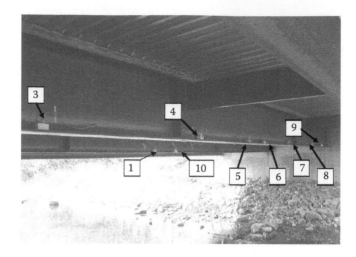

FIGURE 5.60   Wireless sensors layout on bridge. (Courtesy Dr. Jerome Lynch.)

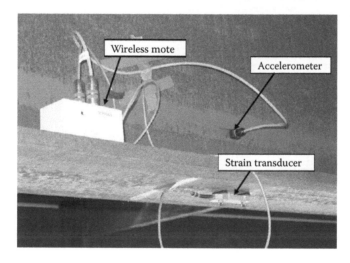

FIGURE 5.61   Details of wireless arrangements. (Courtesy Dr. Jerome Lynch.)

measurements, respectively. The good agreements between the two systems indicated the accuracy of the wireless system.

5.6.2.3   Benjamin Franklin Bridge, PA

The Benjamin Franklin Bridge—connecting Philadelphia, PA and Camden, NJ—was opened to traffic in 1926. It supports seven lanes of traffic and two outboard tracks that support a transit rail service. On the basis of biennial inspection and follow-up structural evaluations, two cantilever floor beam members were suspected to have potential fatigue problems under the train loading of the rail service. The analysis performed by Rong and Cuffari (2004) indicated that stresses were near the fatigue threshold and hence it was decided to use monitoring to ensure that the actual field stresses are below fatigue threshold.

The authors described an SHM system. The system data collection rate, sensor locations, number of sensors, and triggers for data collection (to collect only the data needed when strain exceeds certain level) were designed based on the structural analysis and loading analysis (train speed, type of data, etc.). Actual sensors (strain gages) were wired and were collected to data loggers (nodes)

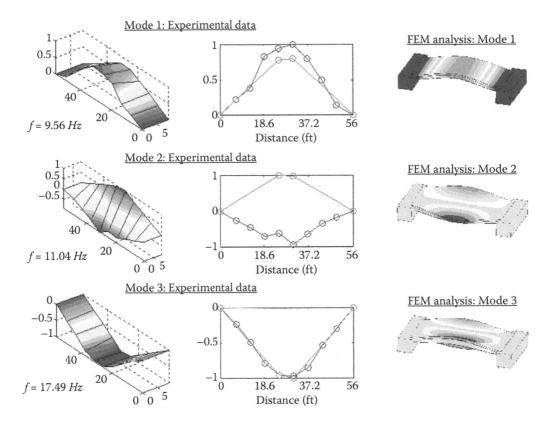

FIGURE 5.62 Measured and analytical results. (Courtesy Dr. Jerome Lynch.)

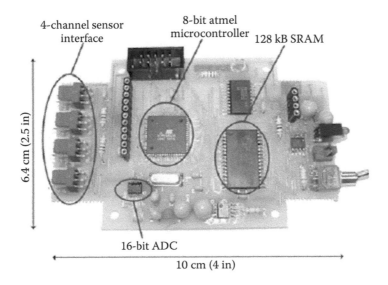

FIGURE 5.63 Wireless sensor printed board. (Courtesy Dr. Jerome Lynch.)

FIGURE 5.64   Assembled wireless unit. (Courtesy Dr. Jerome Lynch.)

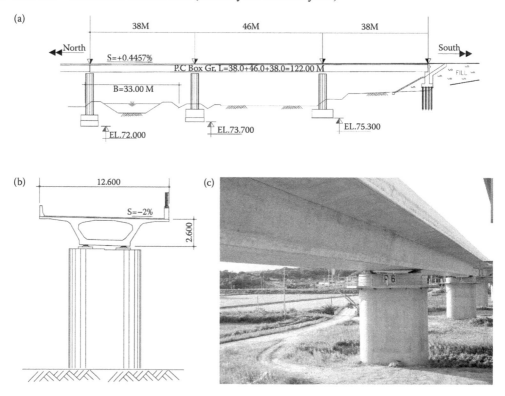

FIGURE 5.65   Views of Geumdang Bridge (a) elevation, (b) cross section, and (c) general view of bridge. (Courtesy Dr. Jerome Lynch.)

within 20 ft of the sensors. The data from these nodes were transmitted to two base stations wirelessly and to personal computer from base stations in similar fashion.

The measured strain data was about 60% lower than the predicted values and thus it was determined that the fatigue is not an issue. The project illustrates the importance of stating an objective, and designing the process around such an objective. It also illustrates how wireless sensing can be used effectively in collecting project data.

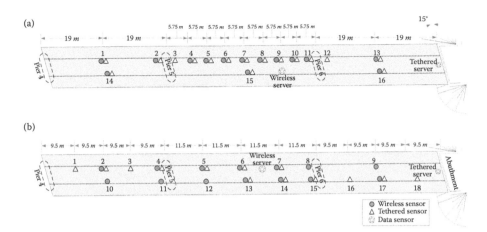

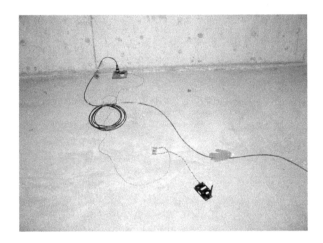

FIGURE 5.66   Wireless and tethered sensor locations. (Courtesy Dr. Jerome Lynch.)

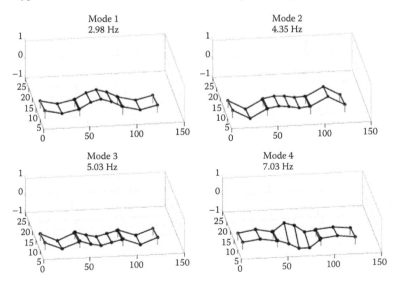

FIGURE 5.67   Typical installation of wireless sensor assembly. (Courtesy Dr. Jerome Lynch.)

FIGURE 5.68   First four modes—wireless sensors. (Courtesy Dr. Jerome Lynch.)

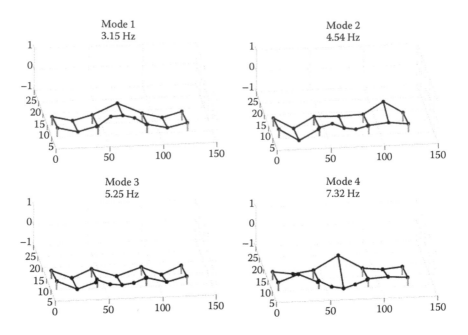

FIGURE 5.69   First four modes—tethered sensors. (Courtesy Dr. Jerome Lynch.)

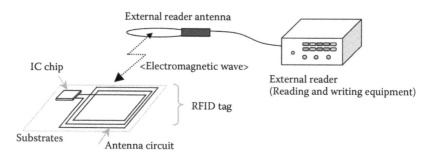

FIGURE 5.70   Basic system of RFID techniques. (Reprinted from ASNT Publication.)

5.6.3   WIRELESS SENSORS–SPECIFIC METHODS AND TECHNIQUES

5.6.3.1   RFID in Civil Infrastructure

Satou et al. (2006) discussed wireless sensors using passive RFID techniques that are perceived to be very useful for bridge management purposes, in the future, and this chapter investigates the fundamental properties of an RFID tag to be attached to a reinforcing bar within the concrete (see Figure 5.70). It also explores some of the possible applications for this technology.

The results indicate the following:

1. If RFID tag is placed inside the structure behind a reinforcing bar, it shortens the communication distance significantly as the bar absorbs and reflects (see Figure 5.71) electromagnetic wave and also changes the external reader's resonant frequency (see Figure 5.72)
2. Concrete has little influence on the communications distance with the RFID tag embedded inside it (see Figure 5.73)

The authors reported that the RFID technology was successfully used, experimentally, in measuring the concrete temperature during the hardening process and detection of duct grouting.

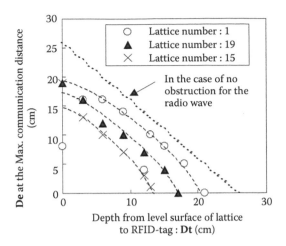

FIGURE 5.71 Influence of lattice frame with reinforcing bars on communication distance. (Reprinted from ASNT Publication.)

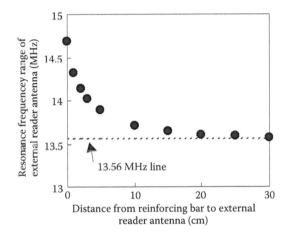

FIGURE 5.72 Influence of reinforcing bar on resonance frequency of external reader antenna. (Reprinted from ASNT Publication.)

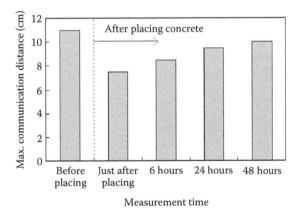

FIGURE 5.73 Influence of concrete on communication distance. (Reprinted from ASNT Publication.)

Another system concept that is based on radio frequency (RF) interrogation was presented by Thompson and Bridges (2002). The sensor is wireless, and require no power. It can measure strains, and can be embedded for continuous monitoring. It belongs to the "smart pebble" family of sensors.

5.6.3.2 Web-Based Systems

Lynch et al. (2009) presented an overview of the potential of using cyber technologies in SHM. The authors advocated the integration of different sensing technologies to collect pertinent data and streaming it through the internet where data are processed. Decision making operations can then be performed to help bridge managers in reaching appropriate decisions.

Shenton et al. (2004) provided an efficient and cost-effective system that is made from off-the shelf components for quick short-term field testing using strain gages. The system known as in-service bridge monitoring system (ISBMS) is a small battery-operated system to measure live loads strains in a girder or slab for up to 3 weeks unattended. It can be accessed via the web for downloading the data using an interface. The system has three modes of operation: peak detect (measures peak strains that exceed a user specified trigger threshold), waveform (dynamic strain history of strain of a triggered event such as passing vehicle), and rainflow (count cycles of randomly varying strain). The system can be set up in hours and utilizes BDI full bridge strain gage or a quarter bridge foil strain gage. 12-V-rechargeable battery powers the modem and a 2-V-battery powers the strain gage. Figure 5.74 shows the ISBMS as mounted to a bridge. Figure 5.75 shows a sample recorded time history.

This approach can be very useful for cost-effective fatigue evaluation, permit vehicle presence checking, and load rating analysis.

5.6.3.3 Structural Identification

One of the potential advantages of wireless sensing over tethered sensing is that wireless sensing offers an almost unlimited potential in the number of sensors that are used in any particular SHM experiment. This brings the question of whether there is really no limit to the number of wireless sensors in experiments, such as the Geumdang Bridge or the Trout Brook Bridge cases that were discussed above. To explore the presence of such limit, let us describe the overall problem first.

Assume that the purpose of the experiment is similar to the two cases presented earlier: to measure dynamic responses of the bridge and identify its modal behavior (mode shapes and natural frequencies). For simplicity, let us assume that the bridge of interest can be assumed to be a simple supported beam. As such, we can express the nth natural frequency of the bridge as (Timoshenko 1955).

$$f_n = \frac{\pi n^2}{2\ell^2}\sqrt{\frac{EIg}{A\gamma}} \qquad (5.45)$$

With E, I, ℓ, A, and γ representing the modulus of elasticity, moment of inertia, length, area, and unit weight, respectively. Next, we assume that the sampling frequency of each of the sensors is f_{SAMPLE}. This implies that the highest usable frequency f_{USABLE} of the project is approximately

$$f_{USABLE} = f_{SAMPLE}/2 \qquad (5.46)$$

Note that we used a factor of 4 to relate usable and sampling frequencies. Such a factor might be increased if higher measuring reliability is required. Using Equations 5.45 and 5.46 above, we can compute the theoretical upper limit of natural frequency/mode that can be identified in this project as

$$n_{Max} = \sqrt{f_{SAMPLE}\frac{\ell^2}{\pi}\sqrt{\frac{A\gamma}{EIg}}} \qquad (5.47)$$

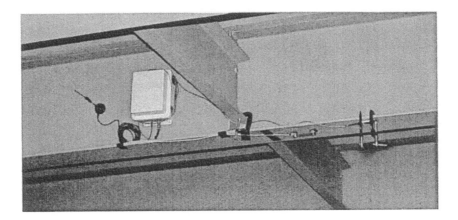

FIGURE 5.74   Photograph of the ISBMS mounted to a bridge. (Reprinted from ASNT Publication.)

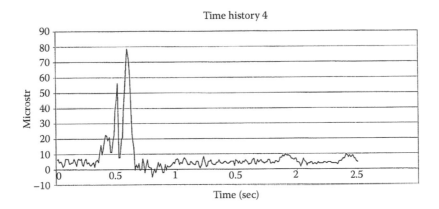

FIGURE 5.75   Sample time history recorded by the ISBMS using the time history program from Bridge I-704. (Reprinted from ASNT Publication.)

To get realistic values on the theoretical limit on the usable number of modes and the sensor's properties (sampling frequency) let us consider a rectangular simply supported reinforced concrete beam. The relations between sampling frequency and n_{Max} are shown in Table 5.12. We reached an important observation that relates sensors and STRID: the theoretical limit of the number of identified modes is governed by the sensor properties.

To continue this line of development further, following Timoshenko (1955), the first few natural modes of a simply supported beam can be illustrated as in Figure 5.76. We propose, rather unconservatively, that we need one sensor for each natural mode to be identified. We can then reach an otherwise interesting and important conclusion: the maximum number of sensors that can be used efficiently for this experiment is n_{Max}. Any sensor that is used in excess of n_{Max} sensors can't be utilized efficiently, since it can't provide additional useful information to the experiment. Of course, additional sensors can improve the accuracy of the identified results.

We should note that the above developments and example applies for both tethered and wireless sensors. However, since the practical limit of the number of tethered sensors is much smaller than the practical limit of the number of wireless sensors, the upper limit that is described by Equation 5.47 will apply mostly to wireless sensors.

Note that the above development was based on assuming that the maximum usable frequency is half of that of the sampling frequency. In practice the ratio might be as larger, that is, the maximum usable

TABLE 5.12
Theoretical Limits on Identified Modes

f_{SAMPLE} (KHz)	Width (in)	Depth (in)	A (in²)	γ (lb/in³)	E (psi)	I (in⁴)	g (in/sec²)	ℓ (in)	n_{Max}
50	18	36	648	0.086806	3.00E + 06	69984	386.4	240	20
100	18	36	648	0.086806	3.00E + 06	69984	386.4	240	28
200	18	36	648	0.086806	3.00E + 06	69984	386.4	240	39

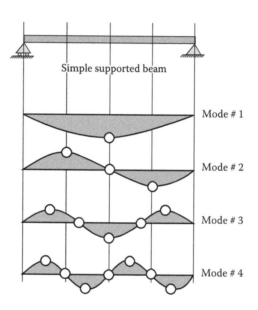

O Location of sensors needed to identify this mode accurately

FIGURE 5.76   Sensors needed for modal identification.

frequency might be taken as one-third of one-fourth the sampling frequency, to ensure better accuracy. Doing so will reduce the maximum efficient number of sensors in the experiment. Another factor is the number of additional sensors needed for each additional frequency. In the simple beam example, we assumed that a single additional sensor is needed for each additional mode shape. For complex structural geometries the additional number of sensors for additional frequencies might be larger than one sensor. This will result in an increase of the maximum efficient number of sensors in the experiment. However, the main conclusion remains that there is an upper bound of the number of sensors that can be used to identify the structure using modal identification methods. Any additional sensor beyond that optimum number will not provide any useful additional information. Figure 5.77 shows the steps that can be used to evaluate this maximum number of usable sensors for general geometry.

5.6.3.4   Damage Identification

As wireless sensing offers an immense potential for increasing the number of sensors, question regarding spatial resolution of those sensors and their arrangements arise. We observe that for every sensor, there is a zone in which the sensor can monitor its ToM. We call this zone as the active sensing zone (ASZ). For example, the ASZ can be global, for a displacement sensor, or local for a strain or AE sensors. In addition, we observe that there is an active damage zone (ADZ). The ADZ is the zone at which a damage, or response, can be felt. The presence of the damage outside an ADZ will not be felt. For large damage sizes the ADZ is large and the reverse is true. Figure 5.78 shows

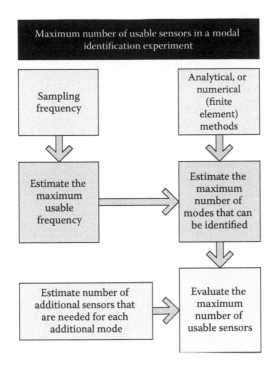

FIGURE 5.77   Maximum number of usable sensors for structural identification SHM experiment.

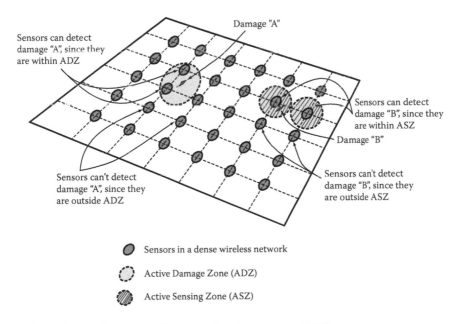

FIGURE 5.78   Active sensing zone (ASZ) and active damage zone (ADZ).

conceptually the ASZ and ADZ. For accurate sensing, the sensor network needs to consider the interrelationships between ASZ and ADZ as shown in Figure 5.78.

Hay et al. (2008) report that AE technique has been used to monitor several bridge locations as identified in Figure 5.79. Authors propose a *conceptual* wireless AE monitoring system to replace wired AE continuous monitoring system. The field systems have 16-channel data acquisition

Hanger connection   Link pin connection

Cope and stringer   Welded stiffener

FIGURE 5.79   Representative areas monitored by acoustic emission. (Reprinted from ASNT Publication.)

capability with pattern recognition and AE data interpretation methodologies. The power management strategies using hibernation mode is described with wake-up mechanism based on the parametric input of a strain sensor.

5.6.3.5 Remote Programmability

Remote programmability of wireless sensors was addressed by Arms et al. (2004). They described a methodology used to develop modular wireless sensor network capable of wireless reprogramming using cell phone network (see Figure 5.80).

The authors then give more details of the system used in the previous chapter. Nodes distributed over the structure report to base station located near the structure. A cellular phone link from the base station allows the sensing network to be controlled and reprogrammed from an office elsewhere.

5.6.3.6 Decision Making

Chen et al. (2004) presented a simple methodology, assuming the bridge behaves similar to a single degree of freedom (SDOF) system, to get the stiffness change in the bridge by estimating change in frequency. Once the stiffness is estimated, the load capacity is estimated by multiplying stiffness with the allowable deflection. A moving car (using wireless drive-by network) is used to induce ambient vibrations with one sensor to estimate the frequency. For this method to work, initial stiffness should be known and also considering lot of other factors influence the frequency. In addition, this research explored the value of utilizing monitoring systems. Figure 5.81 shows sample computational cost results.

5.7 SMART STRUCTURES

5.7.1 What Is a Smart Structure?

"Smart structures" (SS) is a term that has increased in use lately. There are several definitions of the term. For our purpose, we define SS as those structures that can sense changes in their environments

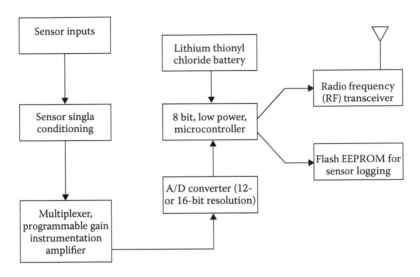

FIGURE 5.80   Wireless sensor node block diagram. (Reprinted from ASNT Publication.)

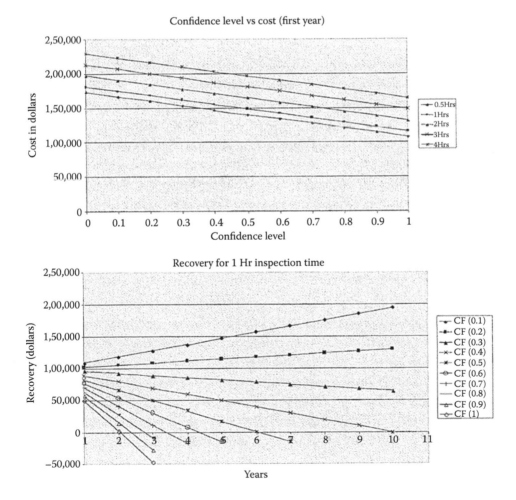

FIGURE 5.81   Cost estimation results. (Reprinted from ASNT Publication.)

(such as loading conditions, environmental conditions, and or damaged conditions). Moreover, a SS might be able to respond to such changes in environment by alerting the stakeholders of the structure, or by actively changing some of the structural properties so that the harmful effects of those changes are mitigated.

Upon closer inspection to the above definition of SS, it becomes clear that a SS is simply a structure that has an implemented SHM/IHCE system. The only requirement is that the SHM/IHCE system should have longer-term objectives, rather than short-term objectives.

5.7.1.1 Smart Structures and Smart Materials

Smart composite materials were categorized into three levels of intelligence by Huston (2007) as shown in Table 5.13. Huston correctly argued that the levels of complexity and capability increases as the material intelligence level increase. We would like to generalize such levels such that they are applicable to SS. First, we note that composite materials, as manufactured materials, are fairly different from other main engineering materials (steel, concrete, masonry) in the following two main points:

- The mass of composite materials are usually less than the mass of other engineering material. This enables miniature actuators (level II) to be effective on material level.
- The chemical and other processes that are used/can be used in material self-healing (level III) are more feasible for composite materials than for steel, concrete, or masonry.

So, it might not be easy to identify steel, concrete, or masonry as potentially feasible smart material. However, we need to remember that a structure is a composite tapestry of different components. There are distinct differences between structural and material behavior; the following are among them:

- The structure contains several substructure components that are assembled together by a set of joints and connections. Good part of the structural damage occur at or near those joints/connections.
- Structural damage also can occur due to geometry (stability problems) rather than material damage.
- Structural failures were reported also due to damages of supporting ground (see Chapter 1 in Ettouney and Alampalli 2012).

Considering all of those differences, we still can identify three intelligence levels for the structure. For example, we propose to consider the relatively small effective structural mass, thus the structure can be made to react to stimuli as in level II. We also propose that a structure can induce self-healing processes in a way similar to a composite material changing its color. We can introduce Table 5.14 as the structural equivalent to Table 5.13.

TABLE 5.13
Levels of Smart Composite Materials

Intelligence Level	Smart Material
Level I	Sensing
Level II	Sensing + reacting so as to change geometry/shape
Level III	Sensing + reacting so as to change properties (self-healing, color, stiffness, etc.)

TABLE 5.14
Attributes of Smart Structures

Intelligence Level	Smart Structure Function	SHM Components
Level I	Sense	Sensors/instrumentations
Level II	React to stimuli: active control measures	All four components of IHCE are needed
Level III	React to stimuli: recommend healing methods	All four components of IHCE are needed

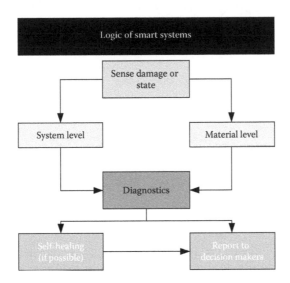

FIGURE 5.82 Smart systems logic.

What we are suggesting here is that a complete IHCE project (four components) is actually a SS system (see Figure 5.82). The structural material need not be a smart material for the structure to be a SS.

We now discuss the revere situation: can every structure that contains smart material be considered a SS? Let us consider one of the most important problems in structural behavior: structural stability. It is well known that structure become unstable because of their geometry. Stability problem is a brittle-type problem: the structure can become unstable suddenly, without major material damage. In such situation, smart materials might not be capable of detecting the impeding instability situation. On the other hand, carefully placed and monitored structural vibration sensors might be capable of detecting the upcoming instability situations (see Chapter 8).

We now reached an interesting conclusion: use of smart material in a structure does not guarantee a safe structural behavior; only a carefully designed smart structural system, that may, or may not take advantage of smart material technologies, can provide such a guarantee.

5.7.2 SMART STRUCTURES AND IHCE COMPONENTS

Sensing: Sensing is a common issue for all levels of SS. We need first to recall the comprehensive sensing description of smart material by Huston (2007). In that work, the smart material sensing main attribute was that it is distributed within the material itself (see Figure 5.83). Huston described several topologies for such distribution such as (a) linear (fiber optic sensing), (b) grid, and (c) random. The sensing technologies described by Huston ranged from electromagnetic to optics. SS sensing will need those smart materials sensing attributes. In addition, more sensing attributes are needed to account for the more complex features of structures as compared to materials, as shown in Table 5.15.

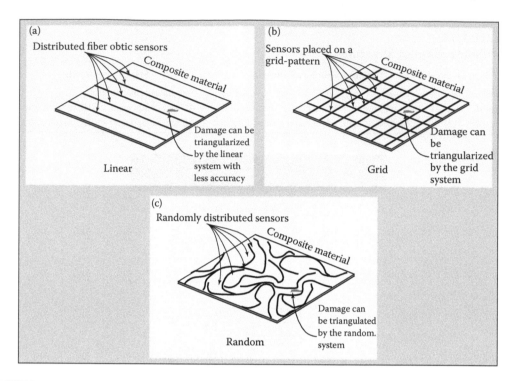

FIGURE 5.83   Geometries of sensors in composite smart materials (a) linearly distributed FOS, (b) grid-based FOS, and (c) randomly distributed FOS. (Parts (b) and (c) are courtesy of Dr. Dryver Huston.)

TABLE 5.15
Sensing Attributes of Smart Materials and Structures

Sensing Attribute	Smart Material	Smart Structure: In Addition to Smart Material, also:
Time	Intermittent and/or continuous	
Locations	Material body	Connections, joints, foundations, decks, local, and global
ToMs	Material strains/stresses, motions, humidity, temperature, etc.	Behavior of Soil, connection/joint global motions, etc. Also demands such as loadings, seismic motion, wind speeds, flood levels, etc.
Scope	Local in nature	Global

STRID: Structural identification tools are needed for levels II and III. Examples of potential uses of STRID in SS are shown in Table 5.16.

DMID: Damage identification is an integral part of any smart system, albeit smart material or SS. Several authors categorize different levels of damage identification as (a) existence of damage, (b) location of damage, (c) extent of damage, (d) effects of damage, and (d) different ways to respond to # 1 through 4. The first four levels are general enough to be applicable to smart material and SSs (we recognize the fifth level to be a decision making issue that is discussed below). However, identification of damage in SS can, in a very general way, be considered as a superset of that in smart material. As was discussed above, identifying material damage is necessary, but not sufficient, in a structural sense. Table 5.17 shows the differences between smart material and SS when identifying some types of damage.

Decision Making: The type of decision making and its consequences is what differentiate between smart structures levels II and III. In general, level III decision making is more advanced and complex than level II.

TABLE 5.16
STRID for Smart Materials and Smart Structures

STRID Method	SS Applications
ANN	ANN to estimate seismic damages in real time immediately after a seismic event and reported to officials
Modal identification	Use of chaos technique to estimate potential loss of structural stability
	Measure natural frequencies to estimate potential loss of structural stability (see Chapter 8)
Parameter identification	Use FEM in evaluating critical scour depths and set up a threshold limits at which the officials will be notified

Note that STRID is utilized for smart structures levels II and III.

TABLE 5.17
DMID for Smart Materials and Smart Structures

Type of Damage	Smart Material	Smart Structure
Fatigue cracks	Yes: require special attention for joints and connection	Yes
Wear and tear	Yes	No
Soil erosion (scour, liquefaction, etc.)	No	Yes
Corrosion	Yes	No
Overload: general	Yes	Yes
Overload: stability conditions	No	Yes

Note that DMID is utilized for smart structures levels II and III.

5.7.3 EMERGING PARADIGMS FOR SMART STRUCTURES

As structures continue becoming "smarter," the need for new methods of sensing becomes even greater. One potential limitation of conventional sensing is the quantification of the reliability of sensing. Bergmeister (2004) discussed the general behavior of different sensors as well as data acquisition techniques. The durability of sensors and probabilistic analysis of results were also of a concern to the author: the reliability of the sensor measurements must be of concern to officials to ensure accurate decision making process. A novel approach to accommodate several sensing limitations was offered by McNiel and Card (2004). These authors used neural network technique for data analysis of sensor measurements. Another technique that aimed at minimizing sensing errors by careful selection of sensor locations was presented by Weber et al. (2004).

The above shows that there are many inherent limitations of conventional sensing. Huston (2002) explains the limitations of regular sensors when used in field validation. The author then gives some practical suggestions.

- Owners lack enough data on structures due to lack of good sensing systems. Sensing system must be easy to use, reliable, and cost effective.
- Issues of geometry and timescale complicate implementation of a sensing system. Structures are large, yet often the key damage is localized.
- Many of the items of interest are inaccessible.
- Their life is very large compared to natural hazard's short time scale, where as some damages such as fatigue and corrosion can occur most of the structures' life.
- A dense array of systems can be expensive, overload of nonessential info in most cases, and hard to maintain.

The author then observes the need for adaptive sensing systems. Sensors with electromagnetic waves as part of their transduction and transmission system offer the possibility of reconfigurable sensing because of the freedom with which the geometry of the system can be readily changed. One approach to adaptive sensing is to arrange sensors to be sensitive to failure modes or other important behavior of structures. If these modes/patterns are predictable, sensors can be placed at appropriate locations. But, one needs good understanding of structural mechanics for this to be successful. Thus:

(a) Sensor arrays with local signal processing: Sensor data is processed locally and send only important/required data/results to central system. An example is passive peak strain sensor, where only the max strain at allocation during a particular timeframe is recorded.

(b) Imaging system with squinting: Here first the whole system is scanned coarsely and then interesting/suspected areas are scanned at higher resolution. An example is beam crawling robotic inspection system. GPR and IRT also fall in the same category.

5.8 OPTIMUM SENSOR SELECTION

5.8.1 OVERVIEW

This chapter shows a very clear message: there are numerous types of sensors, and there are even greater numbers of sensors that can perform similar functions. The different technologies of strain sensors, for example, are very evident. Figures 5.11 through 5.16 show several types of strain sensors. This leaves the professionals in charge of SHM project in a very good position: there are many options to choose from. Unfortunately, this begets the question: of the many sensor choices that are available, which sensor is most appropriate? It is clear that every experimental of monitoring situation is different in scope, objectives and means. The choice of the wrong sensor can lead to a less-than-optimum result. Currently, choosing sensors is mostly a qualitative process that is based on the experience of the professional. As usual, such a qualitative process has the advantage of simplicity; unfortunately it runs the risk of producing an inefficient process and perhaps even missing the goal of the project.

In this section, we explore this situation. First we enumerate different parameters that can have an effect on the sensor performance within a given SHM experiment. We then introduce two optimum sensor selection (OSS) methods that utilize those parameters. One of the methods is quantitative, which is based on decision tree analysis. The second OSS method is qualitative that is based on the judgment of the user. We present examples for the use of both methods

5.8.2 GENERAL PARAMETERS

There are numerous parameters and factors that affect the choice of an appropriate sensor for a given experiment. We will discuss some of the parameters that are general enough to be applicable to most situations. We observe that there can be more special parameters that are applicable only to specific monitoring situations. We will also discuss some of those specific situations at the end of this section. Also, note that we offer more guidelines on sensor selection that are more specific to bridge load testing in Ettouney and Alampalli (2012).

5.8.2.1 Objectives of Sensing/Project

Objectives of sensing are an important factor since it is not always possible to use the same sensor in each test. For example, many of the commercially available inclinometers require longer response time to obtain stable data, therefore, sampling speed or load cycle time is important and should be given special consideration. Sensor selection depends on loading type and expected structural response in each test. In many cases, different installation procedures and adhesive types may be required, depending on whether static or dynamic load effects are measured, so that gages may survive structural response duration. Because of all of the above considerations objectives of sensing is a common and important parameter in any/all SHM experiments. Different types of possible objectives are shown in Table 5.18.

5.8.2.2 Type of Structure

Type of structure to be monitored affects the type of sensor as it may have to match the structural material properties condition. For example, prevailing standards and specifications should be considered. In steel bridges, drilling holes for gage installations in tension zones of bridge girders may not be permitted. Strain gages are made with materials to match thermal conductivity of structural materials to increase accuracy and reliability of measured data.

Table 5.19 shows different structural properties and how they can affect sensor selection process.

5.8.2.3 Physical Parameter for Sensing (ToM)

Type of physical parameter plays a major role in selection of sensor and all other aspects of testing. Some parameters such as deflection can be measured directly where as others can be measured only indirectly. For example, stress in most cases cannot be measured directly where as pressure can be, even though both are similar (see Table 5.20).

5.8.2.4 Monitoring Time Span

Monitoring time span and number of times a sensor is planned to be used also play a role in the selection of equipment/sensors, costs, and other factors; these are briefly discussed in Tables 5.21 and 5.22.

5.8.2.5 Operating Environment

Operating environment plays a major role in choosing a sensor. Specially, when one considers the fact that many of the available sensors are designed for laboratory applications and may not sustain

TABLE 5.18
Objectives of Sensing/Project

Objective	Comments
Type of decision (consequence)	Examples:
	Just a warning to act on at leisure
	Warning requiring immediate attention
	Avoiding life threatening condition
Research	Examples
	Behavior of new material in the system
	Distribution coefficients on different girders
	State of composite action
Validate design assumptions	Examples
	Deflections of a prestressed concrete bridge
	Strains of an innovative connection
	Properties (strength/stiffness) of a newly rehabilitated system
Cost implications	Examples
	Traffic measurements
	Corrosion threshold
	Deterioration (e.g., extent of cracks)
Hazard-specific (earthquake, scour, wind, etc.)	Mitigation efforts, evaluation, prediction, etc.
Security	Deterrence, defense, etc.

TABLE 5.19
Type of Structure

Objective	Some Parameters to Consider
Material type (steel, concrete, frp, aluminum)	Material characteristics (thermal conductivity, coefficient of expansion, weldability, adhesive properties, conductivity etc.)
	Gage length to accommodate strain distribution
Design type (box beam, I sections, slab, etc.)	Accessibility, strain distribution across sections, protection from environment, structural stiffness affecting sensor range, signal issues
Design life of structure	Sensor life, sensor durability, power requirements
Size (large, small)	Number of sensors, sensor size, wired vs. wireless
Connections	Range of expansion, space requirements, access, sensor protection
Construction type	Access, sensor protection, sensor installation phases, wiring sequences, sensor protection
Underground, over ground	Access, sensor protection
Below water or above water	Sensor protection, corrosive resistance, wiring protection
Construction equipment used	Sensor durability, sensor protection, wired vs. wireless sensors, installation costs

TABLE 5.20
Physical Parameter for Sensing

Objective	Comments
Strain	Gage type (Foil strain gages, strain transducers, vibrating wire, strain sensor range, and resolution)
Stresses	Mostly measured from strain gages or by measuring loads (forces)
Temperature	Thermo couples, thermography
Humidity	Humidity sensors
Vibration (acceleration/ velocity/displacement)	Sensor type (accelerometers, velocity transducers, LVDTs), range and resolution, reaction time, frequency range
Pressure	Unlike stress, can be measured directly using pressure sensors
Inclination	Inclinometers, relatively hard to install

TABLE 5.21
Monitoring Time Span

Objective	Comments
Short	Sensor response time, wired vs. wireless
Long	Sensor durability and type, data range, sensor and equipment protection, amount of data and storage, power requirements
Continuous	Sensor type, data thresholds and triggers, environmental protection, sensor and equipment protection, data storage, maintenance
Periodic	Data continuity, sensor durability, repeatability

TABLE 5.22
Durability

Objective	Comments
One time use	Durability is not an issue, need low cost sensors
Multiple use	Environmental durability, installation issues, repeatability, documentation

inclement weather conditions without proper protection. However, there are a few rugged sensors specially manufactured for use in variable or hostile environments that are more suitable in bridge testing. If these sensors are unattainable, proper weatherproofing (against temperature, dust, water, wind, vibration, etc.) should be specified to meet necessary requirements for sensor operation. Even though they are relatively expensive, rugged gages are a must in long-term monitoring. When weatherproofing is required, the entire measurement system, including connecting cables, should be properly protected.

Among operating environment issues is the loading conditions: Some sensors do not operate effectively under certain types of loading. Each sensor takes a brief time (reaction time) to stabilize, depending on its principle of operation. For example, many LVDTs and inclinometers require longer reaction time compared to strain gages. If structural response changes faster than the sensor reaction time, accurate data cannot be collected even when high sampling speeds are used. If transient or dynamic response is to be measured, sensors with short-reaction times should be chosen. Also, when long-term monitoring is planned, possible instrument drift should be determined.

Table 5.23 summarizes many of the important issues in sensing operating environments.

5.8.2.6 Assembly/Maintenance/Disassembling (Excluding Labor Cost)

Preparation for the test, access requirements, disassembly of setup, etc. also play an important role in making decision to proceed with the test or not, test costs, and data quality (see Table 5.24 for some comments on these factors).

5.8.2.7 Cost

Total cost of a test depends on cost of sensors, equipment, processing time, and expertise needed for data analysis, maintenance of traffic required during tests, and other costs as noted in Table 5.25. These costs depend on structural type, complexity, structure location, etc. and all these should be considered from the test planning stages to get more value.

TABLE 5.23
Operating Environment

Objective	Comments
Hot	Operating environment needs to be within manufacturers' specifications. Structural characteristics in different temperatures should be understood and installation issues can be very different. Operator experience could become very important in some cases. Sensor covering and sensor material play a significant role
Cold	
Hot and cold	
Humidity	
Acid-rain conditions	
Chloride environment	
Traffic type and loads	
Lab vs. field	

TABLE 5.24
Assembly/Maintenance/Disassembling

Objective	Comments
Ease of initial erection/assembly	Saves time and cost. Gives more time for actual testing and data verification. Durability of adhesion, etc. can depend of these factors
Ease of disassembling project setup	Saves time and cost
Access	Ease of access of sensors can affect project performance
Maintenance and protection under traffic	Increases cost. Reduces test time and quality

TABLE 5.25
Cost

Objective	Comments
Cost of sensor	Impacts number of sensors
Cost of additional hardware	Impacts costs and need for multiplexing
Cost of initial labor	Can be a major portion for short tests and complex structures, good planning can reduce these costs
Cost of labor during experiment	Mostly skilled expertise with understanding of test subject and instrumentation, critical for test success, can be relatively low for most projects
Other costs	Need careful management to make the benefits outweigh costs

TABLE 5.26
Instrumentation Type

Objective	Comments
Turn-key	Need limited experience and can be used by users with limited experience
Power	Batteries (DC), electric cables (AC), solar, other
Environment needed by instrumentation	Have significant impact on operating costs, e.g., A/C or watertight enclosures
Memory	Higher sensor-level memory capabilities can enhance certain decision making process
Data collection mode	Stored at-site, transmitted to remote site
Interface of sensors with rest of instrumentation	Wired, wireless, individual vs. network
Processing capabilities at site	Sensor-level processing, onsite, offsite, remote

5.8.2.8 Instrumentation Type

It is very important to select sensors that are software and hardware compatible. The excitation voltage, output voltage, wiring scheme, overload handling conditions, and resetting time of the hardware and software should be considered when selecting a sensor. It is possible to obtain sensors of very high accuracy, but if the software and hardware are less reliable, it may be more economical to choose sensors with the same order of reliability.

Table 5.26 illustrates other important instrumentation factors that need to be considered when choosing a sensor.

5.8.2.9 User Knowledge/Experience

Sensor selection also depends on available technical skills. While a large number of sensors can be installed with relative ease, a few installations need more experience and technical skills to obtain reliable and consistent data. For example, some strain gages are manufactured and sold with cables that eliminate the necessity for field soldering and connections. However, they are relatively expensive compared to their counterparts without cables, a trade off for the time saved and the lesser need for gage installation expertise. These parameters are illustrated in Table 5.27.

5.8.2.10 Sensitivity/Range/Resolution

Range of measurements during the entire test period has to be estimated before the sensor selection is made. Accuracy of some sensors varies when used to measure data at different ranges. Some sensors behave linearly only for a small part of their operable range and may behave differently for

TABLE 5.27
User Knowledge/Experience

Objective	Comments
Experienced	NDT Level III or P.E. type of certifications, several years of both direct test experience and test project management experience, etc.
Medium	NDT Level II, several years of direct test and installation experience, etc.
Low	NDT Level I, trainee or limited test experience

TABLE 5.28
Sensitivity/Resolution

Objective	Comments
Loading type	Static vs. dynamic
Processing type	Time domain vs. frequency domain
Decision type	Safety vs. security vs. durability

TABLE 5.29
Sensor Physical Attributes

Objective	Comments
Size	Important to accommodate rapidly varying change in ToM, detail type and size, ToM type, and installation ease
Weight	Not critical for most bridge applications except for some connections
Interaction effects with the structure	Important to obtain the effect due to loading condition alone
Ruggedness	Makes operations easy and a necessity for many civil applications

positive and negative responses. Hence, the sensor's operating range and required accuracy should be chosen with the expected response range in mind.

Resolution is often coupled with data range. Normally, the minimum possible range has to be used to obtain maximum resolution. Most available instrumentation converts analog response to digital response using A/D converters into a fixed number of steps. For example, 20 $\mu\varepsilon$ data range with 12 bit A/D gives a resolution of 0.078 $\mu\varepsilon$. When the same A/D used with 200 $\mu\varepsilon$ range gives a resolution of 0.78 $\mu\varepsilon$. If expected structural response is less than 10 $\mu\varepsilon$, a sensor with a range close to 10$\mu\varepsilon$ should be chosen. At the same time, when the data range is around 200 $\mu\varepsilon$, and if a resolution of 0.078 $\mu\varepsilon$ is required, an A/D with more number of steps has to be used.

Table 5.28 shows some parameters that involves sensitivity/resolution of sensors in an experiment.

5.8.2.11 Sensor Physical Attributes

Physical attributes of sensors can play a major role in an experiment or test, and can affect the accuracy of test results and should be considered carefully (see Table 5.29). For example, selecting the right foil gage that is compensated for the material of the structural member it is adhered to is very important to get the accurate strain the structural member is subjected to. Similarly, to get a better idea of average strain in concrete, a longer gage length is required due to the heterogeneous nature of the material.

5.8.2.12 Formalizing Sensor Choice Parameters

To formalize different methods of sensor choices, let us define the vector$\{SP_j\}$, with SP_j as the jth parameter that affects sensor choices. The counter $j = 1, 2, \ldots N_{SP}$ and N_{SP} is the total number

TABLE 5.30
Sensor Selection Parameters

j^{th}	SP_j
1	Objectives of sensing/project
2	Type of structure
3	Physical parameter for sensing
4	Monitoring time span
5	Durability
6	Operating environment
7	Maintenance (excluding labor cost)
8	Cost
9	Instrumentation type
10	User knowledge/experience
11	Sensitivity/resolution
12	Sensor physical attributes

of parameters. Following the discussion of last section, the different values of SP_j are shown in Table 5.30. This list will be used later in the sensor selection methods.

The above-mentioned guidelines do not cover the entire criteria for sensor selection. Several factors such as cable types, distance between test structure and instrumentation, cable lengths, surrounding environment, and electrical installations are also important in deciding the right sensors for the test. For example, if the cable lengths between instrumentation and sensors are very long, then sensors with built-in amplifiers should be considered to reduce noise/signal ratio. In summary, relying on the above guidelines and good engineering judgment, a variety of sensors can be considered first, and then the most appropriate ones based on application, accuracy, reliability, repeatability, and available instrumentation can be selected.

5.8.3   Decision Tree Method

We can use the sensor choice parameters to help in quantifying processes of choosing the sensors for any SHM experiment. One method is the decision tree method. We present it by an example. Consider an SHM project that calls for sensing strains. The professional found that there are two types of sensors: A and B that can be used in the experiment. It was found that there are four parameters that can affect the decision as follows:

1. Cost: There are two types of costs to be considered in this decision. First: the direct cost of sensors. This is the cost incurred if the sensor fails. The second type of costs is the cost of wrong readings by a sensor. Those types of costs are usually much higher than the direct sensor costs.
2. Environment: Severe environments can have an effect on sensor performance. This is especially true for long-term SHM experiments.
3. Reliability: Sensors can fail due do environmental effects. They can also fail due to other factors.
4. Accuracy: Sensors might give wrong results, thus causing the SHM experiment to fail, with a much higher resulting costs.

Figures 5.84 and 5.85 show the decision trees for sensors A and B, respectively. Each figure also shows the probabilities of occurrence for each branch of each node of the tree. The decision trees also show the different costs of failure, and/or inaccurate readings. For simplicity, we are not considering the benefits of the SHM process in this example: thus, the decision will be based entirely on the above-mentioned parameters and their consequences.

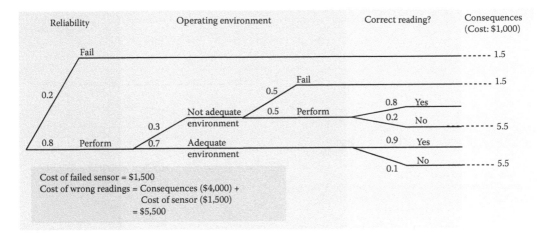

FIGURE 5.84   Decision tree for choice of sensor A.

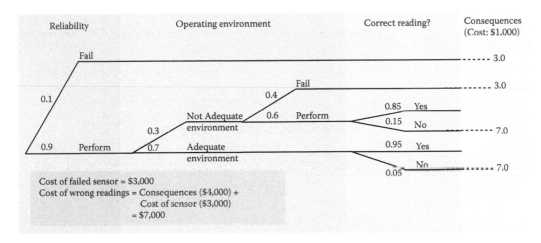

FIGURE 5.85   Decision tree for choice of sensor B.

Using the decision trees approach, the cost of each option is obtained by adding the costs of all paths. Each path cost will be the product of its probability of occurrence times its cost.

Thus, the cost of sensor A can be shown to be $1,800, while the cost of sensor B is $1,900. Obviously, consequences of using sensor A will cost less than using sensor B.

Note that in the above example, the consequences of wrong readings were assumed to be $4,000 (which make the cost of wrong readings = $1,500 + $4,000 = $5,500). Such a cost is fairly close to the cost of the sensors, which biased the total cost towards sensor A, which costs half of that of sensor B. Assume that the consequences of wrong readings were $20,000 instead and that will make it much higher than the direct costs of the sensors. We also assume that the direct costs of sensors A and B are same as before. The new corresponding decision trees are shown in Figures 5.86 and 5.87. Now the cost of using sensor A will change to $3,100. The cost of using sensor B will be $2,900. Now the consequences of using sensor A will cost more than using sensor B. This is due to the more reliable performance of sensor B as compared to sensor A, even though the direct cost of sensor B is twice that of sensor A.

The above example is suitable for use when the number of parameters affecting the sensor choice decision is small. It is also suitable for quantitative decision making. For qualitative decision making, we offer a method that is based on estimating qualitative weights of parameters.

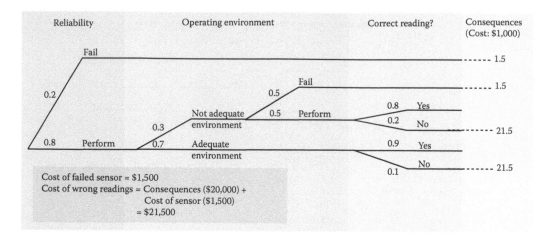

FIGURE 5.86 Decision tree for choice of sensor A, modified example.

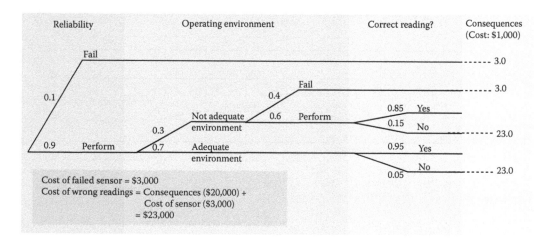

FIGURE 5.87 Decision tree for choice of sensor B, modified example.

5.8.4 Sensor Utility Method

The sensor utility method in choosing sensors relies on the fact that each SP_j has certain relative value or weight WSP_j in the overall project goals. These parameters are dimensionless with $0 \le WSP_j \le 10$. As the suitability of SP_j to the experiment increases, the value of WSP_j increases. If SP_j is not applicable to the experiment $WSP_j = 0$. These weights will differ from one experiment to another. They are assigned by the official, preferably with the advice of all stakeholders of the experiment. Note that the weights are dependent on the type of project, and they are independent from the type of sensors.

We also define another dimensionless factors α_{ij} with $0 \le \alpha_{ij} \le 10$ as weighting factors that measure the suitability of the ith sensor for the jth SP_j parameter in the experiment on hand.

Finally we introduce the general utility $U_S|_i$ of a given sensor i as

$$U_S|_i = \frac{1}{N_{SP}} \sum_{j=1}^{j=N_{SP}} \alpha_{ij} \cdot WSP_j \qquad (5.48)$$

When comparing different sensors for an experiment, the optimum sensor would be the one with the maximum utility such that

$$U_S\big|_{OPTIMUM} = MAX\left(U_S\big|_i\right) \tag{5.49}$$

The advantage of this approach over the decision tree approach is its simplicity. Note that it does not account for consequences of failure, or different probabilities, in a formal fashion: it account for them in a qualitative manner in assigning the different values of α_{ij} and SP_j.

As an example, consider the situation of a load testing of a short-span bridge. The engineer is interested in choosing appropriate strain sensor at the midspan for the test. Table 5.31 shows the different SP_j and the corresponding weights WSP_j that were assigned to the experiment. There were three possible sensors that were available for use in the experiment; we define them as A, B, and C, respectively. After studying different attributes of each sensor carefully, the project management team assigned values of α_{ij} to the three sensors as shown in Table 5.32. From Tables 5.31 and 5.32, and Equation 5.48, the general utilities of the sensors are

$$U_S\big|_A = 35.5$$

$$U_S\big|_B = 28.9$$

$$U_S\big|_C = 30.7$$

It is clear that the three sensors are fairly matched; with the general utility of sensor A is slightly better (higher) than sensor B or C. From Table 5.32, it appears that sensor A has better utility due to its better (optimum) resolution and user familiarity, as well as overall higher scores in other issues even though it costs more than both sensors B and C.

The sensor utility method can be improved by accounting for the inherent uncertainties in some, or all, weights WSP_j and/or α_{ij}. Thus, Equation 5.48 becomes a function of random variables that can be solved using any of the methods of Chapter 8. The uncertainties of sensor utility $U_S\big|_i$ can then be accommodated using the resulting moments. Confidence levels of $U_S\big|_i$ can be used in deciding the

---

TABLE 5.31
Weights of Sensor Choice Issues

j^{th}	SP_j	WSP_j	Comments
1	Objectives of sensing/project	9	Objectives of this project is safety related; thus very important
2	Type of structure	3	The structure is fairly typical, thus this factor is not a governing factor
3	Physical parameter for sensing	3	Strain sensing is fairly typical, thus this factor is not a governing factor
4	Monitoring time span	2	Limited time span
5	Durability	2	Due to the short time span, this factor is not important
6	Operating environment	4	Due to the short time span, this factor is not as important However, bad weather can still have an effect
7	Maintenance (excluding labor cost)	3	Due to the short time span, this factor is not important
8	Cost	9	Important factor
9	Instrumentation type	2	Strain sensing is fairly typical, thus this factor is not a governing factor
10	User knowledge/experience	8	Important factor
11	Sensitivity/resolution	8	Important factor

TABLE 5.32
Factors of Sensor Suitability to Parameters

j^{th}	SP_j	Factors α_{ij} for Each Sensor			Comments
		A	B	C	
1	Objectives of sensing/ project	9	9	9	Generally speaking the sensors must be good match with the project objectives
2	Type of structure	9	3	6	Installation problems, or lack of them need to be considered in this issue. For example: how are the sensors attached to the structure? What are the effects of wireless sensing?
3	Physical parameter for sensing	10	10	10	Must be a good match
4	Monitoring time span	8	8	8	Can the sensor accommodate the different needs of intermittent or long time spans
5	Durability	7	4	6	Durable sensors score higher values. Wireless and remote sensing can affect this issue
6	Operating environment	8	5	5	This can be scored by investigating manufacturer's
7	Maintenance (excluding labor cost)	7	5	6	specifications as well as past performances of the sensor What are the effects of wireless sensing?
8	Cost	2	5	7	Higher costs score lower values
9	Instrumentation type	8	7		How close does the sensor output match the objectives of the project? Are there any extra information does the sensor provide beyond the direct project needs (see serendipity principle, Section 4.7)
10	User knowledge/ experience	9	4	4	How familiar the user is with the sensor under consideration?
11	Sensitivity/resolution	8	6	6	Does the sensor produce optimum resolution? (not too coarse, or unwanted refinement?)

optimum sensor for the experiment on hand. Note that for routine tests, these methods may not be needed, but can be very useful for long-term and/or major projects.

5.9 OPTIMUM SENSOR LOCATION

5.9.1 OVERVIEW

One of the basic tools of SHM is different sensors that are located optimally throughout the structure. In general, the number of these sensors is small when compared with the size and complexity of the structure. Thus, the locations of these sensors have to be chosen optimally, to insure accuracy of detected damage and efficiency in both computations and cost. Even for situations with very large sensor arrays, with large number of sensors, an optimal sensor arrangement is still needed.

The optimal sensor location problem (OSLP) has been studied by many authors. Shah and Udwadia (1977), and Udwadia (1994) presented an OSL algorithm that is based on the optimization of the norm of Fisher information matrix. Papadimitriou et al. (2000) presented an entropy-based OSL formulation. Their method is optimized by the entropy of the information of the structural system. Cobb (1996) presented an OSL technique that is based on the sensitivities of the eigensolutions of the structural system.

Although the above OSL algorithms vary in their mathematical details, they all have at least two commonalties. These commonalties are (a) they are usable in structural damage detection schemes, and (b) they all have two main parts, namely the structural sensitivities and the structural damage. The structural sensitivities parts can be, for example, the Fisher information matrix as suggested by Udwadia (1994), or the eigensolutions derivatives, as presented by Cobb (1996). The structural damage part, which is common to all OSL algorithms, assumes, *a priori*, that the location and magnitude of the sensitive (damaged) elements are known. It was shown by Shah and Udwadia (1977) and Udwadia (1994) that the OSL depends on the assumed locations of the sensitive (damaged) structural elements. Unfortunately, it is not possible to assume this information, *a priori*, that is, before measuring the actual damage, which is what the sensors are supposed to do.

In addition to the above apparent causality problem is that the environmental sources of the structural damage can be varied. For example, structural damage patterns that result from a severe windstorm can be different from the damage pattern of the same structure that result from corrosive effects. Each of these damage patterns would require different OSL. Given that only one layout of sensors is practical, the question is how to resolve this damage profile conflict?

This section presents an attempt to address these issues in a logical and simple manner. That attempt was presented by Ettouney et al. (1999). They first tried to estimate the expected damage location and magnitudes in a given structure. Second, the authors attempted to resolve the conflicting OSL demands resulting from different environmental sources. They then related the structural environment to the OSL algorithms. It was accomplished by using the stress ratio as a damage indicator. Stress ratio is well known and widely used structural design parameters, see McCormac (1992 and 1995). The possible conflicts between environmental sources of damage can then be resolved by applying the stress ratio concept and the load factor concept, AISC (1989 and 1995), to the goal programming approach, Ravindran (1987).

Later in the section, two example problems will be presented to demonstrate the applicability of the solution. The two structures, though realistic, do vary in complexity. The example problems demonstrate the importance of considering the structural environs and damage sources during the solution phase of any OSL computations. It should be noted that the methodologies of this study are applicable to any OSL algorithm. However, the OSL logic proposed by Cobb (1996) was used by Ettouney et al. (1999). We summarize their work in the remainder of this section.

5.9.2 ANALYTICAL TREATMENT

5.9.2.1 General

The OSL development of this section is based on evaluating the eigenvalue and eigenvector derivative for a given structure. These derivatives are computed with respect to any desired structural component. The eigensolutions derivatives will then be used to compute the eigensolutions gradients for any assumed structural damage profile. The gradients of the eigensolutions can be used to prioritize the sensor locations, after colinearity check is performed to eliminate any redundancies in the prioritized sensor locations. For detailed discussion about evaluating eigensolutions derivatives and the use of these derivatives in the sensor prioritization and colinearity checks, the reader is referred to Cobb (1996).

In this section, we will summarize the governing equation leading to the sensor prioritization scheme. The stress ratio S_r based OSL approach will be introduced, and the goal programming methodology will be used to resolve any conflict in the OSL results from different structural loading conditions.

5.9.2.2 OSL Methodology

Consider the eigenvalue problem

$$\left([K] - \lambda_i [M] \right) \{\phi\}_i = [0] \tag{5.49}$$

where $[K]$ and $[M]$ are the structural stiffness and mass matrices. The ith Eigensolutions is represented by $\{\phi\}_i$ and λ_i. Nelson (1976) showed that the derivatives of the Eigensolutions could be expressed by

$$\frac{\partial \lambda_i}{\partial g_j} = \{\phi\}_i^T \left(\frac{\partial [K]}{\partial g_j}\right) \{\phi\}_i \tag{5.50}$$

$$\frac{\partial \{\phi\}_i}{\partial g_j} = c_i \{\phi\}_i + \{V\}_i \tag{5.51}$$

$$c_i = -\{\phi\}_i^T + [M] \{V\}_i \tag{5.52}$$

The vector $\{V\}_i$ is the solution of the equation

$$\left[\tilde{E}\right]\{V\}_i = \left\{\tilde{F}\right\}_i \tag{5.53}$$

The matrix $[\tilde{E}]$ and the vector $\left\{\tilde{F}\right\}_i$ are the rearranged forms of $[E]$ and $\{F\}_i$ where

$$[E] = [K] - \lambda_i [M] \tag{5.54}$$

$$\{F\}_i = \left([M]\{\phi\}_i \{\phi\}_i^T - [I]\right)\frac{\partial [K]}{\partial g_j}\{\phi\}_i \tag{5.55}$$

Rearranging $[E]$ and $\{F\}_i$ to obtain $[\tilde{E}]$ and $\left\{\tilde{F}\right\}_i$ is essential, since $[E]$ is singular, by definition.

Cobb (1996) has developed a rearranging methodology that enables the accurate solution of Equation 5.53.

The parameter g_j is any structural variable that can be associated with damage, that is, changed its value during the operation of the structure under consideration. It can be an axial stiffness of a particular element, a joint rotational stiffness or a prestressing force in a bridge element. The derivative $\partial [K]/\partial g_i$ can be explicitly evaluated for any parameter g_j.

Using $\partial \lambda_i / \partial g_i$ and $\partial \{\varphi\}_i / \partial g_i$ Cobb (1996) showed that the gradients of the eigensolutions $\Delta \lambda_{ij}$ and $\Delta \{\phi\}_{ij}$ can be evaluated as

$$\Delta \lambda_{ij} = \frac{\partial \lambda_i}{\partial g_j}\Delta g_j \tag{5.56}$$

$$\Delta \{\phi\}_{ij} = \frac{\partial \{\phi\}_{ij}}{\partial g_j}\Delta g_j \tag{5.57}$$

The gradients of the expected structural damage Δg_j include information about the type, location, and magnitude of the expected damage in the structural system. Cobb (1996) presented a sensor prioritization scheme using $\Delta \lambda_{ij}$ and $\Delta \{\varphi\}ij$ for as many parameters g_j as required. His method is based on ordering the vector

$$\{D\} = \sum_i \sum_j \left| \Delta \{\phi\}_{ij} \right| \tag{5.58}$$

in a descending fashion. The priority of the sensor location follows the entries of the vector $\{D\}$. The top n locations will be the OSL for n sensors. To eliminate any redundancy in the resulting sensor location, Cobb (1996) employed a colinearity check using $\Delta\lambda_{ij}$ and $\Delta\{\varphi\}_{ij}$. If a pair, or more, of the top n locations are found colinear, this indicates that their measurements are redundant. Only one sensor of each colinear group is chosen in the final OSL profile. For more details on the colinearity checks, the reader is referred to Cobb (1996).

5.9.2.3 Stress Ratio–Based OSL

Two basic types of information in the OSL algorithm outlined above are needed. They are the eigen-solutions sensitivities $\Delta\lambda_{ij}$ and $\Delta\{\varphi\}_{ij}$ as well as the profile and amplitude of the expected structural damage Δg_j. In reality, all OSL algorithms depend on types of available information, structural sensitivities, and expected damage gradients. The developments of structural sensitivities are well established, as shown above. Unfortunately, there is no consistent manner to evaluate the damage gradient Δg_j.

To determine Δg_j, let us ask the question: what are the causes of structural damage? The answer is that different structural environments could cause structural damage. Thus, if a specific structural environment can be specified, the damage gradient Δg_j can be determined.

One of the basic damage caused is structural overstressing, due to different loading conditions. For a given loading condition ℓ it is reasonable to assume that the damage gradient due to this loading condition $\Delta g_j{}^\ell$ is proportional to the stress ratio $(S_r)_{\ell j}$ in the jth structural component. This assumption is consistent with the fact that the stress ratio is the ratio of the actual stress to the allowable design stress in the structural components. It should be noted that $0.0 \leq (S_r) \leq 1.0$. On the basis of this the damage gradient profile can be accurately estimated as

$$\Delta g_j{}^\ell = \left(S_r\right)_{\ell j} \tag{5.59}$$

The number of elements included in the OSL algorithm can be established by determining a damage threshold below which the structural components are assumed not susceptible to damage. All the structural elements will be included if the threshold is taken to be 0.0, conversely, the problem becomes trivial, that is, no elements would be included, if the threshold is taken to be 1.0.

5.9.2.4 Multiple Loading Conditions—Goal Programming

All realistic structures are subjected to multiple environments and loading conditions that promote structural damage. It was shown in the above section that for a given structural loading ℓ there exists a specific OSL. Suppose that there are L structural loading conditions. It is reasonable to expect that each of these loading conditions would have its own OSL profile. These ℓ profiles are not necessarily similar. Furthermore, it is not practical or feasible to change sensor locations to accommodate each of these optimal locations for each loading condition.

The possible conflicting OSL problem can be related to the goal programming problem. The solution of a goal programming problem depends on assigning a weighting factor to each of the loading conditions, and solving the optimization problem. The difficulty here is to find reasonable weighting factors. In any structural design problem, the relative importance of the multiple loading conditions is resolved statistically, and a weighting factor is assigned to each loading condition for design purposes. These statistical factors will be used in the goal programming technique in this study.

5.9.3 Case Studies

To study the importance of the consideration of structural loading conditions (environmental effects) on the optimal sensor locations, two realistic structural cases will be presented. First, a simple structural steel tower is studied, which is subjected to its own weight, as well as, lateral wind loads. Second, a more complex bridge structure is presented, which is subjected to its own weight, as well as, lateral seismic loading (SL).

5.9.3.1   Steel Transmission Tower

Consider the tower of Figure 5.88. The tower is a steel structure that is subjected to both its own weight (dead load [DL]), as well as, wind loading (WL). The DL is acting vertically downward while the WL is acting laterally. It is required to place sensors on this structure to measure the possible damage that might result from either of these two loads, or any combination of them. Only six sensors are available for such a task. First, the OSLs will be estimated independently from the loading conditions. This is achievable by assuming uniform damage pattern for all structural elements. Figure 5.89 shows the OSL for this case. Also, Figure 5.89 portrays the prioritization of the six sensors. Note that the location with the highest priority is at the top of the tower. In addition, no horizontal sensors are required for this OSL approach.

The OSL were then estimated using the applied loads on the tower. The damage levels and profile is estimated using the DL alone. In addition, the damage level and profile due to WL alone is estimated. The OSL due to DL alone were computed, as shown in Figure 5.90a. The OSL due to WL alone was also computed and shown in Figure 5.90b. Finally, the goal programming technique of this study was applied to the two loading conditions. The weighting factors for each of the two optimum conditions were the design load factors applicable to each loading condition. The resulting OSL for the combined case is shown in Figure 5.90c. A comparison between the three solutions of Figure 5.90 shows that the OSL was not sensitive to the change of the loading conditions. It is interesting to note that the OSL of Figure 5.90 is different from that of Figure 5.89. The locations of highest priority are now near the bottom of the tower. In addition, two horizontal sensors are among the top six prioritized sensor locations. The importance of the lower locations of sensors for this type of structures has also been addressed by Shah and Udwadia (1977).

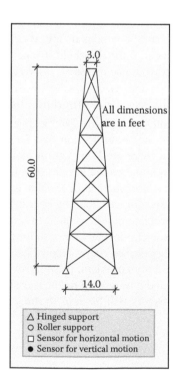

FIGURE 5.88   Tower structure.

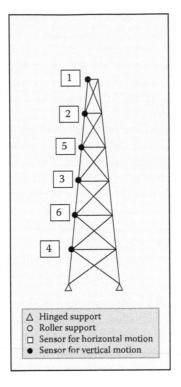

FIGURE 5.89   OSL using a uniform damage profile.

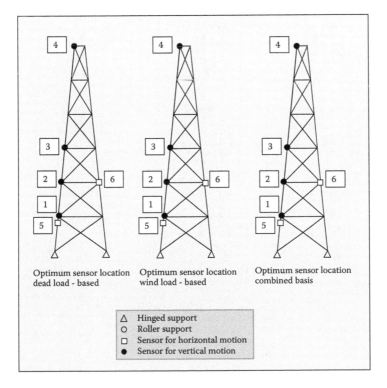

FIGURE 5.90   OSL with a threshold of 0.5: (a) dead load-based OSL, (b) wind load-based OSL, and (c) load combination-based OSL.

The OSL configurations of Figure 5.90 were obtained using a lower damage threshold of 0.5. This means that only structural elements with stress ratio Sr ≥ 0.5 were included in the methodology of this study. To investigate the importance of this limit, the OSL problem was repeated with stress ratio threshold of 0.8. The resulting OSL for DL-only, WL-only, and combined conditions are shown in Figure 5.91. Again, it seems that the OSL is not sensitive to the loading conditions, as in the case of Sr ≥ 0.5. However, the changing of the stress ratio threshold had an effect on the optimum locations of three out of the six sensors.

The insensitivity of the OSL to the loading condition for this example problem, either Figures 5.90 or 5.91, is noted. The reason for this insensitivity lies in the fact that the tower is essentially a vertical cantilever system that is supported at the base. The stress ratio, and hence the damage profile and magnitude, for all studied loads (DL, WL and any resultant combination) increase from top to bottom. Thus, the computed OSL should be similar for all these loading conditions.

In the methodology presented in this study, each of the assumed damaged structural elements is assumed to act independently from the other. If the elements are assumed to be damaged simultaneously, as in $\Delta g_j = 1.0$, the resultant OSL profile is shown in Figure 5.92. This OSL result is less accurate than the results of Figure 5.90c or Figure 5.91c albeit more computationally efficient.

5.9.3.2 Suspension Bridge

We now investigate the more complex bridge structure of Figure 5.93. This bridge is subjected to two loading conditions and their combination. One loading condition is the vertically applied DL. The horizontal seismic load (SL) is the other loading condition. The combination of these two loading conditions is assumed the basis for any damage that might occur in the bridge.

Due to the complexity of this structure, we will assume that 20 sensors are available for damage detection. As in the case of the tower, the OSL will be estimated independently from the loading

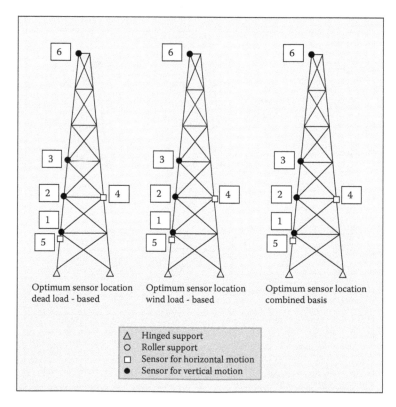

FIGURE 5.91 OSL with a threshold of 0.8: (a) dead load-based OSL, (b) wind load-based OSL, and (c) load combination-based OSL.

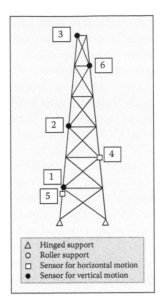

FIGURE 5.92 OSL using a damage-dependent element assumption.

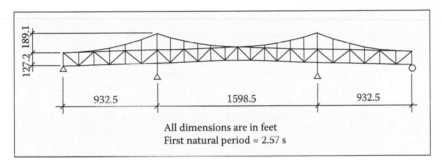

FIGURE 5.93 Bridge structure.

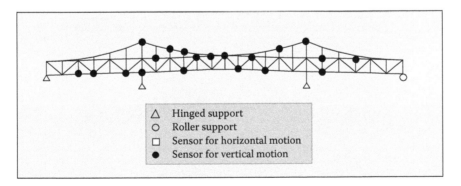

FIGURE 5.94 OSL using a uniform damage profile.

conditions. Figure 5.94 shows the OSL when uniform damage in all the structural elements is assumed. The numeric prioritization of each of the 20 sensors is not reported for the bridge case, since it would complicate the illustrations. It should be noted that the prioritization of the sensors were different for each of the cases studied. Figure 5.94 shows that all the sensors that were computed with the uniformity assumption are vertical sensors.

The effects of the loading conditions on the OSL were then estimated. The damage levels and profile were estimated using the DL alone. In addition, the damage level and profile due to SL alone were estimated. The OSL due to DL alone were computed, as shown in Figure 5.95a. The OSL due to the horizontal SL alone was also computed and shown in Figure 5.95b. The Goal programming technique of this study was applied to the two loading conditions, where the weighting factors for each of the two optimum conditions were the design load factors applicable to each loading condition. The resulting OSL for the combined case is shown in Figure 5.95c. A comparison between Figures 5.95a, 5.95b, and 5.95c shows that the OSL is extremely sensitive to the change of the loading conditions, unlike the tower case. For the DL-base OSL, Figure 5.95a, all the sensors are in the vertical direction. For the horizontally based SL, Figure 5.95b, the OSL are all in the horizontal directions. For the combined loading case, Figure 5.95c, the sensors are all in the horizontal direction, albeit in different locations than those reported in Figure 5.95b. The overpowering effects of the SL on the damage in the bridge are obvious, when compared with the DL.

The OSL of Figure 5.95 was obtained using a lower damage threshold of 0.5. This means that only structural elements with stress ratio Sr ≥ 0.5 were included. The OSL problem was repeated with stress ratio threshold of 0.8. The resulting OSL for DL-only, SL-only, and combined conditions are shown in Figure 5.96. The sensitivity of OSL to the loading conditions is evident. The importance of the SL loading is also clear. Note that varying the stress ratio threshold has an effect on the optimum locations.

Figures 5.95 and 5.96 depict the sensitivity of the OSL to the loading condition for this example problem. The reason for this sensitivity lies in the fact that the bridge is a complex structural system with multiple supports. The stress ratio, and hence the damage profile and magnitude, for all studied loads (DL, SL and their combinations) have differing profiles and amplitudes. Thus, the computed OSL should obviously be different for all these loading conditions.

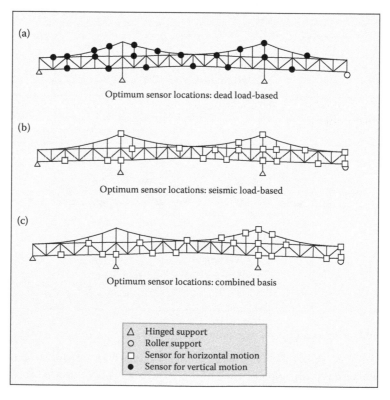

FIGURE 5.95 OSL with a threshold of 0.5: (a) dead load-based OSL, (b) seismic load-based OSL, and (c) load combination-based OSL.

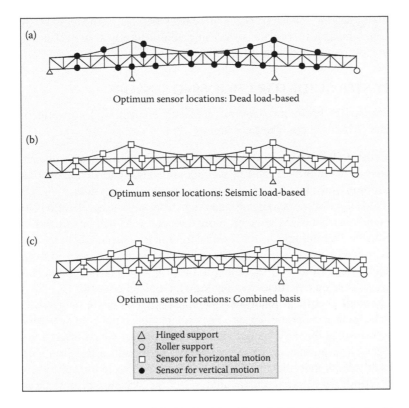

FIGURE 5.96 OSL with a threshold of 0.8: (a) dead load-based OSL, (b) seismic load-based OSL, and (c) load combination-based OSL.

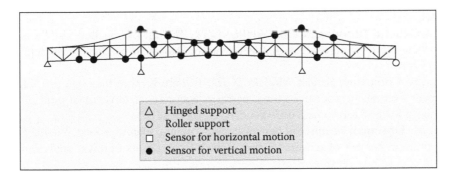

FIGURE 5.97 OSL using a damage-dependent element assumption.

As in the tower case, if the elements are assumed to be damaged simultaneously, as in $\Delta g_j = 1.0$, the resultant OSL profile is shown in Figure 5.97. This OSL result is less accurate than the results of Figures 5.95c or 5.96c, albeit more computationally efficient.

5.9.4 CLOSURE: IMPORTANCE OF OSL FOR SHM

This section showed the importance of optimally placing sensors in structures, especially complex structures. It also explored the importance of accommodating different causes of damage when choosing the location and number of sensors. As of the writing of this volume, there are still many more issues to be researched. For example, applications of OSL to other types of complex structural systems in civil engineering, different types of sensors (strains, velocity, etc.), different damage sources (corrosion, temperature, fatigue, scour), and nonlinear effects are all among the important subjects that can have profound effects on any OSLP.

5.10 STEP BY STEP GUIDE FOR CHOOSING SENSORS

This section summarizes optimum steps for choosing efficient and cost-effective sensors that matches particular SHM projects. The steps are

1. **Identify Project Objectives:** This is perhaps the most important step in an SHM project. The objective(s) of the project must be well identified and understood by all stakeholders. Object definitions must be enumerated in as much detail as possible.
2. **Identify ToM:** Type of measurements can range from strains, accelerations to humidity or corrosion rates. In most SHM projects, more than one ToM will be utilized.
3. **Identify Sensor Technology:** As discussed earlier in the chapter, there are numerous technologies that can be used to sense different ToMs. The advantages and disadvantages of different technologies and their suitability to the project objectives should be studied carefully before settling on a particular technology for a given project.
4. **Perform Needed Technical Analysis:** Technical analysis might be needed at this stage to help in choosing appropriate sensors. For example: the number of structural modes that need to be measured during a modal identification project would affect the chosen sampling limits of accelerometers. Expected seismic drifts will affect limits of location, or displacement sensors that might be chosen to measure expected seismic event. In the first example, preliminary dynamic modal analyses of the structure need to be performed. In the latter, nonlinear seismic analysis of the structure needs to be performed.
5. **Perform Needed Decision Making Analysis:** Decision making analysis might be needed to evaluate cost benefits or utilities of different choices that are available to the decision maker. Examples of decision making for sensor analysis are given in Section 5.8 of this chapter.
6. **Define General Parameters That Might Affect Sensing:** Use Section 5.9 of this chapter to evaluate pertinent parameters that affect sensor choices. If needed repeat steps 4 and 5 above to refine choices.
7. **Compare Competing Sensor Models/Types:** If there is more than one possible sensor or sensor technology that can be used, perform prioritization analysis (similar to those in Sections 5.8.3 or 5.8.4) to help decision making process.
8. **Compute Optimum Number of Sensors:** Use structural identification techniques to identify optimum number of sensors for this project. Experience of professionals can also be used to choose number of sensors.
9. **Identify OSLs:** Use techniques similar to those of Section 5.9.2 to optimize sensor locations.
10. **Estimate Multihazards Needs/Effects:** If there are more than one hazard, estimate its effects and/or needs on the chosen sensor.

Figure 5.98 shows the above steps.

Note that the above steps are for illustration only. The order of executing those steps can change, depending on the particular project on hand.

5.11 REMOTE SENSING IN SHM

5.11.1 Overview

Emerging asset management concepts, in the field of civil infrastructure, call for extending the useful life span of infrastructure. This is a direct result of the increased cost of rehabilitation and

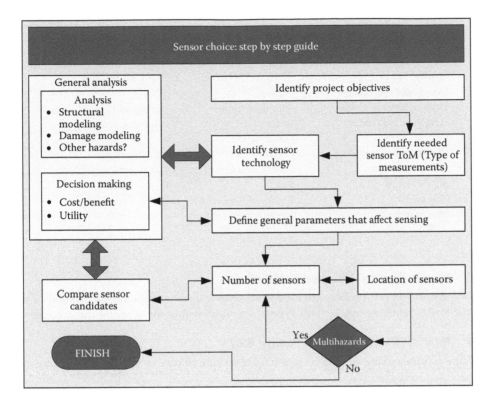

FIGURE 5.98   Steps of choosing sensor in an SHM project.

replacement of systems to accommodate modern hazards and needed safety of the public. One of the rapidly developing tools that can help to extend useful life spans is the field of IHCE and its subset field of SHM. The SHM field calls for monitoring the performance of systems at reasonable intervals. The cost implications of such monitoring are a factor that must be accounted for. One possible way to reduce SHM costs is by the utilization of remote sensing (RS) methods for monitoring performance. It is desirable and may be essential to sense the structural response remotely during a posthazard event to assess potential risks, execute appropriate risk-mitigation programs, and coordinate incident responses. Remote structural sensing also is an emerging field and has potential to be used in conjunction with IHCE.

In this section, we explore different techniques of RS as they pertain to the SHM/IHCE field. We observe that RS technologies have been developed mostly for the NDT field. There are, however, some differences between NDT and SHM/IHCE fields as they pertain to RS, as shown in Table 5.33. For the purpose of this chapter, we define the system, or target system, as the structure to be tested or interrogated. This section will categorize the RS field into four categories, and discuss each of the categories as they pertain to the SHM/IHCE field.

5.11.2   Types of RS

5.11.2.1   Type I: Sensors Attached to Subject

Traditionally, the term RS is used to refer to situations when sensors are attached on the target system and the instrumentation (the sensor results) is placed elsewhere. The sensors and the instrumentations are connected either by wires (tethered) or by wireless signals. This Type I of RS is the conventional sensing method. Many of the basic concepts of Type I sensing is discussed in the section of wireless sensing of this chapter.

TABLE 5.33
Features of RS System for NDT and SHM

Feature	NDT	SHM/IHCE
Sensors and system contact	Needed	
Distance between transmitters, receivers, and system	Short distances are acceptable	Usually need longer distances
Overall Geometry	Less complex	More complex
Exposed surface condition	Less complex	More complex
Ambient vibrations	Usually not a factor	Can be an important factor
Size of system	Small	Large
Accessibility	Usually not a main factor	Can be a major factor
Type of material	Can be demanding	

5.11.2.2   Type II: Sensors Attached to Robots

Another type of RS is when robots are used to sense in difficult situations. An example of this situation is the bomb detecting robots. This type of RS is not within the scope of this volume.

5.11.2.3   Type III: Remote Sensors, Passive RS

Type III of RS is the passive type. This is when a transmitter projects signals onto the target system. The signals will have minimal interaction with the target system. They will mostly just reflect from the system. By processing the reflected signals, some information about the target system can be gathered. Some of the technologies within Type III RS are described next.

Laser Distance Sensing: Laser pulsers have been used for sometime in estimating distances between the source and the target system. This technique has been used successfully to measure distances in several fields, such as surveying, mining, mapping, military, construction, and marine. It also has been used in the vehicle positioning and collision avoidance markets. This technology is well proven and is based on TOF technique. The tolerances of the measurements are extremely low and thus make it suitable to measuring static changes on the surface of the target system. For example, long-term relative motions between both ends of bridge bearings can be observed by this technique. Small swelling of reinforced concrete beams that might result from hidden rusting of rebars that can't be observed by manual inspection may also be measured by this technique. This particular use can differentiate between harmless concrete discoloration and more harmful hidden rusting of steel rebars; the concept is shown in Figure 5.99. In this situation, corrective measures can be taken before the rusting problem becomes more costly. Fuchs and Jalianoos (2008) preformed laser-based experiments that are fairly similar to this concept (Chapter 3 in Ettouney and Alampalli 2012).

Holography: Use of laser pulsers to measure surface response of systems has seen several applications recently. The basic concept is to project one or more laser pulses on the system and observe the reflected pulses. The reflected pulses can be compared to the original transmitted pulses in several manners, such as TOF, intensity, and/or phase. By processing the results, information about the target system emerges. For example, one such technique projects three laser pulses on front and back of a plate and then observes the resulting ultrasonic waves that are generated on the plate surface. By processing interferences between different reflecting pulses, the plate surface waves can be observed in real time. Any defects at or near the surface can easily be observed from this holographic image. Another application was reported by Telscho and Larson III (2003). They used full-field laser measurements at normal video rates to measure modes for plate frequencies in the range of Hz to MHz.

Radar: The use of laser as an RS tool is a very promising technique for SHM/IHCE. However, one of the main disadvantages of lasers is that they might have limited use during foggy days at long

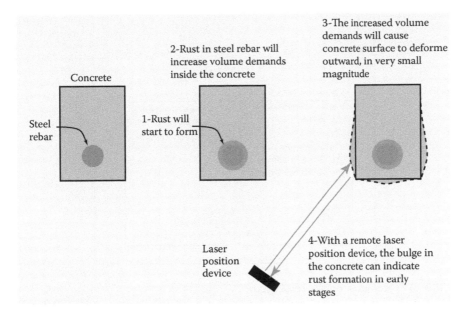

FIGURE 5.99   Conceptual use of RS to detect early rusting stages in reinforced concrete.

distances. A promising emerging application in the field of SHM/IHCE is the use of radar waves. It was shown that Millimeter-wave (MMW) short range radar waves can be used remotely to identify ground-based objects such as grass, asphalt, and snow. Larger objects such as sign posts and guard-rails were also included in the study. Evans et al. (1998) showed that it is possible to achieve higher sensing resolutions by probing in the submillimeter (SMMW) range. For more optimal performance they proceeded to combine SMMW with MMW.

5.11.2.4   Type IV: Remote Sensors, Active RS

Type IV RS is based on sending signals from a transmitter through a low density medium, such as air, into the system that needs to be tested. The signals will then interact with the system generating some type of response. The system response will scatter the incident signal, and the scattered signal will then travel back (or forward) through the low density medium into a receiver that will analyze the received signal. Such an analysis will reveal some information about the remote system.

There are two main difficulties with Type IV RS that distinguish it apart from other types of sensing. At first, there is the impedance mismatch problem. Limiting our interest, for this chapter, to RS through air, let us assume that the impedance of air to ultrasound waves (as an example of input signals) is Z_1 and that of the system material to be tested is Z_2. Then, we can define the strength of the refracted signal through the system T as

$$T = \frac{4Z_1 Z_2}{\left(Z_1 + Z_2\right)^2}$$

The value of T is of the order of $10^{-4} \sim 10^{-5}$ for conventional structural materials. This indicates that most of the incident signal will be reflected from the structural surface and limited signal will penetrate the structure. The same difficulty is present for other types of air borne signals such as electromagnetic waves and laser signals. The other major difficulty of type IV RS is the rapid attenuation of signals within the system, that make it difficult to detect damages that are deep within the thick systems, such as steel or concrete. The two difficulties are conceptually shown in Figure 5.100.

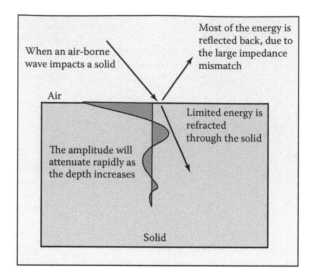

FIGURE 5.100   Difficulties of remote sensing type IV.

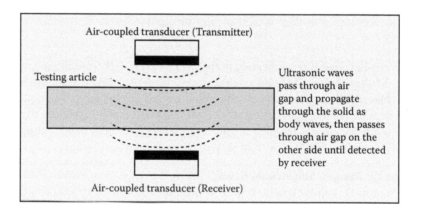

FIGURE 5.101   Air-coupled transmitter and receiver systems (across setting).

On the other hand, Type IV RS offers immense benefits. Among those benefits are the following: (a) no need for contact between the instruments and the system, (b) can be used in difficult or complex conditions, and (c) can detect more types of damages and system properties than Type III RS. Because of these benefits, there has been interest in developing methods for Type IV RS. Those methods are based on ultrasound, electromagnetic, radar waves, AE, and laser methods. We summarize some of those techniques next.

Ultrasound RS Methods: Berriman et al. (2003) used noncontact NDT to evaluate density of plain concrete. They tested a 30-mm-thick specimen, with lateral dimensions of 100 × 100 mm. The objective of the test was mainly to explore the capability of a noncontact setup in detecting the material density of the concrete. They used an air-coupled method as shown in Figure 5.101. The transmitter was a piezoelectric transducer that generated ultrasound waves in the air. These waves created body waves in the concrete specimen that in turn created air waves on the other side of the specimen. By utilizing a piezoelectric transducer on the other side as a receiver, it was possible to detect the ultrasound air waves. By recording the transmitted motions and given the time of arrival and the thickness of the sample, the longitudinal wave speed of the specimen can be computed. In addition to the capability of the noncontact testing method, the experiment had an additional beneficial outcome: the authors observed that due to the noncontact feature of the experiment, the measured results have no artificial by-product of the interaction effects that the conventional contact transducers might have had on the results. Kommareddy et al. (2005) used a similar setting to evaluate embedded defects within an FRP plate. An air-coupled noncontact pair of piezo-ceramic transducers was used in that experiment. The air gap between each of the transmitters and the receiver, and the FRP plate was 2.5 in. Since the transmission of the signal through the FRP plate depended on the presence of defects in the plate, it was possible to detect those defects by observing the transmission of the signals. By performing a C-scan of the plate, the researchers were able to obtain 3D visual plots of the defects.

In a novel technique to increase the transmitted energy T Solovov and Busse (2006a) used an oblique angle of incidence of the ultrasonic waves as shown schematically in Figure 5.102. The slanted transmitter of ultrasound waves through the air would generate ultrasound waves within the solid. These ultrasound waves can be plate acoustic waves (PAW) or surface acoustic waves (SAW). The angle of incidence of the waves can be slanted or normal. The incident ultrasound waves will be either reflected from the surface or transmitted through the solid, if the solid is not too thick. By measuring the reflected or transmitted waves, the defects within the solid can be detected through a TOF technique. The authors compensated for low T by enhancing the transmitted (or reflected) waves through the use of resonance of the air-solid geometry. They showed that by changing the angle of incidence of the waves, they could produce a resonance in the PAW/SAW that will compensate for the low T. They showed that for different materials—such as paper, aluminum foil, and thin wood veneer—gains in the range of 10–23 dB can be achieved through this method. They also showed that the same method can be used for NDT of thick materials through gains of SAW. Another use of the noncontact ultrasound testing is to evaluate classical painting, see Siddiolo and Maev (2006).

Laser RS Methods: Another technique is to use lasers as both the transmitting and receiving signals (see Figure 5.103). Laser RS can generate body waves (shear and bulk) as well as surface (Rayleigh) and plate (Lamb) waves. They can be used in high-temperature, corrosive, and conductive environments. They are also very attractive since the transmitter and the receiver can both be located at long distances from the system of interest. Sanyal (2007) used this technique by using a laser pulser to generate waves across damaged FRP (fiber reinforced polymer) material. The response on the backside of the material is then measured by a 2D scanning of a laser vibrometer. The material properties of the FRP (modulus of elasticity, shear modulus, and Poisson's ratio) were measured through this technique. Also, a C-scan of known defects was performed and the defects were estimated. Both material properties and the presence and extent of defects were shown to have been estimated accurately by this technique.

Combined Laser and Ultrasound RS Methods: Combined laser-acoustic setup is another technology that has been used successfully for RS. It is based on using laser-generated beams that are aimed at the system. Upon impacting the surface of the system, the laser beam will generate acoustic body and surface waves. These waves will propagate through the system and will be scattered through the defects. The scattered waves will generate surface waves that will interact with

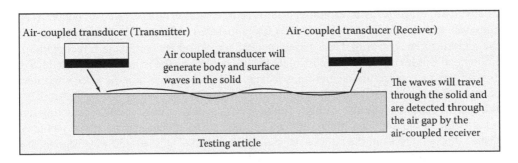

FIGURE 5.102 Air-coupled transmitter and receiver systems (same side setting).

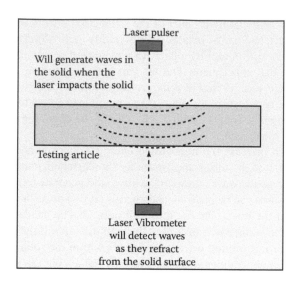

FIGURE 5.103   All-laser transmitter and receiver setting.

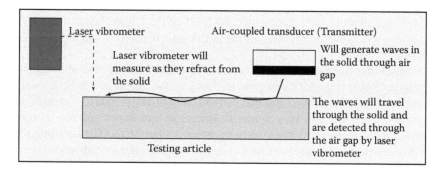

FIGURE 5.104   Combined setting: laser receiver and air-coupled transmitter.

air, which will vibrate accordingly. By measuring the air vibrations by an ultrasonic transducer, it is possible to detect the location and shape of those defects. Kenderian et al. (2006) reported on a successful use of this laser-air hybrid ultrasonic technique (LAHUT) for detecting defects in rail systems (see Figure 5.104). The LAHUT technique was successful on several fronts: (a) It reported a damage detection success rate of 100% in several configurations; (b) It was capable of performing well for a variety of surface conditions, including shiny surfaces; (c) The distances between the transmitters, receivers, and the system were relatively long at 200 mm and higher; and (d) The RS system was used on typical high-density steel rail material. Such a system has great promise for RS of near-surface damage detection of homogeneous dense materials, such as steel or concrete. A similar study was reported by Scalea et al. (2005). They included in their study a comprehensive SHM damage identification algorithm that is based on artificial neural network pattern recognition.

Solodov and Busse et al. (2006b) and Kohler (2006) performed the noncontact coupled RS experiments with the air-coupled ultrasound transducers as the transmitters and the laser vibrometers as the receivers. See Figure 5.105 for this setup.

Radar RS Methods: The use of high frequency radar waves were used in detecting damage of steel structures (Housner and Masri 1996). The radar waves would penetrate the steel construct and scatter through cracks in the base or the welding material. The scattered wave signals are then processed to reveal these defects. The high impedance mismatch between the steel material and air voids that form though the defects would produce high scattering signals. A signal processing

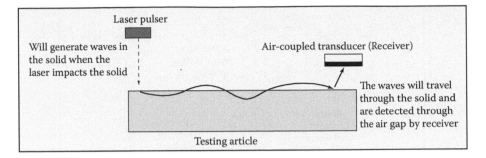

FIGURE 5.105 Combined setting: laser transmitter and air-coupled receiver.

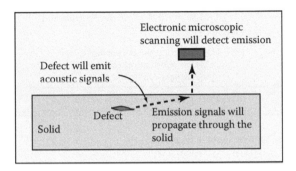

FIGURE 5.106 Acoustic emission concepts in RS.

scheme, such as those used in computer aided tomography scans (CATSCAN) can help in damage identification. Since the radar waves have much higher frequencies than, for example, ultrasound waves, they can penetrate construction materials (such as steel or concrete) deeper than the ultrasound waves.

Electromagnetic Methods: Electromagnetic acoustic transducers (EMAT) can generate acoustic signals in the system of interest as a response to electromagnetic waves generated by a noncontact transducer close to the system. Similarly, noncontact EMAT transducers can also receive signals from the response of the system. EMAT can generate all types of ultrasonic waves within the tested body. Unfortunately, there are some limitations to the use of EMAT as RS devices. The tested system must be electrically conductive. Also, the efficiency of the EMAT decreases rapidly as the air gap between the transducer and the tested system increases. The optimum air gap is usually in the range of 1 mm. As such, it seems that EMAT may not be a practical technology for SHM/IHCE field.

Acoustic Emission: AE is a passive NDT technology. It purports to listen to defects. As such, in the field of RS technologies, there is only a receiver to detect any AE from defects, as shown in Figure 5.106. The AE technique in the field of RS was used to detect emission signals from the crack in stainless steel specimens. The signals were emitted in the frequency range of 9–10 MHz. An electron microscope scanning was performed to detect the AE signals.

5.11.3 TECHNICAL ASPECTS OF REMOTE SENSING

5.11.3.1 Penetration versus Backscatter

All Type IV RS techniques will rely on one of two STRID methods: penetration or backscatter of waves, as shown in Figures 5.107 or 5.108. In the penetration of waves method, the transmitter will generate the wave type of interest (can be laser, ultrasonic, electromagnetic, penetrating radar such as X-rays, etc.). They will impact the structure and penetrate it. It will interact with the structure

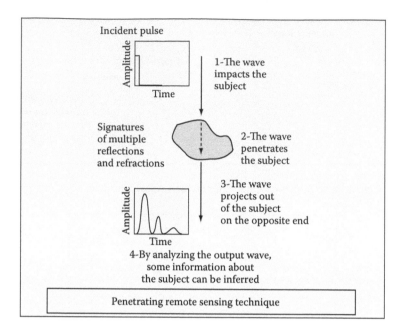

FIGURE 5.107   Penetrating remote sensing technique.

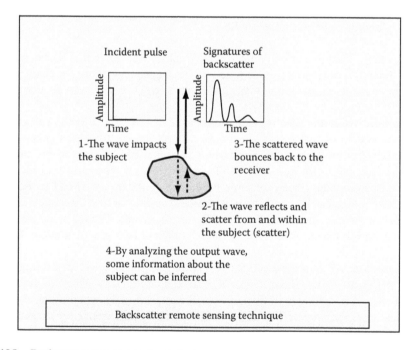

FIGURE 5.108   Backscatter remote sensing technique.

and when the waves exit the structure from the opposite end, it will be detected by the receiver. The received signals have been modified through reflections and refractions by the structure and any structural defects. By analyzing the incident and output wave, information about the structure can be gathered (see Figure 5.109). Perhaps the most popular penetrating RS method is the CATSCAN in NDT and medical fields. Chapter 8 explores the basics of this method.

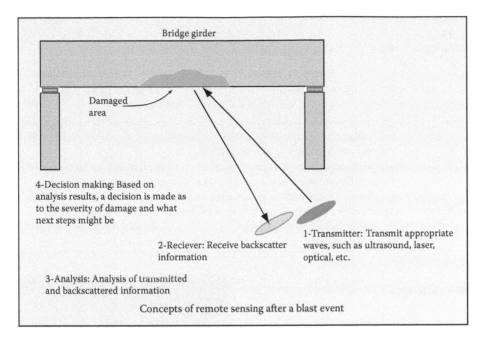

FIGURE 5.109 Remote sensing after a blast event.

In the backscatter method (Figure 5.108), as the name implies, both the transmitter and the receiver are placed on the same side of the structure. When the incident waves impact the structure, it will reflect and refract on the surface, the refracted waves will further penetrate the structure and reflect from other structural boundaries. When the structure responds to this complex wave patterns, it will further affect the reflection/refraction. The process will result in a set of scattered waves from the structure, and some of these scattered waves will find its way back to the receiver. The analysis of incident and backscattered waves can reveal useful information about the structure. The NDT ultrasound methods pitch-catch that was described earlier offer good examples of the backscatter methods. Table 5.34 shows the differences between penetrating and backscattered methods.

5.11.3.2 Decision Making

It was argued by Ettouney and Alampalli (2000) that the decision making process must be an integral part of any SHM/IHCE. This is also true for RS. Several decisions must be undertaken by the officials including whether to embark on an RS approach versus a more conventional sensing system. These can range from budgetary to technological and also require attention as to the frequency of the input. For example, assets in large metropolitan areas in seismic zones may need a more frequent monitoring than those in smaller localities. Clearly, there are several advantages and disadvantages to the RS system. Some of the advantages are (a) usability in complex and difficult geometries, and (b) the process is noncontact in nature. Some of the disadvantages are (a) limitations of the current technologies, and (b) cost disadvantage. As usual in any decision making process, the official should weigh the costs versus benefits of any method before embarking on a particular project.

5.11.4 Current Applications in SHM

5.11.4.1 Thermography

Given the average age of a bridge in the inventory in the United States being about 44 years, there is an increasing focus on the condition assessment, repair and rehabilitation of concrete components in bridges. Quantitative evaluation (location and extent) of the concrete components subjected to spalls and

TABLE 5.34
Penetrating versus Backscattered Methods

Issue	Penetrating Waves	Backscattered Waves
Location of receiver	Opposite side of the transmitter	Same side as the transmitter
Structural size	Larger	Thinner
Type of waves	High energy waves, such as X-rays and Gamma rays are used for deeper penetration, or stronger backscattering of structures. Ultrasonic, electromagnetic and laser waves can be used too, with lesser penetration or backscattering capabilities	
Safety	Use of high energy penetrating waves, such as X-rays or Gamma rays, that can penetrate deep into the structure can pause safety concerns	Same concerns for safety issues if high energy waves are used
Location of defects	Any location within the structure, as long as they are within reasonable proximity of the penetrating waves	Usually close to the surface due to the potentially rapid attenuation of the waves deep inside the structure

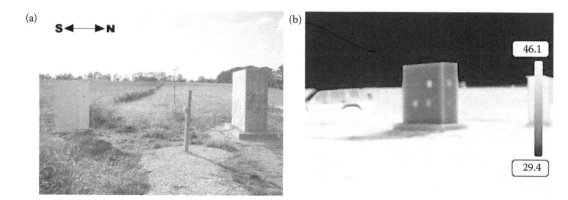

FIGURE 5.110   Test block and its IR image (a) visual image, and (b) IR image. (Courtesy of Dr. Glenn Washer.)

delaminations due to steel corrosion of embedded reinforcing steel is very useful for better bridge management. Traditional inspection methods require arms-reach access to the surface being inspected. This frequently requires lane closures or other traffic disruptions, and can be time consuming for the inspectors. The application of IR thermography for detecting deterioration in concrete bridge components may provide a means for imaging large areas of a structure from a distance, reducing inspection times and averting traffic disruptions. Washer et al. (2008) discussed experimental testing designed to identify the optimum conditions for the detection of subsurface features in concrete bridge components.

The authors erected a test concrete block to evaluate the effect of environmental influences upon a concrete structure (see Figure 5.110). The test block was instrumented with Styrofoam targets of known thickness, area, location, and depth were embedded in the concrete was used to evaluate the effect of environmental influences upon a concrete structure to determine optimum conditions for use of IR method for defect detection. Environmental parameters including wind speed, solar loading, humidity, and ambient temperatures are assessed to determine their influence on the detection of subsurface features in the block. Figure 5.110 shows the IR images of the same block. Note that the depth of the targets is qualitatively apparent by the contrast of the IR images. Figure 5.111 shows the changes in the contrasts of the four targets at different times of the day. The correlation between solar loading and thermal contrast is evident in Figure 5.112.

Initial results of the study suggest that clear skies, resulting in uninterrupted solar loading, provide good contrast in thermal images as the total thermal energy introduced into the block on a

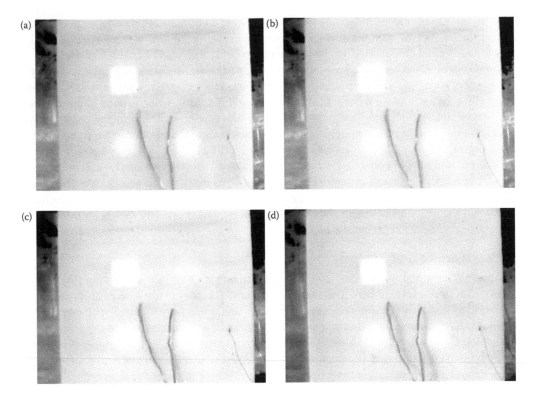

FIGURE 5.111 Thermographic images at different targets taken at different times during the afternoon (a) 1:00 p.m., (b) 1:30 p.m., (c) 2:30 p.m., and (d) 4:00 p.m. (Courtesy of Dr. Glenn Washer.)

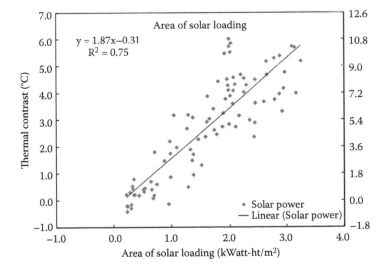

FIGURE 5.112 Scatter plot relating thermal contrast and solar loading. (Courtesy of Dr. Glenn Washer.)

clear day can be greater, establishing a greater thermal gradient in the block and therefore greater contrast in the images.

5.11.4.2 Ground Penetrating Radar (GPR)

Geophysical methods can yield more information than the chain dragging and hammer sounding and if used carefully may have the ability to quantify the degree of deterioration. GPR has been used by several on number of bridges for getting more information on bridge decks with the intention of getting information such as deck thickness, reinforcement/rebar depth (concrete cover), and configuration, potential for delamination, concrete deterioration, and estimation of concrete properties.

Gucunski et al. (2008) provided a comparative study of bridge deck condition assessment by high frequency air (horn) and ground coupled GPR antennas. Horn antennas have been used in the past to primarily provide a fast overview of the condition of the deck, while ground coupled antennas provide more detailed imaging and analysis of the deck condition. The primary objective was to evaluate a new 2.6 GHz ground coupled antenna on bare concrete decks through a comparison with existing high frequency 1.5-GHz ground coupled and 2.0-GHz air-coupled antennas. Figure 5.113 shows the method of data collection in the field. The results showed that the 2.6-GHz antenna provides significantly more detail compared to the 1.5 GHz, so that a strong scatter from the aggregate in the concrete above the rebar level becomes clearly visible. Disadvantages of lower resolution of images from the air coupled antenna are to some extent compensated by the capability to conduct surveys of bridge decks at highway speeds.

5.11.4.3 Thermoelastic Method

A thermoelastic method is described by Fuchs and Jalianoos (2008) for identifying full-field stress distribution on structural members. This method is described as easy to apply, requiring s minimal surface reparation, can be applied at a wide viewing angle, and is relatively immune to vibrations. A thermoelastic stress measurement instrument was used to measure the surface stress of a specimen through thermoelastic effect that takes place when a material's temperature changes as a result of experiencing a bulk compression or expansion. For example, if a solid material is subjected to a tensile force, the material experiences a temperature change proportional to the sum of the principle stresses.

To measure the temperature changes on a specimen (steel light pole, Figure 5.114) that result from stress, the authors took infrared (IR) images of the specimen with a thermal camera. Due to extremely small temperature changes that result from typical stresses and hence the specimens are subjected to dynamic loads in the order of a few cycles per second. The specimen is coated with a flat black paint to make its surface a diffuse emitter of IR radiation.

Testing was done on small specimens as well as a steel light pole and a steel I-girder. The data showed that the color scale in the image is proportional to the sum of the principle stresses in the specimen. Figure 5.115 shows the tests on steel light pole structures loaded dynamically to induce fatigue cracks.

(a)    (b)

FIGURE 5.113 Data collection by GPR (a) leading instrumentation, and (b) data collection vehicle. (Reprinted from ASNT Publication.)

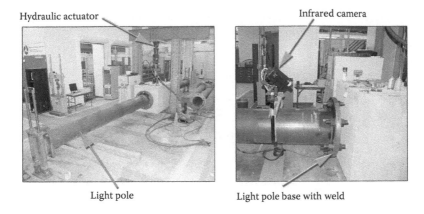

FIGURE 5.114   Steel light pole measurements with the thermographic system. (Reprinted from ASNT Publication.)

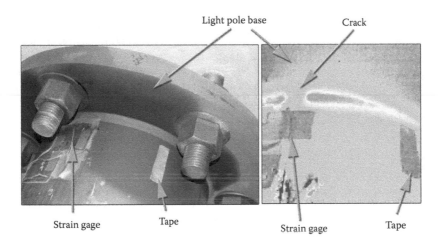

FIGURE 5.115   Thermographic results of crack at the base of the steel pole. (Reprinted from ASNT Publication.)

5.11.4.4   Correlating Load Level with Surface-Wave Laser Measurements

Zoëga and Wiggenhauser (2008) investigated the influence of uniaxial load on the propagation time of surface waves in concrete well below the ultrasonic range. The surface-wave velocity was measured at different load levels along different directions relative to the direction of the applied uniaxial load. Results indicate that the surface-wave velocity is influenced by the applied load and direction; therefore it has a distinct stress-sensitive behavior and it is also possible to determine the load direction from measurements. The different behavior of the velocity curve at different load levels may potentially be used for determining actual stress level and ultimate strength of the concrete specimen. Figure 5.116 shows the experimental setup (a) and the experimental concept (b).

5.11.5   Closing Remarks

This section presented an overview of the field of RS as it can potentially be used in the field of SHM/IHCE. We categorized the RS field in four types. We then described each of the types. Clearly, Types III and IV are the most useful RS types.

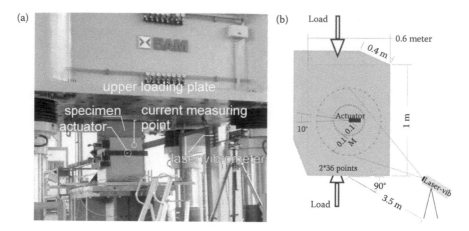

FIGURE 5.116   Laser measurements of surface waves (a) overview of the test setting, and (b) illustration of the concept. (Reprinted from ASNT Publication.)

We also showed several applications of RS in the field of SHM/IHCE. Remote monitoring of early stages of corrosion (rust) of reinforced concrete is one example. Accurate detection of modes is another example. Also, RS of fatigue cracks, or tilting of piers after severe flooding, or hidden but near-surface cracks after an earthquake event are all examples of potentially useful RS. Another important application is protection of first responders after a catastrophic event, such as a terrorist attack, where damage could be monitored remotely. Figure 5.9 shows this concept. Thus, the RS field is evolving rapidly and promises to benefit the SHM/IHCE field and the whole community to a great extent.

REFERENCES

AISC., (1995) *Load and Resistance Factor Design*, American Institute of Steel Construction, Chicago, IL.

AISC., (1989) *Manual of Steel Construction, Allowable Stresses*, American Institute of Steel Construction, Chicago, IL.

American Society of Testing and Materials, ASTM., (1987). Standard Test Method for Half-Cell Potentials of uncoated Reinforcing Steel in Concrete. *ASTM C876.*

Ansari, F., (2002) "Fibre Optic Sensors and Systems for Structural Health Monitoring," Proceedings of 1st International Workshop on Structural Health Monitoring of Innovative Civil Engineering Structures, ISIS Canada Corporation, Manitoba, Canada.

Arms, S. W., Galbreath, J. H., Newhards, A.T., and Townsend, C.P., (2004) "Remotely Reprogrammable Sensors for Structural Health Monitoring," Proceedings, NDE Conference on Civil Engineering, ASNT, Buffalo, NY.

ASNT., (1996) *Nondestructive Testing Handbook*, 2nd Edition, American Society of Non-Destructive Testing, Columbus, OH.

Bastianinin, F., Matta, F., Rizzo, A., Galati, N., and Nanni, A., (2006) "Overview of Recent Bridge Monitoring Applications using Distributed Brillouin Fiber Optic Sensors," *NDE Conference on Civil Engineering*, ASNT, St. Louis, MO.

Benmokrane, B. and El-Salkawy, E., (2002) "Design, Construction, and Monitoring of a Bridge Deck Reinforced with FRP Bars: Wotton Bridge," Proceedings of 1st International Workshop on Structural Health Monitoring of Innovative Civil Engineering Structures, ISIS Canada Corporation, Manitoba, Canada.

Bergmeister, K., (2004) "Reliability Based Structural Health Monitoring of Bridges," Proceedings of 2nd International Workshop on Structural Health Monitoring of Innovative Civil Engineering Structures, ISIS Canada Corporation, Manitoba, Canada.

Berriman, J., Gan, T. H., Hutchins, D. A., and Purnell, P., (2003) "Non-Contact Ultrasonic Interrogation of Concrete," Non-Destructive Testing in Civil Engineering International Symposium, Berlin, Germany.

Cai, S., Hou, S., and Ou, J., (2008) "Settlement and Erosion Monitoring Using FBG Sensors for Soil Under Bridge Approach Slab," *NDE/NDT for Highwaya and Bridges: Structural Materials Technology (SMT)*, ASNT, Oakland, CA.

Chen, G., (2004) "Novel Cable Sensor for Crack Detection of RC Structures," Proceedings, NDE Conference on Civil Engineering, ASNT, Buffalo, NY.

Chen, S-E., Callahan, D., Jones, S., Zheng, L., Biswas, P., Eldina, P., El Yamak, B., Malpekar, A. D., and Lokanath, R., (2004) "Design of a Remote monitoring Techniques of Bridge Integrity Using Wireless Drive-by Network," Proceedings, NDE Conference on Civil Engineering, ASNT, Buffalo, NY.

Cho, S-B. and Lee, J-J., (2004) "Strain Event Detection Using a Double-Pulse Technique of a Brillouin Scattering-Based Distributed Optical Fiber Sensor," *Optics Express,* 12(18), 4339–4346.

Chuang, J., Brown, B., Rivera, E., and Fletcher, D., (2004) "Design and Implementation of an Integrated Fiber Bragg Grating Interrogation System Using Distributed Computing Architecture," Proceedings of 2nd International Workshop on Structural Health Monitoring of Innovative Civil Engineering Structures, ISIS Canada Corporation, Manitoba, Canada.

Cobb, R. G., (1996) "Structural damage identification from limited measurement data," Ph. D. dissertation, School of Engineering, Air Force Institute of Technology, Wright Patterson AFB, OH.

Corrosionsource (2011) "Corrosion Sensors," http://www.corrosionsource.com/technicallibrary/corrdoctors/Modules/MonitorBasics/Types.htm. Site accessed February 23, 2011.

El-Sheimy, N., Taha, M., and Niu, X., (2004) "Next Generation Low Cost MEMS Based Sensors: Challenges and Implementation in SHM Systems," Proceedings of 2nd International Workshop on Structural Health Monitoring of Innovative Civil Engineering Structures, ISIS Canada Corporation, Manitoba, Canada.

Engineersedge (2011). "LVDT Sensors," http://www.engineersedge.com/instrumentation/lvdt_linear_voltage_displacment_tranducer.htm. Site accessed February 23, 2011.

Ettouney, M. and Alampalli, S., (2000) "Engineering Structural Health," Proceedings, Structural Engineers World Congress, Philadelphia, PA.

Ettouney, M. and Alampalli, S., (2012) *Infrastructure Health in Civil Engineering, Applications and Management,* CRC Press, Boca Raton, FL.

Ettouney, M., Daddazio, R., and Hapij, A., (1999) "Optimal Sensor Locations for Structures with Multiple Loading Conditions," Proceedings, International Society of Optical Engineering Conference on Smart Structures and Materials, San Diego, CA.

Evans, K. F., Walter, S. J., Heymsfield, A. J., and Deeter, M. N., (1998) "Modeling of Submillimeter Passive Remote Sensing of Cirrus Clouds," *Journal of Applied Meteorology,* 37(2), 184–205.

Fuchs, P. and Jalianoos, F., (2008) "Thermal Stress Measurements on Structural Defects," *NDE/NDT for Highwaya and Bridges: Structural Materials Technology (SMT),* ASNT, Oakland, CA.

Gage-Technique (2011) "Demountable Strain Gauges" http://www.gage-technique.demon.co.uk/pdf/gti-straingaugetheory.pdf. Site accessed February 23, 2011.

Gangone, M. V., Whelan, M. J., Fuchs, M. P., Janoyan, K., D., and Minnetyan, L., (2006) "Implementation and deployment of a structural health monitoring (SHM) system for bridge superstructures" Presented at Transportation Research Board, TRB, Washington, DC.

Gucunski, N., Rascoe, C., Parrillo, R., and Roberts, R ,(2008) "Comparative Study of Bridge Deck Condition Assessment by High Frequency GPR," *NDE/NDT for Highways and Bridges: Structural Materials Technology (SMT),* ASNT, Oakland, CA.

Hag-Elsafi, O., Kunin, J., and Alampalli, S., (2003) "In-Service Evaluation of a Concrete Bridge FRP Strengthening System," Report 139, Transportation Research and Development Bureau, New York State Department of Transportation, Albany, NY.

Hannah, R. L. and Reed, S. E., (1991) "Strain Gage User's Gandbook." Society for Experimental Mechanics, Bethel, CT.

Hay, T., Jayaraman, S., Ledeczi, A.,Volgyesi, P., and Sammane, G., (2008) "Wireless Technology for Steel Bridges," *NDE/NDT for Highways and Bridges: Structural Materials Technology (SMT),* ASNT, Oakland, CA.

Hecht, J., (2006) *Understanding Fiber Optics,* Prentice Hall, New Jersey, NJ.

Housner, G. and Masri, S. (1996) "Structural Control Issues Arising from the Northridge and Kobe Earthquakes," Proceedings of 11th World Conference on Earthquake Engineering, Mexico, Paper No. 2009.

Huston, D., (2002) "Electromagnetic Interrogation of Highway Structures," Proceedings, NDE Conference on Civil Engineering, ASNT, Cincinnati, OH.

Huston, D., (2010) *Structural Sensing, Health Monitoring and Performance Evaluation,* CRC Press, Boca Raton, FL.

ISIS Canada., (2006), *Civionics Specifications,* ISIS Canada Corporation, Manitoba, Canada.

Kao, T. and Taylor, H., (1996) "High-sensitivity intrinsic fiber-optic Fabry-Perot pressure sensor," *Optics Letters,* 21(8), 615–617.

Kenderian, S., Djordjevic, B., Cermiglia, D., and Garcia, G., (2006) "Dynamic Railroad Inspection Using the Laser-Air Hybrid Ultrasonic Technique," *NDT in Rail Industry, Insight,* 48(6).

Kohler, B., (2006) "Dispersion Relations in Plate Structures Studied with a Scanning Laser Vibrometer," *ECNDT 2006*, We.2.7.1, pp 1–11.

Kommareddy, V., Peters, J., and Hsu, D., (2004) "Air-Coupled Ultrasonic Measurements in Composites," Proceedings, SPIE 3rd International Conference on Experimental Mechanics, Singapore.

Lynch, J. P., Kamat, K., Li, V. C., Flynn, M. P., Sylvester, D., Najafi, K., Gordon, T., Lepech, M., Emami-Naeini, A., Krimotat, A., Ettouney, M., Alampalli, S., and Ozdemir, T., (2009) "Overview of a Cyber-enabled Wireless Monitoring System for the Protection and Management of Critical Infrastructure Systems," *SPIE Smart Structures and Materials*, San Diego, CA.

Lynch, J. and Loh, K., (2006) "A Summary Review of Wireless Sensors and Sensor Networks for Structural Health Monitoring," *Shock and Vibration Digest*, 38(2), 91–128.

Lynch, J. P., Wang, W., Loh, K. J., Yi, J-H., and Yun, C-B., (2006) "Performance Monitoring of the Geumdang Bridge using a dense network of high-resolution wireless sensors" *Smart Materials and Structures*, Institute of Physics Publishing, 15, 1561–1575.

Macrosensors (2011), "LVDT Sensors" http://www.macrosensors.com/lvdt_macro_sensors/lvdt_applications/temperature_effects.html. Site accessed February 23, 2011.

McCormac, J. C. (1992) *Structural Steel Design: ASD Method*, Harper Collins Publishers, New York.

McCormac, J. C. (1995) *Structural Steel Design: LRFD Method*, Harper Collins Publishers, New York.

McNiel, D. and Card, L., (2004) "Novel Event Localization for SHM Data Analysis," Proceedings of 2nd International Workshop on Structural Health Monitoring of Innovative Civil Engineering Structures," ISIS Canada Corporation, Manitoba, Canada.

Mercado, E. J. and Rao, J. R., (2006) "The Pneumatic Scour Detection System," NDE Conference on Civil Engineering, ASNT, St. Louis, MO.

Metje, N., Chapman, D. N., Rogers, C. D. F., Kukureka, S. N., Miao, P., and Henderson, P. J., (2004) "Structural Monitoring using Optical Fibre Technology," Proceedings, NDE Conference on Civil Engineering, ASNT, Buffalo, NY.

Nelson, R. B., (1976) "Simplified Calculations of Eigenvector Derivatives," *AIAA Journal,* 14(9), 1201–1205.

Neubauer, S., Hemphill, D., Phares, B., Wipf, T., Doornik, J., Greimann, L., and Monk, C., (2004) "Use of Fiber Bragg Gratings for the Long Term Monitoring of a High Performance Steel Bridge," Proceedings, NDE Conference on Civil Engineering, ASNT, Buffalo, NY.

Newton, R., (1999), "Construction of a Chemical Sensor/Instrumentation Package Using Fiber Optic and Miniaturization Technology," NASA Report Number TM-1999-209732, Marshall Space Flight Center, Alabama.

Northrop, R. B., (2005) *Introduction to Instrumentation and Measurements*, Taylor and Francis, New York.

Norton, H. N., (1989) *Handbook of Transducers*, Prentice Hall, Englewood Cliffs, NJ.

Ojio, T., Saito, Y, Yamada, K, and Shiina, S., (2004) "Development of 'Strain Probe' By Frictional Type Strain Gauge," Proceedings of 2nd International Workshop on Structural Health Monitoring of Innovative Civil Engineering Structures," ISIS Canada Corporation, Manitoba, Canada.

Oka, K., Ohno, H., Kurashima, T., Matsumoto, M., Kumagai, H., Mita, A., and Sekijima, K., (2000) "Fiber Optic Distributed Sensor for Structural Monitoring," Proceedings of 2nd International Workshop on Structural Health Monitoring, Stanford University, Stanford, CA.

Omega (2011) "OMEGA Engineering Technical," http://www.omega.com/prodinfo/loadcells.html. Site accessed February 23, 2011.

OZOptics., (2006) "Fiber Optic Distributed Brillouin Sensor Applications," www.ozoptics.com. Site accessed February 27, 2011, ON, Canada.

Papadimitriou, C., Beck, J. , and Au, S. (2000) Entropy-Based Optimal Sensor Location for Structural Model Updating, *J. Vib. Control*, 6(5), 781–800.

Park, H., Jung, H., and Baek, J., (2004) "Safety Monitoring Technique for Steel Beams Using Long-Gauge Fiber Optic Sensors," Proceedings of 2nd International Workshop on Structural Health Monitoring of Innovative Civil Engineering Structures," ISIS Canada Corporation, Manitoba, Canada.

PCB (2011) "PCB Piezotronics Tech Support Documents," http://www.pcb.com/techsupport/tech_indaccel.php. Site accessed February 23, 2011.

Perry, C. C. and Lissner, H. R., (1962) "The Strain Gage Primer." McGraw-Hill Book Company, *2nd Edition*.

Pendat, J. and Piersol, A. (2010) *Random Data: Analysis and Measurement Procedures*, Wiley, New York.

Ravindran, A., Phillips, D., and Solberg, J., (1987) *Operations Research-Principles and Practice*, John Wiley & Sons, New York.

Rivera, E., Mufti, A., and Thomson, D., (2004) "CIVIONICS Specifications for Fibre Optic Sensors for Structural Health Monitoring," Proceedings of 2nd International Workshop on Structural Health Monitoring of Innovative Civil Engineering Structures," ISIS Canada Corporation, Manitoba, Canada.

Rong, A. Y. and Cuffari, M. A., (2004) "Structural Health Monitoring of a Steel Bridge using Wireless Strain Gages," Proceedings, NDE Conference on Civil Engineering, ASNT, Buffalo, NY.

Sanyal, D., (2007) "Laser Induced Ultrasonic Characterization of FRP Composites," Proceedings, International Conference on Advanced Materials and Composites, ICAMC.

Satou, T., Kaneda, Y, Nagaoka, S., and Ogawa, S. (2006) "Development of Wireless Sensors Using RFID Techniques for Concrete Structures," NDE Conference on Civil Engineering, ASNT, St. Louis, MO.

Scalea, L., Rizzo, P., Coccia, S., Bartoli, I., Fateh, M., Viola, E., and Pascale, G., (2005) "Non-Contact Ultrasonic Inspection of Rails and Signal Processing for Automatic Detection and Classification," *NDT in Rail Industry, Insight*, 47(6), 346–353.

Sensorland (2011), "LVDT Sensors," http://www.sensorland.com/HowPage006.html. Site accessed February 23, 2011.

Shah, P. and Udwadia, F. E., (1977) "A methodology for optimal sensor locations for identification of dynamic systems," *Journal of Applied Mechanics,* 45(2).

Shenton, H. W., Jones, R., and Howell, D. A., (2004) "A Web-Based System for Measuring Live Load Strain in Bridges," Proceedings, NDE Conference on Civil Engineering, ASNT, Buffalo, NY.

Siddiolo, A. and Maev, R., (2006) "An Air-Coupled Ultrasonic Techniques for NDE of Ancient Paintings," *ECNDT 2006*, Th.2.4.4.

Singer (2011), "LVDT Sensors," http://www.singer-instruments.com/tutorials/lvdt/lvdt_tut.html. Site accessed February 23, 2011.

Solodov, I. and Busse, G., (2006a) "Mapping of Elastic Anisotropy with Air-Coupled Phonon-Focusing of Guided and Surface Waves," *ECNDT 2006*, We.2.7.1.

Solodov, I. and Busse, G., (2006b) "New Advances in Air-Coupled Ultrasonic NDT Using Acoustic Mode Conversion," *ECNDT 2006*, We.2.4.2.

Telscho, K. and Larsen III, J., (2003) "Determination of Lateral Mode Dispersion from Full-Field Imaging of Film Bulk Acoustic Resonator Motioh," *2003 IEEE Ultrasonics Symposium*.

Tennyson, R., Manuelpillai, G., and Cheng, J., (2002) "Application of Fibre Optic Sensors for Monitoring Pipelines," Proceedings of 1st International Workshop on Structural Health Monitoring of Innovative Civil Engineering Structures," ISIS Canada Corporation, Manitoba, Canada.

Tennyson, R. and Morison, W., (2004) "Fiber Optic Structural Health Monitoring System for Pipelines," Proceedings of 2nd International Workshop on Structural Health Monitoring of Innovative Civil Engineering Structures, ISIS Canada Corporation, Manitoba, Canada.

Texier, S. and Pamukcu, S., (2003) "Fiber Optic Sensor for Distributed Liquid Content Quantification in Subsurface," International Symposium on Field Measurements in GeoMechanics, NGI, Oslo, Norway.

Thompson, D. and Bridges, G., (2002) "Smart Aggregate," Proceedings of 1st International Workshop on Structural Health Monitoring of Innovative Civil Engineering Structures, ISIS Canada Corporation, Manitoba, Canada.

Timoshenko, S., (1955) *Vibration Problems in Engineering*, McGraw-Hill, NY.

Udwadia, F. E., (1994) "Optimal Sensor Locations for Structural Identification," *Journal of Engineering Mechanics,* 110(2), 1757–1770.

Washer, G., Bolleni, N., Fenwick, R., and Alampalli , S., (2008) "Environmental Factors for the Thermographic Inspection of Highway Bridges," *NDE/NDT for Highways and Bridges: Structural Materials Technology (SMT)*, ASNT, Oakland, CA.

Weber, B., Paultre, P., and Proulx, J., (2004) "Damage Detection in a Two-Story Concrete Building: Minimizing the Effect of Measurement Errors," Proceedings of 2nd International Workshop on Structural Health Monitoring of Innovative Civil Engineering Structures, ISIS Canada Corporation, Manitoba, Canada.

Webester, J. G., (1999) The Measurement, Instrumentation and Sensors Handbook, CRC Press, Boca Raton, FL.

Wu, Z. and Xu, B., (2002) "infrastructural Health Monitoring with BOTDR Fibre Optic Sensing Technique," Proceedings of 1st International Workshop on Structural Health Monitoring of Innovative Civil Engineering Structures, ISIS Canada Corporation, Manitoba, Canada.

Yamaura, T, Inoue, Y, Kino, H., and Nagai, K., (2000) "Development of Structural Health Monitoring System Using Brillouin Optical Time Domain Reflectometer," Proceedings of 2nd International Workshop on Structural Health Monitoring, Stanford University, Stanford, CA.

Zeng, X., Yu, Q., Ferrier, G., and Bao, X., (2002) "Strain and Temperature Monitoring of A Concrete Structure Using A Distributed Brillouin Scattering Sensor," Proceedings of 1st International Workshop on Structural Health Monitoring of Innovative Civil Engineering Structures, ISIS Canada Corporation, Manitoba, Canada.

Zoëga, A. and Wiggenhauser. H., (2008) "Propagation Time of Elastic Surface Waves on Concrete Specimens under Uniaxial Loads," *NDE/NDT for Highways and Bridges: Structural Materials Technology (SMT)*, ASNT, Oakland, CA.

Zou, L., Ravet, F., Bao, X, Chen, L., Huang, R., and Khoo, H., (2004) "In-Line Inspection of Pipeline Buckling by Distributed Brillouin Scattering Sensor," Proceedings of 2nd International Workshop on Structural Health Monitoring of Innovative Civil Engineering Structures, ISIS Canada Corporation, Manitoba, Canada.

6 Structural Identification (STRID)

6.1 INTRODUCTION

6.1.1 TWO-DIMENSIONALITY OF MODERN STRID TECHNIQUES

Analysis of structures has followed a generalized Newton's 2nd law of motion since early times. Simply stated, the analyst would estimate some properties of the structure under consideration, and the exciting entity (generally in form of a force). Using a generalized equilibrium equation (which is a generalization of Newton's 2nd law of motion), the structural response is computed. Obviously, the accuracy of such an approach depends on the accuracy of the estimation of the structural properties. Given the complexities of modern structures and the great demands of natural and manmade hazards, it is clear that traditional estimates of structural properties needs to be improved. The high costs of construction, rehabilitation, and failure of infrastructure highlights the need for an even higher degree of accuracy of estimating (identification) the structural properties.

A technique of improving the accuracy of estimating structural properties is emerging in the form of structural identification (STRID) through field testing and monitoring. Thus STRID can be considered as an expansion of traditional Newton-based structural analysis into an experimental-theoretical level. In a way, STRID adds another dimension to the traditional structural analysis. To explain further: the original dimension of the traditional structural analysis as based only on theoretical identification of structures. Modern STRID techniques add to this another dimension of experimental analysis (Figure 6.1).

6.1.2 DEFINITION OF STRID

The subject of structural identification has been known in several fields such as aerospace and mechanical engineering for long time. Civil infrastructure community started investigating the subject in the 1970s. STRID is the process of identifying unknown parameters of a system by using some experimental results. There have been many more complex definitions; however this definition is simple, yet comprehensive. We will follow it throughout this volume. The simplest example of a structural identification process is the single spring construct,

$$KU = P \tag{6.1}$$

The structural identification problem here is to identify the structural property K (the stiffness) where P is the force and U is the displacement. Traditionally, the stiffness K is estimated using some kind of theoretical basis, resulting in an estimate of K_{Theory}. Modern STRID provide an experimental approach of estimating K. We can devise a simple experiment where we apply a force P_{ex} on the system, and measure the resulting displacement U_{ex}. The identified unknown property is then

$$K_{ex} = \frac{P_{ex}}{U_{ex}} \tag{6.2}$$

Obviously, the experimental stiffness is more accurate than the theoretical stiffness: it represents the real value of the stiffness. When using Equation 6.2 to estimate displacements with the measured

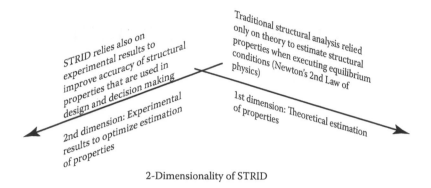

2-Dimensionality of STRID

FIGURE 6.1   2-Dimensionality of STRID.

stiffness value, the improved displacement estimate would naturally be more accurate. The difference between the theoretical and experimental estimations provides the benefit of the STRID effort. Figure 6.2 shows the STRID process.

6.1.3   Objectives of STRID

There are several uses and objectives of STRID, and the following are among them:

Condition Assessment of Structure: This includes design validation of new and existing structures: For new structures, identifying and verifying new techniques might be needed. For existing structures, rehabilitation effectiveness can be verified by STRID efforts. Load testing of bridges and other types of infrastructure sometimes utilize STRID techniques for further verification of bridge condition.

Construction Support: STRID can aid in verifying construction safety and adherence to plans.

Time-Dependent Changes: These include changes in demands such as load and changes in capacity due to environmental deterioration or changes in governing design codes.

Analytical Model Updating for New and Existing Structures: Modern infrastructure are complex and involved; this is reflected by the computer models that are used to analyze the structure. STRID can aid in updating these models so that they are up-to-date and accurate.

Damage Identification of New and Existing Structures: Diligent monitoring of structural health necessitates identifying damages and deterioration as early as possible to reduce costs of rehabilitation, and ensure safety.

Accurate LCA: Life cycle analysis (LCA) is emerging as an essential infrastructure management tool.

6.1.4   General Model of Structural Identification

There are perhaps hundreds of different quantitative methods and techniques that address structural identification issue. Moreover, there are additional countless qualitative personal approaches that handle the issue. One of our objectives of this chapter is to try to have a unified approach to the structural identification issue, both quantitatively and qualitatively, from SHM vintage point.

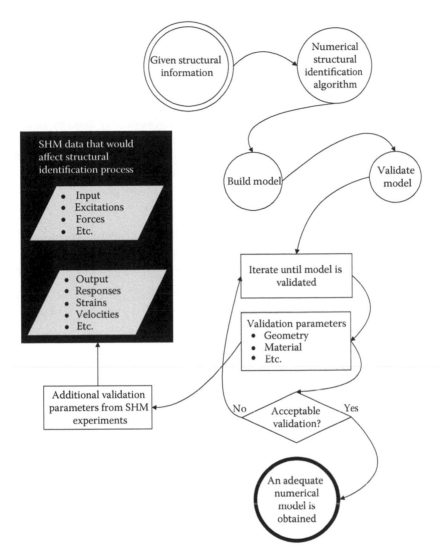

FIGURE 6.2 Process of structural identification with SHM experiments.

Perhaps the *very general* structural scheme of Figure 6.2 can help in devising a general model for structural identification issue. It simply states that for any situation, there is an input *I*, output *O*, and a systematic relationship *S* between them. Formally, the relationship can be written as

$$S I = O \tag{6.3}$$

Either of the input or output, or any combination of the two, can be measured during a nondestructive test (NDT) or SHM experiment. Similarly, they can be estimated by an analytical means. Structural identification problem can now be defined as

Evaluate the components of S that minimize |ΔI| and |ΔO|, subjected to set of constraints C(I_C, O_C, S_C).

$$|\Delta I| = I_{MEASURED} - I_{COMPUTED} \tag{6.4}$$

$$|\Delta O| = O_{MEASURED} - O_{COMPUTED} \tag{6.5}$$

The measured inputs and outputs, $I_{MEASURED}$ and $O_{MEASURED}$ are the inputs and outputs that are obtained as results of an NDT or SHM experiments. Similarly, the computed inputs and outputs $I_{COMPUTED}$ and $O_{COMPUTED}$ are the inputs and outputs that are obtained as results of analytical evaluations. It is clear that structural identification problem, as described above, is an optimization problem. The result of such an optimization problem is the optimized components of $S = S_{OPTIMUM}$.

It was mentioned earlier that the structural identification problem can be either qualitative, or quantitative. Qualitative approach to structural identification is usually done by best guess, engineering judgment, or past experiences. It is usually accurate for simple systems, or when very limited set of measurements are available. When the system becomes more complex, or when the measured data set is large, quantitative structural identification methods are utilized.

6.1.5   General Methods of STRID

Adding the second dimension of experimental basis to the conventional dimension of theoretical basis to form STRID methodology created several categories of STRID. The confluence of time (frequency) and space within the theory and tests helped in creating these STRID categories. They are as follows

- Modal identification (MI)
- Parameter identification (PI)
- Scale independent elements (SIE)
- Dispersion curves (DC)
- Statistical energy analysis (SEA)
- Artificial neural networks (ANN)
- Other miscellaneous methods

We explore each of these methods in more detail in this chapter.

6.1.6   Static versus Dynamic STRID

There are two major categories for STRID: statics based and dynamics based. As the names imply, the static based STRID uses static structural principles, mainly static equilibrium principles, to achieve its objectives. The dynamics based STRID utilize dynamics principles such as modal analysis or frequency domain analysis. Static based STRID is used in day to day bridge management activities such as bridge load rating and different bridge calibration procedures. This is mainly due to the simplicity of static based STRID. Dynamics based STRID offers potential for more accuracy and better insight to the bridge behavior. Of course, there is a case to be made that static based methods are simply a subset of dynamics based methods. For these reasons, we concentrate in most of this chapter on dynamics based methods. Discussion of some static based methods is offered in Section 6.4.4.1. We also note that artificial neural networks (ANN) can be applied to both static and dynamic STRID problems.

6.1.7   This Chapter

This chapter will discuss general aspects of STRID. These include metrics and methods. Of importance is, how space and time (frequency) attributes of the structure can affect choice and accuracy of the appropriate STRID method. We also argue that the metrics of structural response will vary as time and space ranges vary; thus affecting the STRID methodology. The following sections of this chapter will present in detail each of the STRID methods. Modeling techniques and their effects on STRID accuracy are presented next. Finally, decision making issues as they relate to STRID subject are discussed. Specifically: the importance of STRID to

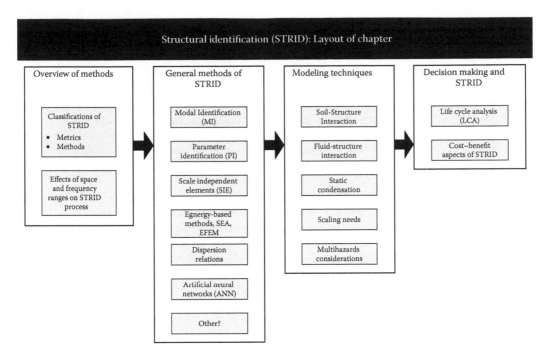

FIGURE 6.3   Structural identification chapter layout.

the emerging issue of LCA is presented. Also, recognizing that STRID efforts can be costly, the cost–benefit of STRID efforts is discussed. Simple methods are offered that can help the decision maker in estimating the value (cost vs. benefit) of an STRID effort. The contents of this chapter are shown in Figure 6.3.

6.2   STRID PROCESSES

6.2.1   GENERALIZED CLASSIFICATIONS OF STRID

Structural identification methods can be categorized into several general methods. The question, of course, is how is the structure identified? Each STRID method purports to identify the structure by using different metrics. Table 6.1 shows STRID general methods and some of the popular metrics each method uses to identify the structure.

The choice of an STRID method obviously depends on the desired metric. The reverse is true. In addition, we note that STRID methods are dependent on time and frequency scales, as discussed next.

6.2.2   STRID: INTERRELATION OF TIME AND SPACE MODELING ISSUES

6.2.2.1   Frequency and Space Scales

6.2.2.1.1   Global Response and Damage

Global response measure of a system can be described by global modes or global metrics such as displacements, natural frequencies of the system. When STRID method is based on subdividing the structure into smaller components (such as in finite element method, or statistical energy analysis method), the size of the components should be related to the frequency range of interest as well as the objective of the identification. For example, let us assume that a STRID project aims at identifying the degree of exposure of piles due to scour. The mode shapes of the piles that are sensitive

TABLE 6.1
Methods of STRID

STRID Method	Identification Metrics
Modal identification (MI)	Mode shapes, natural frequencies, modal damping
Parameter identification (PI)	Material properties, properties of modeling components such as springs, masses and joint/connection properties
Scale independent elements (SIE)	Material properties, properties of modeling components such as springs, masses and joint/connection properties Also, properties of wave mode shapes and wave numbers
Dispersion curves (DC)	Frequency/wave velocity relationships
Statistical energy analysis (SEA)	Modal density, loss factors, and coupling factors
Artificial neural networks (ANN)	Relationships between input and output parameters

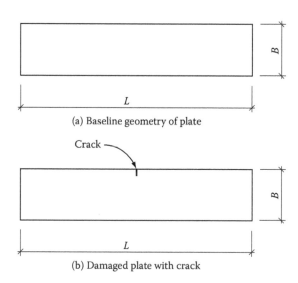

(a) Baseline geometry of plate

(b) Damaged plate with crack

FIGURE 6.4   Local damage example (a) baseline geometry, and (b) damaged plate with crack.

most to scour erosion are those that have the same characteristic length as the scour depth. Assume that the ratio of the total pile height to the exposed pile height is n. For accurate identification the following must be executed:

- The test and the follow-up identification should be performed up to a frequency range of about nf_0. The first natural frequency of the pile is f_0.
- If a structural modeling is used (parameter identification of a finite element model), each pile should include at least n elements.

Failing to follow the above rules at a minimum will result in an inaccurate identification.

6.2.2.1.2   Local Response and Damage

Let us consider the thin rectangular plate shown in Figure 6.4. Figure 6.4a shows the pristine plate and Figure 6.4b shows a plate that is slightly damaged by a small edge crack. The crack has a length of $\alpha\ell$, where α is a scalar constant and ℓ is a length measure. The plate length is L and the width of the plate is B. We are interested in studying the modal behavior of the plate and the assumed crack.

When the two plates of Figure 6.4 vibrate, their vibration modes will be almost identical, with the exception of the small dynamic effects of the crack. If the sources of vibration are located away from the crack the dynamic effects of the crack presence will result in a dynamic pressure bulb as shown in Figure 6.5. The dynamic pressure bulb *DPB* is defined as

$$DPB = \frac{p_1}{p_0} - 1$$

Where p_0 and p_1 are the pressures at a given location for the pristine and the damaged plate, respectively. The idea of dynamic pressure bulb can be visualized as a generalization of the static pressure bulb in soils under building foundations: the extent of the effects of the static (or dynamic) soil-structure interaction (SSI). The *DPB* can be used to explore the extent of the dynamic effects of the crack. The size of *DPB* depends on the frequency of vibration Ω and the wave number λ. Similarly, we can argue that the effects of the trenches in soils (which is similar geometrically to the crack problem of Figure 6.5) are dependent also on the size and shape of the trench. From the above we can claim that

$$DPB = g(\Omega, \lambda, \ell) \tag{6.6}$$

The above relationship is important in the field of structural and damage identification; there are two reasons for this importance:

- Placing any sensors outside the *DPB* would not effectively measure the dynamic effects of the crack (or damage)
- Whenever using modal decomposition in structural identification, the *DPB* that is active during the corresponding modal frequency will be the main clue as to the existence of the damage. As such, it would be of great interest to study the frequency range that maximizes the size of the *DPB*

Let us try to estimate the extent of the *DPB*. It is not an easy task, since there are limited results that can help in such an estimate. However, we can qualitatively assign a maximum size of a *DPB*, L_{DPB_MAX}, to be

$$L_{DPB\_MAX} = \beta(\alpha\ell) \tag{6.7}$$

A value of $\beta = 20$ seems to be reasonable. Such an upper limit of β should be confirmed on a case by case basis for important experiments.

For small size cracks, say a 0.1 in, the estimated maximum dynamic effects of such a crack is about 2.0 in. If such a crack exists in a plate with a depth and length of 25 in and 120 in.,

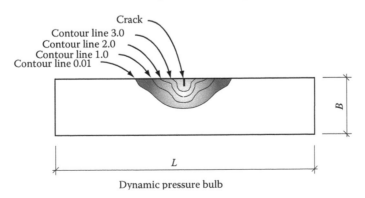

Dynamic pressure bulb

FIGURE 6.5 Example of dynamic pressure bulb.

respectively, its effects will not be easily discernable in most modes in a modal analysis approach. The only time where such an effect can be easily observed is when the mode of vibration is mostly the local damaged area (crack), that is, when the crack vibrates, while the motion of the rest of the plate is negligible. Such a mode is shown qualitatively in Figure 6.6.

Estimating the frequency range of such a local mode is fairly simple. Let us assume that the mass density is ρ, the equivalent vibrating width is W_{equ}, the equivalent vibrating height is H_{equ}, and the shear modulus is G. From Figure 6.5 we can estimate the vibrating mass to be

$$M = \rho \, W_{equ} \, H_{equ} \, t \tag{6.8}$$

and the equivalent stiffness

$$K = G \, \frac{H_{equ}}{2} \tag{6.9}$$

Thus, the resulting frequency is

$$f = \frac{1}{2\pi} \sqrt{\frac{K}{M}} \tag{6.10}$$

For the same 0.1-in crack length example above, we estimate the equivalent width and the equivalent height of the vibrating mass to be 0.07 in and 0.07 in, respectively. Assuming a unit plate width, the local frequency is estimated as 1.57 MHz. At crack length of 10 in the local frequency is 1.57 KHz.

The above example leads to several conclusions

- To detect a local damage experimentally, such as crack, sensors must be within the local dynamic pressure bulb.
- In addition to the above, the sampling rate of such measurements must be consistent with the expected frequencies of the crack size.
- To detect a crack through analytical structural identification methods, the employed analytical or numerical model must have high resolution to accommodate local frequencies. This includes the size of elements (if a finite element method is utilized) and the modal frequency range.
- If the resolution (element size or frequency range) is not consistent with demands of the crack (or damage) size, then the structural identification method that rely on finite element

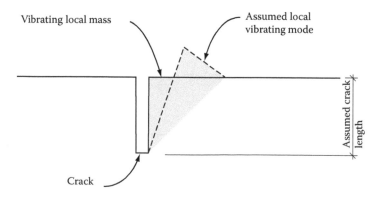

FIGURE 6.6   Local vibration of the crack.

modeling and/or modal analysis will result on averaging the effects of the *DPB*. Such an averaging might give some clues on the existence of damage. Moreover, in the absence of the required resolution, it is impossible to accurately estimate the damage location.

6.2.2.2 Objectives of STRID and Frequency-Space Domain

There are many objectives of STRID and some of those objectives were mentioned earlier. We can summarize those objectives into two basic ones: design verification and damage identification. The interrelationships between STRID objectives and frequency/space scales are discussed next.

6.2.2.2.1 Design Verification

Design verification, which includes parameter identification, system response during construction, or during normal operating conditions, or during abnormal hazards, is a major objective of STRID. Design parameters can be global (displacements, acceleration, overall stiffness, etc.) or local, such as local stiffness of a connection or weld and, local material property (modulus or elasticity, crack length, etc.). Obviously, methods of STRID should change as frequency and space scales change to identify those design parameters in an efficient and accurate fashion. Figure 6.7 shows the effects of frequency and space scales on STRID and design verification parameters.

Damage identification (DMID) is also one of the major objectives of STRID. The interrelationship between STRID and DMID depends on the frequency and space scales, as shown in Figure 6.8. There are three modes of STRID-DMID interrelationship. At lower frequency and larger space scales, to identify a structural damage, two structural states need to be identified. The damage can be identified as the difference between the two states. At high frequency and low space scales, the damage can be identified directly. In these conditions, STRID and DMID coincide. Another category at which STRID can be used for DMID is by using virtual sensing paradigm (VSP): The damage can be related to structural conditions by predetermined set of functions. VSP concepts are discussed elsewhere in this volume. They are most efficient at mid-frequency and mid-space scales.

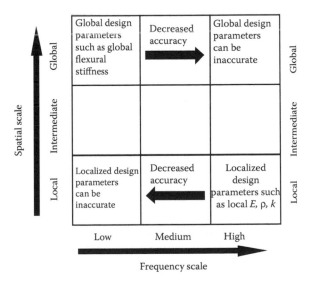

Design verification parameters depend on frequency and spatial scales; this controls the type of STRID method used.

FIGURE 6.7 Design parameters in STRID frequency-space domain.

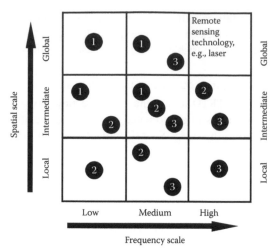

DMID-STRID interrelationship depends on frequency and spatial scales as follows:

1 To identify damage, two states of structure needs to be identified

2 To identify damage, VSP or ANN can be used

3 Damage can be identified directly

FIGURE 6.8   DMID methods in STRID frequency-space domain.

TABLE 6.2
Methods of STRID

STRID Method	Comment
Modal identification (MI)	Modal identification is limited to relatively lower frequency range. As number of modes increase, the accuracy of their identification is reduced. This also limits the method into large space scales
Parameter identification (PI)	Parameter identification is limited to relatively lower frequency range due to computational demands and modeling accuracies. As space scale decreases, those modeling and computational demands increase this limit and the method moves into larger space scales
Scale independent elements (SIE)	Scale independent elements method reduces the frequency and scale demands somewhat, thus permitting the method to be used at mid-frequency and space scale ranges
Dispersion curves (DC)	Dispersion curves can traverse into high frequency ranges and simultaneously to low space scales
Statistical energy analysis (SEA)	SEA is fairly suited to be used at medium and high frequency scales at almost any space scale
Artificial neural networks (ANN)	ANN does not aim to model the physical system properties. It relates input to output parameters directly. Thus it can be used at any frequency or space scales

6.2.2.3   Methods of STRID and Frequency-Space Domain

The efficiency and accuracy of the general categories of STRID methods are directly dependent on the frequency and time scales. Table 6.2 describes the frequency dependency of STRID general categories. Figure 6.9 shows the relative applicability frequency spectrum of STRID methods. Clearly, care is needed when using an STRID method to ensure its compatibility of the desired frequency range.

When we consider the space scale, the interdependency of the STRID methods on the frequency space becomes more intricate as shown in Figure 6.10. As the frequency and space

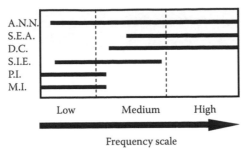

Frequency ranges of STRID techniques

FIGURE 6.9   Frequency ranges of STRID methods.

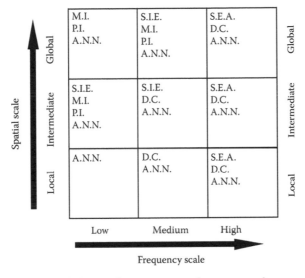

STRID techniques vary as frequency and space ranges change

FIGURE 6.10   STRID techniques in frequency-space domain.

change, the applicable STRID methods change. As of the writing of this volume, the boundaries of applicability has not been sufficiently studied, further understanding of this problem is needed.

6.3   MODAL IDENTIFICATION METHODS

6.3.1   THEORETICAL BACKGROUND

6.3.1.1   Overview

6.3.1.1.1   General

Modal identification (MI) methods aim at utilizing in field measurements to estimate modal values of the structure. The input to these methods is generally time (or frequency)-dependent excitations. The excitation can be ambient (no specific input), single source, or multiple sources. The measurements also vary greatly. They can be single or multiple measurements. Figure 6.11 shows general issues that affect MI efforts. The output of MI efforts will include estimated frequencies, mode shapes, and modal damping. These modal parameters can then be used directly in analysis, design,

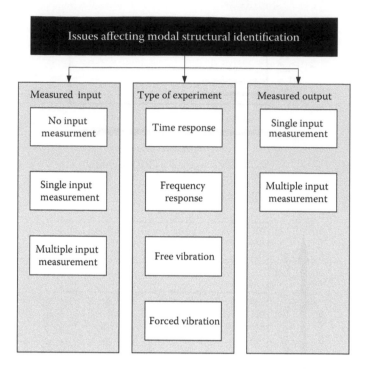

FIGURE 6.11   Issues affecting structural identification using modal methods.

DMID, or other decision making efforts. In some situations, the estimated MI parameters are used in further STRID efforts such as parameter identification.

Among the advantages of MI are that (1) they are well established, (2) they can be used directly in design and analysis efforts of low frequency problems such as seismic designs, (3) there is large body of experience of MI methods, and (4) there are large number of MI methods that can be used in numerous situations. Among the disadvantages of MI methods are the following (1) they are applicable only to low frequency range, (2) nonlinear applications are limited, (3) lack of fluid-structure interaction applications, and (4) limited use for DMID at early stages of damage formation (at early stages, damages usually affect high frequency ranges).

6.3.1.1.2   Different Classifications

There are several types of classifications of MI methods, we explore some of those classifications next.

Experimental versus Operational MI: As the name implies, experimental MI would entail artificially exciting the system on hand by a shaker, for example. Operational MI would utilize the ambient operating conditions as the exciting mechanism, Brinker and Moller (2006). As such, no special exciting equipments are needed. Experimental modal analysis have been around much longer, and thus the MI methods are well established. Lately, however, many operational MI methods have been developed, and their popularity is increasing. For bridge structures, operational modal analysis offer several advantages: (1) They affect traffic much less than experimental MI; (2) Large bridges can be modeled since no mechanical equipment is needed to excite the large bridge mass, (3) They would be more cost effective, and (4) They can be included as a part of long-term SHM projects. The details of an SHM experiment that utilize operational MI are presented later this section.

Order of MI Methods: Single degree of freedom dynamic structural equations are second-order differential equations. For multiple modes, the matrix size N would relate to the number of modes to be identified. It is possible to reduce the size of the matrices while increasing the order of the dynamic

TABLE 6.3
Order of Modal Methods

| Model | Matrix Size N | Order r | Experimental Conditions | | Civil Engineering Examples |
			Spatial Measurements: Number of Sensors	Temporal Measurements: Inertia Effects	
High order	Large	Low	Small	Large	STRID
Low order	Low	Large	Large or complete	Medium to low	STRID
Zeroth order	Large	Zero	Large or complete	Static	Load testing

equation. At the limit, as the dynamic equation is reduced to a scalar equation, the order of such an equation becomes N. There are different MI solution methods for different order of equations. Also, the type of experiment depends on the order of the MI solution method as Table 6.3 shows. Detailed description of model order is explained later.

Frequency versus Time Domain: MI methods can be solved either in the frequency of time domains. Examples of both approaches are given later in this section.

6.3.1.2 Equations of Motion

Equation of motion of free vibration is expressed by

$$[M]\{\ddot{U}\}+[C]\{\dot{U}\}+[K]\{U\}=\{0\} \tag{6.11}$$

The mass, damping, and stiffness matrices are $[M]$, $[C]$, and $[K]$, respectively. The displacements, velocity, and acceleration vectors of the system are $\{U\}$, $\{\dot{U}\}$, and $\{\ddot{U}\}$, respectively. The characteristic equation using Laplace operator s is

$$s^2[M]+s[C]+[K]=\{0\} \tag{6.12}$$

Both Equations 6.11 and 6.12 are used extensively in MI methods.

6.3.1.3 Modal Solution

For a nontrivial solution of Equation 6.11, it needs to have the harmonic form

$$\{U\}=\{\varphi\}e^{\lambda t} \tag{6.13}$$

Substituting into Equation 6.11

$$\left[-\lambda^2[M]+i\lambda[C]+[K]\right]\{\phi\}=\{0\} \tag{6.14}$$

We recognize Equation 6.14 as a quadratic eigenvalue problem, with $2N$ eigensolution of λ_i and $\{\phi\}_i$, where $i=1,2\ldots2N$. The eigenvalue λ_i and the eigenvectors $\{\phi\}_i$ are complex valued, in general.

We note that for an undamped system ($[C]=[0]$) or for proportional damping, Equation 6.14 is reduced to a linear eigenvalue problem. In such a situation, only N eigensolutions are needed to solve the problem. These eigensolutions are real, and the well known equalities

$$[\Phi]^T[M][\Phi]=\langle I \rangle \tag{6.15}$$

and

$$[\Phi]^T[K][\Phi]=\langle \omega \rangle \tag{6.16}$$

apply.

The identity matrix is $\langle I \rangle$. The diagonal matrix $\langle \omega \rangle$ contains the natural frequencies of the system. The matrix $[\Phi]$ contains the vectors of the mode shapes $\{\phi\}_i$.

Returning back to the general problem of Equation 6.14, the displacement of a structural system can be expanded into the series

$$\{U\} = \sum_{i=1}^{i=2N} A_i \{\phi\}_i \, e^{\lambda_i t} \tag{6.17}$$

where A is an arbitrarily constant that is a function of initial conditions. Some methods incorporate the constant A_i with the mode shapes $\{\phi\}_i$ to get the simpler form

$$\{U\} = \sum_{i=1}^{i=2N} \{\phi\}_i \, e^{\lambda_i t} \tag{6.18}$$

Complex natural frequencies λ_k are related to the natural frequency ω_k and modal damping by

$$\mathrm{Re}\,al\left(\lambda_k\right) = -\xi_k \, \omega_k + i\left(\omega_k \sqrt{1 - \xi_k^2}\right) \tag{6.19}$$

$$\mathrm{Im}\,ag\left(\lambda_k\right) = -\xi_k \, \omega_k - i\left(\omega_k \sqrt{1 - \xi_k^2}\right) \tag{6.20}$$

where $i = \sqrt{-1}$.

We changed the mode count from i to k in Equations 6.19 and 6.20 to avoid confusion with the complex constant $i = \sqrt{-1}$. This indicates that by evaluating the complex natural frequency it is possible to evaluate both the natural frequency and the modal damping. The practice is a popular practice in MI efforts.

6.3.1.4 Frequency Response Function (FRF)

In forced harmonic vibrations with a driving frequency Ω the equation of motion becomes

$$\left[-\Omega^2 [M] + i\Omega[C] + [K]\right]\{U\} = \{P\} \tag{6.21}$$

The harmonic force and displacement vectors are $\{P\}$ and $\{U\}$, respectively. The harmonic displacement is obtained as

$$\{U\} = [F]\{P\} \tag{6.22}$$

with

$$[F] = \left[-\Omega^2 [M] + i\Omega[C] + [K]\right]^{-1} \tag{6.23}$$

The matrix $[F]$ is called *FRF*. It relates applied harmonic forces to harmonic displacements of the system. Its components are frequency dependent and complex valued, in general. The components of FRF in the ith row and j^{th} $f_{ij}(\Omega)$ represents the harmonic displacement of the ith degree of freedom due to unit harmonic force in the direction of the jth degree of freedom. Physically, it represents the dynamic flexibility of the system. It is directly related to dynamic influence lines (DIL) discussed in Chapter 8 of Ettouney and Alampalli (2012). FRF is used extensively in MI. Measuring system responses at different points during experiments due to harmonic forces would generate the FRF of the system.

6.3.1.5 Impulse Response Function (IRF)

The inverse Fourier transform (see Chapter 7), of the FRF $f_{ij}(\Omega)$ is a time-dependent signal $h_{ij}(t)$. This signal is called *IRF*. It can be shown to be the time-dependent displacement response of the ith degree of freedom of the system due to unit initial velocity in the direction of the jth degree of freedom. The time-dependent matrix that holds all $h_{ij}(t)$ is also used extensively in MI methods. By applying initial velocity (impulse) to the system and measuring responses, the IRF matrix can be established.

6.3.1.6 Pseudo Inverse

Consider the linear matrix equation

$$[A]_{N \times M} \{x\}_M = \{y\}_N \tag{6.24}$$

If $M \neq N$, the matrix $[A]_{N \times M}$ can't be inverted by conventional means. However, there is an easy way to solve Equation 6.24. First, remultiply by $[A]^T$

$$[A]^T_{M \times N} [A]_{N \times M} \{x\}_M = [A]^T_{M \times N} \{y\}_N \tag{6.25}$$

Rewrite as

$$[B]_{M \times M} \{x\}_M = [A]^T_{M \times N} \{y\}_N \tag{6.26}$$

with

$$[B]_{M \times M} = [A]^T_{M \times N} [A]_{N \times M} \tag{6.27}$$

The square matrix $[B]_{M \times M}$ can be inverted by conventional method. The required solution $\{x\}_M$ is obtained by

$$\{x\}_M = [B]^{-1}_{M \times M} [A]^T_{M \times N} \{y\}_N \tag{6.28}$$

or

$$\{x\}_M = [A]^+_{M \times N} \{y\}_N \tag{6.29}$$

with

$$[A]^+_{M \times N} = [B]^{-1}_{M \times M} [A]^T_{M \times N} \tag{6.30}$$

The matrix $[A]^+_{M \times N}$ is the pseudo inverse of $[A]_{N \times M}$. The pseudo inverse process is used in many structural identification methods, such as the eigenstructure assignment method.

6.3.1.7 Singular Value Decomposition (SVD)

Sometime expansions of matrix $[A]_{N \times M}$ is needed in the form

$$[A] = [U]\langle D \rangle [V]^T \tag{6.31}$$

The matrices $[U]$, $\langle D \rangle$, and $[V]^T$ can be evaluated using popular numerical solutions. They are related to the eigenvalue properties of $[A]_{N \times M}$. These matrices are used in several MI methods. For additional information see He and Fu (2001).

6.3.1.8   Modal Assurance Criteria (MAC)

A very useful estimate of the accuracy of estimated mode shapes is the MAC. MAC is a measure of correlation between the estimated mode shapes $\{\phi\}_{Ei}$ and the measured mode shapes $\{\phi\}_{Mi}$. Note that $i = 1,2, \dots N$ and $j = 1,2, \dots M$. With N and M representing the total estimated mode shapes and the total measured mod shapes, respectively.

The MAC is now defined as

$$MAC\left(\{\phi\}_{Ei}, \{\phi\}_{Mj}\right) = \frac{\left(\{\phi\}_{Ei}^{T}\{\phi\}_{Mj}\right)^{2}}{\left(\{\phi\}_{Ei}^{T}\{\phi\}_{Ei}\right)^{2}\left(\{\phi\}_{Mj}^{T}\{\phi\}_{Mj}\right)^{2}} \tag{6.32}$$

Note that $MAC(\{\phi\}_{Ei}, \{\phi\}_{Mj}$ is a scalar constant. Since it is a correlation measure, its range is between 0 and 1.0. If $MAC(\{\phi\}_{Ei}, \{\phi\}_{Mj})$ is equal to, or close to, 1.0, then the ith estimated mode is well correlated to the jth measured mode. If $MAC(\{\phi\}_{Ei}, \{\phi\}_{Mj})$ is equal to, or close to, 0 then the ith estimated mode is not correlated to the jth measured mode.

MAC is a popular technique that is used by many practitioners and researchers, see, for example, Padur et al. (2002).

6.3.1.9   Matrix and Vector Norms

A norm is a measure of the matrix, or vector, magnitude. Since there are many ways to express such magnitude, there are many ways to define a norm of a matrix or vector. For example, the sum of the magnitude of a square matrix $[C]$ is the norm

$$\|[C]\|_{1} = \sum_{i=1}^{n}\sum_{j=1}^{n}|C|_{ij} \tag{6.33}$$

The size of the matrix is n.

Another important norm that are used frequently in structural identification methods is the Euclidean norm (sometimes is referred to as Frobenius norm) it is expressed as

$$\|[C]\|_{E} = \sqrt{\sum_{i=1}^{n}\sum_{j=1}^{n}C_{ij}^{2}} \tag{6.34}$$

The reader will recognize that this norm is the basic equation in any least square process. The Euclidean norm importance in the SHM field is self evident.

Note that the norm expressions for vectors are similar to the above expressions. In case of vectors, replace the double summation operators with a single summation operator.

6.3.2   Order of System Models

Perhaps one of the most important decisions in the structural identification process, as well as the MI process is the choices of the structural degrees of freedom (DOF) and/or the number of modes of interest to be estimated. For the purposes of this section, let us assume that the number of desired modes to be identifies is the same as the DOF of the structural model N. The general matrix equation of motion of the system is Equation 6.11 with the size of the matrix equation is N. Laplace transform of the equation of motion is Equation 6.12. Note that the power of the Laplace operator of this equation is 2. We can express Equation 6.12 as

$$s^{2}\begin{bmatrix} M_{11} & M_{12} \\ M_{21} & M_{22} \end{bmatrix} + s\begin{bmatrix} C_{11} & C_{12} \\ C_{21} & C_{22} \end{bmatrix} + \begin{bmatrix} K_{11} & K_{12} \\ K_{21} & K_{22} \end{bmatrix} = \{0\} \tag{6.35}$$

Where we expressed the original matrices with size N with individual submatrices with size $N/2$. The new equation can be rearranged as

$$s^4 \left[C_4\right] + s^3 \left[C_3\right] + s^2 \left[C_2\right] + s \left[C_1\right] + \left[C_0\right] = \{0\} \tag{6.36}$$

The size of the above equation is $N/2$ where as the power of the Laplace operator is 4: we reduced the size of the matrices, while increasing the power of the operator.

We can continue this reduction/increase process. After n processes, the generic characteristic equation becomes

$$s^{2n} \left[C_{2n}\right] + s^{2n-1} \left[C_{2n-1}\right] + s^{2n-2} \left[C_{2n-2}\right] + \cdots + \left[C_0\right] = \{0\} \tag{6.37}$$

The size of the matrices is $N/2n$ while the power of the Laplace operator is $2n$. Ultimately, the characteristic equation becomes scalar, with the power of Laplace operator of N such as

$$s^N c_N + s^{N-1} c_{N-1} + s^{N-2} c_{N-2} + \cdots + c_0 = 0 \tag{6.38}$$

The roots of this scalar characteristic equation, or any intermediate reduced equation, are the same eigensolutions of the original system matrix Equation 6.11.

The power of the Laplace operators in any of the above characteristic equations is also referred to as the order of the model. For example, the minimum order of the model represented by Equation 6.12 (the original characteristic equation) is 2, the highest order of the model is N, it occurs for scalar characteristic equations.

There is important significance for the model order representation in structural identification field. Recall that the structural identification purports to estimate modal properties relying only on measurements. The number of stations where the measurements are made P is not always equal to the number of modes of interest. This frequent mismatch necessitates using models with different order for maximum computational efficiency. For example, when P is too small compared with N, the methods with high model order might be more efficient than models with low order. There are numerous MI methods that utilize low- or high-order models.

One of the major challenges to any MI method is the way the characteristic equation is formed, and the relationship between such formation and the measurements. In later sections some popular methods will be presented.

We need to state one final note regarding the reduction/increase of matrix size/power of operators. We demonstrated the concept using Laplace characteristic equations. Similar processes can be made using the differential operators, for example, Equation 6.11 in the time domain. The characteristic equations in this case are differential equations of higher order.

6.3.3 ATTRIBUTES OF MODAL IDENTIFICATION METHODS

The model order is one of the basic attributes of MI methods. There are several other attributes for those methods. For example, the solution domain (time domain or frequency domain). But, some methods do utilize both time and frequency domain solutions. Another attribute is the number of input sources and number of output stations. Input sources are the locations where the structure is excited by an external force, usually to generate an IRF. Output stations are where the structural response is measured, usually in the form of IRFs. If FRFs are needed by the MI method, a simple transform can be performed. There are several input/output combinations that can affect the choice of the MI method. Among those combinations are single input source/single output station (SISO); single input source, multiple output sources (SIMO); and multiple input sources, multiple output sources (MIMO). We also note that there are some methods that do not rely on an artificially generated excitation source, such as the random decrement (RD) method see Asmussen, J.C., (1997). The RD method utilizes random excitation of the structure, which make it ideal for

TABLE 6.4
Attributes of Modal Identification Methods

Method	DOF	Input-Output	Domain	Order
Peak amplitude	SDOF	SISO	FD	
Least squares	SDOF	SISO	FD	
Rational fraction polynomials	MDOF	SIMO	FD	High
Ibrahim method	MDOF	SIMO	TD	
Least squares	MDOF	SIMO	TD	
Random decrement	MDOF	NA	TD	
Autoregressive moving average (ARMA)	MDOF	SIMO	TD	
Least Squares Complex Exponential (LSCE)	MDOF	SIMO	TD	High
Poly-reference	MDOF	MIMO	FD	Low
Ibrahim method-multi-input	MDOF	MIMO	TD	Low
Poly-reference	MDOF	MIMO	TD	Low
Global modal analysis method	MDOF	MIMO	TD	
Eigensystem realization algorithm (ERA)	MDOF	MIMO	TD	High

heavy bridge structures that are difficult to excite by an artificial means. Hence, this method is given in detail with a case study on bridges. Table 6.4 shows the different attributes of some popular MI methods. Some of those methods will be discussed briefly in the next sections. As the number of input sources and/or output stations increase, the cost of experiment as well as the accuracy of the results and completeness of information increase. To realize an optimal balance between cost and benefit of the experiment that aim at MI of particular structure, a study using the theory of experimentation is recommended. See the pertinent example in Chapter 4.

6.3.4 PEAK AMPLITUDE

For the FRF, $\alpha(\omega)$ of an SDOF, define the range of interest. Such a range should contain a peak, or resonance. The peak is defined by $\alpha(\omega)_{max}$ within the range of interest. The natural frequency that corresponds to that resonance is ω_n, where n is the nth mode of interest. The damping ratio ξ_n that correspond to the nth mode can be estimated using the analytical expression for $\alpha(\omega)$ of the SDOF. The modal damping is related to ω_1 and ω_2, which are the two frequencies at both sides of ω_n as the magnitude of $\alpha(\omega)$ drops to $\alpha(\omega)_{max}/\sqrt{2}$. Thus,

$$\xi_n = \frac{\omega_1^2 - \omega_2^2}{4\,\omega_n^2} \tag{6.39}$$

The above completely identifies SDOF parameters. If the measurements involve an MDOF, then the above procedure is applicable for every range of the FRF that contains resonance. In the MDOF case, the identification of the modal amplitude A_n is needed. Recalling the frequency domain modal solution near resonance, the modal amplitude can be defined as

$$A_n = 2\,\alpha(\omega)_{max}\,\xi_n\,\omega_n \tag{6.40}$$

The peak amplitude method is simple, yet can result in very accurate results. For ranges of FRF with close resonances, it might yield inaccurate results. Also, it might not yield accurate results for highly damped systems. Such systems are prevalent in civil infrastructure, such as composite steel-concrete bridges, or soil-structure systems. Nevertheless, due to its simplicity, this method can be used as an approximate estimate for modal parameters.

6.3.5 EIGENSOLUTION REALIZATION METHOD (ERA)

The ERA is one of the most popular MI methods. As usual, it aims at finding modal shapes, modal frequencies, and modal damping of a given system, using a set of measurements, usually displacements. We will summarize some of the essential steps of the method. For more details, the reader should consult other references, such as He and Fu (2001), or Doebling et al. (1999).

We assume that there are N_1 and N_2 measurement points and external forces, respectively. We also assume that the system to be identified has N DOF. The basic step in the ERA method is to form Hankel matrix using available measurement points. Let us define the impulse response vector $\{h\}_{mij}$ to be of size M. Also note that $i = 1, 2, \dots N_1$, and $j = 1, 2, \dots N_2$. Note that the vector represents a time measurement stream at the ith measurement point due to the jth external force. The m subscript represents the time step at which $\{h\}_{mij}$ measurements starts. Thus, the kth element in the vector $\{h_k\}_{mij}$ is the time sample measured at $k+m$ time step at the ith point due to the jth external force. The size of each vector M should be greater than or equal to $2N$. If this condition is not met, then a time shifting scheme can be used to fill the needed spaces in the vector to make it at least equal to.

We can now define the basic Hankel matrix $[H_m]_{Base}$ as

$$[H_m]_{Base} = \begin{bmatrix} \{h\}_{m11} & \{h\}_{m12} & \cdots & \{h\}_{m1N_2} \\ \{h\}_{m21} & \{h\}_{m22} & \cdots & \{h\}_{m2N_2} \\ \vdots & \vdots & \vdots & \vdots \\ \{h\}_{mN_1 1} & \{h\}_{mN_1 2} & \cdots & \{h\}_{mN_1 N_2} \end{bmatrix}_{mN_1 \times N_2} \tag{6.41}$$

The base Hankel matrix $[H_m]$ is then formed by using a time shifted $[H_m]_{Base}$ as submatrices as follows

$$[H_m] = \begin{bmatrix} [H_m]_{Base} & [H_{m+1}]_{Base} & \cdots & [H_{m+m2}]_{Base} \\ [H_{m+1}]_{Base} & [H_{m+2}]_{Base} & \cdots & [H_{m+m2+1}]_{Base} \\ \vdots & \vdots & \vdots & \vdots \\ [H_{m+m1}]_{Base} & [H_{m+m1+1}]_{Base} & \cdots & [H_{m+m1+m2}]_{Base} \end{bmatrix}_{mm1N_1 \times m2N_2} \tag{6.42}$$

The constants $m1$ and $m2$ are arbitrary. They are used to reduce the effects of measurement noise on the accuracy of results. Note that the larger $m1$ and/or $m2$, the larger the amount of measurement information in Hankel matrix, thus, it is expected that the resulting modal parameters will be more accurate. There is a limit, though on how high $m1$ and/or $m2$ can be. Additional data might not improve the accuracy. The values of $m1$ and/or $m2$ can be established by experience, or by iterative processes.

Now Hankel matrix is completely defined, and the realization of the modal properties can proceed. He and Fu showed that if we decompose Hankel matrix that is evaluated at $m = 0$, $[H_0]$ using SVD as

$$[H_0] = [U]^T [\Sigma][V] \tag{6.43}$$

Equation 6.43 contains the singular value decomposition components of $[H_0]$, see Section 6.3.1.7. The matrix $[\Sigma]$ is a diagonal matrix of size $2N$ and contains the singular values of $[H_0]$. The matrices $[V]$ and $[U]$ are the right and left singular vectors of $[H_0]$, respectively. The row size of $[V]$ is $2N$ and its column size is of the same size as the size of the column of $[H_0]$. Also, the row size of $[U]$ is $2N$ and its column size is of the same size as the size of the row of $[H_0]$.

Upon defining another Hankel matrix at $m = 1$, $[H_1]$, we can define the matrix $[A]$ as

$$[A] = [\Sigma]^{-1/2} [U][H_1][V]^T [\Sigma]^{-1/2} \qquad (6.44)$$

Note that $[A]$ is a square matrix with a size of $2N$. The complex $2N$ eigenvectors of $[A]$ are the required system mode shapes. Each of the complex $2N$ eigenvalues of $[A]$ constitutes the ith complex natural frequency s_i of the system. They are related to the natural frequency and modal damping by Equations 6.19 and 6.20. For more details, see He and Fu (2001).

6.3.6 IBRAHIM TIME DOMAIN (ITD) METHODS

As one of the popular MI techniques, the ITD is a low order method that uses the modal equation

$$\{U\} = \sum_{i=1}^{i=2N} \{\phi\}_i \, e^{\lambda_i t} \qquad (6.45)$$

The method is described in several references; see, for example, Ibrahim (1998). We summarize the basic approach of the method. The method aims at computing the complex eigensolutions (complex frequencies) λ_i and the associated complex modes shapes $\{\phi\}_i$. The measurements $\{U(t_j)\}$ include measurements at N locations. Thus, at time t_j

$$\{U(t_j)\} = \sum_{i=1}^{i=2N} \{\phi\}_i \, e^{\lambda_i t_j} \qquad (6.46)$$

The initial sample time is $T_0 = t_1$. Defining $2N$ measurements vector of size N each as

$$[U]_{T_0} = \left[\{U(t_1)\} \{U(t_2)\} \cdots \{U(t_{2N})\} \right] \qquad (6.47)$$

We can write

$$[U]_{T_0} = [\Phi]_{T_0} [\Lambda] \qquad (6.48)$$

The matrix $[\Phi]_{T_0}$ contains the vectors of complex mode shapes $\{\phi\}_i$. The components of the matrix $[\Lambda]$ are

$$[\Lambda]_{ij} = e^{\lambda \, t_j} \qquad (6.49)$$

Note that the matrix $[U]_{T_0}$ is of size $N \times 2N$.

The method defines three more response equations, each with a time shift of T_1, T_2 and $T_1 + T_2$ from the first response equations, thus

$$[U]_{T_1} = \left[\{U(t_1 + T_1)\} \{U(t_2 + T_1)\} \cdots \{U(t_{2N} + T_1)\} \right] \qquad (6.50)$$

$$[U]_{T_2} = \left[\{U(t_1 + T_2)\} \{U(t_2 + T_2)\} \cdots \{U(t_{2N} + T_2)\} \right] \qquad (6.51)$$

$$[U]_{T_1 + T_2} = \left[\{U(t_1 + T_1 + T_2)\} \{U(t_2 + T_1 + T_2)\} \cdots \{U(t_{2N} + T_1 + T_2)\} \right] \qquad (6.52)$$

Furthermore, we can define the time-shifted mode shape vectors as

$$\{\phi\}_{i,T_1} = \{\phi\}_i \, e^{\lambda_i T_1} \qquad (6.53)$$

Defining the matrix $[\Phi]_{T_1}$ to contain the vectors of complex mode shapes $\{\phi\}_{i,T_1}$ the T_1-time-shifter response matrix equation can be written as

$$[U]_{T_1} = [\Phi]_{T_1} [\Lambda] \tag{6.54}$$

The responses of T_0 and T_1 can be combined as

$$\begin{bmatrix} [U]_{T_0} \\ [U]_{T_1} \end{bmatrix} = \begin{bmatrix} [\Phi]_{T_0} \\ [\Phi]_{T_1} \end{bmatrix} [\Lambda] \tag{6.55}$$

In short,

$$[Y] = [\Phi][\Lambda] \tag{6.56}$$

Not that $[Y]$ and $[\Phi]$ are $2N \times 2N$.

Repeating for initial sampling times of T_2 and T_1+T_2.

$$\begin{bmatrix} [U]_{T_2} \\ [U]_{T_1+T_2} \end{bmatrix} = \begin{bmatrix} [\Phi]_{T_2} \\ [\Phi]_{T_1+T_2} \end{bmatrix} [\Lambda] \tag{6.57}$$

or

$$[Y]_1 = \begin{bmatrix} [U]_{T_2} \\ [U]_{T_1+T_2} \end{bmatrix} = [\Phi]\langle\gamma\rangle[\Lambda] \tag{6.58}$$

The ith component of the diagonal matrix $\langle\gamma\rangle$ is

$$\gamma_i = e^{\lambda_i T_1} \tag{6.59}$$

We now eliminate the highly nonlinear matrix $[\Lambda]$ from Equations 6.56 and 6.58

$$[Y]_1 [Y]^{-1} [\Phi] = [\Phi]\langle\gamma\rangle \tag{6.60}$$

Rearranging, we reach the eigenvalue problem

$$[A][\Phi] = \langle\gamma\rangle[\Phi] \tag{6.61}$$

where

$$[A] = [Y]_1 [Y]^{-1} \tag{6.62}$$

Upon solving the eigenvalue problem, Equation 6.61, the resulting mode shapes $[\Phi]$ will represent the structural mode shapes. The eigenvalues γ_i are related to the ith modal damping and natural frequency. See Ibrahim (1998) for details of extracting the modal damping and natural frequencies from γ_i. Techniques for establishing the number of modes that contribute to the response, the mismatch between number of measurements and the unwanted noise effects are also described in detail by Ibrahim (1998).

6.3.7 LSCE (LEAST SQUARE COMPLEX EXPONENTIAL)

Earlier we identified the characteristic equation of a dynamic system, Equation 6.38, as a high-order model for structural identification. This characteristic equation is the basis of the LSCE for structural MI. For convenience, we rewrite Equation 6.38 for an order of $2N$ as

$$\sum_{i=0}^{i=2N} \bar{c}_i \, s^i = 0 \tag{6.63}$$

The roots of Equation 6.63 s_k represent the complex natural frequencies of the system, where $k = 1, 2, \ldots 2N$. To compute the roots of the characteristic Equation 6.63, we first need to compute its constants. Consider the solution for the IRF of Equation 6.38.

$$h_{ij}(t) = \sum_{k=1}^{k=2N} A_{ijk} \, e^{s_k t} \tag{6.64}$$

As usual, we assume that the IRF $h_{ij}(t)$ is the measured responses at the ith DOF due to a force at the jth DOF. Assuming $t = \ell \, \Delta t$, where Δt is the sampling time step and introducing $z_k^\ell = e^{s_k \, \ell \Delta t}$, we can rewrite Equations 6.63 and 6.64) as

$$\sum_{i=0}^{i=2N} c_i \, z_k^i = 0 \tag{6.65}$$

$$h_{ij\ell} = \sum_{k=1}^{k=2N} A_{ijk} \, z_k^\ell \tag{6.66}$$

Note that we changed the constant $\bar{c}_i$ into c_i in Equation 6.65. Simple multiplications and summations of Equation 6.66 produce

$$\sum_{\ell=1}^{\ell=2N} c_\ell \, h_{ij\ell} = \sum_{k=1}^{k=2N} A_{ijk} \sum_{\ell=1}^{\ell=2N} c_\ell \, z_k^\ell \tag{6.67}$$

From Equation 6.67, we write the left hand side of Equation 6.65 as

$$\sum_{\ell=1}^{\ell=2N} c_\ell \, h_{ij\ell} = 0 \tag{6.68}$$

Dropping the i and j subscripts for simplicity, Equation 6.67 becomes a linear combination of a measured IRF set of $2N$ samples. If we assume, arbitrarily without any loss of generality, that $c_{2N} = -1.0$, and repeat Equation 6.67 for $2N$ sets of IRF, we reach the matrix equality

$$[H]\{c\} = \{h\} \tag{6.69}$$

The size of the matrices in Equation 6.69 is $2N$. Vector $\{c\}$ contains the unknown coefficients c_ℓ. The vector $\{h\}$ can be expressed as

$$\{h\} = \begin{Bmatrix} h_{2N} \\ h_{2N-1} \\ \vdots \\ h_{4N-1} \end{Bmatrix} \tag{6.70}$$

Also

$$[H] = \begin{bmatrix} h_0 & h_1 & \cdots & h_{2N-1} \\ h_1 & h_2 & \cdots & h_{2N} \\ \vdots & \vdots & \cdots & \vdots \\ h_{2N-1} & h_{2N} & \cdots & h_{4N-2} \end{bmatrix} \tag{6.71}$$

Equation 6.69 can be solved for the unknown coefficients c_ℓ. The characteristic Equation 6.65 can be solved to find the characteristic roots z_k. Using the relation between z_k and the complex natural frequencies s_k, the complex natural frequencies of the system s_k can be evaluated. Using expressions of complex natural frequencies in Equations 6.19 and 6.20, the natural frequencies and modal damping ω_k and ζ_k are computed.

The mode shapes can be computed by evaluating the modal amplitudes A_{ijk}. Since we know now the values of z_k, we can rewrite Equation 6.67 in matrix form as

$$[Z]\{A\} = \{h\} \tag{6.72}$$

The components of the square matrix $[Z]$ are the computed values of z_k. The right-hand side vector $\{h\}$ contains the first measured $2n - 1$ samples. The unknown vector $\{A\}$ contains the unknown modal amplitudes A_{ijk}. Upon solving Equation 6.72 to find $\{A\}$, the MI of the system is complete. For a comprehensive description of the LSCE method, see He and Fu (2001).

6.3.8 RANDOM DECREMENT

6.3.8.1 General
Reliable bridge-condition evaluation and early detection of bridge-component failures are critical for bridge owners in the United States for better utilization of available resources. Remote bridge monitoring systems (RBMS) have been perceived to assist periodic evaluation of structures to supplement bridge-management systems with quantitative data, and for examining new design techniques. Many RBMS thus far are based on measured bridge vibration. One of the major issues in developing RBMS is the lack of reliable methods to obtain modal parameters using traffic excitation. This section summarizes a signal processing method that was offered by Alampalli and Cioara (2000). The method relies on data measured from one of the RBMS installed in New York State bridges.

6.3.8.2 Overview of Problem
Bridges are relatively complex structures and their structural integrity must be verified through periodic inspection. Most bridges require inspection at least once in every 2 years, by federal mandate, but serious failures can occur between the inspections, FHWA (2004). A visual inspection, often, may not reveal hidden defects, and the quality of the inspection is highly dependent on training, personnel experience, available equipment, and in-service environment, see Moore et al. (2001), Phares et al. (2001), and Phares et al. (2004). Thus other nondestructive methods were developed and used by the inspectors to supplement, as needed, the visual inspection of bridges. See Washer (2002), Tolbert et al. (2004), Siswobusono et al. (2004), Aref and Alampalli (2001), Moore et al. (2004) and Fuchs et al. (2004). Continuous health monitoring of bridge structures has also been a major topic of research due to the aging bridge infrastructure and lack of economic resources to completely renovate all of them. Thus, for maximizing and efficiently using available economic resources, reliable bridge-condition evaluation and early detection of possible failures has been a major subject in transportation research arena.

Remote bridge monitoring systems, using experimental modal investigation where bridge integrity is correlated with structural modification, have received considerable attention in the last decade due to their perceived ability to assist in bridge evaluation and bridge management, see Chong et al. (2001), Alampalli and Ettouney (2003), and Cheung and Naumoski (2002). Recent developments in sensor technology have yielded durable and reliable sensors making use of such systems possible. New York realizing the long-term benefits of RBMS, installed monitoring systems on three bridges on an experimental basis to investigate reliability and performance of these systems for practical application and also to create its own database. New York State systems were equipped with a variety of sensors. Data under known loading and normal traffic loading were collected over a period of 3 years. Reliability and performance of the monitoring systems, sensors, and communication systems were observed.

Most available or developed system identification methods based on measured vibration response require input excitation and the system output, Katkhuda et al. (2005). In practice, for bridge structures, this could be very challenging. If forced excitation, such as an impulse hammer or an electrodynamic shaker, is used, the bridge has to be closed fully or partially to vehicular traffic. This poses public inconvenience, is expensive, and in some cases could be a safety hazard. It is more convenient to use the readily available traffic excitation. But, in these cases, input excitation cannot be measured with any certainty. This requires using output only structural response and requires processing methods that can deal with measured structural vibration due to several sources of unquantifiable excitation. Thus, one of the major issues in developing RBMS has been lack of reliable methods to obtain modal parameters using traffic excitation. In recent days, there has been several studies investigating various signal processing methods to obtain modal parameters from measured ambient vibration data from bridge structures (Katkhuda et al. 2005, Cole 1973, and Ibrahim 1977). This section addresses one such signal processing method and discusses the results obtained using the data measured from one of the RBMS installed on New York State bridges.

6.3.8.3 Theoretical Background

The dynamic model of a bridge can be simplified using a differential equation system,

$$[M]\{\ddot{x}(t)\}+[C]\{\dot{x}(t)\}+[K]\{x(t)\} = \{F(t)\} \tag{6.73}$$

where x is the bridge motion at a given time t, the column vector F represents the general excitation, and the matrices $[M]$, $[C]$, and $[K]$, respectively represent inertia, damping, and stiffness. For a general excitation expressed by the column vector $\{F(t)\}$, the motion $x_s(t)$ of a given point P_s of the discrete structure, including both deterministic and random components, can be written in a modal decomposition form,

$$x_s(t) = \sum_{r=1}^{n} e^{-\sigma_r t}\left[C_{s,Rr}\cos(p_r t) + C_{s,Ir}\sin(p_r t)\right] + \sum_{r=1}^{n}\int_0^t F_{s,r}(\zeta)h_r(t-\zeta)d\zeta \tag{6.74}$$

where p_r and σ_r are circular damped natural frequency and decay rate of rth natural mode (r = 1,2,3 … n). $C_{s,Rr}$ and $C_{s,Ir}$ are n pairs of constants, which can be determined using initial conditions at $t = 0$;

$$x_s(0) = \sum_{r=1}^{n} C_{s,Rr} : \quad v_s(0) = \dot{x}(0) = \sum_{r=1}^{n}\left(p_r C_{s,Ir} - \sigma_r C_{s,Rr}\right) \tag{6.75}$$

and eigensystem,

$$\left([M]\lambda_r^2 + [C]\lambda_r + [K]\right)\{X_r\} = \{0\} \quad (r = 1, 2, 3, … n) \tag{6.76}$$

where the eigenvalues are of the form,

$$\lambda_r = -\sigma_r + jp_r \quad (j = \sqrt{-1} \quad (r = 1, 2, 3 \dots n) \tag{6.77}$$

and the complex eigenvector is given by,

$$\{X_r\} = \{X_{Rr}\} + j\{X_{Ir}\} \quad (r = 1, 2, 3 \dots n), \tag{6.78}$$

The column vectors $\{X_{Rr}\}$ and $\{X_{Ir}\}$ are related to constants $C_{s,Rr}$ and $C_{s,Ir}$ by the following expressions:

$$\{X_{Rr}\} = -\frac{1}{2}\{C_{1,Rr}, C_{2,Rr}, C_{3,Rr} \dots\dots C_{n,Rr}\}^T$$
$$\{X_{Ir}\} = \frac{1}{2}\{C_{1,Ir}, C_{2,Ir}, C_{3,Ir} \dots\dots C_{n,Ir}\}^T \tag{6.79}$$

With $r = 1, 2, 3, \dots n$.

The above formulation yields only $2 \times n$ independent initial conditions, which can be determined using known initial displacement $xs(t = 0)$ and velocity $v_s(0) = \dot{x}_s(t = 0)$. The first sum in Equation 6.74 is influenced by exponential terms $e^{-\sigma_r t}$ decaying in time, and the motion $x(t)$ being modulated by active force components $F_{s,r}(t)$ through the integral convolution with unit impulse functions,

$$h_r(t) = \frac{1}{p_r m_r} e^{-\sigma_r t} \sin(p_r t) \tag{6.80}$$

where m_r is the rth modal mass, dependent on time evolution of excited forces $F_{s,r}(t)$.

Application of structural modal parameter estimation, using structural response to a noncontrolled excitation such as vehicular traffic on the bridge, requires elimination of unknown excitation forces' influence in the measured signal as well as separating the free decay component.

6.3.8.3.1 Selective Random Decrement (SRD) Technique

In the RD technique (Cole 1973; Ibrahim 1977; Asmussen et al. 1999) it is assumed that every transient response time record contains both free decay and forced components. A typical time history data from a structure, which includes both components, is shown in Figure 6.12a. SRD method is used to eliminate the forced component (i.e., the influence of unknown excitation forces) to obtain the free decay component shown in Figure 6.12b. First, on the time history of the signal

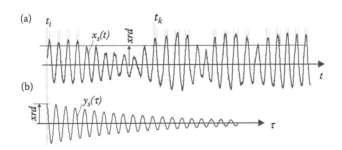

FIGURE 6.12 Illustration of RD processing (a) original signal and (b) damped signal. (From Alampalli, S. and Cioara, T. (2000).)

$x_s(t)$ (see Figure 6.12a) a line of constant magnitude xrd is selected, such that it intersects the signal at N points at discrete times t_k. A new time function is defined,

$$y(\tau) = \frac{1}{N} \sum_{k=1}^{N} x_s(t_k + \tau)$$

(6.81)

with a time shift τ. Using the modal decomposition, Equation 6.74, and knowing that at every tk moment the initial shift time is $\tau = 0$, the following can be obtained:

$$y_s(\tau) = \frac{1}{N} \sum_{k=1}^{N} \sum_{r=1}^{n} e^{-\sigma_r \tau} [C_{s,Rr}^{(k} \cos(p_r \tau) + C_{s,Ir}^{(k} \sin(p_r \tau)] + \sum_{k=1}^{N} \sum_{r=1}^{n} \int_0^\tau F_{s,r}^{(k}(\zeta).h_r(\tau - \zeta)d\zeta\}$$

(6.82)

Writing the rth modal component of excitation force at point Ps of the structure in the form,

$$F_{s,r}^{(k}(t) = X_{s,Rr} \{X_{Rs}\}^T \{F^{(k}(t)\}$$

(6.83)

and changing the order between summation and integral operations in Equation 6.82 yields,

$$y_s(\tau) = \sum_{r=1}^{n} e^{-\sigma_r \tau} \left[\frac{1}{N} \left(\sum_{k=1}^{N} C_{s,Rr}^{(k} \right) \cos(p_r \tau) + \frac{1}{N} \left(\sum_{k=1}^{N} C_{s,Rr}^{(k} \right) \sin(p_r \tau) \right]$$
$$+ \sum_{r=1}^{n} \left[X_{s,Rr} \{X_{Rs}\}^T \int_0^\tau \left(\frac{1}{N} \sum_{k=1}^{N} \{F^{(k}(\zeta)\} \right) h_r(\tau - \zeta)d\zeta \right]$$

(6.84)

In the above when

$$\frac{1}{N} \sum_{k=1}^{N} \{F^{(k}(\tau)\} \Rightarrow \{0\},$$

(6.85)

the function $y_s(\tau)$ represents a free decay of n modal components vibration signal. (Note that Figure 6.12b is the time history of a single mode free decay component signal.) Taking into account the first initial conditions of Equation 6.75, the following can be obtained for $\tau = 0$,

$$y_s(\tau = 0) = \sum_{r=1}^{n} \left(\frac{1}{N} \sum_{k=1}^{N} C_{s,Rr}^{(k} \right) = \frac{1}{N} \sum_{k=1}^{N} \left(\sum_{r=1}^{n} C_{s,Rr}^{(k} \right) = \frac{1}{N} \sum_{k-1}^{N} x_s^{(k}(t_i)$$

(6.86)

If all $x_s(k(t_i))$ samples are triggered at the same level x_{rd} at every time t_i, then

$$y_s(\tau = 0) = x_{rd}$$

(6.87)

The triggering can be also performed at level $x_{rd} = 0$ for a positive or negative slope (i.e., $x_s(t_i + \Delta t) > 0$ or $x_s(t_i + \Delta t) < 0$, where Δt is the sampling time). Mathematically, Equation 6.85 is satisfied for a Gaussian white noise excitation that requires large number of points ($N \to \infty$). In many cases, when vibration responses are obtained from structures such as bridges excited by vehicular traffic, this condition is hard to fulfill (see Figure 6.13).

For a typical multi-span bridge shown in Figure 6.13, with a deck Dk on the metallic girders Gr, more than one moving vehicle will be on the bridge inducing the transient bridge deformations. The deck and girders are supported on piers through a number of bearings Be with foundations Fd

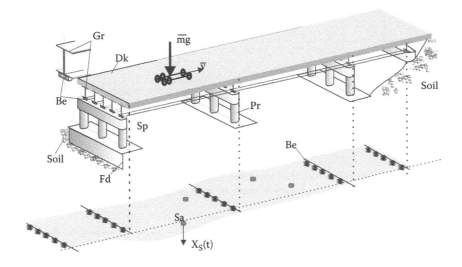

FIGURE 6.13 Typical bridge structure carrying vehicular traffic.

transferring the loads to the soil. The deck is also connected to the highway to ensure the smooth traffic flow. Typical damage in the bridge structure, such as a crack on the midspan of one girder, changes the local area stiffness and modifies the parameters contributing to the static and dynamic equilibrium of the structure. Hence, predicting the change in the measured structural parameters such as natural frequencies and localizing the area contributing to the change are essential for damage detection. For the first step, in the case of a bridge structure, the ambient response triggered by the vehicles passing along the bridge is needed to predict structural modification caused by a localized damage. This response does not depend on the dynamical characteristics of the moving loads, but is excited by these loads. SRD algorithm developed above can be used easily to extract the response to ambient excitation.

A typical time history recorded using an accelerometer Sa on the bridge, six events occur (marked 1 to 6 in Figure 6.14). In the first period of every excitation, unknown forces from moving loads increase the vibration magnitude, which is dominated by a second component of the general form of motion xs in Equation 6.74.

At the end of each period, excitation forces drop in magnitude and the dominant first component of Equation 6.74 is more appropriate to a multi-modal free decay vibration motion. During each such event, a narrow time window T_w can be selected such that triggering start of the RD function (signal) $y_s(\tau)$ cis of length T_s. By positioning the window T_w in the area of the envelope signal decaying slope, it can be made certain that the level of unknown excited forces drop in magnitude and the sum in Equation 6.85 converges quickly to zero, for a limited number (N) of averages. For this particular case, the RD processing can be redefined, more appropriately, as the SRD. The RD function processed (Figure 6.15) using the data shown in Figure 6.14 clearly shows the free decay components.

When events occur at close intervals (Figure 6.16), length T_s of the SRD function $y_s(\tau)$ overcomes the span between events 2–4 and 5–6. End of the function $y_s(\tau)$ will be affected here by deviation from free decay law in processing, that is, the forced component. To overcome this problem, it is necessary to reduce the time trace length T_s of $y_s(\tau)$ to less than the span between the two closed events, and apply an exponential time window for each sequence processed,

$$w_j(t) = e^{-\sigma_0(t-t_j)}$$

$$(6.88)$$

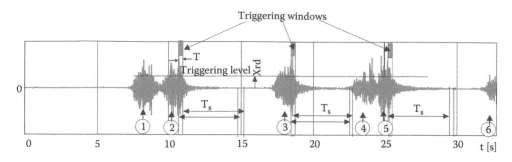

FIGURE 6.14 Typical vibration signal of a bridge due to vehicular traffic. (The markers 1 through 6 indicate separate marked events as measured by an accelerometer. Each event marks increase in live load, such as passage of a truck.)

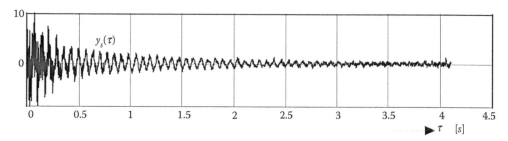

FIGURE 6.15 RD function obtained from the signal shown in Figure 6.14.

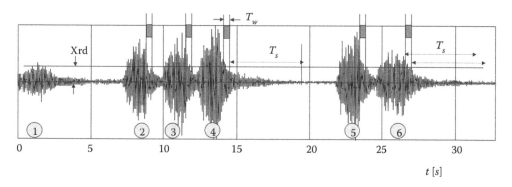

FIGURE 6.16 Typical vibration time history with closely spaced multiple events. (The markers 1 through 6 indicate separate marked events as measured by an accelerometer. Each event marks increase in live load, such as passage of a truck.)

at the start of triggering time t_j, where σ_0 is virtual damping decay rate. In addition, all the modal decay rates will be estimated from modal parameter estimation using a new modal free decay signal,

$$z_s(t) = x_s(t)w_j(t) \tag{6.89}$$

The decay rate σ_0 is calculated such that the end of the signal, $z_s(t = T_o s) \Rightarrow 0$, where $T_o s$ is the new time length of the signal $z_s(t)$. Finally this will yield the following SRD function, ready for modal parameter estimation.

$$R_s(\tau) = \sum_{r=1}^{n} e^{-(\sigma_0 + \sigma_r)\tau} \left[\frac{1}{N} \left(\sum_{k=1}^{N} C_{s,Rr}^{(k)} \right) \cos(p_r \tau) \right] + \sum_{r=1}^{n} e^{-(\sigma_0 + \sigma_r)\tau} \left[\frac{1}{N} \left(\sum_{k=1}^{N} C_{s,Ir}^{(k)} \right) \sin(p_r \tau) \right] \tag{6.90}$$

6.3.8.3.2 Modal Parameter Estimation

The processed free decay signal can be used now for highly accurate estimation of modal parameters using a circle-fit algorithm, see Cioara (1988), and Cioara (1999). First, Fourier transform of the form shown below, is applied to the free decay,

$$F(\omega) = \sum_{r=1}^{n} \left(\frac{U_r + jV_r}{\sigma_r + j(\omega - p_r)} + \frac{U_r - jV_r}{\sigma_r + j(\omega - p_r)} \right) \tag{6.91}$$

where ω is the circular frequency with harmonic steady state excitation. Note that the Nyquist plot in the vicinity of rth damped natural frequency p_r is approximated by circle C_r (see Figure 6.17). Experimentally determining the coordinates of the representative points $P(\omega_i)$, $P(\omega_k)$, $P(\omega_s)$, on the curve C_r and determining the angles β_{ik} between the vector radius $O_rP(\omega_i)$ and $O_rP(\omega_k)$, one can define,

$$T_{ik} = \frac{1}{\tan(\beta_{ik})} \tag{6.92}$$

to build a system of linear equations,

$$p_r(\omega_k - \omega_i) + \sigma_r[(\omega_s - \omega_k)T_{ks} - (\omega_s - \omega_i)T_{is}] = \omega_s(\omega_k - \omega_i)$$
$$p_r(\omega_s - \omega_k) + \sigma_r[(\omega_s - \omega_i)T_{is} - (\omega_k - \omega_i)T_{is}] = \omega_i(\omega_s - \omega_k) \tag{6.93}$$
$$p_r(\omega_s - \omega_i) + \sigma_r[(\omega_s - \omega_k)T_{ks} - (\omega_k - \omega_i)T_{ik}] = \omega_k(\omega_s - \omega_i)$$

where the damped natural frequency pr and decay rate of the rth natural mode are unknowns.

Usually the modal test implies the constant increment of the frequency $\Delta\omega$ and three consecutive points $P(\omega_i)$, $P(\omega_i + 1)$, and $P(\omega_i + 2)$ with,

$$\omega_i = (i-1)\Delta\omega \tag{6.94}$$

Using Equations 6.94 and 6.93), one can obtain

$$2p_r + (T_{i+2,i+1} - T_{i+1,i})\sigma_r = 2i\Delta\omega \quad (i = \dots k - 1, k, k + 1 \dots) \tag{6.95}$$

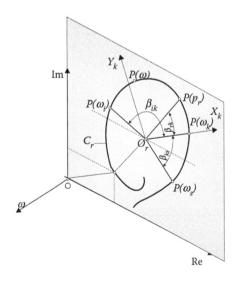

FIGURE 6.17 Illustration of circle fit.

This can be solved in a least squares manner, since it has more than three equations and only two unknowns: p_r and σ_r. The algorithm developed by Cioara (1988) and (1999) allows us to determine mode shapes in terms of modal complex components U_r and V_r.

An illustrative example, to show the high accuracy of this method to estimate modal parameters estimation, is presented in Figures 16.8 and 6.19. These figures contain FFT spectrum and circle-fit algorithm of the signal shown in Figure 6.15. The SRD signal was obtained by processing the ambient vibration data obtained from a bridge (see Figure 6.14). This exhibits two damped natural frequencies $f_1 = 9.267$ Hz and $f_2 = 14.361$ Hz. These are indicated as P1 and P2 in the FFT spectrum (Figure 6.18) and the Nyquist plot (Figure 6.19). Application of the same procedure to two other signals (recorded simultaneously with the signal shown in Figure 6.14) showed very small deviations (<0.01%) relative to frequency values f_1 and f_2, even though the modal components have very different magnitude levels between the three signals.

6.3.8.4   Infrastructure Application

New York State installed structural monitoring systems (STRUMS) on three bridges (Alampalli 1998b; Alampalli and Fu 1994). These systems include sensors for measuring vibration (accelerometers), stress (strain gages), and slope (clinometers). Each vibration data file contains 16,384 data points collected using an 8-bit A/D converter. These systems were named STRUMS20, STRUMS30, and STRUMS31 and data were acquired during a period of 2 years.

The data presented in this chapter is from the bridge located in Rochester, NY, where STRUMS 30 was installed, Alampalli and Fu (1994). It was built in 1963 and carries I490 East Bound over Conrail Mainline in Monroe County. It is a 3-span structure with seven steel girders carrying concrete deck. The structure is about 76-m-long with 44 degree skew. The end spans were supported by steel rocker bearings and were continuous over the reinforced concrete piers; the middle span was

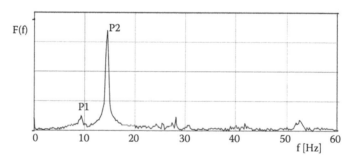

FIGURE 6.18   FFT of signal shown in Figure 6.15.

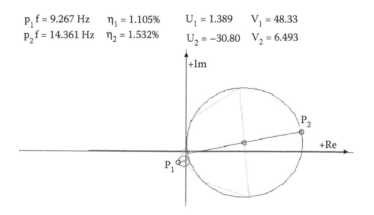

FIGURE 6.19   Nyquist plot of spectrum shown in Figure 6.18.

connected to the end spans by pin-and-hanger. The bridge carries three lanes of traffic. This bridge was rehabilitated in 1993 to replace corroded diaphragms, repair concrete deck with a concrete overlay, install a new joint system, replace steel rocker bearings with elastomeric bearings, and provide continuity between the spans by splicing with steel plates. This bridge was instrumented with STRUMS to examine the consistency of measured structural indices to the in-service environment and changes in the structural conditions.

Structural changes and environmental conditions will influence natural frequencies of bridges over long periods of time. To study these effects, data acquired during March 93 and April 95 from several accelerometers (S2-S15) of STRUMS30 were analyzed using the method illustrated above and results are presented in Figure 6.20 and Table 6.5. The bridge was instrumented with 11 accelerometers, and the data was collected simultaneously from all the accelerometers. The subset of the data, during March 93 and April 95, showing the variation of the first natural frequency is shown in Table 6.5. The plotted vertical bar in Figure 6.20 represents the average values recorded by various accelerometers (S2 to S15) simultaneously (average of columns shown in Table 6.5 during a single event).

To estimate the influence of moving loads (such as a truck) passing on the bridge during an event, several events were recorded on the same day and are designated with an extension (2t or 3t) for second and third events. Data reliability is expressed using 95% confidence intervals, plotted using shadow bars at the top of each frequency bar (Figure 6.20) if more than three corresponding values exists.

The analyzed data records contain 16,384 points collected using an 8-bit digital conversion (±128 units). The levels of signals are located in the lower range of ±15 units, and created major difficulty in data processing using conventional methods. Therefore using dedicated methods such as those developed in this chapter will yield higher accuracy results. Similar results were obtained for other higher order modal frequencies of STRUMS30, STRUMS 20, and STRUMS 31.

6.3.8.5 Concluding Remarks

Analysis of data, for natural frequencies, collected during 3 months from STRUMS20 station (15 data records), 24 months from STRUMS30 station (17 data records), and 20 months from STRUMS31 (12 data records) showed small standard deviations, indicating the reliability of measured data. The first important conclusion from the present study is that specific methods developed for the bridge-structure modal parameter estimation, using vibration signals monitored from ambient excitation, work very well.

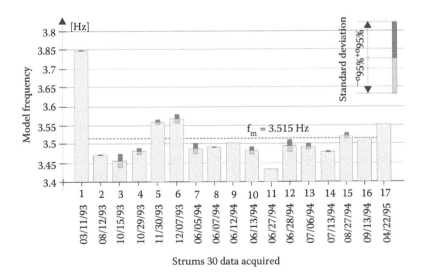

FIGURE 6.20 Variation of first mode during 1993–1995.

TABLE 6.5
Variation of First Modal Frequency of the Bridge Structure during 1993–1995

Sensor No.	Modal Frequency Values [Hz]																
	1	2	3	4	5	6	7	8	9	10	11	12	13	14	15	16	17
S2		3.472				3.629		3.493	3.497	3.489		3.519	3.499				
S2-2t		3.480		3.570	3.534												
S3	3.749			3.516						3.491							
S3-2t				3.477													
S4					3.529	3.571				3.473							
S4-t				3.451													
S5	3.750	3.472		3.478	3.548				3.510	3.484					3.544		
S5-2t		3.473		3.482													
S6	3.749	3.472	3.465		3.568	3.555	3.488				3.434	3.495	3.504				3.548
S6-2t		3.472		3.484													
S7	3.748		3.464	3.485	3.569		3.493					3.502	3.496		3.513	3.516	3.555
S7-2t		3.471		3.482	3.570												
S8	3.749	3.472	3.464	3.502	3.567	3.552		3.493				3.510	3.485	3.487	3.513	3.516	
S8-2t		3.468		3.485	3.565												
S9		3.465			3.567		3.500	3.492				3.483		3.469			
S9-2t		3.472		3.468													
S10	3.748			3.456	3.547		3.473					3.462	3.482	3.482	3.522		
S10-2t			3.434	3.487	3.551												
S14																	
S15		3.475					3.492					3.498					

Sensor no. Extension 2t or 3t indicate second or third events of the same data record.

The natural frequencies do not exhibit negative slope with time indicating that they do not correspond to possible local stiffness reduction due to continuous structural degradation. (Note that the change in frequency between March 1993 and August 1993 was attributed to bridge rehabilitation that includes replacement of steel rocker bearings with elastomeric bearings.) Thus, these variations may be attributed to environmental parameters such as temperature and humidity. These environmental conditions modify the mechanical characteristics near bridge boundary conditions such as foundations (Fd in Figure 6.13). Note that the soil dynamic characteristics and dynamic characteristics of bridge bearings are also dependent on environmental conditions. The same conclusions were reached by Alampalli (1998a), Alampalli et al. (1998), and Farrar and Doebling (1998), after significant experimental efforts.

Comparison of frequencies listed in the columns of Table 6.5, obtained from different events from the same data record show no notable deviations. This indicates that moving loads have small influence on dynamic behavior of the bridge structure. Predominant influence of moving load characteristics should lead to noticeable deviations between modal frequencies estimated from consecutive events.

This study also indicates that practical methods for bridge fault diagnostics using the vibration signals monitored on the bridge structures are necessary for further supplementary research work. Relative influence of various environmental parameters on dynamic response of bridge structure should be determined. It is also necessary to find the natural modes, which are most insensitive to environmental parameter variations, for structural monitoring and damage detection. This can be determined using dynamic model simulation by modifying the parameters of bearing and boundary conditions in the ranges of interest.

6.4   PARAMETER IDENTIFICATION (PI)

6.4.1   PROBLEM DEFINITION

The model, or parameter, update problem in structural identification starts with an analytical or numerical model of the structure. For example, a finite element model of the structure. As such, estimates of the stiffness $[K]$, mass $[M]$, and damping $[C]$ matrices are assumed to be known. It is desired to improve the estimates, *update*, of one, or more, of those matrices. This estimate improving, or update, is performed using an available set of test results. In our case, it would be results of an SHM project. Numerous methods for model updating have been developed and presented. For simplicity, those methods are subdivided into three generic categories. These categories are (1) the optimal matrix update, (2) sensitivity-based parameter update, (3) eigenstructure assignment. Overview of those categories were described by Friswell and Mottershead (1995) and Doebling et al. (1996). We should observe here that vast majority of the model update methods and categories utilize modal information that result from MI methods of the previous section. Note that when a method is based on an assumed composition of the stiffness, mass, and damping matrices, any errors or mistakes in the initial assumption will result in a major errors that will render the whole procedure inaccurate.

Thus, we summarize the problem of parameter identification as follows:

1. Establish, based on theoretical and personal experience, the system matrices: $[K]_0$, $[M]_0$ and $[C]_0$. The details of the numerical model should be consistent with the goals and objectives of the effort (design assessment, local of global behavior, etc.). If damage identification is the goal of the effort, an accurate baseline model should be established first.
2. Select desired parameters to be updated. Note that the analytical model should be consistent in its details with those parameters.
3. Perform testing of the system as appropriate. Collect appropriate data from the test.
4. Apply an appropriate PI method to update the parameters chosen in #2. The updated matrices are matrices: $[K]_1$, $[M]_1$ and $[C]_1$ can now be used to achieve the original objectives of the effort.

Note that the above steps constitute only a STRID effort. The resulting matrices define the structural state *as* tested. No accurate damage identification can be achieved using these steps. The differences between the original matrices, with subscript of 0, and the updated matrices, with subscript of 1, only define the differences between the original estimate of the matrices (based on theory and personal experiences) and the model that is based on test results. The effects of the damage in the tested system only reflects in the updated matrices. No information is available in this process that can help in identifying the already existing damage state in the system. To identify the existing damages in the system, two approaches are possible: (1) Perform testing of a pristine non-damaged system, or (2) Perform direct DMID efforts (Chapter 7).

6.4.2   EIGENSTRUCTURE ASSIGNMENT UPDATE

The method of Eigenstructure assignment also starts by assuming that the pristine (not damaged) structural model is known and is correct. The objective of the method is to identify the damage in that pristine structure. To achieve this objective, a set of measured mode shapes and modal frequencies are obtained, usually through a modal testing and identification procedure. The Eigenstructure assignment method then adds a set of artificial forces in the pristine structural equilibrium equation. Through a set of algebraic reassignments least square optimization, the damaged elements (or DOF) of the structure can then be estimated. There is large body of knowledge and research of the method. One of the earliest works was by Andry (1983). Some of the more relevant applications of the method to SHM were by Lim (1995) and Lim and Kashangaki (1994). We summarize the basic equations of their approach as follows.

Assume that the matrix equation of a given structure models accurately the structure in its pristine condition. We assume that all the system matrices, $[M]$, $[K]$, and $[C]$ are known. Let us assume that the physical structure is now damaged somehow, and it is desired to obtain updated estimates of the damaged structure through identifying the location and the magnitudes of the damage. We also assume that the mode shapes, $\{\phi\}_k$ and the modal frequencies, λ_k of the damaged structure are identified, perhaps through a MI process. For simplicity, we assume that the damage of interest is manifested mostly in the stiffness matrix, $[K]$ and no damage is present in the damping or the mass matrices. By assigning the known modal parameters to the known structural matrices through the structural free vibration, we get

$$([M]\lambda_k^2 + [C]\lambda_k + [K])\{\phi\}_k = \{R\} \tag{6.96}$$

The vector $\{R\}$ is a residual vector that results from the changes in $[K]$ due to damage. As such $\{R\}$ represents additional forces in the structure that occurred due to the unknown structural damage. The eigenstructure assignment method utilizes $\{R\}$ to estimate the locations and magnitude of damages by adjusting the free vibration equation to accommodate the damage as

$$[M]\{\ddot{U}\} + [C]\{\dot{U}\} + [K]\{U\} = [Q]\{F\} \tag{6.97}$$

The vector $\{F\}$ contains artificial control forces, while the matrix $[Q]$ is an influence matrix. The displacements and control forces can be related by

$$\{F\} = -[G][Q]^T\{U\} \tag{6.98}$$

Thus

$$[M]\{\ddot{U}\} + [C]\{\dot{U}\} + ([K] + [Q][G][Q]^T)\{U\} = \{0\} \tag{6.99}$$

Note that the change in the stiffness matrix due to the structural damage is expressed by $[Q][G][Q]^T$.
 The modal parameters can now be applied as

$$([M]\lambda_k^2 + [C]\lambda_k + ([K] + [Q][G][Q]^T))\{\phi\}_k = \{0\} \tag{6.100}$$

Rearrange Equation 6.100 to express the damage effects on the right side

$$([M]\lambda_k^2 + [C]\lambda_k + [K])\{\phi\}_k = -([Q][G][Q]^T)\{\phi\}_k \tag{6.101}$$

Following Lim (1995) assuming that the potential of damage only in the jth element, the influence matrix $[Q]$ is evaluated as $[Q]_j$. Note that $[Q]_j$ is completely defined by the location of the jth element within the overall matrix composition of Equation 6.101. In addition, the unknown gain matrix is reduced to an unknown scalar g_j. Equation 6.101 is reduced to

$$([M]\lambda_k^2 + [C]\lambda_k + [K])\{\phi\}_k = -g_j([Q]_j[Q]_j^T)\{\phi\}_k \tag{6.102}$$

Finally,

$$\{\phi\}_k = [L]_{kj}\{\vartheta\}_{kj} \tag{6.103}$$

with

$$[L]_{kj} = ([M]\lambda_k^2 + [C]\lambda_k + [K])^{-1}([Q]_j[Q]_j^T) \tag{6.104}$$

and

$$([M]\lambda_k^2 + [C]\lambda_k + [K])\{\phi\}_k = -g_j([Q]_j[Q]_j^T)\{\phi\}_k \tag{6.105}$$

$$\{\vartheta\}_{kj} = -g_j\{\phi\}_k \tag{6.106}$$

Since there are potential errors in forming the structural matrices, as well as measurement errors, a new vector, called *best achievable vector* $\{\phi\}_{bkj}$ can be estimated, using a least square approach. Defining $[L]_{kj}^{+}$ as the pseudo inverse of $[L]_{kj}$ (see previous section), the vector $\{\vartheta\}_{kj}$ is eliminated and the best achievable vector is related to the measured mode shape as

$$\{\phi\}_{bkj} = [L]_{kj} \, [L]_{kj}^{+} \, \{\phi\}_{k} \tag{6.107}$$

The relationship between the measured and the best achievable mode shapes as expressed in Equation 6.107 would be an indication for the damage presence in the jth element. If the damage is located in the jth element then the $\{\phi\}_{bkj}$ and $\{\phi\}_{k}$ will be identical; if the damage is not located in the jth element then $\{\phi\}_{bkj}$ and $\{\phi\}_{k}$ will be different. A measure for the presence of damage is the angle α_{kj} between the two vectors

$$\alpha_{kj} = \cos^{-1}\left(\frac{\{\phi\}_{bkj}^{T} \, \{\phi\}_{k}}{\|\phi_{bkj}\|_{E} \, \|\phi_{k}\|_{E}} \right) \tag{6.108}$$

With $\|...\|_{E}$ is the Euclidean norm of a vector. The angle α_{kj} measure the presence of damage in the jth element in the measured k^{th} mode. For more details on the physical interpretations of α_{kj}, the reader can refer to Lim (1995) and Lim and Kashangaki (1994).

The above approach can be extended to estimate the level of damage in the jth element. The details of the analytical approach can be found in Andry et al. (1983), Lim (1995), and Lim and Kashangaki (1994).

6.4.2.1 Summary of Input/Output Measurements/Testing

To utilize the eigenstructure assignment method in an SHM experiment the following steps can be followed:

1. A modal testing setup should be performed first. As such, adequate number of motion sensors need to be installed on the required structure.
2. Adequate number of modal information should be realized next using any of the MI methods that were described earlier.
3. Adequate structural matrices (stiffness, damping, if needed, and mass) are composed. Special care should be given while constructing those structural matrices: their resolution should be large enough to meet the objectives of the SHM experiment. The finite element modeling technique is usually utilized in this step, although other numerical techniques can be used (such as finite difference).
4. Finally one of the many eigenstructure methods can be applied to the results of #2 and #3, above. The method that was summarized, above that was developed by Lim (1995) or any other adequate method can be used. For a comprehensive review of available eigenstructure assignment methods, the reader is referred to Friswell and Mottershead (1995) or Doebling et al. (1996)

6.4.2.2 Advantages/Disadvantages and Applicability to SHM

Among the advantages of the eigenstructure assignment methods are that they can update all structural matrices, including damping matrix. Also, by definition, these methods will satisfy exactly the required measured mode. The disadvantages of the methods are the intensive computational demands as well as the lack of possible physical interpretation of the updated results, since the updates are applied to the whole matrices. In simplified systems, such as simple beams, simple planer pinned trusses, or lumped mass systems, the physical interpretations of the corrected DOF might be possible to interpret physically. Unfortunately, in a realistic structure, such as complex post-tensioned reinforce concrete bridge, physical interpretation of damaged tendons using a 3D finite element model of post-tensioned steel and concrete might be difficult, if not impossible to do.

6.4.3   Optimal Matrix Update (OMU)

6.4.3.1   General Description

Optimal matrix update method is based on updating an assembled system matrices using test results. As most of the methods in this section, this class of methods also depends on the accurate analytical modeling of the system on hand. The method starts by assuming a baseline analytical model of the structure K. On the basis of some test observations, the model information is updated using an optimization scheme. For example, a minimization of the rank of the matrices was used by Zimmerman and Kaouk (1994) to update K. Doebling (1996) introduced a new technique called *minimum rank elemental updates* (MREU) that computes and updates the minimum ranks of the element stiffness parameters. Hemez and Doebling (2001) provided an in-depth review of update methods for nonlinear dynamic systems. Yang and Liu (2007) compared some matrix update methods using a simulate truss system. They observed the importance of measurement noise on the accuracy of matrices updating.

6.4.3.2   Advantages/Disadvantages and Applicability to SHM

Since the OMU methods rely mostly on analytical/numerical modeling solutions, they are not demanding computationally. They do not offer, however, in-depth physical insight to the corrections they offer. This is mainly due to the fact that they operate on full structural matrices. This is a rather limiting disadvantage, especially in case of a complex SHM situation (such as an SHM project dealing with FRP bridge deck, with all the geometric and material complexities it presents). What makes many of these methods even less attractive to SHM is the tendency for many of these methods to spread the updates throughout the entire matrices under considerations as observed by Hemez (1993). This would result in an inaccurate error, or damage, localization. Such inaccuracies would reduce the benefits of any SHM project.

6.4.4   Sensitivity-Based Update

6.4.4.1   Sanayei Static-Based Methods

Most structural identification methods are based on dynamic motions of the structures. On one hand, measuring a stream of time histories provides an abundance of information that would help in identifying the structure. Of course, identifying dynamic properties of the structure includes identifying structural stiffness, damping, and inertial properties. Static-based identification methods would be far simpler, since they involve only the stiffness of the structure. They also involve simpler static measurements. As such they would be suitable for parameter identification of stiffness matrices, design verification, and assessment for static-based loading. We will detail the basics of a static-based parameter identification method that was developed by Sanayei and Saletnik (1996a, 1996b), and Sanayei et al. (1997). The method is elegantly simple, yet fairly powerful and easy to implement. Many of the basic features of structural identification methods (both dynamic-based and static-based) are included in this method; hence it is suited for our immediate task of exploring general characteristics of structural identification methods.

Let us assume that for a given linear elastic structural system the finite element stiffness matrix $[K(p)]$ is well developed. The order of this matrix is N, which represent the total number of constrained DOF of the structure on hand. The functional relationship of some parameters of the structure p are assumed to be known. These parameters include any structural parameters that are used to form $[K(p)]$ such as cross-sectional area, different moment of inertia, or different types of springs. We would like to point out that parameter identification is perhaps most valuable in identifying spring values in finite element modeling. In practice, engineers tend to approximate complex components of the structural system by springs. For example, large swathes of not well-defined soils are approximated by soil springs. Also, steel or concrete connections are modeled by equivalent springs. Using a method such as the current method to accurately identify those otherwise not well-defined springs can prove to be a valuable effort. Some parameters that can yield a highly accurate finite element model when identified are shown in a typical bridge and abutment example of Figure 6.21. We should note that there is no theoretical limit to the number of parameters p that need to be identified. In practice, only limited number of parameters is considered.

The equilibrium equation of the structure can be written as

$$[K(p)]\{U\} = \{F\} \tag{6.109}$$

where $\{U\}$ and $\{F\}$ are the displacement and force vectors, respectively. Both vectors are of the order N. One of the main features of the method is that it defines an elemental strain ε_i for the ith finite element. This definition relates the strains in the finite element model directly to measured strains. Elemental strains can be related easily to the element displacements $\{U\}_i$ by using the strain displacement relation

$$\varepsilon_i = \{B\}_i^T \{U\}_i \tag{6.110}$$

The vector $\{B\}_i$ can be developed easily for any type of finite element.

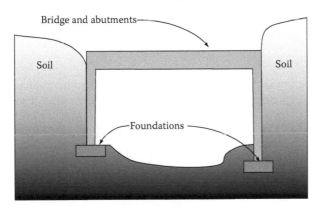

Physical system

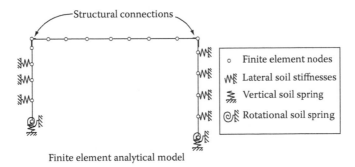

Finite element analytical model

Identifying highly uncertain structural parameters such as soil springs and/or structural connections can result in a highly accurate structural model

FIGURE 6.21   Practical example of parameter identification.

Assembling all elemental strains in the system we get

$$\{\varepsilon\} = [B]\{U\} \tag{6.111}$$

where $[B]$ is assembled from the individual $\{B\}_i$. We eliminate $\{U\}$ from Equations 6.109 and 6.111, to get

$$\{\varepsilon\} = [B][K(p)]^{-1}\{F\} \tag{6.112}$$

Since only those measured strains at locations of measurements are of interest to us, Equation 6.112 above is subdivided such that

$$\{\varepsilon\}_a = [B]_a[K(p)]^{-1}\{F\} \tag{6.113}$$

where $\{\varepsilon\}_a$ is the strain vector whose components are the analytically computed strains at measurement locations. The matrix $[B]_a$ is a subset of $[B]$. Note that Equation 6.113 represent only a single force distribution. The method is originally developed for multiple force distributions, and the generalization is straightforward.

If the force distribution $\{F\}$ is applied to the structure, and the corresponding strains $\{\varepsilon\}_m$ are measured, then it is possible to define the error function

$$\{\varepsilon(p)\} = \{\varepsilon(p)\}_a - \{\varepsilon\}_m \tag{6.114}$$

Note that the computed strains and the error function are both functions of the parameters p. Using Equation 6.113

$$\{e(p)\} = [B]_a[K(p)]^{-1}\{F\} - \{\varepsilon\}_m \tag{6.115}$$

What remains now is to estimate the parameters p that would minimize the error function $\{e(p)\}$. The error function is first expanded in a Taylor series around a base value of p such as

$$\{e(p) + \Delta p\} \cong \{e(p)\} + \left[\frac{\partial\{e(p)\}}{\partial(p)}\right]\{\Delta p\} \tag{6.116}$$

The sensitivity matrix is defined as

$$[S(p)] = \left[\frac{\partial\{e(p)\}}{\partial\{p\}}\right] \tag{6.117}$$

The sensitivity matrix can be evaluated numerically, since the functional relationship between $[K(p)]$ and p is known, in other words, the matrix $\partial/\partial p_j([K(p)])$ can be evaluated. Note that p_j is the jth parameter that needs identification.

The error function can be minimized in a least square sense by minimizing the norm

$$J(p) = \{e(p + \Delta p)\}^T\{e(p + \Delta p)\} \tag{6.118}$$

Assuming that the required minima occur when the gradient of $J(p)$ is set to zero,

$$[S(p)]^T[S(p)]\{\Delta p\} = -[S(p)]^T\{e(p)\} \tag{6.119}$$

The above equation can be solved for the required change in parameters $\{\Delta p\}$ from the assumed initial parameter set $\{p\}$. There are several solution techniques that were suggested by the authors of the method. The pseudo inverse method, the singular valued decomposition (SVD) method, and iterative methods can all be used to solve for $\{\Delta p\}$. Pseudo inverse method and SVD methods are described elsewhere in this chapter. The computed $\{\Delta p\}$ can be used to update $\{p\}$. The method is then repeated in an iterative way until converged. Criteria of convergence have been recommended by the authors.

6.4.4.1.1 Modal Extension of the Method

The aforementioned methods were generalized to take advantage of potential measured modal values, see Sanayei and Santini (1998) and Olson (2002). The modal expression for a free vibrating system is

$$
\begin{bmatrix} K_{aa} & K_{ab} \\ K_{ba} & K_{bb} \end{bmatrix} \begin{Bmatrix} \Phi_a \\ \Phi_b \end{Bmatrix}_i = \varpi_i^2 \begin{bmatrix} M_{aa} & M_{ab} \\ M_{ba} & M_{bb} \end{bmatrix} \begin{Bmatrix} \Phi_a \\ \Phi_b \end{Bmatrix}_i
$$
(6.120)

The ith subscript refers to the mode number. The a subscript refers to the measured entity, while the b subscript refers to the unmeasured entity. A modal-based error function can be expressed using a condensation scheme as

$$
\{e_s(p)\}_i = [(K_{aa} - \varpi_i^2 M_{aa}) - (K_{ab} - \varpi_i^2 M_{ab})][(K_{bb} - \varpi_i^2 M_{bb}) - (K_{ba} - \varpi_i^2 M_{ba})]\{\Phi_a\}_i
$$
(6.121)

Note the similarity between the expressions of error functions of Equations 6.114 and 6.121. Similar developments to Equations 6.115 through 6.119 can be made until optimal unknown parameters are obtained. Details of the method can be found in Sanayei and Santini (1998) and Olson (2005). This method of modal-based parameter estimation was used by Olson (2005) to investigate dynamic evaluation methods for bridges.

6.4.4.2 Advantages/Disadvantages and Applicability to SHM

Among the advantages of the sensitivity-based methods are the following: (1) They can be physically interpreted, since the parameter to be updated are selected *a priori*; (2) Many of these methods use localized updating techniques, making it easier to control, understand, and interpret the results; and (3) If formulated properly, they can be integrated with the commercially available finite element tools. Two disadvantages of these methods are (1) they can be computationally demanding, and (2) using Taylor series would require the derivation of the derivative of the parameter, which might not an easy task. Perhaps the most important limitation of this class of method is that the parameters must be chosen *a priori*. This would require knowledge of the problem that might negate an SHM procedure where the cause of the problem to be monitored might not be known *a priori*. This prior knowledge of the parameters can be a severe limitation of the use of sensitivity-based methods in an SHM project.

6.4.5 OTHER PARAMETER IDENTIFICATION METHODS

There are many other parameter identification methods. For example, Beardsley et al. (2000) investigated a parameter updating method: Generalized least square (GLS). The method relies on computing the displacements vector $\{U(t)\}$ of a nonlinear F.E. model using the initial nonlinear structural F.E. model

$$
M(p)\{\ddot{U}(t)\} + K(p)\{U(t)\} + \{Fi(p,t)\} = (F(t))
$$
(6.122)

Note that $M(p)$ and $K(p)$ are the mass and stiffness matrices that were derived from an F.E. modeling technique. Both matrices are assumed to be linear, and functions of different parameters p.

The parameters p are those that need updating during this structural method. Note that all dynamic nonlinear effects were lumped together in the nonlinear internal force vector $\{Fi(p,t)\}$. The external force vector is $\{F(t)\}$.

If during an experiment, the measured structural displacements are $\{U(t)_{test}\}$, then the updating problem can be defined as minimizing the objective function $J(p)$ where

$$J(p) = \|R(p,t)\| \tag{6.123}$$

and

$$R(p,t) = \{U(t)_{test}\} - \{U(t)\} \tag{6.124}$$

The above approach is similar to the method of sensitivity-based updates in the fact that it updates the parameters p by minimizing the differences between measured and computed displacements. The method, however, applies the minimization scheme (a least square approach) across both the 2-norm spaces (the parameter space p and the time space t).

Another model updating method called principal component decomposition (PCD) method that addresses the problem of Equation 6.122 was developed by Hasselman et al. (1998). They applied the SVD on the computed displacement vector $\{U(t)\}$ to find the displacements, nonsingular modal components, thus increasing the accuracy of the results, as in

$$[U][\Sigma][V(t)]^T = SVD([U(t)]) \tag{6.125}$$

The objective function to be minimized is

$$J(p) = W_U \|R_U(p,t)\| + W_\Sigma \|R_\Sigma(p,t)\| + W_V \|R_V(p,t)\| \tag{6.126}$$

where

$$[R_U(p,t)] = [U_{test}]^T[U] - I \tag{6.127}$$

$$[R_V(p,t)] = [V_{test}]^T[V] - I \tag{6.128}$$

$$[R_\Sigma(p,t)] = [\Sigma_{test}][\Sigma] \tag{6.129}$$

It is interesting to note that SVD basis of the PCD method is similar to the SVD basis of the ERA method for modal realization. Both approaches use the SVD to realize the nonsingular modes of the time samples. The basic difference, of course is that the ERA methods apply SVD on *measured* time samples, while the PCD methods apply SVD on *computed* time samples.

Beardsley et al. (2000) compared the results of GLS and PCD methods as applied to an impact testing. They concluded that the two methods produce comparable results. However, they observed that the PCD approach has two potential advantages when compared with the GLS approach: (1) It is more sensitive to parameter changes, thus less likely to experience numerical problems, and (2) Due to its reliance on an SVD process, it will filter away some of the unwanted measurement noise, or rigid body behavior.

6.5   ARTIFICIAL NEURAL NETWORKS (ANN)

6.5.1   INTRODUCTION

Artificial neural network is a numerical method that relates unknown output parameters to known input parameters. The relationship is evaluated using complex interrelationships between input and output parameters. To aid in computing these interrelationships (called *weights and biases*), a dataset of input-output measurements are used. This makes ANN fairly suited to SHM projects, where pairs of input-output datasets are known. ANN is also utilized in many other fields such as finance and investing, Trippi and Turban (1992). Software specifically designed for modeling ANN problems is popular; see, for example, Demuth et al. (2007). One of the distinguishing features of ANN is that it does not follow symbolic algorithms, same as humans: we will not find any equations inside the human brain. Symbolic algorithms are higher forms of neural algorithms. Other features of ANN are that the (1) connectivity (interrelationships) between basic units can be modified, terminated, or generated during the computing process; (2) the computations are mostly distributed computations, which are suitable for advance computing machines; (3) ANN can be performed in a supervised or unsupervised setting (which is suited for real time SHM); and finally, (4) ANN computations are local, not global, a feature that is also suited for modern computing algorithms. All of those ANN features have parallels with biological neural networks, as shown in Figure 6.22.

This section explores many features of ANN that are pertinent to SHM field. We also describe some of the yet untapped immense potential of ANN in the fields of structural and damage identification. Finally we present some examples to illustrate some of those features.

6.5.2 BASIC DEVELOPMENTS OF ANN

6.5.2.1 Single Layer ANN

The simplest ANN contains single input measure and single output measure. They are related by a single relationship unit. Also, there is a single bias unit that can affect the input-output relationship such that

$$y = f(wx + b) \tag{6.130}$$

where x, y represent the input and output. The weight and bias are w and b. The input unit, bias unit, and output unit are called *neurons*, in recognition to the biological analogy of the ANN. The single neuron ANN is shown in Figure 6.23.

For a given input, which is assumed to be known, and a given output (measurement) dataset pairs, the values of w and b are computed so as to minimize any errors. Such an error is the

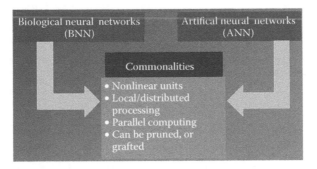

FIGURE 6.22 ANN versus BNN.

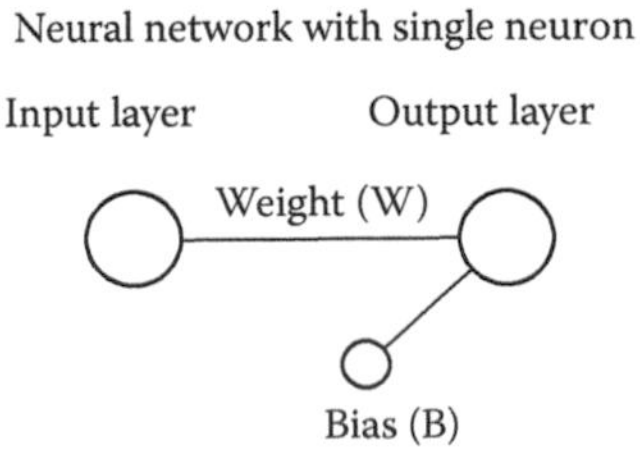

Neural network with single neuron

Input layer   Output layer

Weight (W)

Bias (B)

FIGURE 6.23 Single-neuron ANN.

difference between the computed (using an assumed values for the weight and bias units) and the measured output. The weights and biases are assumed functional relationships between the input and computed output. Those functional relationships are called *activation functions* (sometimes referred to as transfer functions). Some examples of activations functions are shown in Table 6.6 and Figure 6.24.

Note that sigmoid and hyperbolic activation functions have similar forms. The range and the average of the two functions vary. The average of the sigmoid function is 0.5, while the limiting values are 0.0 and 1.0, respectively. The average value of the hyperbolic tangent is 0.0, and the two limiting values are –1.0 and 1.0, respectively. This makes the sigmoid activation function more suited for problems relating to average type behavior. The hyperbolic tangent functions are more suited to problems where the digression from the average is required.

TABLE 6.6
Activation Functions

Activation Function	Expression	Comment
Identity function	$f(y) = y$	Used mostly in linear situations
Step function	$f(y) = A$ for $0 < y$ $f(y) = 0$ for $0 \geq y$ A is a constant, usually assigned to 1.0	Not widely used, since the differential of the function do not exist
Sigmoid function	$f(y) = \dfrac{1}{1+e^{-Ay}}$ A is a constant, usually assigned to 1.0	Popular for nonlinear problems
Hyperbolic tangent function	$f(y) = \dfrac{2}{1+e^{-2y}} - 1$	Special form of the sigmoid function, $A = 2$

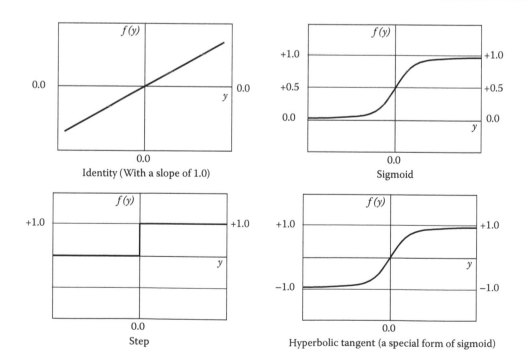

FIGURE 6.24 Examples of activation functions.

The values of the weights and biases are optimized, generally in least square sense, so as to minimize the resulting error. Single neuron ANN has very limited practical use. We will consider more complex ANN models next.

6.5.2.2 Multilayer ANN: Versatile Complex Problems

The two-layer ANN (Figure 6.25), is fairly suitable to solve linear problems. We drop the bias from ANN illustrations for simplicity; however, their presence is implied for the rest of this volume. They can also be used quiet effectively for some classes of nonlinear problems. For example, monotonic type nonlinear problem where the input-output relationships are expected to behave in a monotonic fashion, such as a simple nonlinear spring as in Figure 6.26a. Sigmoid or hyperbolic tangent activation functions can be used effectively in a two-layer network to simulate such a behavior. Consider now a more complex nonlinear behavior, such as the softening spring of Figure 6.26b. It is obvious that sigmoid or hyperbolic tangent functions will not be effective in simulating such a behavior in a two-layer ANN. Because of this, additional layers between the input and the output layers are introduced. These layers are called *hidden layers*. There can be as many hidden layers in an ANN system as needed. We note that one or two hidden layers might be sufficient for most practical SHM problems.

6.5.2.3 Theoretical Background of Multilayer ANN

One way to handle nonlinear problems is to use a nonlinear base function to connect the input layer and the output layer. A popular nonlinear activation function is the sigmoid function, shown in Figure 6.24. In addition to nonlinear activation functions, additional layers in the network are used.

Neural network with single layer

Output layer

Input layer

FIGURE 6.25 Two-Layers ANN model.

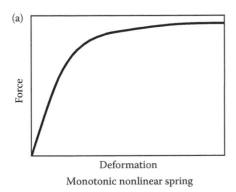

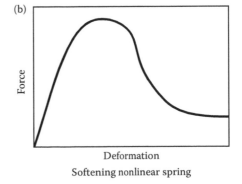

FIGURE 6.26 Examples of nonlinear springs (a) monotonic spring, and (b) softening spring.

These additional layers are called *hidden layers* since they are not seen at the input or output levels. However, the presence of the hidden layers adds immense versatility to the overall neural network. Figure 6.27 shows a neural network with the essential input and output layers as well as with two hidden layers. As usual, the weights in the networks are estimated by minimizing the overall error between the computed results of the network at the output layer, and the measured, or training, data. The adjustment of weights based on an error that is computed at the output layer was a straightforward task for networks with no hidden layers, as was demonstrated above. With hidden layers, the error estimation process is more complicated. Remember that we have no means to compute errors at hidden layers: there are no available measurements or data at the hidden layer. Errors (as defined by a measure of difference between measured and computed data points) can only be estimated at the output layer. The presence of hidden layers presents the obvious question which is: how to distribute errors that are computed at the output layer back into the weights that connects to intermediate hidden layers?

There are several methods that distribute errors in a neural network that includes hidden layers. Among these methods are the delta rule and the generalized delta rule. We will present generalized delta rule below.

Problem Description: We use matrix notations to describe ANN with hidden layer for simplicity. Let the vectors $\{X\}_i$, $\{Y\}_i$ and $\{Y\}_i^m$ represent the input, computed output and measured output pairs for the network, respectively. Note that $i = 1...N_{TRIES}$, where N_{TRIES} is the number of input-output pairs that will be used in the training of the network. We note that the size of the vector $\{X\}_i$ is the number of neurals in the input layer N_{IN} while the size of the vectors $\{Y\}_i$ and $\{Y\}_i^m$ is the number of neurals in the output layer N_{OUT}. For the network topology shown in Figure 6.27, we define the number of layers, including the input and output layers as N_{LAYER}. For such topology, it is implied that all neurals in layer J is connected to all neurals in layer $J + 1$, where $J = 1...(N_{LAYER} - 1)$. The weights connecting layer J and layer $J + 1$ is placed in a matrix. $[W]_j^i$ Similarly, a bias vector is defined as $\{B\}_j^i$. The size of the bias vector is $(N_{LAYER} - 1)$. The number of rows in matrix $[W]_i^m$ is equal to the number of neurals in layer $J + 1$. The number of columns in matrix $[W]_j^i$ is equal to the number of neurals in layer J.

The objective of the analysis is to compute the weights $[W]_j^i$ and the bias $\{B\}_j^i$ that minimize the differences between the computed and the measured output as defined in a least square sense by

$$E = \sum_{i=1}^{i=N_{TRIES}} \frac{1}{2} \left(\{Y\}_i - \{Y\}_i^m\right)\left(\{Y\}_i - \{Y\}_i^m\right)^T$$

Neural network with two hidden layers

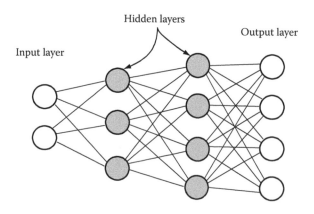

FIGURE 6.27   Neural network with hidden layers.

The steps of the method include feed forward estimations of the responses of different layers, up to the output level. Estimation of the error at the output layers is computed next. The error gradients are propagated backward then. From these error gradients, estimates of the changes in the weights and bias that will minimize the errors are computed. The steps (feed forward, error estimation, and back propagation) are then repeated until the error reaches an acceptable level. The basic equations for each step are summarized below.

Feed Forward: The solution begins by assigning arbitrary initial values for the weights and biases. The response of layer 2 can be estimated as

$$\{H\}_2 = [W]_2\{X\} + \{B\}_2$$

Note that we dropped the indicator i for simplicity. The response of next layer is

$$\{H\}_3 = [W]_3\{H\}_2 + \{B\}_3$$

The feed forward process continues until the last (output) layer

$$\{Y\} = [W]_{NLAYER-1}\{H\}_{NLAYER-2} + \{B\}_{NLAYER-1}$$

Generalized Delta/Back Propagation: The output error at each of the layers (output layers and hidden layers) can be computed by computing the generalized delta vector $\{\delta\}_J$ for each of these layers. Note that the size of $\{\delta\}_J$ is the same as the number of neurons in that layer. It can be shown that for the output layer $\{\delta\}N_{LAYER-1} = \{f\}N_{LAYER-1} \cdot (\{Y\} - \{Y\}^m)$.

For back layers

$$\{\delta\} = \{f\} - [W]_{J+1}\{\delta\}_{J+1}$$

The vector $\{f\}_J$ is the derivative of the activation function with respect to the neural response:

$$\{f\}_j = \frac{\partial f\{u\}_J}{\partial\{u\}_J}$$

Note that $f(u)$ is the activation function as defined in Table 6.6.

The process continues backward for all layers, until layer number 2. There is no error involving input layer since the input is exactly equal to the output for that layer.

Gradients for Weights and Bias: The gradients of the errors with respect to the weights can be computed for the second layer as $\partial e/\partial[W]_2 = \{\delta\}_2\{X\}^T$

For other layers

$$\frac{\partial e}{\partial[W]_{j+1}} = \{\delta\}_J\{H\}_{J+1}^T$$

The gradients of the errors with respect to bias can be computed for each layer as

$$\frac{\partial e}{\partial[B]_j} = \{\delta\}_J$$

Iteration and Training: The above steps are repeated N_{TRIES} times for all available training input-output pairs. Reintroducing the i indicator for each learning pair, the total error gradient for each weight matrix is

$$G([W]_J) = \sum_{i=1}^{i=N_{TRIES}} \frac{\partial e}{\partial[W]_J^i}$$

and for bias vectors

$$G(\{B\}_J) = \sum_{i=1}^{i=N_{TRIES}} \frac{\partial e}{\partial \{B\}_J^i}$$

We can compute the change in each of the weight matrices and bias vectors as

$$\Delta[W]_J^i = \eta G([W]_J),$$

and

$$\Delta\{B\}_J^i = \eta G(\{B\}_J),$$

respectively. While η is the learning step.

The weights and bias are updated as

$$[W]_J = \Delta[W]_J^i + [W]_J$$

and

$$\{B\}_J = \Delta\{B\}_J^i + \{B\}_J$$

The process is repeated with the newly updated weights and bias, until the error E satisfies the condition $E < \epsilon$, where ϵ is a redefined acceptable error limit. It should be noted that there are numerous techniques for the procedure of optimizing the error. Among such procedures is the steepest decent approach (Figure 6.28).

6.5.3 ADVANCED TOPICS

6.5.3.1 ANN versus FEM

Artificial neural networks can be used as a structural identification tool in similar fashion as FEM method. The analogy between the two methods is shown in Figure 6.29. Table 6.7 shows additional comparison between ANN and FEM methods as identification, or analysis tools.

6.5.3.2 Neural Network, Regression Analysis, and Expert Systems

The reader will perhaps recall that another method of formalizing a relationship between a given set of input-output pairs, while using a minimization of errors using a least square approach, is the regression method (see Chapter 8). In this sense, regression and neural networks are fairly similar. In particular, the similarities are that both methods depend on a set of functional forms (activation functions and regression equation for neural network and regression analysis, respectively) and the

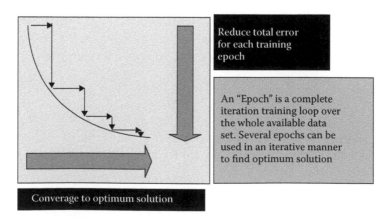

FIGURE 6.28 Method of steepest decent for error optimization.

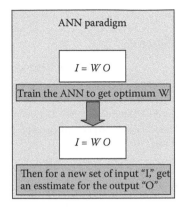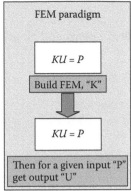

FIGURE 6.29 ANN versus FEM methods.

TABLE 6.7
Detailed Comparisons of ANN versus FEM

Issue	ANN	FEM
Basic component	Neurons, weights, and biases	Finite elements (physical)
Resolution	Number of layers, number of neurons	Size of elements
Behavior	Transfer (activation) function	Stress-strain relationships
Needs	Input-output pairings of dataset	A single set of inputs and a structural description
Objectives	Relates input-output by evaluating weights and biases for future use	Elements are preestablished by user
Can it be hybrid? Mixture of ToMs (displacements, strains, pressures, temperature, etc.)	Yes, without loss of any accuracy	Yes. Special numerical adjustments might be needed
Can it scale local-to-global behavior in single model?	Yes, without loss of any accuracy	Numerical and computational resources can be prohibitive

weights that control those functional forms (weights and bias for neural networks and regression coefficients for regression analysis). However, a basic difference between the two methods is that neural networks of the final functional form is not known *a priori*; it is determined after the network learning process is completed. For regression analysis, the functional form of the regression model is predetermined at the start of the analysis, and it does not change throughout the analysis. For both methods, the weights (for neural networks) and regression coefficients (for regression analysis) are determined from the process.

Another major difference between neural networks and regression models is that the output is affected directly by the input during a regression analysis. That is also true for two-layer (input and output layers) neural networks. However, for neural networks with layers more than two, that is, with one or more hidden layers, these direct input-output relations are lost. The intermediate hidden layers interfere with the input-output direct relationships. This would result in a difficult interpretation of the input-output relationship. This is an interesting balance: added accuracy and complexity from using hidden layers versus increased difficulty of interpreting the input-output interrelationship. Obviously, the added accuracy far outweigh the loss of interpretation capability, since, for complex situations, such interpretation capability is too difficult to begin with.

Another well-known algorithm that has some resemblance with neural networks is the expert system approach. The basic resemblance is that both methods can relate input-output relationships.

TABLE 6.8
Comparison of Expert Systems and Neural Networks

Expert System	Neural Network
Need to know much larger body of knowledge to make a decision	Do not require as much body of knowledge
Very sensitive to presence of noise in data	If trained well, can detect noisy data and produce accurate results
The whole expert system will need readjusting rules, if additional data is acquired	If additional data is acquired, only simple retraining of the network is needed

However, there are several differences between expert systems and neural networks, as shown in Table 6.8.

6.5.3.3 Training versus Validation

It is customary to divide the ANN data set (input-output associations) into two subsets such that

1. Training data subset to realize optimum values (weights and biases) of an ANN
2. Validation data subset to ensure that the ANN performs properly

In general, training data subset is about 67%–75% of the whole data set. For SHM projects that aim to utilize ANN it is essential to be cognizant of this need since it can affect decisions regarding database design and collection.

6.5.3.4 Batch and Online Learning

One of the technical properties of ANN is how the learning/optimization process proceeds. There are two distinct approaches: batch and online learning. The distinction is of importance to SHM usage. Batch learning operates on the whole training dataset, while the online training operates consecutively on each input-output record within the dataset. The rate of convergence of each of the approaches would depend on the size and complexity of the dataset. For simple dataset, it is expected that batch training would offer faster convergence. For smaller size data set, it is expected that the online training would provide faster convergence. Thus it is not an easy task to determine which of the two approaches is more suited for an ANN processing on a generic basis.

For real time, or near real time SHM effort, it seems that an online training ANN process might be the logical approach. In such a situation, we note that data records are continually monitored and added to the main dataset. Such a situation is more suited to an online training than to batch training. More studies are needed to verify this otherwise logical observation.

6.5.3.5 Supervised versus Nonsupervised Learning

Artificial neural networks can be formulated so as to be updated and changed in a nonsupervised mode, Trippi and Turban (1992). This is one of the major advantages of ANN when utilized for a real time, or near real time SHM process. There aren't many other STRID methods that have such a feature. Of course, designing a nonsupervised learning process is not an easy task. Again, this topic (use of nonsupervised ANN in the SHM field) needs further research that is not available as of the writing of this volume.

6.5.3.6 How to Design a Neural Network

Figure 6.30 shows the basic components that need to be addressed when designing a neural network. The types of activation function need to be chosen first. Table 6.6 can be used as a guide in this choice. The important rule to remember when choosing an activation function is that it needs to resemble, as closely as possible to the nature of the problem on hand. Another important issue is the

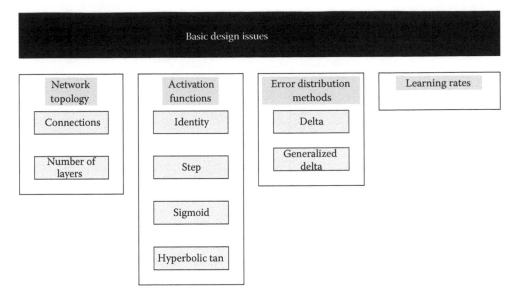

FIGURE 6.30 Neural network design issues.

choice of the error distribution methods. The delta and the generalized delta methods are the most popular and in the following sections, we investigate the generalized delta method. Another issue that must be decided is the topology of the network. This includes the number of hidden layers, the number of neurals in each of the layers and the type of connections between the neural in different layers. Finally, the rate of learning η needs to be addressed.

6.5.4 Applications of Neural Networks in SHM Field

Neural Networks are flexible tools that relates set of inputs to set of outputs in numerous ways, as such, it is expected that there can be many types of Neural Networks applications. We can state two main application types: they are the identification problems and the classification problems. Figure 6.31 illustrates these two main types, and some specific uses of each type. In this chapter, we will discuss the identification application of neural networks. Figure 6.31 illustrates the wide range of identification applications of neural networks. For example, the method can be used to identify the flexibility behavior model of the structure. Flexibility coefficients in a flexibility model is the displacement at a given point and direction that results from applying a force at same (or another) point and direction. Such a model is fairly simple and well identified. Humar et al. (2004) proposed the use of neural networks to define the structure, then to use a modal type algorithm to define the damage. A more complex model is the general structural (or modal) identification model that is discussed extensively in this chapter. A much more complex and not well-defined model is the damage identification application. Using neural network method for damage identification is more complex due to the general lack of numerical or analytical damage models. Examples of the three uses of neural networks are given in this section and the chapter on damage identification. In Chapter 9 of Ettouney and Alampalli (2012), we will discuss the classification application of neural networks.

Laurens et al. (2006) applied neural networks in structural identification (they aimed at identifying the electrical resistivity of concrete, i.e., identifying the properties of materials). This shows the benefits of using neural networks, especially if good database is available for training. The neural network was trained with an experimental database of electrical resistivity of concrete (see Figure 6.32) porosity, water content, and chloride content.

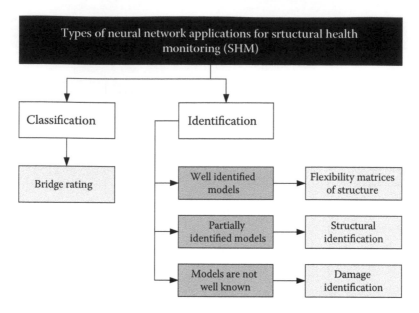

FIGURE 6.31   Applications of neural networks in SHM field.

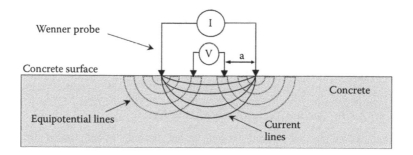

FIGURE 6.32   Wenner device: principle of resistivity measurement. (Reprinted from ASNT Publication.)

6.5.5   EXAMPLES AND CASE STUDIES

6.5.5.1   Single Input, Single Output: Linear Neural Network

To illustrate the use of ANN in a simple structural identification problem, let us consider the stiffness of a linear SDOF. Assume that during an SHM experiment a dataset of 33 force-displacement points is observed. An ANN model with single input-single output neurons and no hidden layers is used to model the observed behavior (Figure 6.23). The ANN model used linear activation function for simplicity. The error in both the weight and bias during the batch iterations are shown in Figure 6.33. Clearly, the ANN model converged quickly. After 50 batch (rather than online) training epochs is found to be sufficient to produce an accurate model. The ANN produced a functional relationship as shown in Figure 6.34. The figure also shows the ANN functional model after different batch iterations. The least square solution is also shown in Figure 6.34. Again, it is clear that 50 batch iterations produced an accurate solution when compared to the least square solution.

6.5.5.2   Nonlinear Spring Behavior

A conventional nonlinear problem in structures is when the force-displacement relationship is a monotonically increasing function. Suppose that a dataset of 55 force-displacement points is observed, as in Figure 6.35. An ANN model with single input-single output neurons and three hidden layers, each with two neurons is used to model the observed behavior. The ANN model used

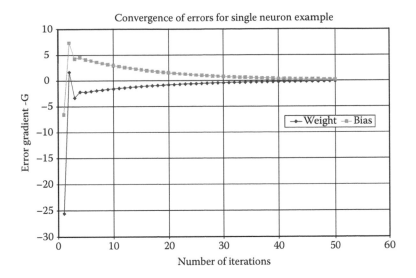

FIGURE 6.33   Convergence of errors for single-neuron example.

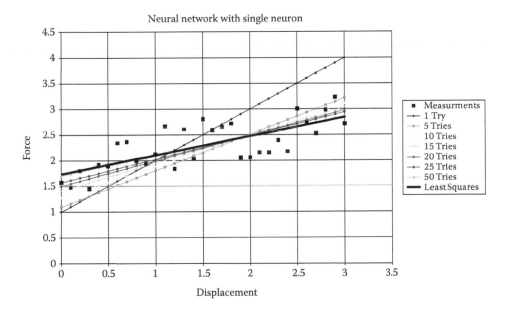

FIGURE 6.34   ANN behavior versus least squares.

hyperbolic activation function. After 100 online (rather than batch) training epochs, the ANN produced a functional relationship as shown in Figure 6.35.

A more difficult nonlinear problem in structures is when the force-displacement relationship is a softening function. Such problem occurs in nonconfined concrete, or in columns that are subjected to large axial forces beyond their stability limits. Suppose that a dataset of 128 force-displacement points is observed, as in Figure 6.36. An ANN model with single input-single output neurons and four hidden layers, each with three neurons is used to model the observed behavior. The ANN model used hyperbolic activation function. After 100 online (rather than batch) training epochs, the ANN produced a functional relationship as shown in Figure 6.36. Note that more information is needed for this ANN model than the previous one due to the higher nonlinearity.

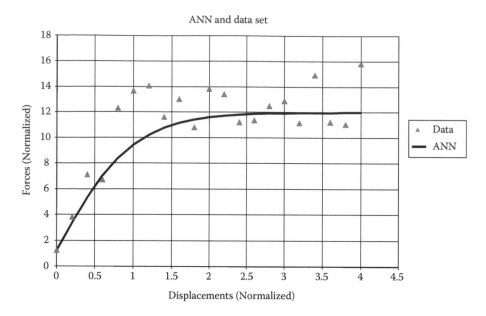

FIGURE 6.35 ANN versus data for monotonic spring behavior.

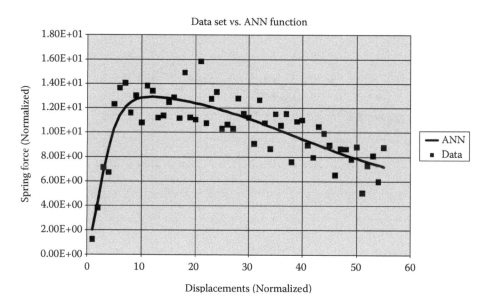

FIGURE 6.36 ANN versus data for softening spring behavior.

The use of ANN in simulating spring behavior can be expanded for even more complex STRID utilization. Consider, for example, a bridge testing experiment. Assume that trucks with a known weight P were positioned at N stations along the bridge. For each station, a set of M displacements were measured at M stations. This produced a load vector$\{F\}$ and an equivalent displacement matrix $[U]$. The order of the load vector is N and the size of the matrix is $M \times N$. If the matrix $[U]$ is rearranged in a vector form $\{D\}$ where the displacement set for the trucks at each loading station, then we recognize the two vectors pair $\{F\} - \{D\}$ set as a potential single data set in an ANN model.

Let us assume that the experiment is repeated N_{ANN} times to produce $\{F\}_i - \{D\}_i$ data sets with $i = 1, 2 \dots N_{ANN}$. Using ANN, this data set can be used to produce functional stiffness, flexibility, or hybrid approaches as follows:

ANN-Based Stiffness Approach: If we assume that the input layer for an ANN model is the forces $\{F\}$ and the output layer is the displacement $\{D\}$, this would make the ANN model a stiffness model. The input and output layers would have an N and an M neurons, respectively.

ANN-Based Flexibility Matrix: If we assume that the input layer for an ANN model is the displacement $\{D\}$, and the output layer is the forces $\{F\}$ this would make the ANN model a flexibility model. The input and output layers would have an M and an N neurons, respectively.

ANN-Based Hybrid Approach: Given the versatility of the ANN method, different hybrid models can be built. Such models can include smaller subsets of either forces and/or displacements as input, or output. This can easily be done without impacting the overall accuracy or resolution of the resulting ANN models.

Advantages/Limitations of Using ANN: The advantages of using ANN to produce functional stiffness/flexibility matrices are as follows

- Accounts as built conditions
- Accommodates any nonlinear effects in an accurate manner
- Resolution can be improved by adding more data sets with minimal efforts (grafting/ pruning)
- Avoids the inherent numerical and size limitations of conventional finite elements approximations
- Changes in trends can be monitored in real time
- Can model both local and global behavior within the same ANN model

In addition to all of above advantages, we note that there is no reason to expand the vector $\{D\}$ such that it includes other ToM than just displacement. For example, in an experiment, the vector $\{D\}$ can include displacements, strains, rotations, and stresses. There should be no loss of accuracy of the ANN model; a similarly hybrid FE model can experience loss of accuracy, Zienkiewicz (1971).

The main limitation of using ANN as a stiffness/flexibility tool is that it is a functional representation, that is, the results are difficult to relate to actual physical parameters. Such limitation might be overcome by taking advantage of the potential of measuring local as well as global responses of varying ToMs. Further research on this intriguing aspect of ANN is needed.

6.5.5.3 Bridge Scour

An SHM application of the use of ANN is discussed here. The problem involves the identification of the equilibrium scour depth around bridge piers. Scour is one of the most dangerous hazards that afflict bridges (see Figures 6.37 and 6.38). Scour problem is discussed extensively in Chapter 1 by Ettouney and Alampalli (2012). The scour problem results when rapid flowing waters around bridge foundations and piers causes soil erosion. When the degraded soil properties reach a threshold, the foundation/piers fail. It is highly nonlinear and involves several parameters including soil and foundations properties. It also involves the hydraulics of the water flow around and near the foundations. Jeng et al. (2005) used ANN to study damaging effects of scour near bridge piers. The remainder of this section describes their procedure and findings. We note that this study is more of damage identification study, rather than a structural identification study; it aims directly to

FIGURE 6.37 Scour damage. (Courtesy of New York State Department of Transportation.)

FIGURE 6.38 Scour damage. (Courtesy of New York State Department of Transportation.)

identifying the scour damage, rather than identifying the system properties. The authors identified five variables that might affect the scour process as follows

$$
\begin{aligned}
U_c &= \text{Critical flow velocity} \\
D &= \text{Pier diameter} \\
U &= \text{Average velocity of approach flow} \\
Y &= \text{Depth of approach flow} \\
d_{50} &= \text{Median grain parameter}
\end{aligned}
\tag{6.131}
$$

Thus, the ANN model has five input neurons. A single ANN output that describe the scour damage was chosen to be the equilibrium scour depth d_{se}. Figure 6.39 illustrates the problem parameters. Figure 6.40 shows the ANN model for the problem. The model has five input neurons and one output neuron. There is a single hidden layer with four neurons. A sigmoid transfer (activation) function was chosen for the problem. The authors chose a 187 data points that relates the five input parameters to the desired output d_{se}. A learning rate of 0.4 was chosen. A total of 9000 epochs were used to formulate the ANN.

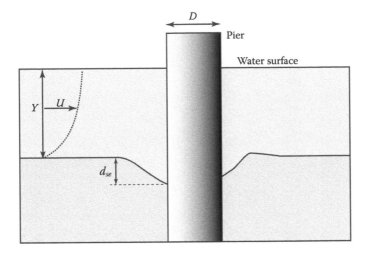

FIGURE 6.39 Simplified scour problem.

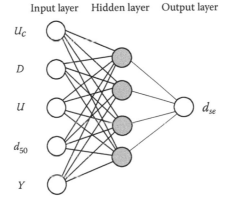

FIGURE 6.40 Scour ANN model.

Figure 6.41 shows the results of the ANN model when compared with actual measurements of d_{se}. The ANN model is well behaved. It can be used to predict scour behavior in similar settings. Such a model can also be used during a real time SHM experiment as follows:

- It can be used in a nonsupervised mode to improve its parameters as more observation data sets are monitored
- If there is a predetermined threshold limit state of d_{se}, an automatic warning signal can be sent to bridge officials whenever the ANN model predicts such a threshold in real time

6.6 OTHER METHODS

6.6.1 FREQUENCY RANGES

Earlier in this chapter, we subdivided qualitatively the frequency spectrum into three ranges: low, medium, and high ranges. We need to discuss this concept further. Low frequency range can be further defined as the range where wave lengths are longer than the typical length of the component of interest. Typically, few wave modes exist in the low frequency range. Also, in this range, the coupling between

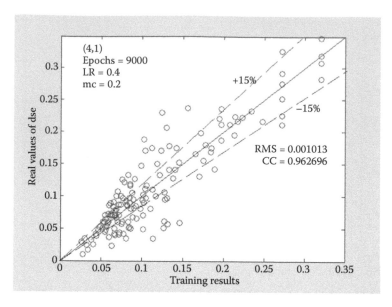

FIGURE 6.41 ANN results. (Courtesy of Department of Civil Engineering, University of Sydney: Jeng et al. [2005].)

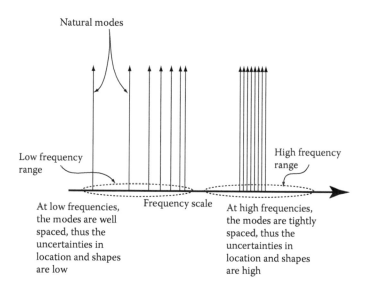

FIGURE 6.42 Mode distributions at low- and high-frequency ranges.

the wave modes is not significant, due to the relatively longer frequency steps between the different wave modes. Thus uncertainties due to coupling between modes are not high. Because of this traditional finite element methods can be used as the basis of STRID efforts with reasonable accuracy.

High frequency range is defined as the range where wave lengths are shorter than the typical length of the component of interest. Numerous wave modes will exist in this frequency range. Because of this, the frequency spacing between the modes becomes small. Thus the coupling between the wave modes can be significant, due to this short frequency spacing. The uncertainties due to the coupling between modes are high. Because of this, the accuracy of traditional finite element methods becomes unacceptable. Intermediate frequency range is the range where some short and long wave modes exist. Figure 6.42 shows the concept of modal coupling at low- and high-frequency ranges.

Because of these physical uncertainties, there is a need for different STRID methods in the high and medium frequency ranges. Several methods accommodate those needs. Some of them are discussed next.

6.6.2 Energy-Based Methods

6.6.2.1 Statistical Energy Analysis (SEA)

Lyon (1975), Lyon and Maidanik (1962), and Keane and Price (1969) observed that at high frequency range, the vibration energy of a system can be averaged using representations of its modal densities. The modal density in a particular vibration entity n is an estimate of the number of resonances (modes) that exist in a prescribed frequency range. Thus, by estimating the energy losses in the system, which can be shown to relate to modal densities, and the energy input to the system, an energy balance equation can be established.

In SEA, the system is subdivided into physical components. The physical components in a civil infrastructure can be plates, beams, shells, or columns. If a physical component can excite more than one type of waves, energy flow for each wave type need to be considered separately. For example, beams need to have four subsystems for each of the four wave types that can propagate within it (axial, torsional, flexural, and shear). For plates, shear, flexure, and axial waves need to be represented by three subsystems. The energy flow within each of the subsystems can be characterized by the component length, shape, and material properties (which include damping). This subdivision is one of the basic tenets of SEA: evaluating properties of subsystems (either by testing or analytically) is feasible at high frequencies. Analytical expressions of many of the parameters for the subsystems are available. Thus, the energy flow between the subsystems can be analyzed in a formal fashion.

As an example, let us consider a vibration of a system that contains several components; each component contains its own subsystem, as appropriate. The total number of subsystems is N. The energy balance equation of the system is expressed as follows:

$$\Omega[H]\{E\} = \{P\} \tag{6.132}$$

The driving frequency is Ω, the energy loss matrix is $[H]$, the subsystem energy vector is $\{E\}$, and the vector of the power input to the system is $\{P\}$. The order of Equation 6.132 is N.

The components of the matrices in Equation 6.132 was formalized by Koizumi et al. (2002) as follows

$$\{P\} = \begin{Bmatrix} P_{i1} \\ P_{i2} \\ \vdots \\ P_{iN} \end{Bmatrix} \tag{6.133}$$

The input power to the jth subsystem is P_{ij}. The subscript i indicates input power. Also:

$$\{E\} = \begin{Bmatrix} \bar{E}_1 \\ \bar{E}_2 \\ \vdots \\ \bar{E}_N \end{Bmatrix} \tag{6.134}$$

with

$$\bar{E}_j = \frac{E_j}{n_j} \tag{6.135}$$

The average energy and modal density of the jth subsystem are E_j, and n_j, respectively. Energy loss matrix [H] is expressed as

$$[H] = \begin{bmatrix} \bar{\eta}_{11} & \bar{\eta}_{12} & \cdots & \bar{\eta}_{1N} \\ \bar{\eta}_{21} & \bar{\eta}_{22} & \cdots & \bar{\eta}_{2N} \\ \vdots & \vdots & \ddots & \vdots \\ \bar{\eta}_{N1} & \bar{\eta}_{N2} & \cdots & \bar{\eta}_{NN} \end{bmatrix} \qquad (6.136)$$

Koizumi et al. (2002) proved that the diagonals of [H] are

$$\bar{\eta}_{jj} = n_j \left(\eta_j + \sum_{i \neq j}^{i=N} \eta_{ji} \right) \qquad (6.137)$$

and for off-diagonal $(i \neq j)$ of [H],

$$\bar{\eta}_{ij} = -n_i \eta_{ij} \qquad (6.138)$$

The energy loss within the ith subsystem is η_i. The coupling loss factor from i^{th} subsystem to j^{th} subsystem is η_{ij}.

Equations 6.132 through 6.138 provide the basis of SEA method. The properties of the ith subsystem, η_i, n_i, and the coupling loss factors to the jth, η_{ij}, provide complete description of the system. For any know input power distribution to the whole system {P}, the energy flow in the whole system {E} can be estimated. At a subsystem level, the energy can be related to the average velocity v_i of the subsystem as

$$E_i = M_i v_i^2 \qquad (6.139)$$

The mass of the subsystem is M.

SEA is a popular method in the fields of automotive, aeronautics, and naval architecture. Its simplicity and applicability at high frequency range account for this popularity. Limited applications are known in the field of civil infrastructure. Koizumi et al. (2002) used it to explore noise flow in buildings. One potential use is in estimating remaining fatigue life evaluation. Instead of using strains as metric for the evaluation, energy content in frequency band is used (see Lyon 1975). Since incipient damage is felt first by higher modes before it is felt by lower modes, detecting damage using SEA holds high potential. López-Díez et al. (2005) used SEA to detect damage in spacecraft structures. However, the use of SEA in the fields of STRID and/or DMID in civil infrastructure remains largely untapped as of the writing of this volume.

6.6.2.2   Energy Finite Element Analysis (EFEA)

Another energy-based method was developed to model both high and medium frequency ranges in an analogous manner to traditional finite element method. The EFEA is based on computing an average of the energy in space and frequency range in a similar manner to the computation of displacements by traditional finite element method. For example, the traditional flexural beam equation is (Timoshenko 1955)

$$EI \frac{d^4 y}{dx^4} + m\ddot{y} = p(x,t) \qquad (6.140)$$

The displacements and accelerations for the beam are y and $\ddot{y}$. The flexural rigidity and mass of the beam are EI and m, respectively. The applied load is p in the space and time variables x and t. The equivalent average flexural equation can be written as (Zhao and Vlahopoulos 2004)

$$\frac{-c_g^2}{\eta\Omega}\frac{d^2\langle e\rangle}{dx^2}+\eta\Omega\langle e\rangle=Q_{in} \qquad (6.141)$$

The flexural group wave velocity is c_g. The hysteretic damping is expressed by η and the driving frequency is expressed by Ω. The unknown energy density is $\langle e\rangle$. The averaged input power is Q_{in}. As before, the averaging is over the time (frequency range) and space. Equation 6.141 is expressed in matrix form for a single i^{th} beam element as (Bernhard and Huff 1999)

$$[E^e]_i\{e^e\}_i=\{F^e\}_i+\{Q^e\}_i \qquad (6.142)$$

The element matrix is $[E^e]_i$. The input power vector is $\{F^e\}_i$, and the internal power flow is $\{Q^e\}_i$. The average nodal energy density vector is $\{e^e\}_i$. The superscript e indicates element level in the equation. The analogy between Equation 6.142 and traditional element equation is obvious. The general equation of energy flow of the system under consideration is then assembled using elements equations similar to Equation 6.142. The power flow between the ith and j^{th} elements are accounted for using special energy coupling equation in the form (Cho 1993)

$$\begin{Bmatrix}Q_{ic}^e\\Q_{jc}^e\end{Bmatrix}=[J]_j^i\begin{Bmatrix}e_{ic}^e\\e_{jc}^e\end{Bmatrix} \qquad (6.143)$$

The matrix $[J]_j^i$ includes power transfer values between the ith and j^{th} elements. The subscript c indicates common nodes between the ith and j^{th} elements. Finally, the joint energy flow equation between the ith and j^{th} elements is expressed as (Zhao and Vlahopoulos 2004)

$$\left(\begin{bmatrix}[E^e]_i&\\&[E^e]_j\end{bmatrix}+[JC]_j^i\right)\begin{Bmatrix}\{e^e\}_i\\\{e^e\}_j\end{Bmatrix}=\begin{Bmatrix}\{F^e\}_i\\\{F^e\}_j\end{Bmatrix} \qquad (6.144)$$

The joint matrix $[JC]_j^i$ is a rearranged version of $[J]_j^i$ to accommodate the nodal sequencing in the model. The resulting equations can then be used to estimate energy flow at high- and medium-frequency ranges for structural systems.

Vlahopoulos et al. (1999) used EFEA to evaluate energy flow at high frequency of welded plates as shown in Figure 6.43. A comparison between test measurements and results from EFEA computations are shown in Figure 6.44. The accuracy of analytical results at such high frequency is evident.

Energy finite element analysis holds great potential in the field of STRID in civil infrastructure. It can identify structural properties at high and medium frequencies, thus compliment low frequency STRID methods. When compared to SEA, the EFEA can model more system details and EFEA can utilize the vast numerical solution algorithms that are available for conventional FEM, Vlahopoulos et al. (1999).

6.6.2.3 Concluding Remarks

There are other energy-based methods for analyzing system behavior at high frequency ranges. Sarradj (2004) provided a discussion of those methods. High frequency boundary element method (HFBEM) is based on boundary integral formulation that uses energy instead of displacement variables. Sarradj (2003) compared the results of SEA, EFEM, and HFBEM. An energy source is

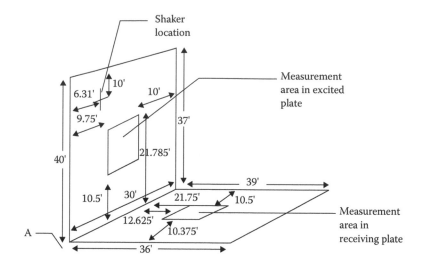

FIGURE 6.43 EFEA test setup. (Reproduced with permission from Academic Press.)

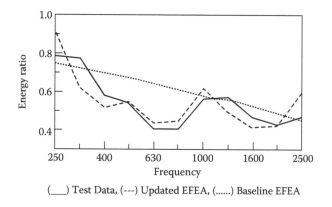

(___) Test Data, (- - -) Updated EFEA, (......) Baseline EFEA

FIGURE 6.44 EFEA test results. (Reproduced with permission from Academic Press.)

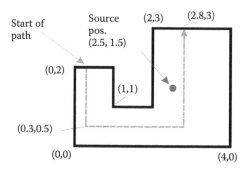

FIGURE 6.45 Steel plate experiment. (Courtesy of Dr. Ing. Ennes Sarradj.)

applied to a large steel plate using a shaker as shown in Figure 6.45. The energy along the path indicated in Figure 6.45 was computed using the three methods at different driving frequencies (1 and 10 K.Hz) and two damping levels (0.1% and 1%). The results are shown in Figure 6.46. The EFEM and HFBEM show more path-dependent distributions than the SEA, as expected. Also, HFBEM seems to produce higher fidelity results than EFEM.

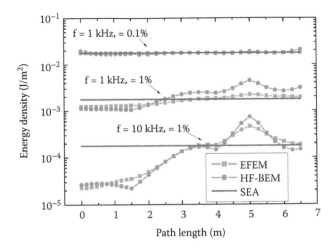

FIGURE 6.46 Comparison of energy-based methods. (Courtesy of Dr. Ing. Ennes Sarradj.)

The energy-based STRID methods offer potential in identify systems at high frequencies. Thus complimenting traditional low frequency MI methods. Their capability in directly identifying small size damage is an additional obvious advantage.

6.6.3 DISPERSION ANALYSIS

When the speed of wave propagation in a system varies as a function of frequency, the system is called *dispersive system*. The relationship between wave phase velocities in solid systems and frequencies of vibration is called *dispersion relations*. Such relationships depend on geometry and material properties of solid. When a system is dispersive, the wave propagates at different speeds at different frequencies. By understanding dispersive relations, the behavior of systems can be understood. Numerous studies regarding dispersive systems and the nature of wave propagating through them. A simple way of illustrating the dispersion relations is by plotting such relationship in a diagram, such diagrams are called *dispersion curves*. Detailed descriptions and derivations of dispersion relations are given by Rose (2004).

Figure 6.47 shows dispersion curves of axial waves in solid infinite cylinder (Rose 2004). The dispersion relations of circumferential waves in hollow cylinder are shown in Figure 6.48 (Rose 2004). Dispersion curves can be obtained analytically for simple systems. For example, the development of the dispersion relations for plates is shown in Section 7.10.

As the system geometry becomes more complex, analytical solutions of dispersion relationships becomes difficult. Numerical solutions, such as finite element methods can be used. Scale independent element (SIE) method can also be used to evaluate dispersion relationships (Chapter 8).

Since dispersion relations are fundamental part of system behavior, we consider their identification as an STRID process. The knowledge of a dispersion relation for a given system is similar to the knowledge of modal properties of a system: both will help in understanding certain aspects of the system, albeit at different space and frequency scales. In addition, given the applicability of dispersion relations to high frequency ranges, it is possible to use their measurements directly to identify damage presence, location, shape, and extent (see Chapter 8).

6.6.4 GENETIC ALGORITHMS

Another class of STRID uses genetic algorithms (GAs). Koh and Htun (2004) described a GA methodology in STRID. As of the writing of this volume GAs are not as widely used in STRID civil infrastructure applications as other methods described in this chapter.

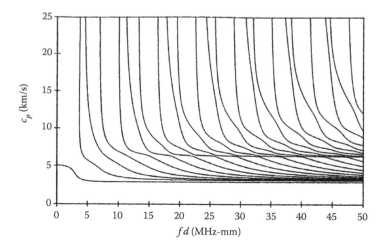

FIGURE 6.47 Dispersion curves for axial waves in solid cylinder. (Rose, J. *Ultrasonic Waves in Solid Media*, Cambridge University Press, Cambridge, UK, 2004. Reprinted with the permission of Cambridge University Press.)

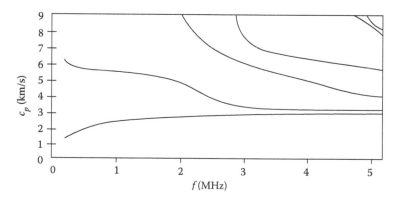

FIGURE 6.48 Dispersion curves for circumferential waves in hollow cylinder. (Rose, J. *Ultrasonic Waves in Solid Media*, Cambridge University Press, Cambridge, UK, 2004. Reprinted with the permission of Cambridge University Press.)

6.7 MODELING TECHNIQUES

6.7.1 OVERVIEW

Many times the accuracy of STRID solutions depends on modeling techniques of system components other than the structural system itself. In particular, modeling of soils and water can significantly influence results of STRID problems. We discuss next some modeling issues with regards to soils and water. We conclude the section with discussion of some general modeling issues that can affect STRID activities such as model scaling, static condensation, and molding problems in a multihazards environment.

6.7.2 SOIL-STRUCTURE INTERACTION ROLE IN STRUCTURAL IDENTIFICATION PROBLEM

The interaction between the structure and its supporting soil is an important factor in the overall behavior of the system in many, but not all, situations. Because of this, several authors pointed to the importance of SSI in STRID problems. In general, whenever we encounter a structural identification

situation, it is important to (1) decide whether the Soil Structure Interaction (SSI) effects are important in that particular situation, and (2) the most effective way to account for such an effect.

There are some general rules that can help in assessing the importance of SSI in any structural
identification situation. The importance of SSI can be ascertained based on type of hazard, magnitude of hazard, and/or overall relative structure-to-soil stiffness. Some hazards would require the
consideration of SSI in any structural identification problem. For example: it seems logical to consider SSI effects while addressing the effects of scour and earthquakes. On the other hand, SSI effects
are negligible when addressing corrosion effects. Higher levels of some hazards can increase the
importance of SSI effects. Impact of barges on a bridge pier might require the consideration of SSI
effects. Perhaps the most important factor that controls the importance of SSI effects is the relative
stiffness of the structure and the underlying soil. If the structure is supported by a rigid foundation,
the SSI effects are negligible. Such a situation arises when the structure is supported directly by
rock. When the structure becomes stiffer, relative to the supporting soil, the SSI effects need to be
considered. Note that this relative stiffness effects must be considered in a static and dynamic sense,
depending on the kind of problem under consideration. For slowly varying loads, such as a bridge
load testing situation, the condition can be assumed as a static condition. In such situation, the relative
structure-to-soil stiffness can be evaluated using the relevant static stiffness parameters.

As a guide for evaluating the relative importance of SSI effects for the rotation of the bent foundation, if the soil rotational stiffness is K_{SOIL_ϕ} and if the bent column rotational stiffness is K_{STR_ϕ},
then the scalar ratio

$$\alpha = \frac{K_{STR\_\phi}}{K_{SOIL\_\phi}} \tag{6.145}$$

should indicate the need for accommodating SSI effects for such a problem.

For dynamic situation, Equation 6.145 must be generalized to account for time, or frequency,
effects. In the time domain:

$$\alpha(t) = \frac{K(t)_{STR\_\phi}}{K(t)_{SOIL\_\phi}} \tag{6.146}$$

and in the frequency domain

$$\alpha(\Omega) = \frac{K(\Omega)_{STR\_\phi}}{K(\Omega)_{SOIL\_\phi}} \tag{6.147}$$

The time and driving frequency are t and Ω, respectively. Evaluating 6.146 in the time domain is not
a trivial endeavor, since it requires solution of iterative nonlinear problem. Evaluating Equation 6.147
in the frequency domain is a bit easier. There are some frequency-dependent soil stiffness expressions that are available (Ettouney 1978 and 1979). Also, some frequency-dependent structural stiffness were reported in some cases, see, for example, Ettouney (1979). Note that α alpha in Equation
6.146 is a real function while α in Equation 6.147 is a complex number.

Soil-structure interaction effects need to be evaluated for more than just rotation of foundations.
They need to be evaluated for horizontal, vertical, and tensional motions. In some situations, coupled SSI motions should also be investigated.

We should emphasize that the above are merely general rules. The analyst should study the specific condition of the problem under consideration, and make the decision on a case-by-case basis.

Let us assume that the analyst decided that SSI effects are important to accommodate in the
STRID problem under consideration. In such a case, the analyst must answer two questions: (1)
What is the acceptable resolution of the soil model, and (2) What type of nonlinear material model
is needed for the soil representation?

TABLE 6.9
Some Soil Modeling Techniques

Model	Advantages	Disadvantages
Linear spring	Simple suited for modal analysis methods	Accurate only for static, or low frequency zone
Nonlinear spring	Can simulate certain aspects of soil nonlinearity	Amount of energy dissipation is neither controllable, nor verifiable. Can produce unconservative results
Frequency-dependent spring	Better representation of damping	Requires frequency domain analysis
Multiple degree of freedom (finite elements)	Suited for modal analysis methods (if soil elements are linear) Accommodates higher modes Better geometry representation	Analytically demanding

There are many types of soil models that can be used in a soil-structure coupled modeling method. Some of the popular models are as follows: (1) single linear spring, (2) nonlinear spring, (3) single frequency-dependent spring, or (4) multiple DOF (finite elements soil representation). The advantages and disadvantages of each model are shown in Table 6.9.

The above discussed only coupled soil-structural modeling issues. It must be mentioned that there are several other methods that are based on decoupled modeling of SSI situations. Decoupled methods are usually based on certain assumptions. We believe that the magnitude of errors that are introduced by those assumptions might not be acceptable in a structural identification effort. Until a detailed study on the adequacy of decoupled soil-structure modeling for STRID problems is performed, it is recommended that only coupled soil-structure methods as discussed above is used for structural Id situations.

Finally we discuss the nonlinear soil behavior issue. We observe that soil is a nonlinear material. We immediately observe that most STRID methods that were discussed in this chapter, and elsewhere, are based on linear modeling techniques. This poses the following apparent paradox: if SSI effects are important, and since soil behavior is nonlinear in general, then we should expect to use nonlinear methods for the STRID problem. However, most of the structure STRID methods are linear. There are two possible solutions to this apparent paradox: either use linear approximation of the soil properties, or use a true nonlinear STRID method. Again, nonlinear STRID methods are not well developed as of the writing of this volume. Further research is needed in this field.

6.7.3 SOIL-STRUCTURE INTERACTION ROLE IN DAMAGE IDENTIFICATION PROBLEM

Sometimes, it is desired to utilize STRID techniques to identify damage directly. Olson (2005) used parameter identification techniques to identify scour damage (as simulated by certain level of exposure of piles). When soil and structure are components in an STRID effort, attributes such as soil-structure interface, soil nonlinearities (especially near the interface), and the potential importance of higher modes need to be considered in such models. If water is present, such as submerged foundations during a scour process, the water-structure interaction behavior should also be considered carefully.

6.7.4 FLUID-STRUCTURE INTERACTION (FSI) IN STRID PROBLEM

Structural identification for potential use in scour problems will naturally concern itself with local scour; other types of scour damage occur away from the bridge structure where STRID efforts would not be of particular value. The distinguishing feature of local scour, from STRID view point, is that it should involve the interaction of three very distinct systems: the structural foundation, the

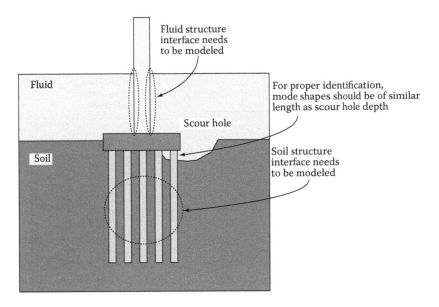

FIGURE 6.49 STRID modeling needs for scour problem.

surrounding soil, and the fluid. In a MI test, the frequency characteristics of the combined system should be considered. We observe that the presence of the fluid would have an effect of an added mass to the system, assuming low frequency range, see Ettouney and Daddazio (1978). The added hydrodynamic mass can have large effect on the dry modes of systems, Ettouney et al. (1990). Similarly, soil effects should be considered in any scour-related STRID effort. In particular, we note that scour holes near foundations may result in slight changes of low modes (both frequency and eigenshapes), until the system approaches the state of instability. A way to increase sensitivity of MI of scour hole presence is to identify higher response modes. Mode shapes with specific length that is close to the scour hole depth, or shorter, offer better chance of being identified (see Figure 6.49).

6.7.5 STATIC CONDENSATION MODELING TECHNIQUES

Conventional SSI problems, such as seismic, or scour problems, would require modeling both soil and structural system. Such a model can be computationally demanding. A popular method is to subdivide the model into soil and structure parts, as shown in Figure 6.50. The governing matrix equation for the soil is

$$[K]^* \{U\} = \{P\}^* \tag{6.149}$$

The soil stiffness is $[K]^*$. The soil displacement vector is $\{U\}$ and the applied loads on the soil is $\{P\}^*$. Equation 6.149 can be subdivided as

$$\begin{bmatrix} K_{11} & K_{12} \\ K_{21} & K_{22} \end{bmatrix} \begin{Bmatrix} U_1 \\ U_2 \end{Bmatrix} = \begin{Bmatrix} P_1 \\ P_2 \end{Bmatrix} \tag{6.150}$$

The subscripts $i = 1$ represent the DOF at soil-structure interface. The remainder of soil DOF is represented by subscripts $i = 2$. Operating on Equation 6.150 we get

$$[K]^* \{U_1\} = \{P\}^* \tag{6.151}$$

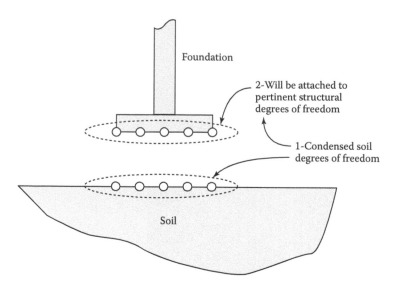

FIGURE 6.50   Static condensation concept.

where

$$[K]^* = [K_{11}] - [K_{12}][K_{22}]^{-1}[K_{21}]$$

(6.152)

Equation 6.151 can now be considered as multi-DOF soil spring. The equivalent soil spring $[K]^*$ can be added to the pertinent structural foundation DOF. The soil is now completely represented in the structural model. The resulting soil-structure model can now be either solved using conventional finite element techniques, or utilized efficiently in an STRID parameter identification problem.

Such an approach in dynamic problems would lead to inaccurate results, since it produces modal relationships that do not resemble the actual modal relationships. Note that the soil dynamic equation is

$$[M]\{\ddot{U}\} + [K]\{U\} = \{P\}$$

(6.153)

The soil mass matrix is $[M]$ and the soil acceleration vector is $\{\ddot{U}\}$. The condensed matrix $[K]^*$ does not have any information regarding the inertial properties of the soil; thus it is incapable of reproducing accurate soil modal behavior. It also truncated the number of soil modes arbitrarily, so that higher soil modes can't be present in the system. The conservatism, or the lack of it, can't be assured in such an approach. In parameter identification, the problem is compound, since the static condensation matrix $[K]^*$ is not a physical matrix: it is a numerical approximation. Trying to identify any of its components might produce meaningless results: we are trying to update a nonphysical entity to a physical one.

6.7.6 Scaling Needs

When MI is used for the purpose of STRID, the professional usually tries to verify a numerical model of the structure by the identified modes. In parameter identification, the numerical model itself is updated using the experimental results. In both situations the resolution of the numerical model should be adequate for accurately modeling the frequency range of the dynamic test. For example, it is common practice to model structural columns by few, usually single, finite elements. Dynamically, the dynamic range of such an element is limited to single flexural and single axial modes. Higher modes can't be captured by single element. For higher modes, additional elements are required, to ensure accurate dynamic representation. As the dynamic frequency range of interest is increased, the number of elements increases. At certain level, this can add unacceptable numerical demands on the model. Scaling techniques, such as scale independent element (described in the next section), can help in reducing such demands, while preserving needed accuracy.

6.7.7 MULTIHAZARDS CONSIDERATIONS AND STRID

There is a tight relationship between STRID modeling and multihazards considerations. However, the three major STRID techniques vary with regards to their multihazards interactions, as follows:

6.7.7.1 Modal Identification

Identifying modal information is related to hazards that are dynamic in nature. Chapter 8 presents qualitative ranges of some important hazards. It is logical to relate the frequency range of desired identified modes to the range of hazard of interest. If there is a mismatch in the indentified modal range and the hazard of interest, the value of the identified modal properties is reduced. This implicit interrelationship can control the type of MI project.

6.7.7.2 Parameter Identification

If the system under consideration is numerically modeled (by finite elements method, for example) then, according to multihazards theory, such a numerical model can be utilized to analyze all hazards of interest. The physical system accommodates all interdependencies of all the hazards, and hence an analytical model should be capable of handling all hazards.

Unfortunately, this is far away from the truth. In translating the physical system into an analytical model, numerous simplifying assumptions must be made and these assumptions would affect how the analytical model analyzes and estimates the response to different hazards. Specifically, these assumptions can be classified into three categories: spatial, temporal, and response level (amplitude). Spatial assumptions include the size of discretization (element size in finite element method), extent of the analytical model and boundary conditions (see Figure 6.51). Temporal assumptions include frequency range of analytical model and time integration schemes, if applicable. Response level (amplitude) assumptions include types and levels of nonlinearities. Obviously, the assumptions made would depend on the hazard of interest and the expected behavior of the structure to that hazard. Any hazard has to be described spatially, temporally (frequency), and quantitatively (amplitude), as shown in Figure 6.51. Figure 1.26 in Chapter 1 shows schematically how different hazards command different ranges in the response level (amplitude)-temporal (Frequency) space. Figure 4.25 in Chapter 4 shows spatial characteristics of various hazards.

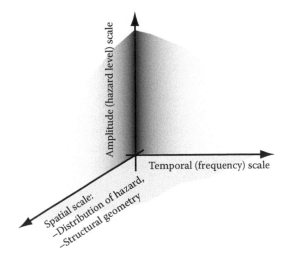

FIGURE 6.51   Hazards attributes.

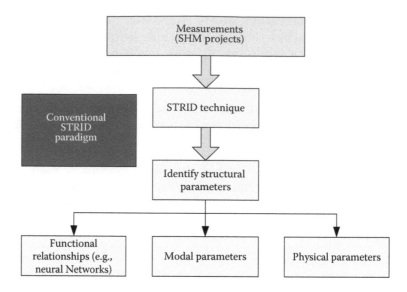

FIGURE 6.52   Conventional STRID paradigm (without accommodating multihazards).

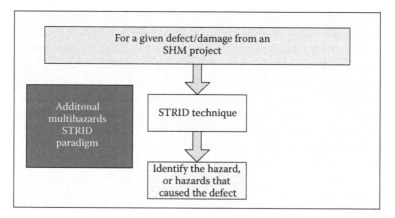

FIGURE 6.53   Multihazards-based STRID paradigm.

On the basis of the previous discussion, it is clear that different hazards demand different numerical assumptions and numerical analysis techniques to ensure accurate analytical results. So, we reach the following important conclusion: even though the physical system can accommodate different hazards, including their interdependencies, numerical analysis and modeling of different hazards would require separate analysis and modeling considerations. Ignoring this would produce inaccurate results. This can be simply exemplified through multihazard parameter identification of bridge piers for purpose of seismic or blast load investigations. For seismic hazard, the design and analysis model can be a simplified beam/column model. On the other hand, a higher precision FEM model may be required for analysis and design for blast load cases. Figures 6.52 and 6.53 show the conventional and multihazards-based STRID paradigms, respectively.

6.7.7.3   Neural Networks

By definition, neural networks represent direct informal relationship between input and output. Different hazards, such as traffic loads, scour, and earthquakes are generally the input to the network. The output can be any desired response function, such as displacements, and scour damage. Because of the large variations in hazard special and temporal characteristics, having a network that can span over different hazards is almost impossible.

6.8   SCALE INDEPENDENT METHODS (SIM)

6.8.1   SCALE INDEPENDENT ELEMENT

Discrete deterministic methods such as finite elements offer great flexibility in analyzing the dynamic response of vibrating systems. However, these methods can easily grow beyond available computer resources as frequencies of interest grow higher. In this chapter we present a new approach for the frequency domain dynamic analysis of structures. A theory is developed for the analysis of systems that are uniform along a single coordinate axis but otherwise arbitrary in geometry and material composition. This approach, termed the Scale Independent Element (SIE), is shown to be an accurate, efficient, and general method for the analysis of vibrating systems. This technique extends the applicability of discrete deterministic finite element based modeling to higher frequencies and is capable of bridging the gap to frequency regimes where statistical energy methods become applicable. Ettouney et al. (1995) offered a detailed description of the technique. We reiterate their development next.

6.8.1.1   Overview

Three frequency regimes are usually distinguished in the analysis of structural vibrations in complex systems. At very low frequencies, deterministic modal methods (Leung 1988) are well established. At high frequencies, asymptotic methods such as ray acoustics and frequency response smoothing by geometric mean values (Girard 1990) are applicable. The midfrequency regime, spanning a substantial range of structural behavior, has several competing solution methodologies. On one end, discrete deterministic methods such as finite elements offer great flexibility in analyzing problems that are intractable analytically. These methods can easily grow beyond available computer resources as frequencies of interest grow higher. At the other end, where resonances are hard to distinguish, statistical energy methods (Lyon 1975) are appropriate. However, a gap remains in the midfrequency regime where no method has a clear advantage.

In this chapter, the SIE is introduced as a new super element designed for the frequency domain dynamic analysis of finite structures (or substructures) that are uniform (or periodic) in the length direction, say x, and arbitrary in the other directions. The heterogeneity of the cross-section, whether geometric or material, calls for a discretization technique such as the finite element method. On the other hand, the uniformity along x calls for some analytical or modal expansion of the solution in that direction. This is especially desirable in the mid-frequency regime where frequencies of interest are high enough to make the cost of wavelength resolution by discretization substantial, but too low for asymptotic theories to govern. The SIE approach reconciles these two demands by conceptualizing the finite structure of length Nh as an assembly of N cells of length h. For each SIE, a single cell needs to be modeled by finite elements having for length dimension h. The characteristic equation resulting from this periodic discrete structure is a quadratic eigenvalue problem. The cell eigensolutions form the complete basis for the representation of the SIE dynamic stiffness matrix. Even though, exactly equivalent in its dynamic characterization to a finite element model of the entire finite structure, the SIE computations only involve cell matrices. In other words, the computational effort is independent of the length scale N, and thus the name SIE. Furthermore, since the SIE displacements are expressed in an expansion basis consistent with conventional finite elements, SIEs can easily be connected to conventional finite elements used to model nonuniform regions.

The SIE approach differs from better known substructuring techniques in two respects:

1. The SIE approach does not truncate the modal basis and therefore maintains the frequency resolution inherent in the definition of the unit cell. In this, the SIE is akin to an exact dynamic condensation procedure.
2. The modeling and computational efforts in an SIE formulation are independent of the length scale. On the other hand, conventional substructuring or super-element techniques involve at least the meshing of the entire substructure and sometimes manipulation of the assembled substructure matrices.

6.8.1.2 Theory

The formulation of the SIE, presented below, is based on a series of ideas that have their origin in the development of the consistent energy transmitting boundary for semi-infinite layered soil strata by Lysmer (1970), Waas (1972), and Kausel (1974). The concept of an eigenfunction expansion in the horizontal direction was extended with the "hyperelement" to finite layered continua by Kausel and Roesset (1977). The "Cloning Algorithm" of Dasgupta (1982) for unbounded homogeneous continua constitutes a further extension in that the coordinate system inherent to the problem geometry is no longer assumed separable. Subsequently, the "hyperelement" concept was further generalized to both semi-infinite and finite truss and beam structural systems by Ettouney et al. (1990). The SIE concept was introduced by Ettouney et al. (1995a and 1995b). The concept extends the "hyperelement" idea to structural systems that are uniform in one direction but have arbitrary cross-sections. In what follows, we will summarize the essential equations of the method.

6.8.1.2.1 Equations of Motion

Consider a finite structure, that is uniform in its length direction x. We introduce a finite element approximation, with uniform discretization h along x of the equation of motion. This translates into a periodicity along x that we track with a cell counter i. Thus, cell i is bounded by the $(i - 1)^{th}$ and i^{th} sets of nodes. To each set of nodes i corresponds M degrees of freedom.

The discretized time-harmonic equations of motion for cell i may be written as

$$\begin{bmatrix} S_{11} & S_{12} \\ S_{21} & S_{22} \end{bmatrix} \begin{Bmatrix} u_{i-1} \\ u_i \end{Bmatrix} = \begin{Bmatrix} -P_{i-1} \\ P_i \end{Bmatrix}$$

(6.154)

where u_i is the displacement solution vector for node set i and S is the cell dynamic stiffness given by

$$S_{kl} = K_{kl} + j\omega C_{kl} - \omega^2 M_{kl} \quad k = 1, 2, \ l = 1, 2$$

(6.155)

K_{kl}, C_{kl}, and M_{kl} are the stiffness, damping, and mass matrices, respectively, ω is the forcing circular frequency and $j = \sqrt{-1}$.

The assembly at node set i of contributions from the ith and $(i - 1)^{th}$ cells yields the following governing stencil

$$S_{21}u_{i-1} + (S_{11} + S_{22})u_i + S_{12}u_{i+1} = 0$$

(6.156)

It can be shown that the stencil, being a recurrence relation with constant coefficients, admits solutions of the form

$$u_i = A\lambda^i = Ae^{-jkx_i}, \quad x_i = ih$$

(6.157)

where k is the complex wavenumber.

Substitution of Equation 6.157 into Equation 6.156 yields the following characteristic equation

$$[\lambda^2 S_{12} + \lambda(S_{11} + S_{22}) + S_{12}]A = 0$$

(6.158)

6.8.1.2.2 Characteristic Equation: A Quadratic Eigenvalue Problem

Equation 6.158 is an $M \cdot M$ (M × M) quadratic complex symmetric eigenvalue problem with λ for eigenvalue. For a given forcing frequency ω, it can be shown that Equation 6.158 admits for solution an 2M eigensolutions (λ_s, V_s) representing the full spectrum of waves (including evanescent waves) that the system admits. It can further be shown that the 2M eigensolutions can be segregated into two sets of M eigensolutions: the first, denoted by (λ_s, V_s) for $s = 1,...,M$, is characterized by $|\lambda_s| \le 1$ and corresponds to waves decaying in the positive x direction (outgoing waves); the second half, denoted by $(\tilde{\lambda}_s, \tilde{V}_s)$ for $s = 1,...,M$, is characterized by $\tilde{\lambda}_s = 1/|\lambda_s| \ge 1$ and corresponds to waves decaying in the negative x direction (incoming waves). Furthermore, the eigenvectors of each of the two sets can be related by

$$\tilde{V}_s = TV_s \tag{6.159}$$

where T is a simple diagonal transformation matrix that was discussed by Waas (1972) for plane strain/plane stress problems, and by Kausel (1974) for 3D axisymmetric continua. In the general case of 2D and 3D structural elements, the expression for T must be generalized to properly account for rotational degrees of freedom.

To formulate the dynamic stiffness matrix of the finite structure that is represented by the assembly of N identical cells, both of the above sets of eigensolutions will be required (Kausel and Roesset 1977 and Ettouney 1990). This is in contrast to semi-infinite structures that require only one of the two sets; see, for example, Waas (1972), Kausel (1974), Dasgupta (1982) and Ettouney (1990).

6.8.1.2.3 Dynamic Stiffness Matrix

Consider the finite structure of length $L = Nh$ and consisting of N identical cells. The dynamic stiffness matrix relating forces and displacements at $x = 0$ and $x = L$ can be derived using the eigensolutions of the characteristic equation. Following Kausel (1977), the displacement vectors u_0 and u_N at node sets $i = 0$ and $i = N$, respectively can be expanded as follows

$$\begin{Bmatrix} u_0 \\ u_N \end{Bmatrix} = [D_0] \begin{Bmatrix} \Gamma_0 \\ \Gamma_N \end{Bmatrix} \tag{6.160}$$

where

$$[D_0] = D_0(V, \tilde{V}, \Lambda) \tag{6.161}$$

with

$$V = [V_1,...,V_M] \tag{6.162}$$

$$\tilde{V} = [\tilde{V}_1,...,\tilde{V}_M] \tag{6.163}$$

$$\Lambda = \mathrm{diag}[\lambda_1,...,\lambda_M] \tag{6.164}$$

$$\Gamma_i = [\Gamma_{i1}...\Gamma_{im}]^T, \quad i = 0 \quad \text{or} \quad N \tag{6.165}$$

where Γ_{0s} and Γ_{Ns} are the sth mode modal participation factors for outgoing and incoming waves, respectively.

Enforcement of equilibrium, by virtue of Equation 6.154, yields for the consistent boundary force vectors P_0 and P_N, at $x = 0$ and $x = L$, respectively, in terms of the modal participation factors

$$\begin{Bmatrix} P_0 \\ P_N \end{Bmatrix} = [D_1] \begin{Bmatrix} \Gamma_0 \\ \Gamma_N \end{Bmatrix} \tag{6.166}$$

where

$$D_1 = D_1(V, \tilde{V}, \Lambda, S_{kl}) \tag{6.167}$$

Elimination of the modal participation factors from Equations 6.160 through 6.167 yields the expression for the dynamic stiffness matrix D of the SIE

$$\begin{Bmatrix} P_0 \\ P_N \end{Bmatrix} = [D] \begin{Bmatrix} u_0 \\ u_N \end{Bmatrix} \tag{6.168}$$

where

$$[D] = [D_1][D_0]^{-1} \tag{6.169}$$

$[D]$ is a complex symmetric matrix of dimension $2M \cdot 2M$.

6.8.1.3   Efficiency of SIE Method

In studying the relative efficiency of the SIE methodology as compared to the traditional finite element technique, we assume that the frequency considerations of the problem require that the model is subdivided into N identical cells and having M degrees-of-freedom per set of boundary displacements. A simple one-beam system, as well as a more realistic compound systems will be studied.

6.8.1.3.1   One-Beam System

Consider the one beam system (Figure 6.54). Table 6.10 shows a comparison of the computational requirements of both SIE and traditional finite element methodologies.

Note that the formation of the dynamic stiffness matrix of the scale independent element requires the solution of the quadratic eigenvalue problem, Equation 6.157, which is an operation of order M^3, irrespective of N. Thus, for large N, the reduction in computational cost is substantial.

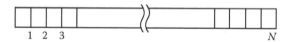

Number of cells = N

Number of degrees of freedom per cell = M

FIGURE 6.54   One beam system.

TABLE 6.10
Comparison of the Two Methods for Simple Structural System

Attribute	Finite Element	SIE
Number of equations	$(N+1)M$	$2M$
Band width	$2M$	$2M$
Order of computation time	NM^2	M^3

6.8.1.3.2 Compound System: Computational

Now consider a more realistic compound systems. Figure 6.55 shows a planar truss that is built up from individual beams. The truss has an N_s subspans. All the members of the truss are assumed to have identical cross-section. Such an assumption is consistent with practical engineering practices. Table 6.11 shows a comparison of computational attributes of the truss if it is modeled using the traditional finite element and the SIE methodologies.

The ratio of the computational requirements of the two methods is N^3. The computational advantage of the SIE is even greater for more realistic systems than in simple systems.

6.8.1.3.3 Compound System: Modeling

Table 6.11 also shows the relative modeling requirements of both traditional finite element method and the SIE method for realistic systems. For the structure under consideration, traditional finite element method requires much more modeling effort than the SIE methodology. Actually, the SIE method will require the minimum amount of geometric information to represent the structure. Any added modeling resolution that is needed in the traditional finite element method due to the driving frequency requirements are not needed by the SIE approach.

It is also of importance to note that the SIE modeling will result in a significant reduction in memory/storage.

6.8.1.4 Examples

The purpose of this section is to demonstrate some practical applications of the SIE methodology. The wave mode shapes of a thin walled, box shaped beam will be presented first. The torsional and bending behavior of a single beam is studied next. Finally, the behavior of a complex framed deck structure is studied.

6.8.1.4.1 Wave Modes

Understanding the behavior of systems in different frequency ranges is an important aspect of analysis. The SIE methodology provides an automatic mean to such an understanding. Evaluation

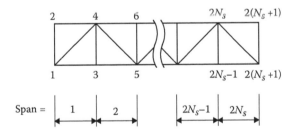

FIGURE 6.55 Compound truss system.

TABLE 6.11
Comparison of the Two Methods

Attribute	Finite Element	SIE
subspans	N_s	N_s
Additional degrees-of-freedom per node, n	NM	M
Number of equations	$2N_sNM$	$2N_sM$
Band width	$4NM$	$4M$
Order of computation time	M^3N^3	M^3

of the eigenvectors V and $\tilde{V}$ of Equations 6.162 through 6.163 can be used for that purpose. Consider, for example, the beam with a box cross-section of Figure 6.56. The material of the section is steel. A unit load is applied symmetrically at one end of the beam. It is of interest to understand the behavior of the beam when the cross-section is no longer rigid.

Figure 6.57 shows the dominant eigenvector of the cross-section plotted at resonant frequencies in the range of 60 to 850 Hz. Even at the lowest bending resonant frequency, 60 Hz, the local deformation of the cross-sectional plating is observed. Care should be taken when modeling such a box cross-section, using conventional beam theory that assumes that the beam cross-section remains rigid throughout the frequency spectrum.

6.8.1.4.2 Torsion and Bending of Box Beams

Figure 6.58 shows a free standing one-beam system. The beam has a tubular cross-section of $6 \times 6 \times \frac{1}{4}$. The moment of inertia of the section is 30.30 in⁴, the cross-sectional area is 5.591 in², while the shear area of the section is 3.00 in². The torsional moment of inertia is 48.5 in². These specifications have been published by The Steel Company of Canada (1966). The thickness of this tubular section is 0.25 in. The radius of the corners is specified at 0.75 in. The manufacturing tolerance of the corner radius has a limit of 10%.

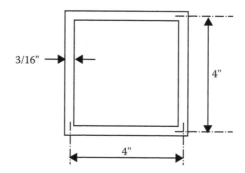

FIGURE 6.56   Box cross-section.

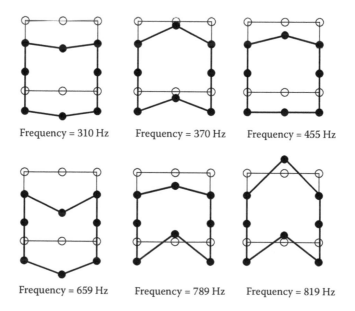

FIGURE 6.57   Wave mode shapes of cross-section.

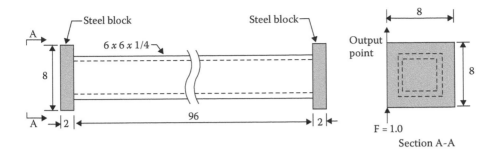

FIGURE 6.58 Box beam configurations.

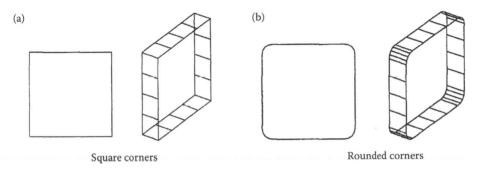

Square corners   Rounded corners

FIGURE 6.59 Rounded and square box corners (a) model with square corners, and (b) model with rounded corners.

To simulate the end joints of the system, heavy steel blocks are welded to the ends of the beam (Figure 6.58). The system is excited by a vertical eccentric unit force applied at one of the steel blocks. This application of force on the system is designed to excite both bending and torsional responses, simultaneously.

Several analytical models were considered for this problem. The SIE methodology was used in all of those models.

First, the beam was modeled using a Timoshenko et al. (1974) approximation, where the plane of the tubular cross-section remains rigid during vibrations. The SIE technology was also used to model the cross-section of the beam as a square tube (Figure 6.59a). A more refined modeling of the cross-section, which account for the rounding of the corners, Figure 6.59b was also considered.

Figure 6.60 shows a comparison between the Timoshenko beam modeling and the square cross-section modeling. Both models simulate the first two bending modes accurately, around ≈100 Hz and ≈330 Hz, respectively. The Timoshenko model predicts the first torsional mode at ≈430 Hz, while the more accurate box model predicts the first torsional mode at around 370 Hz A difference of almost 15%. For higher bending, and torsional modes, the two models predict a progressively divergent frequencies of response, as expected.

Figure 6.61 shows the effect of accurate modeling of the rounded corners on the first torsional ≈370 Hz and the second bending modes ≈330 Hz of the beam. The rounding of the corners will result in a slight decrease of bending impedance, while increasing the torsional impedance in a more pronounced manner.

6.8.1.4.3 Framed Deck-Beam Structure

Consider the structure of Figure 6.62. It is a deck system mounted on four columns. The cross-section of each of the columns is a wide flange section of HP10 X 42, while the deck system is composed of a deck plating and three supporting stringers, as shown in Figure 6.62. The material of the

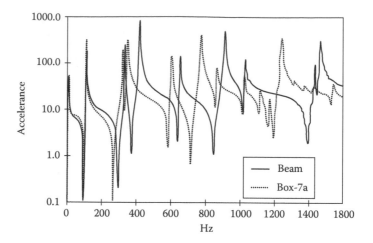

FIGURE 6.60   Effects of elements type.

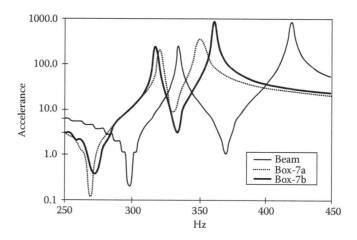

FIGURE 6.61   Effects of corner modeling.

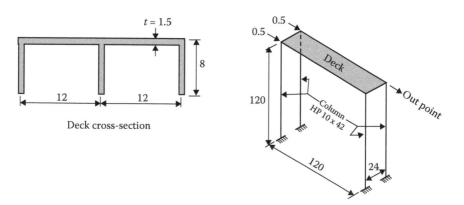

FIGURE 6.62   Framed deck system.

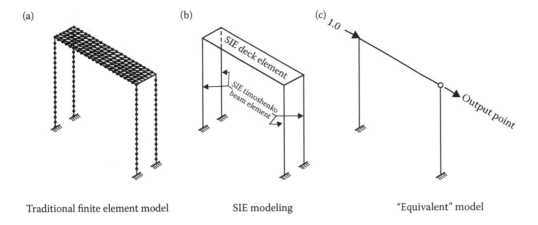

Traditional finite element model   SIE modeling   "Equivalent" model

FIGURE 6.63 Different models for the framed deck system (a) traditional model, (b) SIE model, and (c) equivalent model.

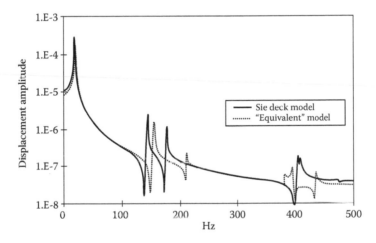

FIGURE 6.64 Comparison between SIE and conventional models.

frame is steel, and the loading on the system is assumed to be a unit steady state horizontal force. The frequency range of interest is 0–500 Hz.

Traditional finite element modeling of this deck-beam system for the frequency range of interest is shown in Figure 6.63a. The SIE modeling for the same system is shown in Figure 6.63b. An equivalent, simple model of the system is also shown in Figure 6.63c. This equivalent model is based on simulating the deck by an equivalent Timoshenko beam. Each of the pair of columns at each end of the frame is simulated by one Timoshenko beam. The modeling and computational efficiencies of the SIE model when visually compared to the traditional finite element model is clear.

Figure 6.64 shows a comparison between horizontal displacement amplitude of the output point (Figure 6.62) as computed by the SIE model and the equivalent simple model. As expected, the equivalent simple model produces good results at lower frequencies. At higher frequency range the need for the detailed deck modeling become apparent.

6.8.1.5 Closing Remarks

SIE represents a new approach for the frequency domain dynamic analysis of structures. We developed a theory for the analysis of systems that are uniform along a single coordinate axis but otherwise arbitrary in geometry and material composition. We showed that this approach, termed the *SIE*, is an accurate, efficient, and general approach for the analysis of vibrating structures. This technique extends the applicability of discrete deterministic finite element based modeling to higher frequencies and is capable of bridging the gap to the frequency regime where statistical energy methods are applicable.

6.9 CASE STUDIES

6.9.1 GENERAL

STRID process is utilized in almost all IHCE projects in varying and innovative ways. To illustrate the variety of the utilization of STRID procedures, we offer few case studies in this section. Section 6.9.2 illustrated the use of modal testing and 3D finite elements modeling. Similarly, Sections 6.9.3, 6.9.4, and 6.9.5 showcase the use of different analysis techniques in support of NDT procedures. Section 6.9.6 illustrates combining signal processing techniques, NDT, and STRID methods. Section 6.9.7 shows the essential importance of STRID methods (modal identification) for optimally placing sensors during an experiment. Finally, Section 6.9.8 shows the use of modal testing methods in some IHCE projects.

6.9.2 GENERATING BASELINE MODELS

An example of monitoring and generation of a baseline model for Maysville Bridge in Kentucky was given by Harik et al. (2002). It is 2100-ft-long cable stayed bridge, with two anchor spans, flaking span, and a center span (1050 ft). A two-plane semi-harped system was utilized in the bridge. Stays were placed at 50′ intervals along each edge of the deck consisting of four sets with 20 cables in each set covered by coextruded high-density polyethylene pipes with white outer layers. The outer layers had small spiral beads around the pipe to break up airflow in cases of light rain and wind thus preventing cable galloping. Soft neoprene collars were employed to connect damping cables to the stay cables. A field ambient vibration test was conducted by the authors using two loaded trucks and data was collected from about 80 locations at 1000 Hz rate. Locations were chosen based on a preliminary FEA model. Two complementary SI methods known as Peak Picking (PP) in the frequency domain and the stochastic subspace identification (SSI) method in time domain were used. A good agreement was found in identified frequencies by both methods. Then a 3D finite element model was developed, calibrated, and analyzed. Test data and FEM results were in good agreement.

Multireference modal testing was conducted on a five-span r/c deck on steel-stringer bridge in Dayton, Ohio by Kangas et al. (2002). The deck was replaced with an interlocking system of FRP deck panels. Modal testing was conducted before and after the rehabilitation, and changes in the bridge characteristics were analyzed using modal flexibility. The changes were attributed the loss of composite action.

6.9.3 SCOUR STRID

STRID methods were also used in bridge scour problems. Over 100,000 bridges in National Bridge Inventory have unknown foundation conditions. Olson (2002) summarized research on surface and borehole methods, including stress wave, vibration, magnetic, and radar NDE methods. These studies were part of NCHRP 21–5 and 21–5(2) projects.

Parallel Seismic Method: NCHRP studies found that borehole-based PS to be more accurate and most applicable NDE method for determination of the depth of unknown bridge foundations for bridge scour safety evaluation purposes. Thus, it was suggested for any project that at least one PS test be performed with other methods so that the reliability of other methods can be validated. This method requires impacting the exposed portion of the substructure and monitoring the response of the foundation and surrounding soil with a receiver in a cased borehole drilled next to the foundation. As the wave energy is monitored by the receiver at depths parallel to and below the bottom of the foundation, it becomes typically weaker and slower below the foundation bottom. The change in the received signals with depth indicates the foundation depth.

The method requires the borehole to be drilled next to the foundation to be tested, preferably within 5 ft or less, cased 2 in ID PVC casing, and extend at least 15-ft-deeper than the minimum required foundation depth.

Surface Ultraseismic (US) method: This was developed to overcome difficulties encountered by sonic echo/impulse response method and the bending wave tests on noncolumnar and complex columnar bridge substructures. US is a broad application in that the initial arrivals of both compression and bending waves and their subsequent reflections can be analyzed to predict unknown foundation depth.

US vertical profiling test is used for one-dimensional imaging of the foundation depth by tracking the upgoing and downgoing events. Bridge columns or abutment is hit from the top or bottom (both vertically and horizontally) and the output wave motion is recorded at regular intervals down the substructure element. Typically, three component record of the wave field is taken to analyze all types of ensuing wave motion.

For 2D imaging of foundation depth, horizontal profiling (HP) US test is used where there is flat, horizontal access for testing, such as top of an accessible pier or abutment. The reflection echoes from the bottom are analyzed to compute the foundation depth. The source and receiver(s) are located horizontally along the top of accessible substructure, or any accessible face along the side of the substructure element, and a full survey is taken along the top of the element.

6.9.4 FOUNDATION IDENTIFICATION

There are different applications for validating information regarding foundations, such as slabs, piles, and so on. These applications are examples of using Condition assessment efforts in support of STRID. For example, as part of the "Re-use of Foundation for Urban Sites" program, Niederleithinger and Taffe (2006) reported on a planned test center, with a slab of varying thicknesses and reinforcements and 10 bored piles of varying length, was built as a test bed (see Figures 6.65 and 6.66) for experimental and validation of methodologies. One of the tasks was to evaluate thickness of slabs using NDT methods. Some of the conclusions show that

1. Thicknesses of r/c slabs are possible up to 125 cm with about 3% uncertainty. Up to 75 cm discrimination between lower reinforcement and back wall seems possible

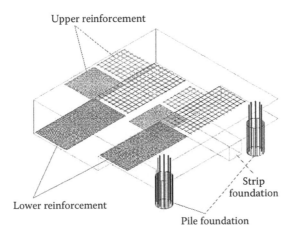

FIGURE 6.65 Illustration of the RuFUS test slab. (Reprinted from ASNT Publication.)

FIGURE 6.66 Detail of the reinforcement and geometrical features of the slab. (Reprinted from ASNT Publication.)

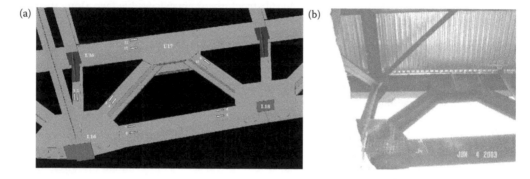

FIGURE 6.67 Instrumented downstream truss members: (a) schematic view of sensors, and (b) photograph of actual sensors placed on bridge.

2. Localization of elements as piles and strips beneath a slab are possible, at least if there are no horizontal sealing layers
3. Radar (500 mHz to 1.5 GHz) was able to locate upper reinforcement, but not the lower one or the back wall
4. The IE results have been unclear

Commonly used low strain integrity testing methods (Pile Integrity Testing, hammer impulse testing, etc.) were also evaluated to find that pile lengths that were under or overestimated in most cases (accuracy is less than 10%).

6.9.5 Condition Assessment for Existing Bridges

Truck Loads: Hag-Elsafi et al. (2004) discussed instrumentation and deck pour monitoring of Court Street Bridge in Tioga County, New York. The bridge is a six-span continuous steel structure, about 338-m-long and 14.45-m-wide, consisting of stringers, floor beams, two trusses (upstream and downstream), and a light weight concrete deck. The deck was built composite with the stringers and the top chords of the trusses. The motive for the instrumentation and monitoring (Figure 6.67) was to determine service load axial forces and secondary moments in the truss main members during deck pours. The authors concluded that the members' actual service dead load axial forces and moments were overestimated by about 20% in the design. Regarding moments, the authors also concluded that service dead load moments were within 20% of those used in the design. The differences between actual and theoretical axial forces and moments for service dead load (Figure 6.68) were attributed to the manner in which the monitoring data was corrected for temperature effects and the possible presence of construction loads on the deck during the pours monitoring.

Temperature Distribution: Mondal and DeWolf (2004) monitored a segmental concrete bridge for Connecticut DOT to learn how different bridges behave, performance evaluation, renovation guidance, and show the value of long-term monitoring. The bridge is a large, multi-span, continuous, post-tensioned box-girder bridge, shows the cross-section of the bridge. It is a 11-span, 250-ft-long, and 178–275- ft-wide bridge. Figure 6.69 shows the cross-section of the bridge. The designers recommended monitoring temperatures over time. The bridge experienced significant cracking and monitoring has offered an opportunity to explore the influence of the temperature variations on the performance (see Figure 6.70). About 16 thermocouples were used with 14 in one cross-section. Other two measured the air temperature inside the box and the temperature inside the instrument cabinet.

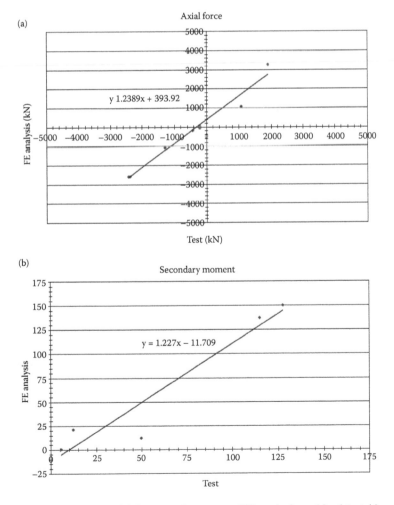

FIGURE 6.68 Service dead load axial forces and moments: FE analysis and load test (a) axial force comparisons, and (b) secondary moment comparisons.

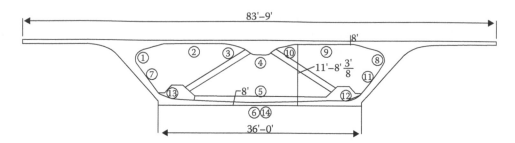

FIGURE 6.69   Bridge cross-section and locations of thermocouples. (Reprinted from ASNT Publication.)

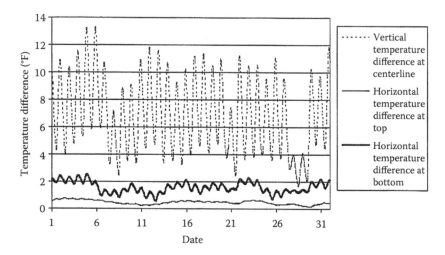

FIGURE 6.70   Vertical and horizontal temperature differences during July 2002. (Reprinted from ASNT Publication.)

Among the conclusions of that landmark research were (1) vertical variations of temperatures were observed to be higher than horizontal temperature variations, (2) temperature gradient profile is similar to AASHTO specifications, even though the magnitudes were different (see Figure 6.71) and (3) the magnitude of stress-induced temperature gradients were low.

6.9.6   Signal Processing

Signal processing techniques can also be used, in conjunction with other NDT techniques, to identify structural properties. For example, Algernon and Wiggenhauser (2006) discussed some of the suggested improvements to data analysis during impact-echo method (see Chapter 7) for nondestructive testing of concrete structures. Impact-echo signals from concrete structures can be very noisy due to geometry effects and may lead to misinterpretation of signals. Hence, autocorrelation methods and Hilbert-Huang transform methods were used to analyze the impact-echo data to improve the accuracy of the analysis.

Using autocorrelation helps to amplify the effect of multiple reflections and minimizes the effects of white noise. Cross-correlation is also used to eliminate variations in the excitation pulse, and is then treated as the original pulse. This was found to be useful in increasing the intensity of the thickness from the enhanced frequency spectrum.

Use of Hilbert-Huang transform was also explored as it has ability to localize an event in time and frequency making it possible to identify short transient signals within longer time sweeps. This was found to be very useful to analyze data from a highway bridge during heavy traffic and obtain a

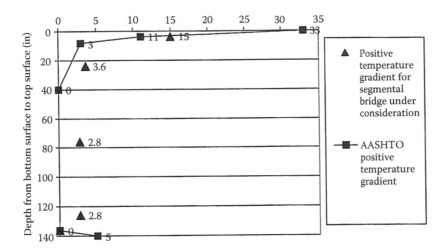

FIGURE 6.71   Observed thermal gradient. (Reprinted from ASNT Publication.)

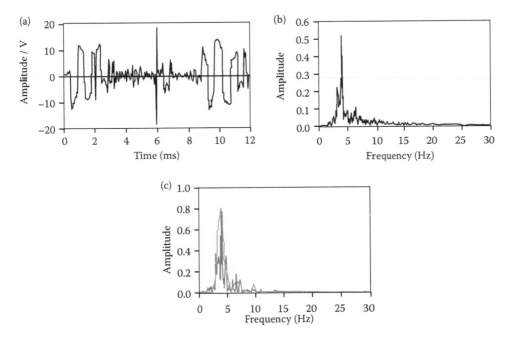

FIGURE 6.72   Comparison of different signal processing schemes (a) time signal, (b) Fourier spectrum, and (c) Hilber-Huang marginal spectrum. (Reprinted from ASNT Publication.)

clear indication of the thickness (see Figure 6.72). Due to its iterative nature, HHT is highly efficient but should be carefully applied as a theoretical performance evaluation is only partly possible. With instantaneous frequency it becomes possible to identify short transient signals within the data as well as frequency fluctuations, which remain hidden for other methods.

6.9.7   Optimal Sensor Locations (OSL)

Papadimitiou et al. (2000) combined an optimum sensor location methodology and a structural identification algorithm. Their approach is based on minimizing the uncertainties involved in experiments. Their method addressed only a single hazard (loading condition). They showed that the optimal location of sensors is dependent on the number of structural modes that are considered in the experiment, as well as the number of available sensors.

6.9.8 STRID AS A COMPONENT IN SHM

Examples of the use of structural identification methods as an integral part of an SHM as applied to bridges was reported by Stubbs et al. (1999), Sikorsky et al. (1999), and Bolton et al. (1999 and 2000). Some of the goals of their experiments/structural (modal) identification were (1) assess instrumentation techniques with minimal interference with the operations of the bridges, (2) assess testing techniques, (3) apply MI method (MEScope software, version 2.0) on measurements to estimate modal responses (frequencies, damping, and mode shapes), and (4) compare with numerical solutions such as finite elements.

The authors stated clearly that the whole effort is based on global NDE paradigm, that is, identify damage from properties of the whole structure. To achieve this, they tested two bridges: a reinforced concrete and a steel bridge. We summarize their experiments, point to some of the highlights of their results and show the value of such experiments, with some cautionary notes.

Used two different field tests to analyze the RC 2-span bridge: the differences between the two tests were mainly on how the data were collected and handled. The first field test collected the data in two sets: (1) vertical bending and torsional behavior, and (2) lateral bending behavior. A single axis accelerometers were used for the first field test. The analytical MI were performed separately. MI of the results of the second field test were performed using unified sets of measurements. This was accomplished by using triaxial accelerometers to measure combined behavior of the bridge. Table 6.12 shows the MI results as reported by the authors.

The results of those tests are very illuminating in that they demonstrate the many potential benefits of using structural identification methods in SHM. For example, note that the estimated modal damping for different modes vary from 3.5% to 1.01%. This information would be invaluable to bridge analyst and designer, since the common practice is to assume a constant modal damping when analyzing/designing a bridge. If the assumed constant damping is in the range of 3.5%, then the resulting analysis/design will be *not* safe. On the other hand, if the assume constant modal damping in analysis/design is in the range of 1.01%, then the resulting analysis/design will

TABLE 6.12
Modal Identification Results

Mode #	Field Test I		Field Test II	
	Frequency (Hz)	Damping (%)	Frequency (Hz)	Damping (%)
1	3.09	3.5	3.35	3.51
2	3.21	3.1	4.82	3.30
3	4.42	3.3	6.73	2.34
4	6.78	2.0	8.75	2.57
5	8.32	2.7	10.63	2.10
6	10.58	1.9	12.93	3.65
7	11.64	4.0	14.62	2.33
8	14.45	1.7	20.55	1.69
9	20.72	1.3	20.93	1.30
10	23.35	1.1	21.71	1.58
11	24.42	1.8	24.42	1.87
12	Not reported	Not reported	27.04	1.95
13	Not reported	Not reported	29.01	1.01

TABLE 6.13
Modal Properties of Steel Bridge

Mode	Frequency (Hz)	Damping (%)
1	2.21	0.520
2	3.08	0.439
3	5.01	0.407
4	6.18	1.548
5	7.57	0.487
6	8.94	0.403
7	9.86	0.262
8	10.31	0.480
9	11.35	0.279
10	13.01	0.233
11	13.89	0.246

---

not be economical. Clearly, using a combination of testing/structural identification can result in a safer yet more economical structure.

Another important observation that was reported by the authors of this experiment is that it shows the conventional use of FE methods can adequately simulate the estimated mode shapes. Since the maximum reported acceleration was about 0.050 g, with most of the measured responses in the range of +0.025 g, the bridge responded in an elastic fashion. Thus, the conventional assumptions of elastic responses while using FE modeling techniques is validated. The high frequency range of the experiment ($\cong$30 Hz.) would make such results fairly adequate to analyze the response of the bridge to earthquake motions.

The *value* of the information gained from such an experiment can easily be determined using a cost/benefit approach as described in several parts of this volume. For example, we can compute the *benefit* from the experiment by computing the cost savings from using a more realistic damping than either too high or too low damping levels. Computing the *cost* of the experiment is straightforward. Thus the value of the experiment is the computed benefit to cost ratio.

We should also point out to the fact that the results of such a successful experiment should not be extrapolated or interpreted beyond its intended goals. For example, since the experiment is based on measuring motion (accelerations, velocities, or displacements), using the results to interpret deformation measures (such as strains, or stresses) would risk leading to erroneous results.

The author also reported a similar MI experiment on a single span-steel truss bridge. The results of the MI scheme are shown in Table 6.13. The value of this experiment is similar to the first one. It is of interest to note that the modal damping is more evenly distributed than that of the reinforced concrete bridge, and the damping of mode number 4 (1.548%).

6.10   LIFE CYCLE ANALYSIS AND STRID

Life cycle analysis requires large amount of information about the system on hand. As more information become available, the accuracy of LCA increase. STRID techniques can be of help in the process of LCA. Recall that a generic model of estimating life cycle cost (LCC) is

$$LCC = \iint p(H) \; c(H) \mathrm{d}H \; \mathrm{d}t \tag{6.170}$$

The probability of hazard level H is $p(H)$ and the cost of the hazard is $c(H)$. The double integrals in Equation 6.170 are over the hazard space and the life span of the structure. The cost of hazard is a function of system properties as they respond to h. These properties include system stiffness K, system mass M, system damping β, system strength R, and system response U. Note that all of these system properties and responses include all system components that include the structure, the supporting soil, and the surrounding water, if present. Also, note that K, M, β, R, and U are general functions that can be scalar, matrices, or analytic functions, as appropriate. As such, we have the relationship

$$LCC = f(K, M, \beta, R, U, ..., H, ..., x_i)$$

(6.171)

The variables x_i represent the many other nonstructural related factors that might affect LCC, such as consequences of failure, or management costs. Some expressions of the function $f(\ )$ are discussed in Chapter 10 of Ettouney and Alampalli (2012).

What concerns us here is the implied relationships in Equation 6.171. Such relationships imply that the accuracy of LCC depends directly on the accurate estimations of K, M, β, R, and/or U. Since the main goal of STRID is to improve the accuracy of some or all of those parameters, we can conclude that STRID can have a major role in the accurate estimation of LCC. To gain an understanding of the nature of the relationship, let us consider a discrete form of Equation 6.170 as

$$LCC = \sum_i p_i \, c_i$$

(6.172)

Note that Equation 6.172 is for only a unit period of time for simplicity. The sum in Equation 6.172 is over the hazard space. The cost of the ith hazard is c_i. This cost is a function of the damage due to the ith hazard, D_i such that

$$c_i = D_i \, \bar{c}_i$$

(6.173)

The cost per unit damage is $\bar{c}_i$, assuming a linear relation between damage and cost. We can further assume the structural response U_i to H_i can be evaluated using

$$KU_i = H_i$$

(6.174)

The representative stiffness of the system is K. For simplicity, we assume a linear structural performance. For nonlinear performance, the logic of this development will not be affected. The final step in the development is to relate the damage D_i to the response U_i such that

$$D_i \sim U_i$$

(6.175)

Thus

$$D_i \sim K^{-1} H_i$$

(6.176)

or

$$D_i \sim F \, H_i$$

(6.177)

The flexibility of the system is F.

Equation 6.172 becomes

$$LCC = \sum_i p_i F \, H_i \bar{c}_i$$

(6.178)

or

$$LCC = F\,A \tag{6.179}$$

with A being a constant such that

$$A = \sum_i p_i\,H_i\,\bar{c}_i \tag{6.180}$$

Equation 6.179 proves that there is a direct relationship between the flexibility, or stiffness, measures of the system, and the estimate of its LCC. In this case, the relationship is linear. For more complex systems, the relationship can be different; however, the direct correlation will still be there.

Let us now consider the effects of accurate estimation of flexibility on the estimation of LCC. A conventional LCC is estimated using conventional (not based on experiment) flexibility, F_0 such that

$$LCC_0 = F_0\,A \tag{6.181}$$

Let us assume that experimentally base flexibility (estimated using some form of STRID method) is F_1. The corresponding LCC is

$$LCC_1 = F_1\,A \tag{6.182}$$

It is reasonable to assume that LCC_1 is more accurate than LCC_0, since it is based on experimental results. The dimensionless improvements is

$$\Delta LCC = \frac{LCC_1 - LCC_0}{LCC_0} = \frac{F_1 - F_0}{F_0} \tag{6.183}$$

We reach an interesting conclusion. The dimensionless improvement of LCC due to experimentally based STRID is proportional to the dimensionless improvement of flexibility estimates. In case of stiffness, the proportionality is an inverse one.

The above shows an important STRID use: the accurate estimation of LCC. As such, *STRID should be considered as an essential infrastructure management tool.*

6.11 COST–BENEFIT ANALYSIS OF STRID

The effort of STRID will have costs as well as benefits to asset manager. The costs, as usual, are fairly easy to estimate. They include the following:

1. Cost of the experiment (sensors, instrumentation, etc.)
2. Cost of management, including costs of decision making as well as owner's costs
3. Cost of labor
4. Cost of analysis

The first three are the usual costs of SHM. The cost of analysis includes costs of performing different STRID methods, data analysis, validation, reporting, and so on.

Estimating benefits of STRID is a difficult task. Usually, the benefits are estimated on a qualitative manner. We try for the remainder of this section to present few examples of estimating the benefits of STRID in a quantitative manner. Thus the value of STRID becomes easier to establish for decision makers. Our general approach will be to estimate the benefits as the cost savings. For example, if the costs of certain situation without the STRID effort is estimated to be C_0 and the costs of the same situation, given the information of STRID effort, is estimated to be C_1, then the benefit of the STRID effort is

$$B_{STRID} = C_0 - C_1$$

(6.184)

Given the above, let us consider some cases, as follows:

Analytical Model Updates: Let us assume that a large analytical model of a bridge structures is available. The model is formed using analytical estimates and *best guesses engineering judgment.* Let us assume further that a major retrofit is planned to mitigate a potential earthquake hazard. The seismic retrofit is based on a modified model from the original model, again based on *best guesses and engineering judgments.* Let us define the differences between the estimated analytical state and the actual state as ΔS_0. In an actual earthquake event, the cost of the damage C_0 is going to be proportional to ΔS_0 such that

$$C_0 = A \, \Delta S_0$$

(6.185)

with A as a constant that relates the cost of damage to the errors in the model that was used in design. If an STRID effort is done to update the design-based analytical model, then the differences between the estimated STRID-based analytical state and the actual state will be ΔS_1. In an actual earthquake event, the cost of the damage C_1 is going to be proportional to ΔS_1 such that

$$C_1 = A \, \Delta S_1$$

(6.186)

Since it is reasonable to assume that

$$\Delta S_0 > \Delta S_1$$

(6.187)

The benefit of STRID will be real (positive) in the sense

$$B_{STRID} = A(\Delta S_0 - \Delta S_1)$$

(6.188)

A detailed cost estimate analysis of Equations 6.185 and 6.186 can yield a quantitative estimate of the benefit of STRID effort. Figure 6.73 shows conceptual benefits of STRID for model updating efforts.

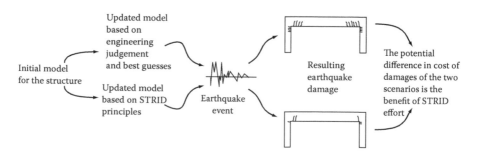

FIGURE 6.73   Benefits of STRID in model updating.

REFERENCES

Alampalli, S., (1998a) *Effects of Testing, Analysis, Damage, and Environment on Modal Parameters.* Proc. Modal Analysis & Testing. NATO-Advanced Study Institute, Sesimbra, Portugal.

Alampalli, S., (1998b) "Remote Bridge-Monitoring Systems: New York's Experience." Structural Materials technology: An NDT Conference, San Antonio, TX, SPIE Volume 3400.

Alampalli, S. and Cioara, T., (2000) "Selective Random Decrement Techniques for Bridge Vibration Data," Structural Materials Technology: An NDT Conference, Atlantic City, NJ.

Alampalli, S. and Ettouney, M. (Eds.), (2003) Proceedings of the Workshop on Engineering Structural Health. New York State Department of Transportation, New York.

Alampalli, S. and Fu, G., (1994) "Remote bridge monitoring systems for bridge condition." *Clinet Report 70*, Engineering Research and Development Bureau, NYS Department of Transportation, Albany, NY.

Algernon, D. and Wiggenhauser, H., (2006) "Impact-Echo Signal Processing," NDE Conference on Civil Engineering, ASNT, St. Louis, MO.

Andry, A. N., Shapiro, E. Y., and Chung, J. C., (1983) "Eigenstructure Assignment for Linear Systems," IEEE Transactions on Aerospace und Electronic Systems, Vol. AES-19, No. 5.

Aref, A. and Alampalli, S., (2001) "Vibration Characteristics of a Fiber Reinforced Polymer Bridge Superstructure." *Journal of Composite Structures*, Elsevier Science, 52(3–4).

Asmussen, J.C., (1997) "Modal analysis based on the random decrement technique. application to civil engineering structures." Ph.D. Thesis, Aalborg University, Department of Building Technology and Structural Engineering, Denmark.

Asmussen, J.C., Brincker, R., and Ibrahim, S.R., (1999) *Modal Analysis based on the Random Decrement Transform.* Proc. Modal Analysis & Testing, NATO-Advanced Study Institute, Sesimbra Portugal.

Beardsley, P., Hemez, F., and Doebling, W., (2000) "Updating Nonlinear Finite Element Models in the Time Domain," Proceedings of 2nd International Workshop on Structural Health Monitoring, Stanford University, Stanford, CA.

Bernhard, R. and Huff, J., (1999) "Structural-acoustic design at high frequency using the energy finite element method", *Journal of Vibration and Acoustics, Vol 221*, Academic Press.

Bolton, R., Stubbs, N., Park, S., Choi, S., and Sikorsky, C., (1999) "Measuring Bridge Modal Parameters for Use in Damage Detection and Performance Algorithms," Proceedings 17th International Modal Analysis Confrence & Exposition, Kissimmee, FL.

Bolton, R., Stubbs, N., Park, S., Choi, S., and Sikorsky, C., (2000) "Field Measurements of Modal Parameters for Non-Destructive Damage Detection Algorithms," Proceedings of 2nd International Workshop on Structural Health Monitoring, Stanford University, Stanford, CA.

Brinker, R. and Moller, N., (2006) "Operational Modal Analysis – a New Technique to Explore," *Sound and Vibration,* Volume 40, No 6, pp. 5–6.

Cheung, M. S. and Naumoski, N., (2002) "The First Smart Long-Span Bridge in Canada – Health Monitoring of the Confederation Bridge.' Structural Health Monitoring Workshop, ISIS Canada Research Network, Manitoba, Canada.

Cho, P., (1993) "Energy flow analysis of coupled structures," Ph.D. Dissertation, Mechanical Engineering Department, Purdue University.

Chong, K. P., Carino, N. J., and Washer, G. A., (2001) "Health Monitoring of Civil Infrastructures." Proceedings of the SPIE Health Monitoring and Management of Civil Infrastructure Systems, Plenary Paper, Newport Beach, CA.

Cioara, T. Gh., (1988) "On a Curve-Fitting Algorithm for Modal Parameter Estimation." Sixth International Modal Analysis Conference, Kissimmee, FL.

Cioara, T. Gh., (1999) *Vibration Monitoring of Mechanical System – Fault Diagnosis and Vibration Level Reduction.* USC-ME-LAMSS-99–101, University of South Carolina.

Cole, H., (1973) *On-Line Failure Detection and Damping Measurement of Aerospace Structures by Random Decrement.* NASA CR-2205, Washington, DC.

Dasgupta, G., (1982) "A finite element formulation for unbounded homogeneous continua," *Trans. ASME Journal of Applied Mechanics*, 49, 136–140.

Demuth, H., Beale, M., and Hogan, M., (2007) *Neural Network Toolbox 5: User's Guide*, The Mathworks, Inc., Natic, MA.

Doebling, S., (1996) "Minimum-rank optimal update of elemental stiffness parameters for structural damage identification," *AIAA Journal.* 34(12), 949–957.

Doebling, S., Farrar, C., Prime, M., and Shevitz, D., (1996) "Damage Identification and Health Monitoring of Structural and Mechanical Systems from Changes in Their Vibration Characteristics: A Literature Review," Los Alamos National Laboratory *Report, LA- 13070-MS*, Los Alamos, NM.

Ettouney, M., (1978) "Compliance Function of Footing in Two Layered Medium," ASCE Geotechnical Engineering Division Specialty Conference, Pasadena, CA.

Ettouney, M., (1979) "Empirical Expressions for Frequency Dependent Soil Springs for Machine Foundation Design," Proceedings, the 1979 Annual Convention, American Concrete Institute, Milwaukee, Wisconsin.

Ettouney, M. and Alampalli, S., (2012) *Infrastructure Health in Civil Engineering, Applications and Management*, CRC Press, Boca Raton, FL.

Ettouney, M., Benaroya, H., and Wright, J., (1990) "Wave propagation in hyper-structures," in *Space 90: Engineering, Construction and Operations in Space*, Editors: Johnson, S. J. and Wetzel, J. P. ASCE, New York.

Ettouney, M. and Daddazio, J., (1978) "Hydrodynamic Interaction Between the Walls and the Floor of Cylindrical Fluid Containers," Proceedings, Conference or Structural Analysis, Design and Construction in Nuclear Power Plants, Port Allegre, Brazil.

Ettouney, M., Daddazio, R., and Abboud, N., (1995a) "Scale Independent Elements for Dynamic Analysis of Vibrarting Systems," Proceedings Third International Conference on Mathematical and Numerical Aspects of Wave Propagation Phenomena, Mandelieu - La Napoule, France.

Ettouney, M., Daddazio, R., and Abboud, N., (1995b) "Some Practical Applications Of the Use of Scale Independent Elements For Dynamic Analysis of Vibrating Systems," *Computers and Structures, Vol 65*, No. 3, Pergamon Press.

Ettouney, M., Daddazio, R., and DiMaggio, F., (1990) "Wet Modes of Submerged Structures," Technical report prepared for David Taylor Research Center, Bethesda, MD. Contract No. N00167-86-c-0100, Weidlinger Assoc., Report No. WA 9008, NY.

Farrar, C. R. and Doebling, S. W., (1998) *Damage Detection - Field Application to a Large Structure.* Proc. Modal Analysis & Testing, NATO-Advanced Study Institute, Sesimbra, Portugal.

FHWA., (1995) "Recording and Coding Guide for the Structure Inventory and Appraisal of the Nation's Bridges," *Report No.* FHWA-PD-96–001, Office of Engineering, Bridge Division, Washington, DC.

FHWA., (2004) *National Bridge Inspection Standards (NBIS)*, Federal Regulations 69 FR74436, Federal Highway Administration, USA.

Friswell, M. and Mottershead, J., (1995) *Finite Element Model Updating in Structural Dynamics*, Kluwer Academic Publishers, Norwell, MA.

Fuchs, P. A., Washer, G. A., Chase, S. B., and Moore, M., (2004) "Applications of Laser Based Instrumentation for Highway Bridges." *Journal of Bridge Engineering, ASCE*, 9(6), 541–549.

Girard, A. and Defosse, H., (1990) "Frequency smoothing, matrix assembly and structural paths: a new approach for structural dynamics up to high frequencies," *Journal of Sound and Vibration*, 137(1), 53–68.

Hag-Elsafi, O., Kunin, J., and Alampalli, S., (2004) "Monitoring of Court Street Bridge for Deck Load Stresses," Proceedings, NDE Conference on Civil Engineering, ASNT, Buffalo, NY.

Harik, I., Zatar, W., Herd, D., Goodpaster, S., Hu, J., Givan, G., and Crane, R., (2002) "NDE / NDT of Bridges in Kentucky," Proceedings, NDE Conference on Civil Engineering, ASNT, Cincinnati, OH.

Hasselman, T., Anderson, M., and Wenshui, G., (1998) "Principal Component Analysis for Nonlinear Model Correlation, Updating and Uncertainty Evaluation," Proceedings, 16th IMAC, Santa Barbara, CA.

He, J. and Fu, Z., (2001) *Modal Analysis*, Butterworth Heinemann, Boston.

Hemez, F. M., (1993) "Theoretical and Experimental Correlation between Finite Element Models and Modal Tests in the Context of Large Flexible Space Structures," Ph.D. Dissertation, University of Colorado, Department of Aerospace Engineering Sciences.

Hemez, F. and Doebling, S., (2001), "Review and Assessment of Model Updating for Non-Linear, Transient Dynamics," *Mechanical Systems and Signal Processing, Vol 15,* No. 1, Academic Press.

Humar, J., Xu, H., and Bagchi, A., (2004) "Application of Artificial Neural Networks in the Detection of Structural Damage," Proceedings of 2nd International Workshop on Structural Health Monitoring of Innovative Civil Engineering Structures, ISIS Canada Corporation, Manitoba, Canada.

Ibrahim, S. R., (1977) Random Decrement Techniques for Modal Identification of Structures. *Journal of Spacecraft and Rockets*, 14(11), 696–700.

Ibrahim, S., (1998) "Fundamentals of Time Domain Modal Identification," in *Modal Analysis and Testing*, Editors: Silva, J., and Maya, M. Nato Science Series, Kulwer Academic Publishers, The Netherlands.

Jeng, D., Bateni, S., and Lockett, E., (2005) "Neural Network assessment for scour depth around bridge piers," University of Sydney, Department of Civil Engineering Research *Report No.* R855, Sydney, Australia.

Kangas, S., Helmicki, A., Hunt, V., Shahrooz, B., Brown, D., and Lenett, M., (2002) "Modal Test-Based Condition Assessment," Proceedings, NDE Conference on Civil Engineering, ASNT, Cincinnati, OH.

Katkhuda, H., Flores, R., and Haldar, A., (2005) Health Assessment at local Level with Unknown Input Excitation. *Journal of Structural Engineering, ASCE,* 131(6), 956–965.

Kausel, E., (1974) "Forced vibrations of circular foundations on layered media," *Research Report R74–11,* Soils Publication No. 336, Structures Publication No. 384, Massachusetts Institute of Technology, Cambridge, MA.

Kausel, E. and Roesset, J. M., (1977) "Semi analytic hyper element for layered strata," *Journal of the Engineering Mechanical Division, ASCE,* 103(EM4), 569–588.

Keane, A. and Price, W., (1969) *Statistical Energy Analysis,* Cambridge University Press.

Koh, C. and Htun, S., (2004) "Adaptive Search Genetic Algorithm for Structural System Identification," Proceedings of 2nd International Workshop on Structural Health Monitoring of Innovative Civil Engineering Structures, ISIS Canada Corporation, Manitoba, Canada.

Koizumi, T., Tsujiuchi, N., Tanaka, H., Okubo, M., and Shinomiya, M., (2002) "Prediction of the Vibration In Buildings Using Statistical Energy Analysis," Proceedings of the 20th International Modal Analysis Conference (IMAC), Los Angeles, California.

Laurens, S., Shartai, Z. M., Kacimi, S., Balayssac, J. P., and Arliguie, G., (2006) "Prediction of Concrete Electrical Resistivity using Artificial Neural Networks," NDE Conference on Civil Engineering, ASNT, St. Louis, MO.

Leung, A. Y. T., 1(988) "A simple dynamic substructure method," *Earthquake Engineering and Structural Dynamics,* 16, 827–837.

Lim, T. W., (1995) "Structural Damage Detection Using Constrained Eigenstructure Assignment," *Journal of Guidance, Control and Dynamics,* 18(3), 411–418.

Lim, T. W. and Kashangaki, T., (1994) "Structural Damage Detection of Space Truss Structures Using Best Achievable Eigenvectors," *AIAA Journal,* 32(5), 2310–2317.

López-Díez, J., Torrealba, M., Güemes, A., and Cuerno, C., (2005) "Application of Statistical Energy Analysis for Damage Detection in Spacecraft Structures," *Engineering Materials, Vol 293–294,* Trans Tech Publications, Switzerland.

Lyon, R., (1975) *Statistical Energy Analysis of Dynamical Systems,* M.I.T. Press, Cambridge, MA.

Lyon, R. and Maidanik, R., (1962) "Power Flow between Linearly Coupled Oscillators," *Journal of the Acoustical Society of America,* 34, 623–639.

Lysmer, J. (1970) "Lumped Mass Method for Rayleigh Waves," *Bulletin of the Seismological Society of America,* 60, 89–104.

Mondal, P. and DeWolf, J. T., (2004) "Long-Term Monitoring of Temperatures in a Segmental Concrete Box-Girder Bridge in Connecticut," Proceedings, NDE Conference on Civil Engineering, ASNT, Buffalo, NY.

Moore, M., Phares, B., Graybeal, B., Rolander, D., and Washer, G., (2001) Reliability of Visual Inspection for Highway Bridges, Volume I: Final Report. *Report No. FHWA-RD-01–020.* Federal Highway Administration, Washington, DC.

Moore, M., Phares, B., and Washer, G., (2004) *Guidelines for Ultrasonic Inspection of Hanger Pins.* Federal Highway Administration, *Report No. FHWA-HRT-04–042,* Washington, DC.

Niederleithinger, E. and Taffe, A., (2006) "Improvement and Validation of Foundation NDT Methods," NDE Conference on Civil Engineering, ASNT, St. Louis, MO.

Olson, L., (2002) "Determination of Unknown Depths of Bridge Foundations for Scour Safety Investigation," Proceedings, NDE Conference on Civil Engineering, ASNT, Cincinnati, OH.

Padur, D. S., Wang, X., Turer, A., Swanson, J., Helmicki, A., and Hunt, V., (2002) "Non-Destructive Evaluation/ Testing Methods – 3D Finite Element Modeling of Bridges," Proceedings, NDE Conference on Civil Engineering, ASNT, Cincinnati, OH.

Papadimitiou, C., Katafygiotis, L., and Yuen, K., (2000) "Optimal Instrumentation Strategies for Structural Health Monitoring Applications," Proceedings of 2nd International Workshop on Structural Health Monitoring, Stanford University, Stanford, CA.

Phares, B., Graybeal, B., Rolander, D., Moore, M., and Washer, G., (2001) Reliability and Accuracy of Routine Inspection of Highway Bridges. Transportation Research Record, *Journal of Transportation Research Board,* National Research Council, Issue Number 1749, 82–92.

Phares, B., Washer, G., Rolander, D., Graybeal, B., and Moore, M., (2004) Routine Highway bridge Inspection Condition Documentation Accuracy and Reliability. *Journal of Bridge Engineering,* ASCE, 9(4), 403–413.

Rose, J., (2004) *Ultrasonic Waves in Solid Media,* Cambridge University Press, Cambridge, UK.

Sanayei, M., Imbaro, G., McClain, J., and Linfield, B., (1997) "Structural Model Updating Using Experimental Static Measurments," *Journal of Structural Engineering, ASCE,* 123(6), 792–798.

Sanayei, M. and Saletnik, M. J., (1996a) "Parameter Estimation of Structures From Static Strain Measurements, Part I: Formulation," *Journal of Structural Engineering, ASCE,* 122(5), 555–562.

Sanayei, M. and Saletnik, M. J., (1996b) "Parameter Estimation of Structures From Static Strain Measurements, Part II: Error Sensitivity Analysis," *Journal of Structural Engineering,* 122(5), 563–572.

Sanayei, M. and Santini, E., (1998) "Dynamic Bridge Substructure Evaluation and Monitoring System," *Report* of subcontract to Tufts University from FHWA grant to Olson Engineering.

Sarradj, E., (2003) "High frequency boundary integral method as an alternative to SEA," Proceedings of 10th International Congress on Sound and Vibration, Stockholm, Sweden.

Sarradj, E., (2004) "Energy-based vibroacoustics: SEA and beyond." 30th German Convention on Acoustics (DAGA) and 7th Congrès Francais d' Acoustique (CFA).

Sikorsky, C., Stubbs, N., Park, S., Choi, S., and Bolton, R., (1999) "Measuring Bridge Performance Using Modal Parameter Based Non-Destructive Damage Detection," Proceedings of the 17th International Modal Analysis Confrence & Exposition, Kissimmee, FL.

Siswobusono, P., Chen, S., Zheng, L., Yamak, B., Jones, S. L., and Callehan, D., (2004) *Dynamic Load Rating of Rural Bridges.* International Modal Analysis Conference XXII, Michigan, MI.

Stubbs, N., Park, S., Choi, S., Sikorsky, C., and Bolton, R., (1999) "A Methodology to Nondestructively Evaluate the Structural Properties of Bridges," Proceedings of the 17th International Modal Analysis Conference & Exposition, Kissimmee, FL.

The Steel Company of Canada Limited., (1966) *Hollow Structural Sections*, Hamilton, Ontario, ON.

Timoshenko, S., (1955) *Vibration Problems in Engineering*, Van Nostrand Reinhold Company, New York.

Timoshenko, S., Young. D. H., and Weaver, W. Jr., (1974) *Vibration Problems in Engineering*, 4th Edition, John Wiley & Sons, New York, Chapter 5.

Tolbert, R., Zheng, L., and Chen, S., (2004) *Modal Testing of a Unique Cantilevered Structure above the Smith Lake.* International Modal Analysis Conference XXII, Michigan, MI.

Trippi, R. and Turban, E., (1992) *Neural Networks in Finance and Investing*, Probus Publishing Company, Chicago, IL.

Vlahopoulos, N., Zhao, X., and Allen, T., (1999) "An Approach For Evaluating Power Transfer Coefficients For Spot-Welded Joints In An Energy Finite Element Formulation", *Journal of Sound and Vibration*, Vol. 220, No. 1, 1999, pp. 135–154.

Waas, G., (1972) "Linear Two-dimensional Analysis of Soil Dynamics Problems in Semi-infinite Layered Media," Ph. D. Thesis, University of California at Berkeley, CA.

Washer, G. A., (2002) "Developments in NDE for Highway Bridges in United States." Eight European Conference on Nondestructive Testing, European Society of Nondestructive Testing, Barcelona, Spain.

Yang, Q. and Liu, J., (2007) "Structural Damage Identification Based on Residual Force Vector," *Journal of Sound and Vibrations,* 305(1–2), 298–307.

Zhao, X. and Vlahopoulos, N., (2004) "A basic hybrid finite element formulation for mid-frequency analysis of beams connected at an arbitrary angle," *Journal of Sound and Vibration*, *Vol 269*, No 6, pp. 135–164.

Zienkiewicz, O. C., (1971) *The Finite Element Method in Engineering Science*, McGraw-Hill, New York.

Zimmerman, D. and Kaouk, M., (1994) "Structural damage detection using a minimum rank update theory," *Journal of Vibration and Acoustics*, 116(2), 222–230.

7 Damage Identification (DMID)

7.1 INTRODUCTION

7.1.1 MATHEMATICAL REPRESENTATION OF DAMAGE

We start this chapter by trying to define damage. A general definition that would fit our purposes is as follows: *Damage is the change of structural properties; such a change usually results in a degradation of performance.*

Since damage is a change in an initial state, then to identify the damage, we have the following two approaches:

1. Identify the damage directly, by observing it
2. Identify the change, that is, by observing a baseline state, and a changed state, then quantifying the differences between the two states as the damage of interest

Methods that identify damages directly include most conventional nondestructive testing (NDT) methods. In most of these methods, no baseline state is needed. Methods that identify the change in structural state would need a baseline (initial state) and a current state. Both states can be analytical or experimental. Of course, if both initial and current states are described analytically (by finite elements method, for example), the DMID becomes an analytical exercise. We recognize that this practice is followed in many communities in the civil infrastructure field. However, it is not explored in this volume. For our purpose, we consider only situations where experiments are part of the overall project.

Methods that identify damage as a change from one state to another would require a minimum resolution in the method used for detection. Such a resolution can be temporal, spatial, or both. In this situation, we introduce the analogy of changes of functions of some independent variables. We recognize immediately that to accurately compute change in a function, in a finite difference sense, for example, there need to be certain resolutions in the computations of the functions. DMID as a change of state is similar: there need to be certain resolutions in computing the states (both initial and current) of the system. Lacking such resolutions would result in an inaccurate damage definition. When the states (initial and/or current) are evaluated either analytically or experimentally, different types of resolutions are needed, depending on the type of evaluation. These points will be explored in detail in later sections of this chapter.

7.1.2 IDENTIFICATION ADEQUACY, UTILITY, AND PRIORITIZATION OF DMID METHODS

Any DMID method needs to be adequate in three issues: (a) technical, (b) utility, and (c) cost–benefit. Unfortunately, most NDT/structural health monitoring (SHM) experiments concentrate on technical adequacy, and neglect utility, or cost–benefit issues. All three issues must be considered carefully, before embarking on any DMID project.

Technical: There are many technical issues that relate to DMID. Among these issues are:

- Scale (size) of damage
- Failure modes

- Defect modes
- Temporal considerations, which includes before, during, and after hazard occurrence: also includes frequency ranges of interest
- Spatial considerations

Utility: Among utility parameters of DMID are:

- Size of equipment
- Simplicity
- Environmental effects
- Maintenance needs
- Labor needs
- *In situ* versus laboratory settings
- Cost

Methods for computing the utility of any event are discussed in Chapter 8. Extending utility analysis to DMID methods can be easily done using most of those methods.

Cost–Benefit: Decision making methods provide many methods for computing the cost benefits of different methods of DMID. Those methods and numerous examples are given in many parts of this volume. They will not be discussed in this chapter.

7.1.3 THIS CHAPTER

This chapter will explore in detail different damage parameters; we will argue that there are several parameters that define damage and to accurately identify damage, we need to know as much as possible about these parameters, their causes, and their effects on the structure on hand. Next section will be devoted to differentiating between DMID and structural identification (STRID). There are some who consider DMID as an extension of STRID. We show that the two fields are not extensions of each other, and they should be approached with different mindset. We argue that damage is the rate of structural change, both in time and in space. It is analogous to a given function of some variables and its partial derivatives with respect to these variables. To obtain the derivatives of the function, given the function itself, there should be certain numerical rules. We follow such an analogy and provide rules for DMID from popular STRID methods. If such rules are not followed, the accuracy of the identified damage would be in question.

We devote the next section to exploring popular NDT methods and their roles in DMID. We always discuss those NDT methods from SHM in civil infrastructure viewpoint. The following section in this chapter features the acoustic emission (AE) method. We use AE for an in-depth discussion because of its many special properties and its larger potential in SHM applications. Vibration-based DMID and signal processing applications are also discussed. We consider signal processing as an integral part of DMID methods since most of them are time dependent. This chapter will conclude by presenting the role of DMID methods and techniques in Structural Health in Civil Engineering (SHCE).

7.2 DAMAGE PARAMETERS

Damages in a bridge structure, or any type of civil infrastructure, vary "depending on the cause of the damage. In addition to the cause of the damage, several other parameters need to be identified. We will call these parameters damage parameters." Generally speaking, the damage parameters are as follows:

- Cause of damage
- Type of damage

- Temporal characteristics of damage
- Location of damage
- Extent of damage
- Severity of damage

The causes of damage vary immensely in a bridge structure. These causes can be natural disasters such as earthquakes or accidental such as impact. Types of damage include small or large cracks, degradation of material properties, or loss of structural material. Temporal characteristics of damage can range from very slow such as in corrosion damage to very fast such as damages that result from accidental impact. Location of damage can be almost anywhere in the vicinity of the bridge; it is obviously directly related to the cause of the damage. The extent of the damage can range from a microscopic extent to a global extent. Finally, damages vary in their severity: some damages are of minor effect, while others have a severe effect, and must be addressed immediately. Table 7.1 describes some hazards and the parameters of damages that result from them.

TABLE 7.1
Examples of Damage Parameters

Hazard	Type of Damage	Temporal Characteristics	Location	Potential Extent	Type of Potential Failure
Fatigue	Cracks (can be hidden)	Can be slow or sudden processes	Usually at structural connections	Localized damage	Can be catastrophic
Corrosion	Degradation of properties due to changes in material chemical compositions	Slow process	Where concrete or steel is subjected to moisture, running water, or salt (such as de-icing materials)	Can be extensive	Due to the slow process, effects can be mitigated
Concrete Cracking (Other than corrosion)	Cracks, e.g., due to freeze and thaw	Can be slow due to environmental, or sudden due to mechanical loads	Damage location will depend on structural configurations and its response to environmental or mechanical effects	Can be extensive	Due to the slow process, effects can be mitigated
Scour	Soil erosion	Can be slow (sandy soil) or sudden (rocks)	Foundation-soil interfaces	Can be extensive	Due to submergence of damage (soil erosion), the extent of damages can vary from minor to major
Impact	Large deformations, delamination, etc.	Sudden due to mechanical loading	Structural components that are exposed to collision (e.g., bridge underpasses)	Ranges from local to global damage effects	Extent of damage varies from minor to major
Deterioration (wear and tear)	Degradation of material properties	Slow processes	Damage location will depend on structural configurations and its response to environmental effects	Ranges from local to global damage effects	Due to the slow process, effects can be mitigated

FIGURE 7.1   Deterioration damage. (Courtesy of New York State Department of Transportation.)

FIGURE 7.2   Impact (collision) damage. (Courtesy of New York State Department of Transportation.)

It is obvious that all of the damage parameters in Table 7.1 should be directly related to the method of DMID in an SHM experiment. If the DMID method is not consistent with the parameters of damage it aims to identify, the efficiency and accuracy of the experiment can be compromised. For example, Figure 7.1 shows general deterioration and corrosion damages, while Figure 7.2 shows impact damages. Clearly, types of DMID methods for each of these two situations need to be different. Karbhari et al. (2005) studied the pertinence of numerous DMID methods as they relate to fiber-reinforced polymers (FRP) wrapping. Table 7.2 shows examples of pertinence of DMID methods in identifying hazard-related damage.

7.3   STRID, DMID, AND SHM

7.3.1   OVERVIEW

There are numerous STRID methods and techniques. The purpose of STRID in the civil infrastructure arena include the following: (a) design validation of new structures, (b) condition assessment of

TABLE 7.2
Relationship of Damage Sources and Damage Identification Methods

DMID Method	Hazard			
	Fatigue	Corrosion	Earthquakes	Scour
Vibration methods	Usually ineffective, except for very extensive fatigue damages	Usually ineffective, except for very extensive corrosion damages	Useful due to the global nature of earthquake motion	Limited use due to the brittle nature of scour damage
Ultrasound	Useful when damage location is known. Labor intensive	Usually not used due to the extensive labor needs	Effective only after the occurrence of an earthquake (after event).	Effective only after the occurrence of a scour damage (after event)
Strain measurements (at selected locations)	Useful when damage location is known, or suspected	Does not measure corrosion extent, measure only the structural response to damage. As such, it has limited value except when corrosion damage is extremely large	Effective only after the occurrence of an earthquake (after event)	Effective only after the occurrence of a scour damage (after event)
Penetrating radiation	Useful when damage location is known. Labor intensive. Logistical and safety issues can be problematic			Not practical

existing structures, (c) analytical model updating of existing and new structures, and (d) DMID of existing structures. Many STRID researchers utilized STRID methods for DMID. However, not all STRID methods are suited for DMID and not all damages are possible to be identified by popular STRID methods. This chapter investigates the relationship between STRID methods and DMID. We investigate which STRID method is suited for DMID for the type of damages that may affect civil infrastructure. It will be shown that some damages can be identified by STRID methods, while some other damages need specific methods that are not within the conventional realm of STRID methods. Some guidelines for choosing STRID methods that are appropriate for DMID are given.

Structural health monitoring is an important emerging engineering field for infrastructure applications. This is due to its potential of saving costs, while improving safety. SHM was shown (see Ettouney and Alampalli 2000) to involve several fields: measurements, STRID, DMID, and decision making. Several STRID methods have been presented in the past (e.g., see Doebling et al. 1996; He and Fu 2001). Many of the STRID methods have the potential of identifying damage in the structure, thus fulfilling one of the most important objectives of SHM, and bringing to the decision maker valuable information that can help in managing their assets more effectively. Thus, the use of STRID methods in DMID is attractive, since there is a plethora of STRID methods that are fairly familiar to researchers and practitioners.

Unfortunately, the use of STRID methods for DMID must be done with extreme care. Using the wrong STRID method for a particular DMID situation might lead to an erroneous result, which might lead the decision maker to make the wrong decision. This chapter studies the interrelationship between STRID and DMID. First, we identify different objectives of STRID and DMID. Next, different methods of STRID are briefly discussed. We then discuss different pertinent damage classifications and introduce a unifying damage classification that is appropriate for the purpose of this chapter. Theoretical issues that can affect the results of STRID methods as tools for DMID are then presented. Finally, a step-by-step guide to check for the adequacy of STRID methods in detecting damage is presented. Following such steps will ensure accurate DMID when using different STRID methods.

7.3.2   Objectives of STRID and DMID

7.3.2.1   Objectives of STRID

STRID in civil engineering applications has several objectives: condition assessment, design validation, analytical model updating, and DMID. We discuss each of these objectives next.

Condition Assessment of Existing Structures: Condition assessment includes estimating values
of system properties that might be of importance to the performance of the system on hand. For
example, it might be of interest to estimate the effective moment of inertia of a concrete beam,
Poisson's ratio of underlying soil, or effective axial pile stiffness. Parameter estimation methods can
fulfill such an objective directly.

Design Validation of New Structures: Since design of new structures relies on numerous assumptions, validation of some of these assumptions using experimental methods is needed. For example,
modal damping or modal frequencies that are used in modern seismic designs may need verification
or validation. Such validation can improve safety, reduce costs, or both.

Analytical Model Updating for New and Existing Structures: Complex analytical models for structures are used for purposes other than just design. For example, they can be
used for weight estimations, cost estimation, determining the feasibility of construction
procedures and techniques, and developing repair or retrofit needs. Different STRID methods would have potential to provide accurate analytical structural models to accomplish
these objectives.

Damage Identification of New and Existing Structures: Numerous STRID methods are offered
as capable of meeting the above objectives and detecting the damage. The capability of STRID
methods in detecting the damage is the focus of this chapter.

All of the above objectives of STRID are of use in SHM. However, our immediate objective is to
evaluate the DMID potential for different STRID methods.

7.3.2.2   Objectives of DMID

Damage identification is one of the basic components of SHM (see Ettouney and Alampalli 2000).
It involves several objectives as given below (see Farrar and Jauregui 1996):

Identify Occurrence: Did the damage occur? For example: the start of corrosion, the start of
   soil scour, and so on.
Identify Location: Where did the damage occur? For example: the location of a void in a
   post-tensioned tendon.
Identify Extent: How extensive is the damage? For example: the amount of rust in the cross-
   section of a steel rebar.
Identify Remaining Useful Life of Structure: Given the above damage properties, what is
   the remaining expected service life of the structure?
In addition to the above, we add another potential objective of DMID:
Expected Cost: Given the above damage properties, what are its effects on the life cycle cost
   of the structure?

7.3.3   Definition and Classifications of Damage

Before proceeding further into the discussion of the suitability of STRID methods for DMID, we
have to first define "damage." A simple and encompassing definition would be that the damage is
an unwanted change in the system property or properties that occurs due to a natural or manmade
environment.

7.3.3.1 Physical Classifications

The above definition is too general and would not serve us well in our attempt to explore the inter-relationship between STRID and DMID. We need to be more specific. We submit that a single definition of damage, though attractive for its simplicity, might not be well suited to our immediate task. Because of this, we propose the following physical classifications: cause, material, and extent. Causes for damage are many and include corrosion, fatigue, impact, fire, and normal wear and tear. Materials include conventional steel, reinforced concrete, aluminum, post-tensioned or prestressed concrete as well as modern construction materials such as FRP. Damage extent can range from a hairline fatigue crack in a welded steel connection to large sudden (or gradual) soil erosion under a large size foundation.

Due to the large variability in classification and type of damage, it is reasonable to deduct that no single STRID method can identify all types of damage. This is an important observation, but not too beneficial. Let us try to address the apparent complex STRID-DMID interrelationship by introducing new classifications of damage: the detectability classification.

7.3.3.2 Detectability Classification

Instead of classifying the damage itself, we classify the method of identifying the damage. We propose this classification of DMID method: All DMID methods in any structural system can be classified in one of two methods: direct or inferred. Direct DMID methods are those methods that purport to identify the damage directly. Direct methods, generally speaking, do not require base-line measurements, that is, they can detect the damage using only information from the damaged state. Inferred DMID methods are those methods that require information from a baseline state in addition to information from the damaged state. The STRID methods that estimate damage using inferred methods are the subject of this chapter. Those STRID methods can be categorized in to three categories: modal identification, parameter identification, and nonphysical (functional) methods. Detectability classifications of DMID methods are shown in Figure 7.3. We will discuss only STRID methods that detect damage in an inferred mode.

7.3.4 STRID METHODS AND INFERRED DAMAGE IDENTIFICATION

7.3.4.1 Overview

The first STRID category that can identify damage in an inferred mode is the modal identification method category (see He and Fu 2001). This method estimates modal properties by utilizing dynamic measurements of the structure of interest. Modal properties are mode shapes, natural frequencies, and modal damping. Since that damage is not measured directly, it must be detected by inference, that is, its properties must be inferred from the identified modal properties.

Parameter identification is another class of STRID that uses inference to detect damage (see Doebling et al. 1996). The method is based on modeling the structure of interest using a suitable numerical technique, such as finite elements. Some of the main parameters that are used in modeling (such as material properties, element properties, etc.) are then identified using the results of testing. This realization of a numerical model is consistent with tests. The damage is related to the realized model by inference, since the damage is not measured or detected directly—only the changes in structural parameters are realized, not the damage itself.

The final class of STRID methods is nonphysical (functional) methods. The most popular technique in this classification is neural networks (see He and Fu 2001). Neural networks compute relations between a set of inputs and the corresponding set of output. These relationships are then evaluated by minimizing an error function in a least square sense. Neural networks are suited for nonlinear systems. However, the physical interpretations of the realized relationships in the network are usually difficult. Because of this, damage can only be identified by inference.

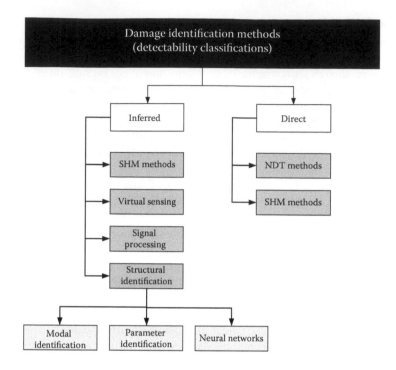

FIGURE 7.3   Damage identification detectability classifications.

7.3.4.2   Factors Affecting Inferred Damage Detection

For any STRID method that is to be used as a DMID tool, there are four factors that need to be satisfied for successful DMID. These four factors are discussed below.

7.3.4.2.1   Error Scale

One of the basic steps in any STRID method is that it relies on optimization. The objective functions of such an optimization is usually associated with some measure of error, E, as computed by the difference between the computed results of the estimated structural model X_i and the measured experimental results Y_i where i is a counter that represents the input-output pairing. The error function can be defined conceptually as

$$E = f\left(\left|X_i - Y_i\right|\right)$$

By definition, the units of such an error function are the same as the measured experimental results.

Examining a damaged structure using a STRID method that detects damage in an inferred mode requires a baseline and a damaged state data to detect the damage. The baseline can be experimental or analytical. Similarly, the damaged state data can be experimental or analytical. We can then define the evaluated damage D as

$$D = g\left(U_D - U_B\right)$$

Or

$$D = g\left(\Delta U\right)$$

where U_D and U_B represent responses (measured or analytic) of the damaged structure and the baseline structure, respectively. Note that U_D and U_B are evaluated with an error of E.

This means that for the estimated damage, the following inequality must be satisfied

$$O\left(D\right) >> O\left(E\right)$$

For example, if the error used in identifying a particular structural parameter is 0.1%, then the effect of the damage must be much larger than 0.1%. A convenient measure of damage effect might be the damage ratio expressed as $\Delta U/(U_D - U_B)$. Admittedly, this is a strict limitation. However, if it is not accommodated, then the STRID method of interest is not adequate in identifying the damage of interest. The error scale is applicable to all STRID methods when used for DMID.

7.3.4.2.2   Dynamics Scale

Any damage of a structure would generate a dynamic effect zone (DEZ) beyond which its effects would be negligible. The size of the DEZ depends on the type and extent of the damage as well as the frequency of excitation and the material properties. Furthermore, it is reasonable to state that the DEZ is largest at specific frequencies of excitation. These specific frequencies are those frequencies where resonances that the damage causes occur. It is possible to estimate the relationships between all of those parameters and the DEZ. Unfortunately, this is beyond the scope of this chapter. Qualitatively, for example, if we can state that for a steel crack of a length of order of 1.0 in, the effective frequency of excitation would be in the order of 1.0 KHz, or higher. Then, for a STRID method to successfully identify such damage, it should be accurately validated at 1.0 KHz, or higher.

The error scale is applicable to most STRID methods when used for DMID. The only exceptions are the statics-based parameter identification methods (see Sanayei and Saletnik 1996a and 1996b).

7.3.4.2.3   Spatial Scale

Damage, as defined earlier, is a change in the structure condition. In an SHM situation, the change should be small, in spatial sense. If the damage (change) is large, then there is no need for SHM in the first place: the size of the damage makes it identifiable by manual (e.g., visual) methods. The small spatial size of the damage would require using smaller size finite elements that are compatible with the expected damage size. For example, if the expected damage is a 1.0-in crack in a steel girder, it is prudent to have the finite elements in the area of a damage of size that is smaller than 1.0 in. If the element sizes are larger than 1.0 in, then the damage effect would be averaged over a larger area. Such an averaging might result in erroneous conclusions. Let us consider the example shown in Figure 7.4. The beams and columns are modeled using conventional axial-flexure elements. In addition, the beam-to-column connections are modeled by simple springs. The spring parameters, in a linear analysis parameter identification method would be the spring stiffness. The axial-flexure parameters would be the area, inertia, and material properties. Such a model will produce average results, such as a prediction of damage in the spring (reduced spring stiffness). However, detecting the exact damage type (loss of bolt tightening, weld cracking, slippage, etc.), is not possible without a more detailed finite element modeling.

The error scale is applicable to most STRID methods when used for DMID. The only exception is the statics-based parameter identification methods (Sanayei and Saletnik 1996a and 1996b).

7.3.4.2.4   Geometric Modeling

Spatial scale referred to the size or resolution of finite elements that are appropriate for parameter identification methods when used for DMID. The geometric modeling issue refers to the accuracy of the model itself. It is essential to accurately model the structural geometry, utilizing appropriate finite elements, in the vicinity of the expected damage. For example, it is popular to model beams and columns in structures by using the popular axial-flexure element shown in Figure 7.4. We should remember that such elements are based on the assumption that the cross-section of the element remains rigid, that is, does not deform except as a plane (Connor 1976). Due to this basic assumption, if the expected damage is in the form of a partial change along the depth of the element (a partial crack, or limited corrosion, for example), modeling the system as an axial-flexure element

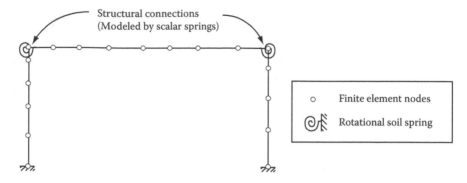

FIGURE 7.4   Example of spatial scale in parameter identification problem.

TABLE 7.3
Factors Affecting STRID Methods for DMID in Inference Mode

STRID method Category	Error Scale	Dynamics Scale	Spatial Scale	Geometric Modeling
Modal identification	Error tolerances during modal or parameter identification realization must be smaller than damage ratio	Frequencies that define dynamic effect zone (DEZ) for the damage of interest must be within the range of the test and/or identified frequencies	N.A.	N.A.
Parameter identification			Size and resolution of finite elements, especially in the damage area, must be adequate for the expected damage size	Geometric and material details in the damage area must be sufficient to model the expected damage
Neural Networks	Acceptable errors during network training must be smaller than damage ratio		NA	Can be of importance when designing the network topology

will not be accurate. A detailed 3D element is needed to model such an expected damage. An example of the need to accurately model the geometry is if the damage of interest is the corrosion of rebars in a given reinforced concrete beam. In such a situation, modeling the reinforced concrete beam as an equivalent homogeneous beam is not adequate. The steel rebars need to be modeled explicitly to be capable of identifying the corrosion damage in the rebars.

Similar to the spatial scale, the geometric modeling issue is applicable only to parameter identification methods. It is not pertinent to modal identification. For neural networks, it might be applicable to the topology of the network. However, this subject is beyond the scope of this chapter.

7.3.4.3   Summary

Table 7.3 summarizes the different requirements of STRID methods when used as a DMID tool in an inferred mode.

7.3.5   GUIDE FOR USING STRID AS A DMID TOOL

On the basis of the previous discussions, it is clear that using STRID methods to identify damage in an inferred mode needs careful consideration. Figures 7.5 through 7.7 show step-by-step guidelines and help a user to assess the adequacy of three STRID methods as a DMID tool in the inferred mode.

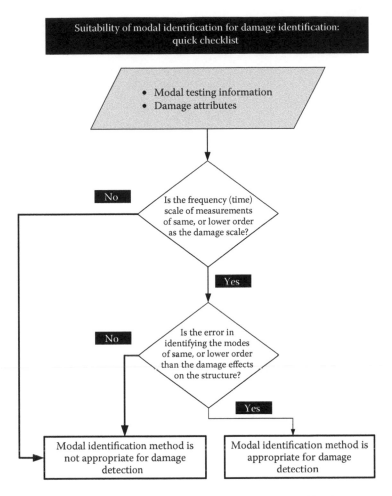

FIGURE 7.5 Guide to using modal identification methods for damage identification.

7.3.6 CLOSING REMARKS

We just explored different methods of DMID. Specifically, we were interested in the applicability of a certain class of STRID methods that are used for DMID. First, we classified DMID methods into direct and inferred classes. Of the inferred class of DMID, we discussed STRID methods that might detect damage in inference mode. The four factors that need to be considered, if the STRID method is to be successful in detecting the damage in an inferred mode, are discussed. We presented a guideline for three of the STRID methods that need to be followed to ensure a successful DMID. Similar guidelines need to be developed for other DMID methods. Developing these guidelines will help professionals to choose the right DMID method for any given SHM project.

7.4 NDT TECHNIQUES

7.4.1 INTRODUCTION TO NDT CONCEPTS AND METHODS

Traditional NDT methods have been used extensively for DMID in civil infrastructure. They are regarded as a mainstay in the emerging civil infrastructure SHM field. Because of this, the decision makers need to understand their basics, strengths, and weaknesses. Without such an understanding, those methods can't be used efficiently in a cost-effective manner in SHM.

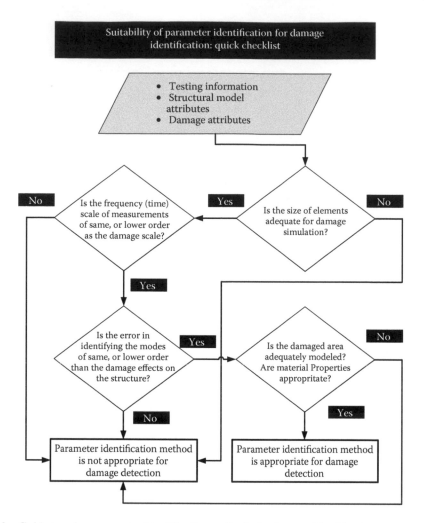

FIGURE 7.6 Guide to using parameter identification methods for damage identification.

Nondestructive Testing methods are numerous, and vary immensely in cost, complexity, scientific basis and techniques; we aim first to find a single basic rule that might help in bring most of NDT methods into focus. Such rule might be the basic wave length relationship:

$$c = \lambda f \tag{7.1}$$

With $\lambda, f,$ and c representing the wave length, the frequency, and speed of light. The rule is perhaps the single underlying common rule that governs most popular NDT methods. The wave length changes inversely as the frequency changes. Different types of waves occur at different wave lengths, as shown in the wave spectrum (Figure 7.8). The relationship between the frequency and wave length is shown in Figure 7.9. This figure also shows the spectrum used by various NDT methods.

We note that different NDT methods target different ranges of the wave spectrum. For example, ultrasonic (UT) methods target lower wave frequency range at higher wave frequency range, thermal radiation (thermography methods) resides in the infrared waves range. Next, we find the visual light frequency range that is the basis of visual and laser-based NDT testing methods. At higher frequency range the microwave and radar frequency range form the basis of electromagnetic wave NDT methods. At even higher ranges we find the penetrating radiation NDT X-rays and gamma rays. Figure 7. 9 illustrates these relationships.

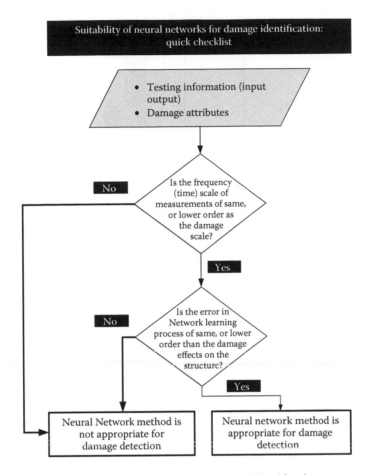

FIGURE 7.7 Guide to using neural networks methods for damage identification.

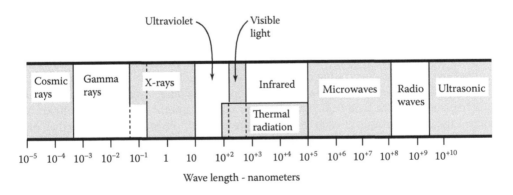

FIGURE 7.8 Wavelength spectrum.

As the wave length and the type of waves change, the properties of these waves change. The interaction of the different waves with test objects also changes. The state of the test object (properties, geometry, or damage) can be detected by observing the interactions of waves of different types and the test object. NDT methods aim at identifying such interactions in quantitative or qualitative manner. The NDT technique changes as the type of wave changes. It is the objective of this section to study how different NDT techniques vary in relationship with the types of pertinent waves, and how the tools, analysis, and processing of such information can lead to identifying the desired state of the test object.

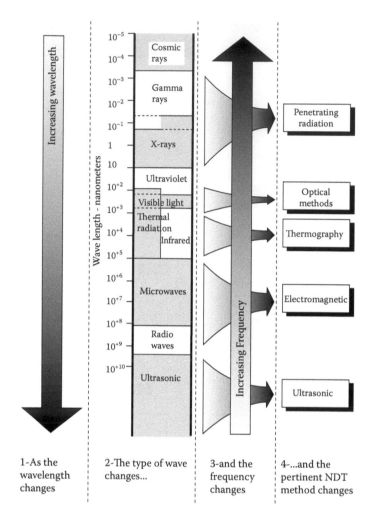

FIGURE 7.9 NDT method dependence on type of waves.

This section will investigate some popular NDT methods such as UT, electromagnetic waves (MV), thermography, and penetrating radiation. More NDT methods are briefly discussed at the end of the section (such as visual, liquid penetrant, and optical methods). In discussing all the methods, we try to present the basic technical grounds of the method and the tools of the method. Also the advantages and disadvantages of each method are presented. If the method has more than one basic technique, we will highlight the different techniques. For each of the methods we will relate them to the field of SHM and explore how different SHM issues are related to NDT techniques.

7.4.2 ULTRASOUND METHODS

7.4.2.1 Overview

Ultrasonic testing is very popular technique in NDT field. It is based on generating UT waves in the test object. As the waves propagate through the object, it changes its form through reflections, refraction, and scattering. By using a listening device to the signals, these signals can be processed and the state of the test object can be detected. There are numerous NDT UT methods such as impact-echo, pulse echo, sonic vibrations, guided waves, and laser-based UT. We will explore the basics of these methods in this section. There are other types of sonic/UT testing. Those types of testing methods are beyond the scope of this document, the reader should consult Birks et al. (2007).

Another major UT NDT method: AE is discussed in more detail elsewhere in this chapter. We note that the term ultrasound is also used to describe UT NDT methods; we use both terms interchangeably in this volume.

7.4.2.2 Theoretical Background

The basic equations of wave propagation in solids are described in the Appendix in Section 7.9 (Lamb waves), the Appendix in Section 7.10 (dispersion curves), and the Appendix in Section 7.11 (Helmholtz equation). More in-depth discussion of waves in solids can be found in Rose (1999). Table 7.4 discusses the basic technical issues in ultrasound waves in solids as they apply to SHM field.

7.4.2.3 Impact-Echo

7.4.2.3.1 General

This is the most basic of ultrasound NDT techniques; see Fisk (2008). It is based on generating normal waves (compression or tension) within the testing object by impacting it mechanically. The signal will travel in the body, and if there are any defects of damages, the resulting sound will reveal such defects.

7.4.2.3.2 Tools and Instruments

The impacting source can be a hammer or mechanically applied pulses. The responses can be detected by microphones or other suitable sensors.

7.4.2.3.3 Uses in NDT Applications

Since this type of testing is mostly qualitative, the resolution and accuracy of the test is also relatively very low when compared with other ultrasound tests. The tests can reveal flaws, debonding, or general qualitative damages. The resolution of the damages is in the order of tens of millimeters.

7.4.2.3.4 Use in Condition Assessment

The traditional approach in condition assessment of bridge decks by impact-echo is based on review of individual test point records. A new automated approach based on three dimensional

TABLE 7.4
Ultrasonic NDT Technical Issues as Applies to SHM

Ultrasound Technical Issues	Effects in SHM
Phase/particle velocities	Can be detected by comparing theoretical dispersion curves with measured frequency-time spectra
Scattering	Can be beneficial if theoretical solutions can be validated with experimental measurements
Wavelengths	Some ultrasonic methods allow for scaling wavelengths, such as ultrasonic vibration methods. This can be invaluable in SHM experiments
frequency range	For successful general SHM experiments, frequency range must be as wide as possible. This is needed to cover both global and local modes of structural behavior
Attenuation	As frequencies increases, waves tend to attenuate faster. This necessitates higher density of sensors. Some ultrasound methods cover wider areas naturally, such as laser-based ultrasonics (LBU)
Medium between probe and test article	Due to the large impedance mismatch between air and engineering solids (steel, concrete, etc.) ultrasonic transducers need to be in full contact with the solids. In certain situation a medium needs to be present between the transducers and the test object. Such needs can make SHM experiments more complicated. Remote sensing methods do not have such a requirement

data visualization was presented by Gucunski et al. (2008). The IE test was conducted using an impact source and a single nearby receiver. The return frequency together with the previously determined P-wave velocity can be used to measure the depth of the reflector, in this case the deck thickness using a simple relationship T = Vp/2fr, where fr is the return frequency. Depending on the extent and continuity of the delamination, the partitioning of energy of elastic waves may vary and different grades can be assigned to that particular section of a deck as a part of the condition assessment process. In the case of a sound deck (good condition), a distinctive peak in the response spectrum corresponding to the full depth of the deck, can be observed (see Figure 7.10). In the case of a delaminated deck, reflections of the P-wave occur at shallower depths as well as deck bottom, showing two distinct peaks in the response. At the same time, a progressed delamination is characterized by a single peak at a frequency corresponding to a reflector depth indicating that little or no energy is being propagated toward the bottom of the deck. In a very severe case of a wide delamination, the dominant response of the deck to an impact is characterized by a low frequency response of the upper delaminated portion of the deck.

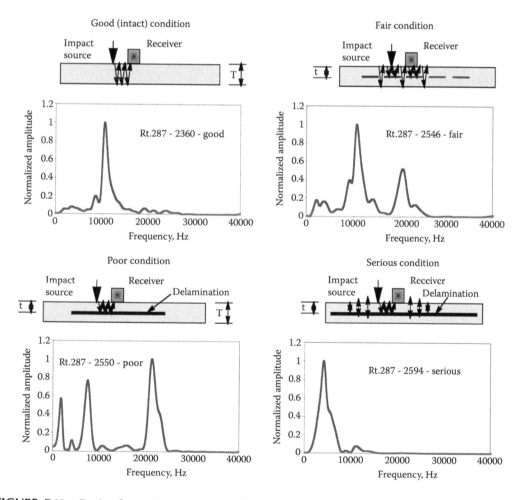

FIGURE 7.10 Grades for various degrees of deck delamination. (Reprinted from ASNT Publication; Gucunski, N. et al., Automated interpretation of bridge deck. Impact echo data using three-dimensional visualization, NDE/NDT for Highways and Bridges: Structural Materials Technology (SMT), ASNT, Oakland, CA, 2008. With permission.)

Using the reflector depth relationship, a unique thickness or depth can be correlated to every component of the frequency spectrum as illustrated in Figure 7.11 by a frequency spectrum and to it corresponding thickness spectrum.

Advanced data visualization techniques were then used for data interpretation through 3D translucent imaging of IE test data and separation of important reflector layers (see Figure 7.12). The presented visualization and to it associated interpretation algorithms allows both the overall assessment of the condition of the deck and identification of deteriorated zones of the deck.

7.4.2.3.5 Impact-Echo in SHM

Colla and Wiggenhauser (2000) discussed the advantages and disadvantages of IE for detecting concrete issues. They observed that IE has shown to be effective to evaluate concrete for locating voids and discontinuities, thickness measurements, duct cover, and delaminations. They noted the following IE disadvantages/limitations:

1. Results in literature have not always been conclusive and reproducible
2. For complex element geometry, such as multiple ducts and reinforcements layers crossing each other, interpretation of the single frequency spectra is not always unequivocal. In those cases, numerical modeling is used along with experimental data, which will limit its use for generalization

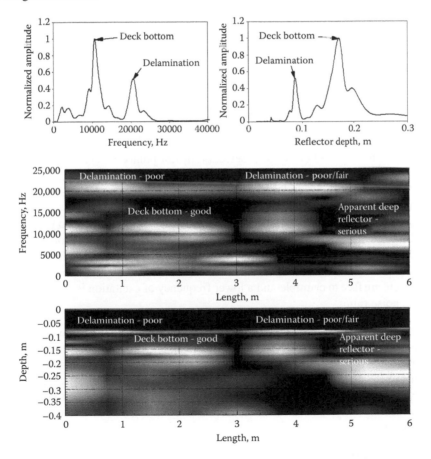

FIGURE 7.11 Frequency and thickness spectra and spectral surfaces. (Reprinted from ASNT Publication; Gucunski, N. et al., Automated interpretation of bridge deck impact echo data using three-dimensional visualization, NDE/NDT for Highways and Bridges: Structural Materials Technology (SMT), ASNT, Oakland, CA, 2008. With permission.)

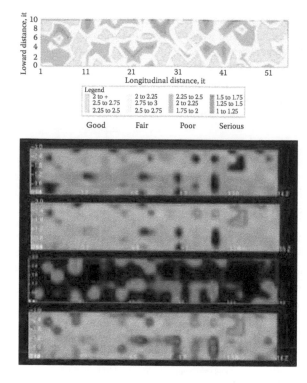

FIGURE 7.12 Condition assessment of Carter Creek Bridge deck. (Reprinted from ASNT Publication; Gucunski, N. et al., Automated interpretation of bridge deck impact echo data using three-dimensional visualization, NDE/NDT for Highways and Bridges: Structural Materials Technology (SMT), ASNT, Oakland, CA, 2008. With permission.)

3. IE is a point testing method. Hence, selection of test points is very important to identify damage.
4. Output of these punctual readings is single frequency spectra, whose interpretation and comparison is difficult.
5. Unfavorable factors include variations in spectra peaks amplitude due to uneven coupling of the transducers, small frequency shift of the peak due to influence from element vibration, and due to geometrical location of the measurement point on the element.
6. The frequencies of interest may not be excited. Bad selection of the hammer can cause the concrete surface to crumble and a lower frequency of excitation is achieved due to bad signal to noise ratios.

In order to avoid some of the above and to eliminate the need for skilled personnel to plan and test using IE, the use of scanning IE was suggested by the authors instead of point measurements. An automated procedure was developed for data collection to reduce influence of transducer coupling. Cross-sections of the element can be plotted. The method showed potential degree of success.

We observe that similar points can be made about most NDT methods, that is, making it a turnkey system with reliability can lead to increased use of NDT methods due to the cost–benefit it offers by doing so.

7.4.2.4 Pulse-Echo (Normal Waves)

7.4.2.4.1 General

The main difference between impact-echo and pulse-echo techniques is the source of the signal. In pulse echo, the signal is generated mechanically, or electrically, using a probe. The signal travels

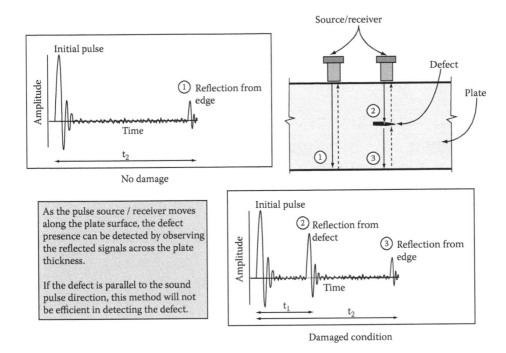

FIGURE 7.13 Pulse-echo principles.

within the media in the form of compression/tension (normal) wave. A source receives the signal, and the pertinent information is then processed from the signal. Figure 7.13 shows the principles of pulse-echo technique.

7.4.2.4.2 Modes of Scanning

There are several techniques to apply the pulse-echo method. They vary in complexity and resulting amount of information. Figure 7.14 illustrates some of these concepts, which are briefly described below.

A-Scan: The UT probe is used to inspect the in-depth state of a plate at a single point on the plate surface. The results of the scan reveal in a 1D scale the size and location of the damages across the plate thickness.

B-Scan: The probe is moved along a line on the plate surface. Thus the ultrasound waves cover a surface that is bound by the depth axis and the line traveled during the scan. The results are displayed in a 2D diagram that shows the size and location of damages in that surface.

C-Scan: The probe is moved along series of parallel lines on the plate surface so as to cover certain surface area. Thus the ultrasound waves cover a volume that is bound by the depth axis and the surface that was covered during the scan. The results are displayed in a 3D diagram that shows the size and location of damages in that volume.

M-Scan: This technology aims to scan moving objects. Biomedical and manufacturing applications are some of the more popular examples of M-scan technology. Obviously, the technology has little importance in SHM applications of civil infrastructure: Civil infrastructure do not move.

7.4.2.4.3 Use in NDT Applications

Pulse echo is used for most civil infrastructure materials such as steel, concrete, masonry, or FRP. It can be used to detect properties and state of as built material or component properties. Defects in materials such as cracks or debonding can be detected using this technique. Also, pulse echo can be

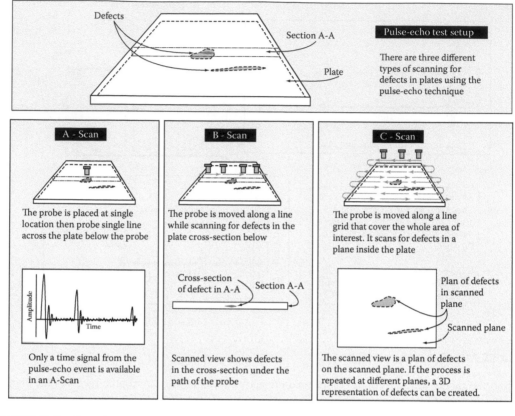

FIGURE 7.14   Scanning techniques for pulse-echo method.

used to measure thickness and density of the material. Most of the applications would be plate-like materials or near surface conditions.

7.4.2.5   Guided (Lamb) Waves

7.4.2.5.1   General

In this technique, also called *transmission UTs*, the UT waves are lamb waves, which are also known as plate waves. Thus, the waves are combinations of flexural and or shear waves. Rayleigh surface waves can also be generated using this technique. Lamb waves travel along the plate axis. Because of this, their detection capabilities cover much wider ranges than normal waves that are utilized in the pulse-echo technique. Because of the versatility of guided waves technique, they can be used in several modes as discussed next.

7.4.2.5.2   Modes of Testing Guided Waves

7.4.2.5.2.1   Pulse-Echo (Transmission Mode)

In this mode, the UT pulse sender would generate a Lamb wave(s) that travels along the plate long dimensions until it is reflected either a natural boundary condition (plate edge, for example), or a defect. Upon reflecting back, the receiver would capture the signal that can be analyzed to obtain pertinent information (see Figure 7.15).

7.4.2.5.2.2   Pitch-Catch

In this mode the sender and receiver are located far apart. This would enable an even longer detection surface than in the pulse-echo technique. This is due to the fact that the wave does not have to travel back from the reflection surface. Figure 7.16 shows the concept of pitch-catch of Lamb waves.

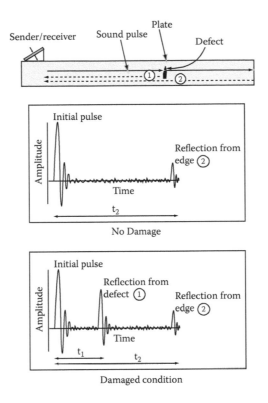

FIGURE 7.15 Pulse-echo concepts for guided waves method.

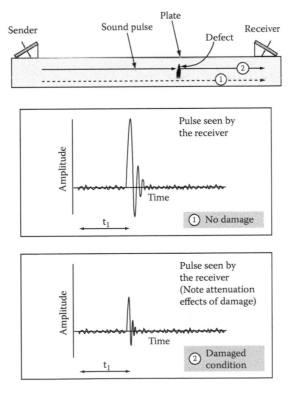

FIGURE 7.16 Pitch-catch concept for guided waves method.

7.4.2.5.2.3 Pulse-Reflection

This method is similar to the pulse-echo method, except that a separate sender and receiver are used during the experiment as shown in Figure 7.17

7.4.2.5.3 L-Scan and Guided Waves

L-scan technique is based on angular spectrum method in which the guided waves in plates will result in propagation modes that involve numerous planar waves. This will result from reflections and refractions from different boundary conditions, as well as any different damages or anomalies in the system being tested. It is desirable to decompose those 3D waves to a set of planar wave. The angular spectrum method can help in identifying different planar waves that are present in a complex wave field. See the Appendix in Section 7.12 for angular spectrum method at end of this chapter for more details on the method.

7.4.2.5.4 Phased Arrays

Phased-array technique is based on producing N UT signal pulses with specific amplitude and time delays. The pulse amplitudes and time delays are designed so as to maximize the range, angle, and overall defect detection of the transducers. The number of the probes within a transducer varies. Some popular transducers include 16, 32, or more probes. The phased-arrays synchronize both the transmitting pulse properties (amplitudes and delays) and the receiving delays and summation laws as shown in Figure 7.18. Those synchronized delays can produce an efficient pulse echo, or pulse-reflection methods. There are numerous probe geometries including linear, 2D, or annual (Figure 7.19).

The phased-array transducers offer numerous detection techniques. The simplest of these techniques are the linear scanning method, where the probes are moved in linear motions along the test object. The dynamic depth focusing (DDF) method changes the focal length of the array along the depth of the specimen, thus producing a high efficiency (range and SN ratio) pulse-echo mechanism. By changing the focal length at a given angle, the phased-array produces a sector scan S-scan. When the S-scan is coupled with a B-scan method, a 3D picture of the object can be produced.

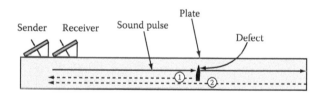

FIGURE 7.17 Pulse-reflection concept in guided waves method.

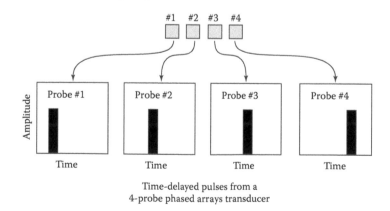

FIGURE 7.18 Time delays in phased arrays.

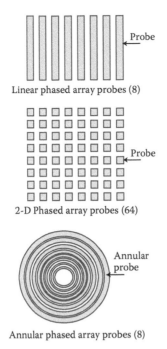

Linear phased array probes (8)

2-D Phased array probes (64)

Annular phased array probes (8)

FIGURE 7.19   Geometry of some popular phased-array probes.

7.4.2.5.5   Use in NDT Applications

Guided waves are very attractive in NDT because of their superior range, and versatile theoretical underpinnings: they can accommodate several types of elastic waves. They are suitable for detecting cracks (as results from fatigue or corrosion) or voids (in FRP laminates, for example). They can also be used to measure as built geometries and properties, such as material densities, wave velocities, or plate thicknesses. One potentially attractive use in FRP bridge decks is to investigate degree of homogeneity of the composite matrix, ASNT (1999). The method can also be used for quality assurance (QA) purposes in construction. The method can detect damages up to 0.2 mm.

7.4.2.6   Ultrasonic Vibration

7.4.2.6.1   General

Ultrasonic vibrations technique generates harmonic signals with varying frequencies in the test object. The response of the object is monitored such that any detected system resonances can be identified and the required system state is inferred. For example, the thickness of a plate can be measured by sweeping UT signals in the plate (see Figure 7.20). When the driving wave length is twice the plate thickness, the response is maximized. This will indicate the desired plate thickness. The experiment can be used to estimate as built plate thickness of steel beams, or FRP bridge deck plating systems.

A potential, yet not fully explored use, of the UT vibrations is to monitor voids in a tendon ducts in a post-tensioned (PT) bridge system. By imparting a sweeping UT harmonics at the surface of the PT beam, the response of the local system can be monitored. In a nonbonded situation, the resonating system will react as if its support is stress free, which will have different resonance frequency than the perfectly bonded system, as shown in Figure 7.21.

We note that the UT vibrations technique can be considered as the higher frequency scale counterpart of the popular global vibration identification techniques that are discussed elsewhere in this volume. A potential complete-scale frequency SHM technique can potentially include both global vibration and UT vibration methods.

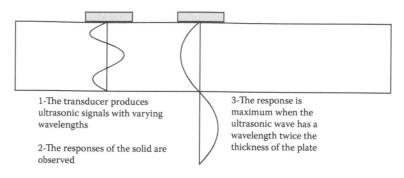

1-The transducer produces ultrasonic signals with varying wavelengths

2-The responses of the solid are observed

3-The response is maximum when the ultrasonic wave has a wavelength twice the thickness of the plate

Using ultrasonic vibrations to measure plate thickness

FIGURE 7.20   Thickness measurements using ultrasonic vibrations.

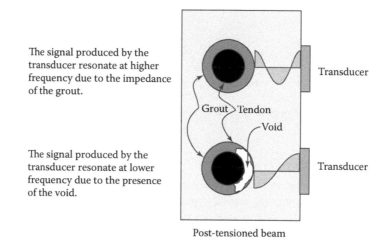

The signal produced by the transducer resonate at higher frequency due to the impedance of the grout.

The signal produced by the transducer resonate at lower frequency due to the presence of the void.

Transducer

Grout Tendon

Void

Transducer

Post-tensioned beam

Using ultrasonic vibrations to detect voids in a post-tensioned bridge beams

FIGURE 7.21   Ultrasonic vibrations and potential PT damage identification.

7.4.2.6.2   Use in NDT Applications

The idea behind UT vibrations is the same as conventional forced vibrations, only with at a higher frequency range. The frequency range can be as high as 20 KHz. Because of this, the frequency scaling from low frequencies (where lower global modes are excited) to high frequencies (where high local modes are excited) holds great promise. Spatial resolutions can be as large as the whole bridge volume and as small as 1.00 mm, which represent smaller size corrosion or fatigue cracks. A careful design of SHM experiment can reveal general changes in dimensions and geometry. The existence of damages can also be revealed. To evaluate exact locations of small damages and size of such damages, pulsers might have to be close in distance to each other, or other types of sensors, such as AE, might be needed.

7.4.2.7   Laser-Based Ultrasound (LBU)

Conceptually speaking LBU is fairly similar to using piezoelectric transducers in a pulse-echo UT setup. The main differences are that the UT waves are generated using impacting laser. Thus, LBU combines two physical phenomena: ultrasound and optics. It is based on generating ultrasound waves in solids by impacting the solid with a pulsed laser source. The laser source will generate UT waves in the solid by either (a) heating the impacted portion of the solid, where the sudden thermal expansion will cause an outward propagating waves, or, (b) the sudden inertial movement caused by the impacting laser would also cause outward propagating waves, ablation process. The UT waves include compression, shear and Rayleigh surface waves in the test object. They can then be received by adequate receiver, such an adaptive interferometer. The motion can reveal defects in the solid as in conventional pulse-echo process (Figure 7.22). The efficiency of detection of LBU is much less than similar conventional transducer detection, however, laser is very efficient generator of UT (see Figure 7.23).

Among the potential advantages of LBU in civil infrastructure is that it can scan larger surface areas than the stationary traditional transducers. The relative costs (both generators and receivers) are more expensive that the traditional transducers. Note that LBU has the flexibility of applying the impacting laser; it can be applied as an array with varying energy intensity, in a phased-array type setting. This will provide flexibility in DMID region. The LBU would generate broadband UT frequencies up to 100 MHz, or even higher if needed.

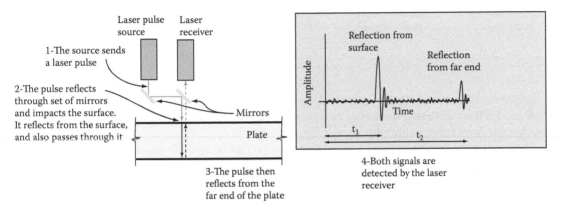

FIGURE 7.22   Concepts of laser-based ultrasonic (pulse-echo setup).

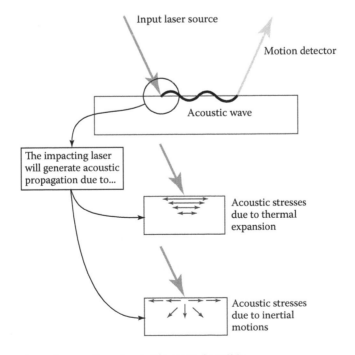

FIGURE 7.23   Laser impact generation of acoustic waves in solids.

TABLE 7.5
Comparison between Some Ultrasound Testing Methods

Issue	Pulse Echo	Impact Echo	Ultrasonic Vibration	Guided (Lamb) Waves	Laser-based Ultrasonic (LBU)
Type of damage?	Cracks, debonding. Also properties such as thickness or mass density		Cracks, debonding. Also properties such as thickness or mass density. Smaller frequencies can detect global changes in the structure	Cracks, debonding. Also properties such as thickness or mass density	
Size of flaws (mm)	0.01	O (10.00)	1.00	0.2	Less than 0.01
In-situ?	Yes				
Predominant waves during test	Propagating: Normal (compression/tension)		Propagating/standing	Propagating: Lamb, flexural, shear, Rayleigh	Propagating: normal, lamb, Rayleigh, flexural, shear

Applications in civil infrastructure include weld or other defect identification, surface delaminations of FRP wrappings in reinforced concrete (RC) installations. It can also be used for measuring thicknesses and temperature responses.

Some general advantages of LBU are that they (a) can be applied remotely, (b) can be used in hostile environment, (c) very flexible for complex geometries, (d) can provide very high resolution, (e) suitable for real time measurements: for example, can monitor real time responses and damages due to truck motions on bridges, and (f) can be used for curved surfaces (Table 7.5).

7.4.2.8 Applications in SHM

7.4.2.8.1 General

Ultrasonic NDT techniques can be utilized in many ways in SHM field. Table 7.6 summarizes some of the potential ways of these applications.

7.4.2.8.2 Choice of Appropriate UT Method

Sometimes in the field of SHM, we are faced with a dilemma: two available choices, with opposing advantages and disadvantages. The Heisenberg principle as applied to signal processing (where higher resolution of frequency sampling is only possible with lower time sampling resolution and higher time sampling resolution is only possible with lower frequency sampling resolution*) is a good example of such a dilemma. Another example of such a dilemma is in the field of UT measurements of defects (mainly in piping systems).

Consider, for example, the UT system presented by Owens et al. (2005). It is a high frequency system (up to 1.0 MHz), that is capable of exciting longitudinal, flexural or torsional waves in the pipes. The echo of those excitations can reveal location of defects along a pipe length of 155 in, as reported by the authors. The advantage of the use of high frequency is that smaller defects can be observed. They reported as small defects as 0.36% CSA (cross-sectional area). Higher sensitivity of detecting defects can safeguard against brittle failures in pipes.

On the other hand, Demma et al. (2005) reported a low frequency (in the range of 0–100 kHz) UT guided wave system that can detect CSA as low as 5%. The main advantage of such a system is the longer distance of the pipe that the system can cover. The authors reported a distance as long as 100 m. A comparison between the two approaches is shown in Figure 7.24. As mentioned above: the dilemma that

---

* For more discussion, see Section 7.7 on signal processing in this chapter. Also refer to Hubbard (1998).

TABLE 7.6
NDT Ultrasonic Utilization in SHM Civil Infrastructure Field

Method	Direct Sensing	Virtual Sensing Paradigm (VSP)	Remote Sensing	Local vs. Global—Scaling Issues	Cost-Benefit Issues
Impact-echo	Mostly direct sensing local capability	Results can be easily used with other STRID techniques to provide DMID insights in a VSP setting	NA	Potential for efficient scaling from local to global SHM is minimal due to the labor demands of this technique	Labor intensive. Cover smaller areas. Mostly local applications
Pulse echo Guided waves (General)	Provide local and semilocal direct sensing capabilities		The techniques can be used in remote sensing settings (see Chapter 5)	In a well-designed experiment or SHM setting, the potential of spatial scaling (local to global) exists	Can provide cost–benefit advantages in well-designed SHM settings
Guided waves (phased arrays)				A series of phased-arrays probes that are places optimally on (or inside) that structure can scale time and space demands, thus providing both local and global identification potential within single setting	Given the potential of local to global scaling, this technique can be optimized to provide optimal cost–benefit returns. In real time, or near real time SHM settings
Ultrasonic vibrations	Works well in a local setting. Can be expanded to larger domains when using several online probes. In such a case this technology can scale from local to global SHM domains	Results can be easily used with other STRID techniques to provide DMID insights in a VSP setting	No direct uses exist as of the writing of this volume. The potential of using the technology in remote sensing does exist	The potential of scaling the time (frequency) scale so as to cover both local and global SHM issues exists	If used in a local-global setting, the cost–benefit of this method becomes more attractive
Laser-based ultrasonic (LBU)	Has the potential of covering large area of detection when used in a remote sensing setting		Using laser as an ultrasonic exciter and high resolution photography as a sensing mechanism can provide remote sensing capability	Offer potential of spatial scaling (local-to-Global) SHM settings	As the costs of equipments are reduced, the cost benefit of this technology becomes more attractive

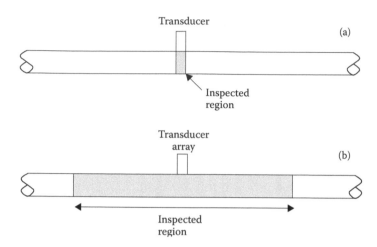

FIGURE 7.24 Two piping inspection solutions: (a) high accuracy but short distance, (b) lower accuracy but longer distance. (From Owens, S. et al., A new high frequency long range ultrasonic guided wave inspection system for pipes, Proceedings of the 2005 ASNT Fall Conference, Columbus, OH, 2005. With permission. Reprinted from ASNT Publication.)

faces the decision maker is: shorter distance and higher accuracy, or longer distance, and lower accuracy. The answer, of course depends on the situation on hand. For (a) brittle pipes, (b) pipes that carry hazardous material, or (c) newer pipes where the defect sizes tend to be small, it would be advisable to use the higher frequency approach; otherwise, the lower frequency approach would be adequate to use.

Since this document is interested mainly in bridge structures, it is of interest to consider potential situations where the above examples might be applied to SHM of bridges. We note that piping systems in general are simple (mostly cylinders) and uniform (mostly single material) constructs, especially when compared with the complex bridge constructs. The high frequency but short distance UT testing is an NDT-type solution. Such solutions were discussed in many parts of this document. Their usability in an SHM project is limited by the fact that the approximate location of the damage must be known in advance. On the other hand, the low frequency but long distance UT approach is more suited for SHM projects. It reduces the need for knowing, *a priori*, the location of the damage. Unfortunately, because of the complexity of the bridge constructs, the needed technology for such a solution is not available. We expect that if/when such technology becomes available, it can be used first in simpler bridge constructs.

7.4.3 THERMOGRAPHY

7.4.3.1 Overview

Measuring the manner structural temperature changes, or heat flow to, from or through materials or structures, is the basis of thermography-based NDT. Heat flow is related to material and structural properties. So, by observing how heat flows, it is possible to obtain information regarding the state of the structure. Such information can be, but not limited to, surface-connected flaws, internal flaws, measures of thickness or densities, mechanical or thermal properties, and certain surface properties such as delamination, see ASNT (1999). Thermography has shown, and holds, great promise in the field of NDT and SHM for civil infrastructure. This section presents some of the basics of thermography, and its uses in SHM for civil infrastructure.

7.4.3.2 Categories of Thermography

There are several was to categorize thermography (Figure 7.25). Methods of heating, methods of sensing and heat flow mechanisms are three basic categories.

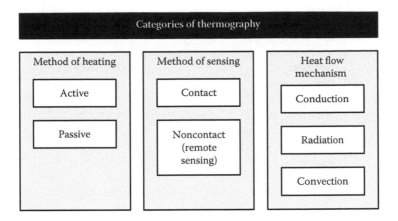

FIGURE 7.25 Categorization of NDT thermography.

Method of Heating: Temperature rises, or drops, in a structure can result either naturally (passive) or from an external source (active). Passive temperature stimulation can occur in manufacturing processes or by simple exposure to atmospheric heat. Active temperature stimulation occurs by subjecting the system of interest to a heating source. Both types of heating have been used in SHM projects.

Mechanisms of Heat Flow: There are three basic methods temperature (or heat) can flow: conduction, radiation, or convection. Conduction heat flow occurs when heat flows from warmer to cooler parts of materials. Radiation occurs because warm surfaces generate MW that radiates from the warm surfaces, De Witt (1988). Such waves travel in near-infrared wave lengths. Infrared radiated waves behave similar to visible light waves, except for the differences in wave lengths and frequencies. Because of this, it can be captured by special infrared cameras. Convection occurs when heated mass is transferred. Conduction and radiation have many applications and uses in SHM while convection has limited use.

An important metric in the field of thermography is the emissivity of the surface ε. Emissivity is a measure of the efficiency of the surface to radiate or absorb electromagnetic radiation. A perfect absorbing, or radiating, surface is when $\varepsilon = 1.0$. Thus, $0.0 \leq \varepsilon \leq 1.0$. It is desired to have as high emissivity as possible during the test. To improve the accuracy of an infrared imaging device in detecting thermal radiation, it is important for the surface to be clean and free of foreign material. Table 7.7 shows emissivity of some engineering materials.

Method of Sensing: Methods of sensing temperature flow are as varied as the mechanisms of heat flow. Sensors can be placed directly on the object (contact). Sensing temperature can also be done remotely (noncontact). Remote sensing thermography is of special interest in the field of SHM because of all of the advantages that remote sensing offers (see Chapter 5). Figure 7.26 shows remote sensing steps when utilizing thermography method.

7.4.3.3 Geometric-Based: Heating Methods

When the temperature of the system is stimulated actively, the spatial (geometric) stimulated surface varies as follows:

Point Temperature Stimulation: a single point on the surface is stimulated. This is generally done by a laser point device. This technique is similar to A-Scan in UT NDT.

Line Temperature Stimulation: a line on the surface is stimulated. This is generally done by a linear florescent lamp. This technique is similar to B-Scan in UT NDT.

Area Temperature Stimulation: an area on the surface is stimulated. This is generally done by a lamp, or set of lamps. This technique is similar to C-Scan in UT NDT. In SHM applications, this method can be most efficient, since it cover more area on the surface of interest.

TABLE 7.7
Emissivity of Some Popular Materials

Material	Emissivity ε
Polished steel	0.03–0.08
Dull steel	0.08–0.20
Steel mill products	0.15–0.25
Heavily oxidized and rough iron or steel	0.60–0.85
Bright brick, tile of plastics	0.80–0.95
Concrete	0.92
Wood	0.90
Sand	0.90

Source: From Botsko, R., Introduction to ultrasonic testing, Section 13, Part 2 in *Nondestructive Testing Handbook: Vol. 10,* ANST, Columbus, OH. Reprinted from ASNT Publication.

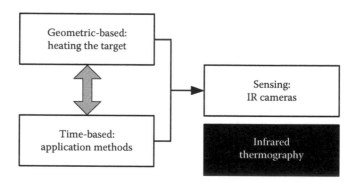

FIGURE 7.26 Steps of remote sensing infrared thermography.

7.4.3.4 Time-Based Thermal Application Methods

The temporal wave form of the thermal stimulation is also of interest when the temperature of the system is stimulated actively. The different possible input temporal heat wave forms are described in Table 7.8.

7.4.3.5 Sensing

7.4.3.5.1 Contact Sensing

Contact thermometry has many uses in numerous fields such as manufacturing and processing. In SHM contact thermometry is used primarily to measure ambient temperature that can be important to correlate with such hazards as corrosion, fatigue, or thermal behavior of the global structural system. For these general uses, thermoelectric devices such as thermocouples would provide adequate performance.

7.4.3.5.2 Noncontact/Remote Sensing

Noncontact (remote sensing) thermal sensors vary in their temperature range resolution as well as distance from the test object. Hand-held thermal scanners offer the portability advantage with the disadvantage of low imaging quality. Perhaps the most popular thermal imaging tool is the high resolution infrared cameras, Hardy and Bolen (2005). They tend to offer high imaging resolution,

TABLE 7.8
Temporal Wave Forms of Temperature Stimulants

Method	Description
Pulse	The stimulus is applied with a very short time duration. The decay properties of the temperature are measured and the needed information is deduced from this information, Shepard et al. (2001)
Step	The stimulus is applied with a finite duration. The rise of temperature is observed, and the needed information is deduced from this observation, Maclachlan-Spicer et al. (1991)
Lock-in thermography: modulation heating/ phased pulses	The thermal waves are generated and detected remotely. The waves are harmonic. The frequencies can be changed such that maximum efficiency of detection can be had, Maldague (2002).
Vibration/mechanical induced heating	Vibrothermography induces heat at defects (by friction, for example) by externally vibrating the system at certain driving frequencies. By observing the heat rise, the defects can be detected. See Moore (2001) and Maldague (2002)

with wide range of temperature range. Infrared cameras also have good response time that makes them versatile for measuring fast and slow changing temperatures.

7.4.3.6 Applications in SHM

Infrared thermographic NDT techniques can be utilized in many ways in SHM field. Table 7.9 summarizes some of the potential ways of these applications. Thermography has been used in several SHM applications. Figure 7.27 shows results of thermographic inspection of the surface of an FRP bridge deck. FRP has also been used to retrofit bridge beams and columns. Thermography was used to monitor status of FRP wrapped bridge beams (Figure 7.28) and columns (Figure 7.29). Table 7.10 illustrates the applicability of thermography techniques in infrastructure applications.

7.4.4 ELECTROMAGNETIC METHODS

7.4.4.1 Overview

Electromagnetic NDT methods utilize the interaction between the electromagnetic flux and waves and the flaws in the test object. Numerous electromagnetic NDT methods are used. Among the more popular methods are: Eddy current (EC), static magnetic field, magnetic particle, and microwaves (radar) techniques. Some basic properties of the four methods are shown in Table 7.11. We also discuss the properties of these methods as they relate to SHM field in Table 7.12. Figure 7.30 shows different aspects of each method that need to be considered while studying the applicability of the method for use in an SHM project. The rest of the section will explore those methods in more details, especially as they relate to SHM field. Note that there are other types of electromagnetic testing such as electric current perturbation (ECP) method that can detect damage in different metals such as aluminum alloys, Beissner (2005). Another modern technique is the megaabsorption technique, Rollwitz (2005a).

7.4.4.2 Eddy Current

7.4.4.2.1 General

Eddy current is a popular NDT method for detecting surface or near surface anomalies (cracks delaminations, etc.). It is also used to detect thickness of objects as well as some material properties such as conductivity, corrosion, and permeability. EC method is based on the following steps

1. An alternating current from an electrical circuit is induced in a circular coil in a portable probe (Figure 7.31).
2. The current will induce a voltage field in the nearby test object.

TABLE 7.9
NDT Thermography Utilization in SHM Civil Infrastructure Field

Method	Direct Sensing	Virtual Sensing Paradigm (VSP)	Remote Sensing	Local vs. Global—Scaling Issues	Cost–Benefit Issues
Conduction	Yes	Yes	No	Mostly local	Can be useful for SHM experiments for correlation purposes, e.g., fatigue, corrosion, thermal expansion problems, etc.
Convection	Yes	NA	No	NA	
Radiation	Yes	Yes	Yes	Can be global and/ or local	High potential for a higher benefit to cost ratio

FIGURE 7.27 Thermographic inspection of FRP bridge deck. (Courtesy of New York State Department of Transportation.)

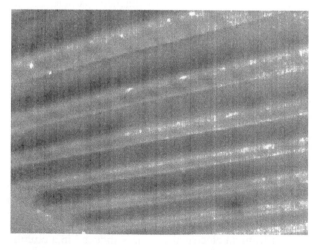

FIGURE 7.28 Thermographic inspection of FRP bridge wrapping of beams. (Courtesy of New York State Department of Transportation.)

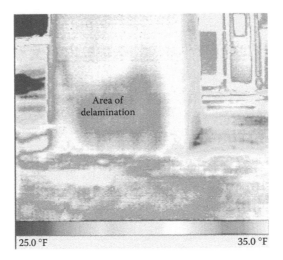

FIGURE 7.29 Thermographic inspection of FRP bridge wrapping of columns. (Courtesy of New York State Department of Transportation.)

TABLE 7.10
Applicability of Thermography Techniques to SHM Hazards

Hazard	Comments
Fatigue	Can be used on a local level for surface cracks
Corrosion—General	Can be used for surface rust-produced cracks
Corrosion—PT/PS	Not promising due to the relative embedment depth of the PT tendons
Scour	NA
Earthquakes	Can be valuable in detecting near surface damages on a global scale after seismic events
FRP wrapping	Very useful due to the surface nature and small thickness of the wrapping materials
FRP decks	Very useful for skin top (or bottom) plates

TABLE 7.11
Comparison between Different Electromagnetic Testing Methods

Issue	Eddy Current	Static Magnetic Field	Magnetic Particle	Electromagnetic Waves (Microwaves)
Type of damage	Surface or subsurface flaws	Surface or subsurface flaws	Surface flaws	Surface, subsurface, interior, or length of cables or solids
Size of flaws	Up to 0.2 mm	0.03 mm	Up to 0.5 mm	10^{-3} mm
Pretest preparations	None. Requires reference samples	Require inducing magnetic field in the test object	Extensive	None
Post-test preparations	None	Require demagnetizing the test object	Extensive	None
In-situ?	Yes, manual. Can be automated	Can be difficult	Yes, mostly manual	Can be automated
Detection methods	Computing electrical impedance properties	Measure magnetic field perturbations	Visual	Measuring electromagnetic wave reflections and scatter can reveal damage states

TABLE 7.12

Electromagnetic NDT Technical Issues as Applies to SHM

Electromagnetic Technical Issues	Effects in SHM
Type of material	Materials are limited to ferromagnetic or metals, depending on the type of test. Electromagnetic waves (microwaves) can be used with other materials such as plastics, concrete, soil, or elastomers
Global vs. local	Some methods are strictly surface-based, or near surface-based, such as magnetic particles. Microwaves can penetrate deep into materials, as well as travel fairly long distances, e.g., steel cables
Surface preparation	Some surface preparation is needed before and after the experiment for magnetic particle testing, which would limit its use in general real time SHM setups
Type of surface	Eddy current and magnetic particle need somewhat smooth surfaces. Electromagnetic waves testing can operate on variety of surfaces

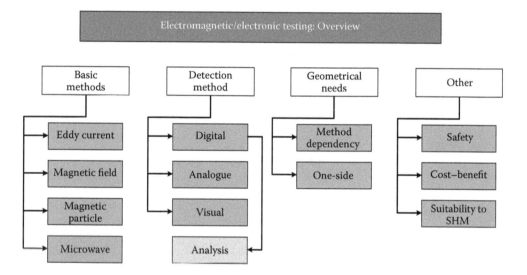

FIGURE 7.30 Overview of electromagnetic/electronic testing.

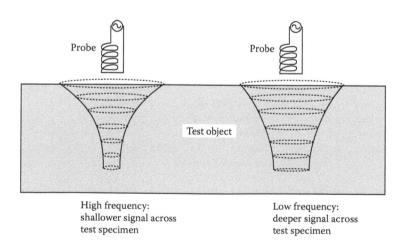

FIGURE 7.31 Effects of frequency on Eddy current signal depth.

3. The circular nature of the coil in the probe will cause electrons in the test object to move in an eddy-like circular fashion. The electrons will in turn generate a circular magnetic field in the test object.
4. The circular magnetic field can then be detected and measured in real time. The detection can be either by the driving coils themselves or by another set of coils in the probe.

The depth of generated magnetic field is a function of the test object material properties, such as geometry and lift off electrical properties. It is also a function of the excitation frequency. For low frequencies, the detection depth is large and as the frequency increases, the depth decreases (Figure 7.31). The depth of EC penetration S of an object is computed by

$$S = (\pi f \mu \mu_0 \sigma)^{-0.5} \tag{7.2}$$

In Equation 7.2 the driving frequency is f, the test object electric conductivity is σ, the relative magnetic permeability is μ, and the permeability of free space is μ_0

7.4.4.2.2 Tools and Instruments

The probes in EC testing can be either single or double probes. The single probe function is illustrated in Figure 7.32. As the probe moves along the surface of the test object, the measured voltage is recorded on display. The real part of the voltage (resistance) is represented by the horizontal axis, while the vertical axis represents the imaginary part of the voltage (inductance reactance). As the probe passes over a defect, going from point 2 to 3, the voltage changes as shown in the image. When the probe moves past the defect, going from point 3 to 4, the voltage traces its footprint, as shown in Figure 7.32. Finally, the voltage reverts back to its baseline at point 5, away from the defect.

A more detailed image can be produced by using a two-probe setup as shown in Figure 7.33. As the double-probe passes over a defect, going from point 2 to 3, the voltage changes as shown in the image. Since each of the probes is recording the voltage with slight space shift (as the first probe moves into the defect the second probe is still in pristine material), a hysterics in the voltage space is formed. Thus, when the probe moves past the defect, going from point 2 to 3, the voltage does not trace its footprint, as shown in Figure 7.33. When the double probe continues its movement across the defect, going through points 3-4-5, an antisymmetric hysteresis is formed. This type of signal is Lissajous* signal.

The Lissajous signal serves more than just providing additional image details. It was found that the signal characteristics can provide information as to the type of defect. Thus, by having a base line of defect characteristic images, it is possible to discern not only the presence of damage, but also the type of damage that is being observed. Figure 7.34 shows different signals that vary according to the type of damage being detected.

7.4.4.2.3 Advantages–Disadvantages in NDT Applications

The advantages of EC method are as follows:

- Very simple and accurate to detect surface defects
- Most efficient in metallic materials (good electrical conductivity)
- Works well for carbon FRP materials
- Near contact between the probe and the test object is not needed
- Efficient locally

---

* Joules Antoine Lissajous (1857) studied the family of curves that are represented by $x_1 = A_1 \sin(\alpha_1 x + \Delta)$ and $x_2 = A_2 \sin(\alpha_2 x)$. The shape of the curves depends on the parameters, α_2, A_1, A_2, and Δ.

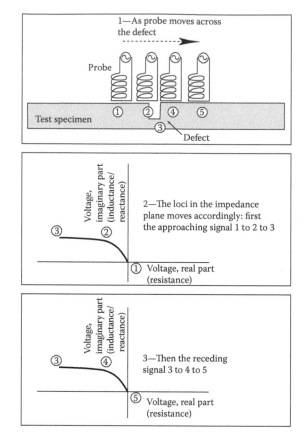

FIGURE 7.32   Generation of impedance curves of a defect: single probe.

The disadvantages are as follows:

- Somewhat limited depth
- Sensitivity is dependent on direction of cracks. Maximum accuracy when the crack is normal to flux lines
- Not global
- Not sensitive for nonconductive materials

7.4.4.3   Magnetic Particle (MP)

7.4.4.3.1   General

Fine ferromagnetic particles are first spread over the test object. When these particles are magnetized, the pattern will follow the magnetic field as shown in Figure 7.36. If there is a surface, or near surface defect or crack that is aligned roughly normal to the magnetic flow, the pattern will show visually; thus revealing the presence of defects. Obviously, this NDT method is capable of detecting defects in a visual manner. The extent or depth of such defects might not be detectable by this method. The method of application is shown in Figure 7.37. Figure 7.38 shows a plate with a preinserted longitudinal crack; the crack is not visible. After applying the magnetic field process, the crack becomes visible as shown in Figure 7.39.

7.4.4.3.2   Tools and Instruments

The needed tools for this method are (a) the ferromagnetic particles themselves that are needed to spread over the test object, and (b) the yoke that is needed to magnetize the particle. The yoke, of course will need power supply for operations.

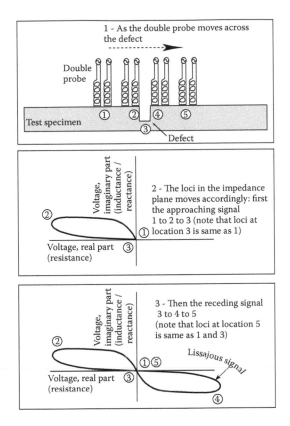

FIGURE 7.33   Impedance curves of a defect: double probe (Lissajous signal).

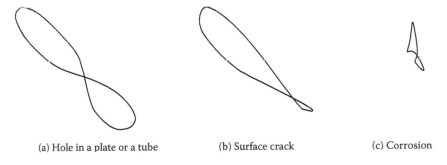

(a) Hole in a plate or a tube   (b) Surface crack   (c) Corrosion

FIGURE 7.34   Patterns of Lissajous signals as a function of type of damage (a) pattern for a hole in plate or tube, (b) pattern for surface crack, and (c) pattern for corrosion.

7.4.4.3.3 Advantages–Disadvantages in NDT Applications

The advantages of MP method are as follows:

- Very simple to detect surface defects
- No need for electrical circuits

The disadvantages are as follows:

- Ferromagnetic materials only, such as steel
- Needs surface preparation before and after the test
- Direction of defects needs to be very close to normal to the magnetic flux
- Only surface defects

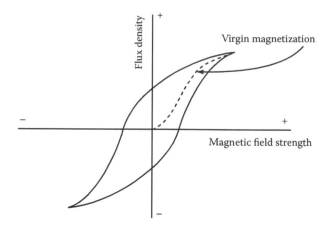

FIGURE 7.35   Magnetic hysteresis.

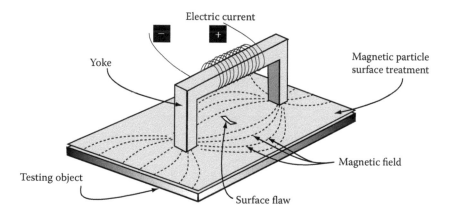

FIGURE 7.36   Components of magnetic particle testing.

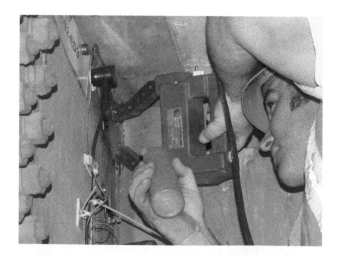

FIGURE 7.37   Magnetic particle yoke and particles in use. (Courtesy of Nondestructive Evaluation Labaratory, FHWA.)

FIGURE 7.38	Magnetic particle testing, baseline test object. (Courtesy of New York State Department of Transportation.)

FIGURE 7.39	Magnetic particle testing, surface crack revealed. (Courtesy of New York State Department of Transportation.)

### 7.4.4.4	Static Magnetic Field (SMF)

#### 7.4.4.4.1	General

Unlike MP method, in SMF the test object itself is magnetized, see Beissner (2005). The normal magnetic field lines are interrupted by the presence of a flaw in the test object, thus creating a flux leakage. Figure 7.40 shows the basic concepts of static magnetic field NTD with and without flaws. A scanning tool is then used to detect the state of magnetic field in the object. If there is a defect near the surface, it will cause disturbance to the field and the scanner will react accordingly. During the relationship between magnetic field strength and flux density can be displayed (Figure 7.35). The shape of the hysteresis can give insights on location, type, and size of the defect in the object. The method can be used to detect flaws, as well as mechanical properties (such as material hardness) in ferromagnetic materials (see Jiles 1986). The technique was applied in several SHM applications such as detecting cracks in steel rebars, see Klausenberger and Barton (1977), and Beissner and McGinnis (1984).

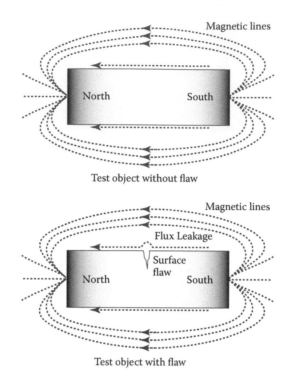

Test object without flaw

Test object with flaw

FIGURE 7.40   Basics of static magnetic field method.

7.4.4.4.2   Advantages–Disadvantages in NDT Applications
The advantages of SMF method are as follows:

- Can be applied to whole objects with varying sizes and shapes
- Can detect both material properties and flaws in an SHM test

The disadvantages of the method are as follows:

- Results are mostly qualitative
- Requires near contact between the scanner and the object
- In-field magnetization can be difficult

7.4.4.5   Electromagnetic Waves: Microwaves or Radio Waves
7.4.4.5.1   General
Microwaves exist in frequency range of about 300 MHz to about 325 GHz, which is an equivalent to a wave length range of about 3×10^9 to 10^5 nanometers (Figure 7.8). These waves are also known as short-pulse radar, impulse radar, or ground penetrating radar (GPR). Radar wavelengths usually exist in the higher range of the microwave bands. Both microwaves and radar waves have been used extensively in NDT field, specifically; they have numerous applications for both NDT and SHM in civil infrastructure, see Alkhalifah (2005).

Microwaves can be generated in several forms. Among these forms are pulsed, constant frequencies, or modulated frequency. They are aimed at the test object, and upon interaction with it, they would reveal needed information about the state of the object. They interact with the object in one, or more, of the following forms:

Scattering: Microwaves scatter from discontinuities or other boundary conditions can reveal important information.

Reflection: When the microwaves reflect from surface, the reflected signals can reveal flaws, and delaminations. This is fairly similar to the pulse-echo technique in ultrasound.

Radiation: Microwave can radiate/attenuate through solids, such as pavements, or soils. Information regarding composition of the interior of the test object can be revealed by measuring this radiation, or attenuation.

In addition to the above-mentioned uses of microwaves, they can also be used to measure thickness of objects, discontinuities, cracks/flaws, moisture, and material anisotropy. They can also be used in the more important stress-corrosion measurements; such an application is important for high strength cables (suspension bridges, cable stayed bridges, and prestressed/post-tensioned bridges), see Rollwitz (2005b).

7.4.4.5.2 Tools and Instruments

The tools needed for a microwave testing include (a) a microwave generator, (b) antennas (transmitting, and/or receiving), and (c) a phase detectors (receiver).

7.4.4.5.3 Advantages–Disadvantages in NDT Applications

The advantages of SMF method are as follows:

- No need for contact with test object
- No safety concerns
- Have many applications in civil infrastructure

The disadvantages of the method are as follows

- Results can be difficult to interpret
- Equipment sizes vary, and can be unwieldy

7.4.4.6 Applications in SHM

7.4.4.6.1 General

Electromagnetic NDT techniques can be utilized in many ways in SHM field. Table 7.13 summarizes some of the potential ways of these applications. Table 7.14 illustrates some potential uses of Electromagnetic NDT in detecting the effect of hazards.

7.4.5 PENETRATING RADIATION

7.4.5.1 Overview

Penetrating radiation is an attractive type of testing for civil infrastructure, but with severe limitations. Different issues that involve these classes of tests are shown in Figure 7.42. These tests involve subjecting the tested object to a type of radiation waves, as shown in Figure 7.41. The radiation waves will either pass through the object, or backscatter from the object. The passed or scattered radiation waves can then be interrogated for information regarding the test object. Generally speaking, there are three types of radiation waves that are used in testing: X-rays, gamma rays, and neutron beams. X-rays are generated by subjecting a metal object to high-energy electrons (Figure 7.41). This will generate photons that will propagate in a predetermined direction through the tested object. Gamma rays are generated from the nuclei of the atom. They are emitted from unstable radioactive materials, such as cobalt, cesium, or thulium. Neutron beams are generated from reactors or isotopes. They tend to produce better resolution than X-rays or gamma rays. They also penetrate deeper into solids than other methods. Table 7.15 shows the

TABLE 7.13
NDT Electromagnetic Utilization in SHM Civil Infrastructure Field

Method	Direct Sensing	Virtual Sensing Paradigm (VSP)	Remote Sensing	Local vs. Global—Scaling Issues	Cost–Benefit Issues
Eddy current (EC)	Yes	As of the writing of this volume, there is no body of work that relates any of these methods to damages within the object in a VSP mode	Even though all these methods do not require direct contact with the test object, they require close proximity. Thus, their usefulness in remote sensing applications is limited	Only local, due to the surface, or near surface, nature of the method	Average for local detection
Magnetic particle	Yes				Average for local detection. Covers larger area than the EC, but requires more effort in surface preparation
Static magnetic field (SMF)	Yes			Due to practical pre- and postpreparations issues, this method can be very difficult to scale from local to global SHM	Can be expensive for SHM applications
Microwaves/radar	Yes		Fairly suited to remote sensing	Given the radiation property of microwaves, this method has a potential of scaling local to global detection of state	For appropriate cases, this method can provide attractive cost–benefit value

TABLE 7.14
Applicability of Different Electromagnetic Techniques to SHM Hazards

Hazard	Applicable Technique	Comments
Fatigue	EC, microwave, magnetic particle, and SMF	All electromagnetic methods can detect fatigue cracks. The optimal method will depend on geometry, and extent of expected damage. Careful decision making analysis is needed to identify the optimal method
Corrosion—General Corrosion—PT/PS	Microwave inspection	Stress corrosion can be detected by microwaves, Rollwitz (2005b)
Scour	Microwave (GPR)	Ground penetrating radar might be suited to detect scour damage
Earthquakes	EC, microwave, magnetic particle, and SMF	All methods can be used locally to detect earthquake damage. Radar waves have been used remotely to inspect postearthquake damage, Housner and Masri (1996)
FRP wrapping FRP decks	EC	Limited on to carbon FRP
Bridge security	Microwaves	Radar waves might be used remotely to detect post-event damages

TABLE 7.15
Differences between X-Rays and Gamma Rays (Figure 7.43)

Issue	X-Rays	Gamma Rays
Nature during testing	Generated by X-ray machines during tests	Naturally emitted
Projected power	Vary	Constant
Generation	High-energy electrons impacting metallic objects and generating photons	Unstable radioactive isotopes
Useful life	Long	Limited by the half-life of the isotope
Wave length	See Figures 7.8 and 7.9	

basic differences between X-rays and Gamma rays from testing view points. Other types of penetrating radiation testing are beyond the scope of this document, the reader should consult Greene (2005), Emigh et al. (2007), and Bossi et al. (2007).

Penetrating radiation testing is attractive for NDT since the X-rays or the gamma rays can penetrate through the object, or backscatter from the object with minimal loss of intensity leading to accurate identification of small flaws in the object. Generally speaking, the intensity of X-rays can be controlled such that it can be used to test thicker and denser objects. Gamma-ray intensity is constant, so the size of the test object can be limited.

7.4.5.1.1 Tools and Instruments
The setup of penetrating radiation testing is fairly simple (see Figure 7.41). There are three basic components of the test: the radiation-generating devices, the testing object itself, and the colleting devices. In addition, some testing methods rely on computational processing components.

The generating devices for X-rays vary in size and portability. The stationary machines are suited for NDT of relatively small objects. However, they are not suited for in-site SHM projects. Portable X-ray machines can be used in site, however, safety precautions must be taken. The energy demands of X-ray machine also generally vary in the range of 150–450 kV. Higher energy X-ray machines can generate more than 450 kV. As mentioned before, gamma-ray machines uses constant energy input. They house the isotope (radiating material) inside a shielded, S-shaped housing. When needed for use, the isotope is pushed out of the housing for the prescribed period of time, then pulled back for safe keeping.

The objects that are suitable for penetrating radiation testing, in the field of civil infrastructure, can be steel, reinforced concrete, or FRP constructs. Both X-ray and Gamma-ray methods can be used with all of those materials. Types of damages that can be detected include voids, cracks, delaminations, rusting, thickness, or density.

After the radiating waves pass through the tested object, or scattered back from the tested object, the intensity of these waves is then collected and processed using different techniques. The oldest technique and most popular is to use radiographic film. The films are covered with thin layers of chemicals that changes when subjected to the photons emanating from X-rays. Such changes are proportional to the intensity of the photons. This intensity is function of the internal composition of the test object. If there are any flows, defects, or any other changes in the composition, such changes will affect the exposed film. Damages can be identified by observing the film. Radiographic film collectors are suitable for static tests: no time-dependent observations can be made.

CRT (TV screens) is also used to observe the intensity of X- or Gamma rays after passing through, or scattering back from, an object. Such a method is also called *radioscopy*. It has the advantage of measuring dynamic, even real time, tests.

In addition to the above visual collecting procedures, more complex analytical methods were developed to infer even more information about the composition of the tested object. These analytical methods process the digital information about backscatter, or passed radiation intensity, then compute more accurate representation of the internal composition of the tested object. The two main computational methods are computed backscatter (CB) and computed tomography (CT). Note

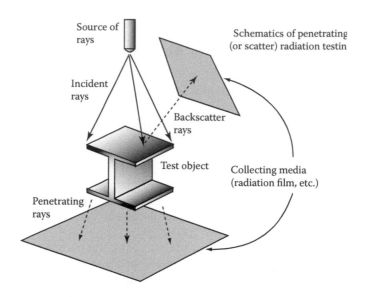

FIGURE 7.41   Elements of penetrating radiation testing.

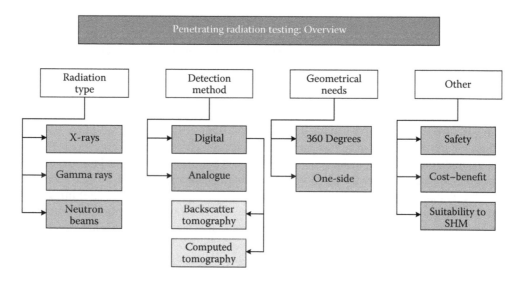

FIGURE 7.42   Penetrating radiation testing overview.

also that CT is sometimes known as computer aided tomography (CAT). These methods will be presented in more detail in the next section.

7.4.5.1.2   Advantages–Disadvantages in NDT Applications

Penetrating radiation testing has been used extensively for NDT applications. As usual, NDT applications involve laboratory testing or for spot testing in site. For those situations, penetrating radiation techniques offer unmatched accuracy for detecting small flaws, delaminations, or cracks. Changes in density detection accuracy make it ideal for detecting rusting or corrosion advances deep within reinforced concrete components. Using CT can result in a 3D representation of the whole object. In addition to the accuracy advantage, there are several other basic advantages to penetrating radiation technologies. Both X-rays and gamma rays travel in straight lines and are not sensitive to magnetic or electrical fields; yet they can be scattered at surfaces of objects or edges of flows.

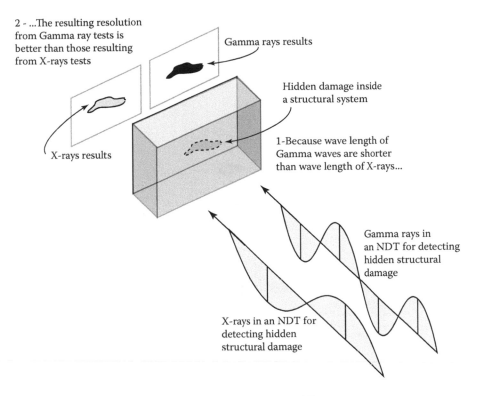

2 - ...The resulting resolution from Gamma ray tests is better than those resulting from X-rays tests

Gamma rays results

Hidden damage inside a structural system

X-rays results

1-Because wave length of Gamma waves are shorter than wave length of X-rays...

Gamma rays in an NDT for detecting hidden structural damage

X-rays in an NDT for detecting hidden structural damage

FIGURE 7.43   Comparison between X-rays and Gamma rays in NTD tests.

Unfortunately, such advantages come with steep price. CT, radioscopy, or radiographic films need access to at least front and back side of the test object. This imposes immense practical limitation for on-site testing. Another practical limitation is the size and portability of the projectors. Perhaps the most important limitation to this class of methods is the safety concerns. Heavy shielding from the radiating waves is needed to protect workers. All of those limitations add to large cost–benefit ratio.

7.4.5.2   Computed Tomography

Inside the laboratory or in situation when the radiation generator can rotate around the test object, CT can be used to generate a digital illustration of the state of the interior of the object in a given plane. By moving the plane of the radiation generator along the test object, a 3D image of the test object can be generated. The concept is shown in Figure 7.44. The concept of CT is based on the absorption of an incident X-ray beam with an intensity of R_0 after it penetrates an object to an intensity of R. The incident and penetrated intensities are related by Lambert's Law

$$R = R_0\, e^{-\mu r} \qquad (7.3)$$

With r as the traveled distance and μ as an attenuation coefficient of the test objet. The coefficient μ is a material property. For cross-sections that contain different materials, such as steel rebars and concrete, for example, the distribution of the attenuation coefficient in a cross-section is $\mu(x,y)$ that is a function of the Cartesian coordinates of that cross-section x and y. Equation 7.3 can be generalized to

$$R = R_0 e^{A} \qquad (7.4)$$

$$A = -\int_{r_0}^{r_1} \mu(x,y)\, dr \qquad (7.5)$$

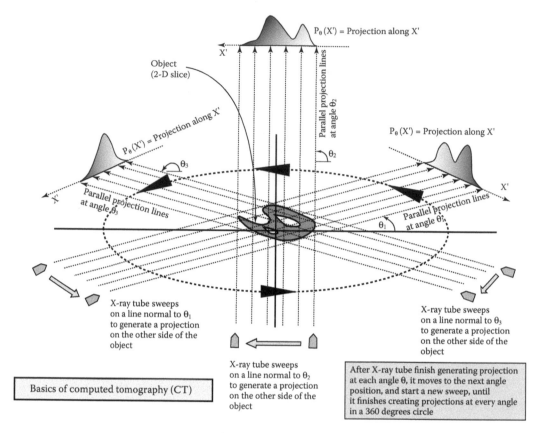

FIGURE 7.44   Basics of computed tomography (CT).

The limits of the integral in Equation 7.5 range over the point of entry and the point of exit of the beam. Operating on Equations 7.4 and 7.5 yield

$$P = \int_{r_0}^{r_1} \mu(x, y)dr \tag{7.6}$$

$$P = \ln(R_0/R) \tag{7.7}$$

The transform in Equation 7.6 of $\mu(x,y)$ is the Radon transform. The object of CT is to estimate $\mu(x,y)$ given the test results that produce P at different stations around the test object. Several analytical techniques have been proposed to estimate the transform of Equation 7.6. Among those techniques are direct polar Fourier transform (FT). It is reported that such solution does not produce accurate resolutions, Dennis (2005). More accurate resolutions for the transform can be obtained following filtered back projection techniques, which utilizes filtering of the measured intensities by processes of convolution; see Dennis (2005) for in-depth description of the method.

What concern us immediately is the potential uses of CT in SHM projects. Given the capabilities of resolving interior states of solid such as steel or concrete, the process should be of great interest. Since the radiation generator needs to revolve around the object, the method obviously is most applicable to linear-type geometries such as columns, truss members, and suspension cables. Two basic obstacles remain to be resolved:

- **Physical setup:** The mechanics of revolving of the radiation generator around the *in-situ* test object and recording the data accurately needs to be designed.
- **Safety:** Penetrating radiation in the field can produce safety concern.

As of the writing of this volume, the above two issues have not been resolved in a practical sense. However, we believe that the promise of CT method in SHM applications will prompt the resolution of the above obstacles in a practical and acceptable manner.

7.4.5.3 Computed (Compton) Backscatter

When an X-ray, or gamma ray hit a target, photons will scatter from the target with a scattering angle θ. This process changes the incident wave length of the incident waves λ_i into scattered wave length λ_s. This effect was discovered by Arthur Compton in 1923. In 1927, he earned the Noble prize in physics for his discovery. The relationship between the incident and scattered wave lengths are

$$\lambda_s - \lambda_i = \frac{h}{m_e c} (1 - \cos \theta) \tag{7.8}$$

Where h is Planck's constant ($h \cong 6.626 \times 10^{-34}$ J. s); it is a measure of energy. Note also that c is the speed of light. The wave length shift as predicted by Equation 7.8 is used to measure surface and internal voids and anomalies in the tested object.

Again, we are mainly interested in the potential uses of CB in SHM projects. Given the capabilities of resolving interior states of solids such as steel or concrete, the process should be of great interest. The other feature of the CB approach in penetrating radiation testing is that, unlike CT, it needs radiation exposure only from one side of the tested object, thus offering great practical advantage. Because of this, the method is applicable to most infrastructure geometries. Again, two basic obstacles remain to be resolved:

- **Analytical Accuracy:** The numerical resolution of backscattering problem for general, online, and in-field setups are not popularly available as of the writing of this volume.
- **Safety:** Penetrating radiation in the field can produce safety concern.

7.4.5.4 Applications in SHM

Penetrating radiation NDT techniques can be utilized in many ways in SHM field. Table 7.16 summarizes some of the potential ways of these applications. We only cover in-field, and near real time SHM projects in this table. Table 7.17 illustrates some potential uses of penetrating radiation in detecting effects of the hazards.

7.4.6 OTHER METHODS

7.4.6.1 Visual Methods

Visual inspection is the simplest and most straightforward method of NDT. It is based on observing the condition of the object of interest visually, and trying to detect its current state based on personal experience and existing guidelines, if any. Visual NDT methods should be an integral part of any NDT, or SHM project; they can save costs and time. They also can guide the progress of the project and help in adjusting its direction midway, if any adjustments are needed.

The advantages of visual method are as follows:

- Very simple and fast
- Can scale from local to larger areas without any additional efforts
- Can apply remotely

The disadvantages are as follows:

- Qualitative
- Might need special equipment for access and other reasons

TABLE 7.16
Radiation NDT Utilization in SHM Civil Infrastructure Field

Method	Direct Sensing	Virtual Sensing Paradigm (VSP)	Remote Sensing	Local vs. Global— Scaling Issues	Cost–Benefit Issues
X-rays, Gamma rays, and neutron beams	Difficult, given safety and physical sizes of infrastructure	VSP might be needed on a case-by-case basis if the testing is performed on isolated or localized components	Can be used to detect state of remote components. Safety concerns can be an issue	Safety issues can be of concern if applied to more than localized area	Cost–benefit issues need careful studies before embarking on an SHM testing
Computed tomography (CT)	Possible for linear-type geometries, such as columns, truss members, and suspension cables. Safety concerns must be addressed	Since the method produces 3D images of the interior state of the test object, no need for VSP applications.	Due to the need to revolve around the test object, remote sensing in traditional sense is not applicable	Can be designed to detect the state of whole components, such as whole column, or entire truss members	Can be cost effective if limitations are resolved
Computed backscatter (CB)	Can have numerous applications in SHM. Safety and numerical issues need to be resolved		CB is a remote sensing method, by default	Can be designed to detect the state of entire components, such as steel connections, or bearing setup	

TABLE 7.17
Applicability of Different RT Techniques to SHM Hazards

Hazard	Applicable Technique	Comments
Fatigue	CB	Film radiography can be difficult in-situ
Corrosion—General	CB/CT	Depending of geometry of reinforced concrete, CT can be used, if safety can be
Corrosion—PT/PS	CB/CT	assured
Scour	NA	Due to presence of soil and water, penetrating radiation is not practical
Earthquakes	CB/CT	After earthquake event, CB or CT can be used to assess damaged connections and welds
FRP wrapping	CB/CT	CB can be used to assess conditions of FRP wraps in most situations. CT can be used for columns wraps, since it is possible to access the column from all sides
FRP decks	CB	The most effective technique is CB, since it is difficult access the deck from all sides

7.4.6.2 Liquid Penetrant

This method is similar in concept to MP method. It is based on applying a liquid to the surface of the object being tested. The liquid will penetrate the cracks (as small as 0.001 mm in width). After the excess fluid is removed, a suitable powder coating is applied to the surface. If there are any cracks in the surface, the residual liquid will interact with the powder coating and becomes visible causing the crack to be detected.

There are several types of liquid penetrant, such as florescent and/or visible types. They can be removed either by washing them (water), or by using solvents. The choice of the penetrant type depends on several factors such as size of test area and the access. More details of the methods is presented by Tracy (2007).

The advantages of liquid penetrant method are as follows:

- Very simple and quick to detect surface defects
- Applicable to most engineering materials
- Can be adapted to large as well as small surfaces

The disadvantages are as follows

- Surfaces need to be within reach
- Needs surface preparation before and after the test
- Inside of cracks needs to be clean, for example, corrosion free
- Porous materials need special procedures
- Only surface defects
- Might produce false indications of cracks for rough surfaces

7.4.6.3 Optical Methods

There are numerous NDT optical methods. We already discussed some laser-based techniques earlier. Similar to the fiber optic sensors (Chapter 5), optical NDT methods are mostly based on optical interferometry. Generally speaking, interferometry concerns itself with the interaction of two waves, in both amplitude and phase. Interferometer is the tool that controls such an interaction. When the resultant of the interacting two waves is known, it is possible to discern the attributes of one of the original waves. At its simplest mode of behavior, when the original two waves are in phase, they will reinforce each other; when the two waves are out of phase, they will cancel each other. Recall that Fabre Grating Fiber Optic sensors operate on this principle, as explained in Chapter 5. By using interferometer principles, state of strains (deformations) and displacements at the surfaces of the test object can now constructed both visually and quantitatively. There are three basic NDT optical methods: Grid, holography, and shearography, see Botsco and Jones (2007). Generally speaking, there are two major differences in the attributes of the three methods: first difference is in the way the two light signals that are to be interrelated are generated. Second difference is in the complexity of the setup.

Moiré Method: Two light signals are split, then are beamed the object. Upon reflection, their interferences are analyzed and surface displacements and strains are computed. See Post et al. (1994) for more details.

  Holography: This type of optical NDT is based on applying a laser beam (reference beam) to the test object. The reflected light from the object (called *object beam*) is then compared. with reference beam. The two beams would show information about the object, including surface displacements and strains. This information is revealed in real time. This general holographic technique would yield a low-resolution images. For higher resolution, holographic interferometry is used. In such a situation, the images are generated from the interference of the two wave fronts emanating from the deformed object, instead of the two simple laser beams. This would result in high resolution of strains and displacements. For more details on this subject, see ASNT (1995).

  Shearography: The setup in shearography is simpler than the other optical methods. Only one beam is directed at the test object. Upon reflection from points at different surface coordinates, those reflected signals are then sheared using a birefringent crystal so as to make them nearly collinear. The resulting interference patterns between these two collinear signals can produce the desired surface displacement or strain images. Generally speaking, shearography does not require special environmental requirements that are needed to generate the reference beam in holography. It is also known to produce higher resolution (sensitive) results, see ASNT (1995).

The advantages of optical methods are as follows:

- Remote sensing
- High resolution
- Scales from local to global detection
- In general, faster than other comparable NDT techniques
- Faster in application than similar ultrasound methods
- Can be portable

The disadvantages are as follows:

- It may not be possible to detect interior defects
- Direct line of vision required; hidden, or interior surfaces can't easily be tested
- Fairly recent and thus, body of knowledge is still being developed

7.4.6.4   Applications in SHM

Different NDT techniques discussed above can be utilized in many ways in SHM field. Table 7.18 summarizes some potential applications.

TABLE 7.18
Use of Different NDT Methods in SHM Civil Infrastructure Field

Method	Direct Sensing	Virtual Sensing Paradigm (VSP)	Remote Sensing	Local vs. Global— Scaling Issues	Cost–Benefit Issues
Visual methods	Qualitative results	Due to qualitative nature of the method, VSP utilization is not too promising	Visual inspection can be done remotely, either by naked eye, or by optical tools	Can visually inspect both local and global zones/ objects	Inexpensive. However, given the limited benefit, the cost–benefit rations may not be too effective
Liquid penetrant	Surface cracks. Mostly qualitative		Remote sensing is not consistent with this method	Can apply the penetrant both on limited and large areas	Limited expenses. For limited goals, cost–benefit can be attractive. For general SHM project, the qualitative nature of the method limits its cost–benefit attractiveness
Optical methods	Surface and subsurface stains. Can also reveal displacements	Holds great potential. For example, STRID using boundary element method instead of finite element method	Yes, by default	Yes, by default. Local zones need to be in the direct line of sight, not hidden	Cost–benefit can be very attractive for SHM projects, if decision making analysis is performed appropriately

7.5 ACOUSTIC EMISSION

7.5.1 OVERVIEW

The concept of AE is fairly simple: damage events in materials and structures generate stress waves that propagate throughout the structure. The properties of these stress waves are dependent on the properties of damage as well as the properties of the structure. These stress waves are called *AE*. By monitoring the AE and analyzing the AE signals, information can be gathered about the damage as well as the state of the material and the structure.

The frequency range of AE signals depends on the kind of triggering events. They can be in the range of 1.0 Hz for example, if we consider ground seismic motions as a form of AE. However, in the field of NDT, the frequency range of AE is similar to other UT waves (20 KHz ~ 1.0 MHz). There are numerous differences between ultrasound and AE methods. One of the basic differences is that AE is a passive NDT method; it relies on listening to damage signals. Ultrasound methods, on the other hand, are an active method; it generates sound signals and monitors the interaction of those signals with the test object.

This section will explore theoretical and practical attributes of AE in NDT field. We then will discuss the potential applications of AE in SHM field. Several examples will be presented. Many of the examples are from fields other than civil infrastructure; however, we believe that those applications can be utilized in civil infrastructure with minimal additional research effort. Figure 7.45 shows some basic components of AE field.

7.5.2 THEORETICAL BACKGROUND

7.5.2.1 Basics of AE

Acoustic emission testing relies on observing AE signals. These are elastic wave forms that propagate in the frequency range of 20 KHz to 1.0 MHz. By definition, AE waves are those waves resulting form energy release (due to nonlinear deformations in material). As such, earthquake motions can be considered as a form of large amplitude AE. AE also can have very small amplitude. For example, corrosion processes in reinforced concrete can be considered as low amplitude AE, with amplitude in the range of 30 dB*. Figure 7.46 shows some of the basic concepts of AE that will be discussed next.

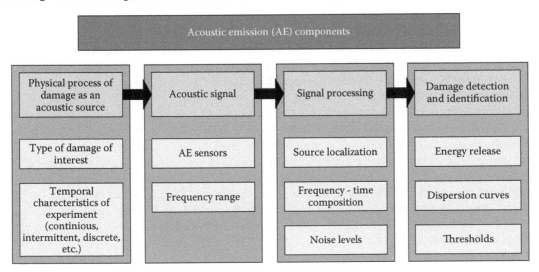

FIGURE 7.45 Components of acoustic emission.

---

* dB = 20 log (A_1/A_0), with $A1$ as the measured signal and A_0 as a suitable reference baseline.

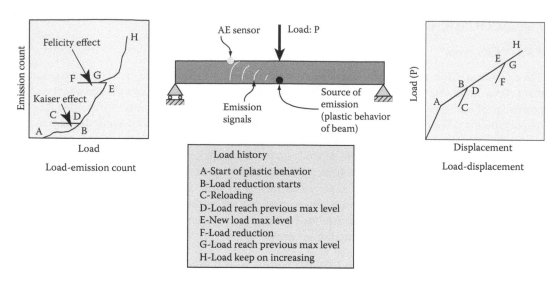

FIGURE 7.46   Basics of acoustic emission.

7.5.2.2   Kaiser, Dunegan, and Felicity Behavior of AE

The load-emission count relationship is such that as the loading increases, the emission count increases. However, there are several effects that can be observed during the loading and unloading of a system. Three main effects govern the AE behavior, and understanding those effects can help in performing an accurate and efficient AE experiment.

Kaiser Effects: Kaiser effects happen when materials start behaving in a nonlinear form (a form of failure, or damage), as it starts emitting stress waves that originate from the damaged location. The emission will stop if the loading stops, or if there is a reduction of the load level. The emission will only start again when the load exceeds the previous loading level. For example, in Figure 7.46, when the loading P increases, the emission count increases, as shown from pint "A" to point "B." If the load decreases to point "C," the emission count stays constant. The emission count will remain constant when the load increases again, until it reaches the previous level at point D = B. If the load increases beyond that, the emission count starts increasing again.

   Dunegan Corollary: Dunegan et al. (1970) noticed that in many situations the Kaiser effects need to be modified. Emission can occur when the load is kept constant, which appears to be a deviation from Kaiser effect. The discovery of this corollary led to the successful applications of AE in many field NDT proof tests, Miller (2007).

   Felicity Effects: At higher stress levels the Kaiser effect does not hold true. In such a situation, emission can occur at load levels that are lower than the previous maximum load level. For example, Figure 7.46 shows that if at a higher load level, "E," the load decreases to level "F," and the emission count will remain same. When the load starts increasing again from "E" to "G," and the emission count remains same. However, at level "G" that is less than the previous maximum load level "E," the emission count start increasing. Such an effect is called "Felicity Effect" The Felicity ratio is defined as P_R/P_M with P_R as the load level at which the emission count start increasing and P_M as the maximum previous load level. Obviously, if the Felicity ratio is 1.0, we have Kaiser effects.

   Felicity effects occur in time-dependent material behavior (such as FRP and concrete). Several civil infrastructure situations can experience Felicity effects such as Hydrogen embrittlement (high strength tendons), corrosion (reinforced concrete rebars), and FRP bridge decks. For more detail, see Miller (2007). The effects have been incorporated in standard load tests and were devised to generate AE signals in different structures such as for FRP pressure vessels, and bucket truck and lift inspections. We observe the many similarities between the proof testing of those situations and the proof testing of bridges, (see Chapter 8 by Ettouney and Alampalli 2012). On the basis of those similarities, and the success of AE in providing satisfactory proof testing methodology, it seems that developing a rigorous AE proof testing guide, or even standard, for bridge proof testing is a worthy future effort for the bridge community.

7.5.2.3 Emission Initiation and Propagation

Emission is initiated when energy is released from a local source due to the state of stresses or deformations within the material. The emission from the stressed source behaves as a pulse. This pulse will cause elastic waves to propagate throughout the body in all directions. Depending on the shape of the damage, some preferred propagation paths might be followed. Preferred propagation paths might also occur in inhomogeneous materials such as FRP. The stress wave propagation in an AE situation will follow conventional rules of wave propagation (Rose 1999).

There are two types of AE waves: burst and continuous. Burst AE, as the name implies occur when the energy is released suddenly in a form of pulse in short time span. It occurs during proof load tests, for example. Continuous AE signals occur when the damage in the material occur continuously, as in corrosion processes in metals. We explore examples of both types later.

7.5.2.4 Important Characteristics of AE Wave

There are several important characteristics of AE signals: amplitude, frequency content, and rate of decay are the most important. However, the overall wave forms of AE signals have been used to discern valuable DMID information: signal processing of AE signals is discussed later this section. See Figure 7.47 for illustration of AE wave form characteristics. In addition to the above AE wave forms attributes that are shared with many other wave forms (such as seismic, impact-echo, etc.), there are specific AE wave forms attributes that are used in AE experiments such as the following:

AE Counts: The number of passages of amplitude over a given threshold during an AE signal is called *AE count* (see Figure 7.47). Clearly, the AE count can be used as a measure for the strength of an AE signal: the larger the count, the stronger the signal. The relationship between elapsed time and AE count is used extensively in AE test. In addition to revealing the strength of AE signal, it can also reveal the overall condition of the damage in the test object. If the rate of increase of the AE count is high, it indicates a more severe damage state than slower rate (Figure 7.48).

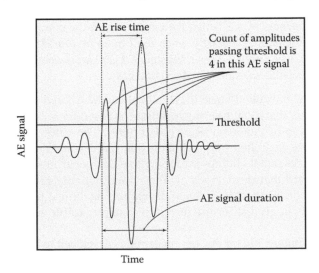

FIGURE 7.47 Acoustic emission signal characteristics.

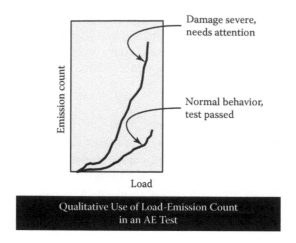

FIGURE 7.48 Use of AE count.

AE Energy: AE energy is measured by RMS (root mean square) or TMS (total mean square, which is the square of RMS) of the signal. It is a measure of the strength of the signal. Similar to AE counts, its rate can also be used to qualitatively infer the level of damage, or the lack of it in the object. Using AE energy as a metric is particularly useful for continuous AE Tests, which would be valuable for SHM experiments. In general the resulting energy measures are smoother results than the counting technique. Energy results are also less sensitive to the choice of threshold value; which is another advantage of using AE energy in SHM tests.

AE Signal Duration: Signal duration is evaluated as in Figure 7.47. It can be used as a measure of the strength of AE signals.

AE Transforms: Signal processing using FT, WFT, WT, or HH techniques can be used to infer damage presence, location, and extent. In general, those techniques are limited by the sampling rate of the signals. Thus only frequencies up to 5 MHz can be analyzed.

More details on AE signal attributes and their usefulness during NDT/SHM experiments are discussed by Pollock (2005).

7.5.2.5 AE Data Filtering

One of the important features of AE is that the emission acoustic energy can be extremely small. This can be a severe limitation, especially when used in an *in-situ* SHM experimentation where the background noise can interfere with the AE signal. There are several practical solutions to this limitation. Two of the popular solutions are discussed next.

Swansong Filters: The unwanted noise in AE signal can be filtered analytically by Swansong filtering technique. This analytical filtering technique is based on a set of rules that were developed by the Association of American Railroads (1999). The rules are based on eliminating unwanted noise that are generated by specific outside sources that are common in the railroad community such as mechanical sliding or rubbing that can occur in rail cars. A research study by Endo (2000) reported that those types of filters were not very effective in removing noise from moving vehicles on bridges. The idea of developing analytical filters to remove unwanted noise from AE signals in civil infrastructure, for example, traffic noise, should be explored further.

Guard Sensor Technique: Guard sensors concept was developed to help the AE sensors record only relevant signals. The concept is shown in Figure 7.49. Carlyle and Ely (1992), Carlyle and Leaird (1992), and Carlyle (1993) successfully used guard sensors in AE bridge testing to filter unwanted noise from AE signals. Guard sensors can be incorporated in an AE sensor network.

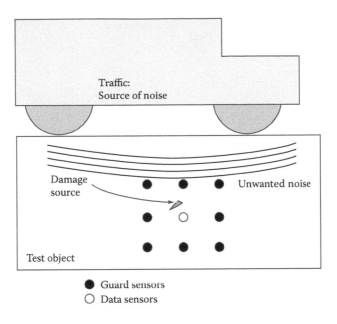

Guard sensors

Data sensors

FIGURE 7.49   Guard sensor concept.

A careful study of potential sources of noise and potential crack or damage locations can help in optimizing the performance of an AE-based SHM experiment.

7.5.2.6   Calibration of AE Sensors

Sensors need to be calibrated prior, or during the experiment. A popular calibration technique for AE sensors and experimental setup is the pencil lead breaking technique. It was found out that the fracture of pencil lead produces AE signals that are similar in amplitude and frequency range to the process of fracture in many materials, such as FRP material. Thus, the AE sensors can be calibrated by this process, which is also known as Hsu-Nielsen pencil technique.

7.5.2.7   Source Location

One of the advantages of AE method is its capability to locate damage locations accurately. When the AE signal has a definitive arrival time, as in the burst mode AE, for example, the damage location can be estimated in linear, 2D, or even 3D geometries. The process is based on the well known time of flight (TOF) technique. For example, consider the linear situation in Figure 7.50, where the source of the damage is located on the same line between two AE sensors, S_1 and S_2, that are spaced with a distance L. Assume that the arrival times of the AE signals were recorded as T_1 and T_2 at the two sensors. The distance between the damage and sensor S_1 can be shown to be

$$L_1 = \left(\frac{(T_1/T_2)}{1+(T_1/T_2)} \right) L \tag{7.9}$$

The distance from the damage source to sensor S_2 can be computed in a similar fashion. The linear geometry has an important application in the field of high strength wires, such as suspension bridges, cable stayed bridges, or post-tensioned/prestressed bridge girders. A well-positioned AE sensors along the cables can detect wire breakage (due to deterioration), and can also locate the damage source fairly accurately. A more involved, but similar expressions can be developed for 2D and 3D geometries. The 2D geometry is shown in Figure 7.51. Note that at least two AE sensors are

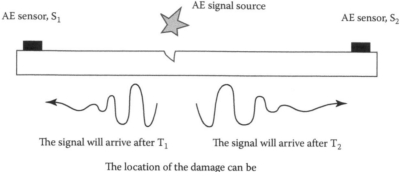

The location of the damage can be
estimated from different times of arrival

FIGURE 7.50   Linear setup for AE damage location.

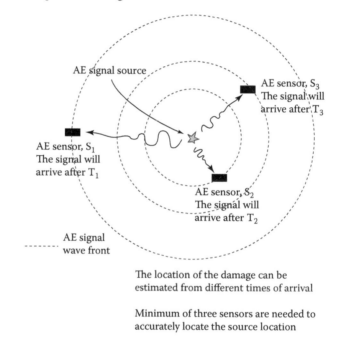

The location of the damage can be
estimated from different times of arrival

Minimum of three sensors are needed to
accurately locate the source location

FIGURE 7.51   2D setup for AE damage location.

needed for linear geometries. A minimum of three AE sensors are needed for 2D geometry. For 3D geometry, a minimum of four AE sensors are needed.

When the AE signal is continuous, as in corrosion processes, TOF method is not appropriate. Time averaging techniques are used to estimate the damage location using the zone location method, Baron and Ying (1987). The method is based on observing the AE sensor recording the largest observed AE amplitude A_{AE_0}, within a given AE sensor array. It is logical to assume that the source of damage is closest to that sensor. Assuming a reasonable attenuation rate can place the damage on the circumference of a circle with that sensor at its center. The source location can be narrowed by observing the sensor with the next highest amplitude A_{AE_1}. Using the same attenuation rate, the damage source can be pinpointed using technique that is similar to the TOF technique. More involved techniques for continuous AE signals can reveal even more information. For example, a cross-correlation evaluation (Bendat and Piersol 1971) between the two continuous signals can reveal the time difference between the two signals. When the time differences are know, the TOF method can be used to estimate the source location. A coherence analysis between the two signals can also be made to detect the presence of damage. For more detail see Baron and Ying (1987).

The simplicity of evaluation of damage source location, both for burst and continuous AE modes, using AE can be utilized so scale real time damage detection in an SHM setting using an array of AE sensors that are optimally located (both in number and location).

7.5.2.8 Properties of AE as Related to SHM

Civil infrastructure SHM projects have been gaining popularity lately. For AE to be successful in any SHM project, several of its attributes must be utilized efficiently. Table 7.19—AE attributes in civil infrastructure SHM—shows some of the special features of AE phenomenon, and how they can be used in civil infrastructure.

7.5.3 AE versus Other NDT Methods

AE method and its underlying physical behavior are very different from other NDT methods, as shown in Table 7.20. Since the UT and AE methods operate mostly within similar frequency range, it is of interest to compare the two approaches, as shown in Table 7.21. Finally, Table 7.22 shows a general comparison between AE method and the popular DMID vibration-based technique.

Duke and Horne (2006) provided an example of the effectiveness of AE in identifying the existence and location of the crack in metal components of bridge structures, and in detecting if the crack is actively growing. Hence, monitoring AEs from the formation and growth of cracks in steel commonly used for bridge structures was reviewed by the authors. The emphasis of the review was on cracks that penetrated the thickness of the plates, commonly used in bridge girders, and propagating in the plane of the plate. Figure 7.52 shows the relative amplitude of AE signals due to mechanism responsible for AE during material deformation. Figure 7.52 also shows that all detectable AE can not be immediately attributed to the final generation of new crack surfaces.

AE wave forms were collected during the overload tests of specimens containing cracks introduced by fatigue in A36, 1018, A588, HPS70W, and A514 steels, with each specimen overloaded numerous

---

TABLE 7.19
AE Attributes in Civil Infrastructure SHM

Acoustic Emission (AE) Issues	Effects in SHM
Noise Levels	Traffic and other ambient noise will need special considerations in an AE testing, such as guard sensors, or special types of numerical filters. Noise effects become important for continuous AE signals such as AE generated by corrosion processes
Distance from source—attenuation	When sources of damage are known, then AE sensors can be located near those sources; this is an NDT-type experiment. In an SHM type experiment, where the sources/locations of damage are not known, or when it is desired to have a continued monitoring of the structural state; several AE sensors must be used. The distances between those sensors must be optimized so as the AE signals will not attenuate be for they reach at least one AE sensor
Damage localization	AE sensors need to be close to each other to detect damages reliably
Frequency range	AE sensors need to cover as wide frequency range as possible. This is needed to accommodate as many types and sizes of damage sources as possible
Passive vs. Active AE Testing	AE is a passive type of testing: the sensors detect the signals emitted from damage sources. Some situations pulsers might be used to send acoustic pulses throughout the test object. Those acoustic signals will interact with defects and their reflections, or scattering from the damage sources can then be detected by AE receivers; this is similar to pulse-echo technique in ultrasonic testing
	A summary of the difference between active and passive emissions was given by Butt and Limaye (2002). They described both approaches as damage identification methods

TABLE 7.20
AE versus Other NDT Methods (General Comparison)

Issue	AE	Other NDT
Role	Listener (passive)	Initiator (Active)
Scope	Localized, or whole structure	More localized
Main limitation	Background noise	Complex geometries

TABLE 7.21
AE versus Ultrasonic

Issue	AE	UT
Physics	Stress waves	Stress waves
Frequency range	Higher	Lower
Techniques	Listens to waves created from anomalies	Generate pulses into the media, then looks for responses
Sensors	Use multiple sensors to localize the damage effects	Can use single or multiple sensors
Detects	Location, size of damage	Location, size of damage
Accuracy	Less	More

TABLE 7.22
AE versus Vibration-Based Testing

Issue	AE	Vibration-Based DMID
Physics	Propagating waves	Standing waves
Frequency range	20 kHZ –1.2 MHz	Up to about 100 Hz
Techniques	Listens to waves created from anomalies. Emphasis on analyzing wave forms and relating them to damage	Measures displacements or accelerations. Emphasis and generating natural frequencies and mode shapes
Issue	AE	GV
Sensors	Use multiple sensors to localize the damage effects. Higher the number of sensors, higher the possibility of damage detection	Can use single or multiple sensors. Higher number of sensors does not necessarily mean higher number of detected mode shapes
Detects	Location, size of damage	Useful for STRID, Limited DMID, value

times at different crack lengths. On the basis of these studies they concluded that the amount of AE detected were insufficient to provide a reliable means of monitoring crack extension.

7.5.4 APPLICATIONS IN SHM—GENERAL

7.5.4.1 General

Constructed infrastructure includes nuclear structures, bridges, (including special types of bridges such as suspension, cable stayed, or movable bridges), pressure vessels, dams, tunnels, pipelines, transmission towers, and storage tanks. AE testing has been used extensively in most of those infrastructure. In this section we look closely at issues that involve the use of AE in civil infrastructure such as bridges, tunnels, or buildings.

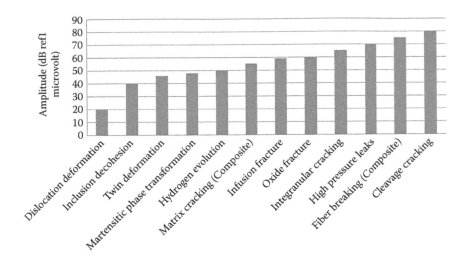

FIGURE 7.52 Mechanisms responsible for AE during material deformation and their typical AE signal amplitudes. (Courtesy of Physical Acoustics Corporation.)

7.5.4.2 Types of AE Tests in Infrastructure

There are two modes of AE behavior that can be utilized in civil infrastructure's DMID efforts. They are burst and continuous AE modes as shown in Figure 7.53. The main difference between the two modes is in the temporal behavior of the material. In burst mode AE, the energy is released in bursts or pulses as the material undergoes plastic deformations. Burst mode AE can be with high amplitudes, for example, when a high tension cable in a post-tensioned beam breaks suddenly, the AE signal can be very large. Lower amplitude burst AE modes can also occur in proof tests of bridges. Proof tests involve increasing the loading demands until a preset loading level. If the bridge undergoes a plastic deformation, a well-placed AE sensors can detect those deformation effects. Also, when bearing plates move against each other, thus generating friction, the friction energy would produce a burst mode AE signals. Continuous AE signals occur when the changes in material are continuous, such as corrosion process, or continuous scour process under bridge foundation.

7.5.4.3 Applications in SHM

Acoustic emission techniques can be utilized in many ways in SHM field. Table 7.23 summarizes some of the potential applications. Table 7.24 and 7.25 presents specific potential uses of AE in the field of civil infrastructure.

7.5.5 Corrosion Rate Measurements (Continuous AE)

Corrosion is one of the costliest hazards in the field of infrastructure. Detecting corrosion rates or corrosion damage in reinforced concrete, or steel structures is always an issue in SHM/structural health in civil engineering (SHCE). It has been observed that corrosion as a chemical process can generates AE signals. The AE due to corrosion is a good example of continuous AE; since corrosion is a continuous process, it is reasonable to assume that the resulting emission is also a continuous signal; thus, all techniques and procedures of continuous AE might be used to detect corrosion.

Riahi and Khosrowzadeh (2005) tried to answer the question: can AE methods detect corrosion rate in steel? They devised a test that induced corrosion process using different solution molarities. The sampling AE frequencies during the test were in the range of 100–400 KHz. One of the problems they tried to address during the test was the relative ambient noise level and the actual AE amplitude level. For a successful AE detection, a good compromise has to be made between noise at low frequencies and fast attenuation at high frequencies. The results of the test showed that the

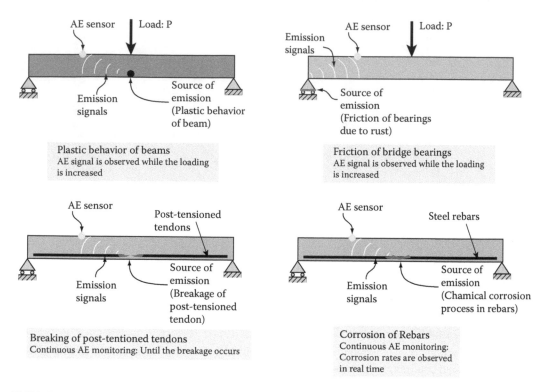

FIGURE 7.53   Modes of AE in civil infrastructure.

TABLE 7.23
Acoustic Emission Utilization in SHM Civil Infrastructure Field

AE Method	Direct Sensing	Virtual Sensing Paradigm (VSP)	Remote Sensing	Local vs. Global—Scaling Issues	Cost–Benefit Issues
Continuous AE	Yes	Relationships between AE count, amplitude, or energy and damage needs to be developed. For example, correlation between AE count and corrosion rates	No: Sensors need to be in direct contact with the test object	Can be local or global, depending on the number and distribution of AE sensors	For a well-planned experiment, benefits to cost ratio may be attractive
Burst AE	Yes	Relationships between AE count, amplitude, or energy and damage needs to be developed. For example, correlation between AE signals, dispersion curves and wave velocities	No: Sensors need to be in direct contact with the test object	Can be local or global, depending on the number and distribution of AE sensors	For a well-planned experiment, benefits to cost ratio may be attractive

TABLE 7.24
Applicability of AE Technique to Selected Problems in Infrastructure

Hazard	Comments
Yielding of materials	When materials reach post-elastic state, AE are generated. Thus, any scenario at which the material reaches or operates at yielding or post-yielding state, AE method of testing offers a reasonable choice. Such scenarios include postseismic behavior, soil scour due to flooding, fatigue, or impact conditions
Cracking of concrete	When concrete cracks, either by design or as a result of excess deformation, such cracks cause AE. When such AE is detected, it can offer insight on the causes and nature of these cracks
Friction movements	Many components in civil infrastructure are designed to move against each other, such as movable bridge bearings. Some components are not designed to move relative to each other, such as bolted or riveted connections or plate girders. AE resulting from relative movements, whether by design or as a result of malfunction, can be recorded. A decision as to the nature of such AE (within design limits, or a nonacceptable design condition) can help the system to operate safely
Welding	State of welding has been inspected by AE in many industries as well as in the civil infrastructure field
Failure of matrix or resin in composite materials	State of matrix or resin has been inspected by AE in many industries as well as in the civil infrastructure field

TABLE 7.25
Applicability of AE Technique to SHM Hazards

Hazard	Comments
Fatigue	A very popular application of AE testing. The sensors can be temporarily or permanently placed. Also the experiment can be either local of global, depending on the objective of the test
Corrosion—General	Corrosion process itself in the wires might be monitored by AE (see example below)
Corrosion—PT/PS	A popular application of AE is to detect breakage of PT/PS wires. Corrosion process itself in the wires might be monitored by AE (see example below)
Scour	AE was shown to detect changes and deformations in soils and rock as they occur, (Koerner et al. 1981 and Koerner and Lord 1982). AE might be useful to detect emission from soil erosion due to rapid flow of water, however, ambient flow noise might be of concern
Earthquakes	Can be useful to detect postyield deformations of systems after an earthquake. The governing behavior in such a situation would be Dunegan Corollary
FRP wrapping	Can be useful, see Karbhari et al. (2005)
FRP decks	Can be useful, especially for proof testing of deck
Bridge security	Can be useful to detect post yield deformations of systems after a blast event. The governing behavior in such a situation would be Dunegan Corollary

mean AE amplitude due to corrosion process is about 40 dB, and thus they decided to use a threshold of 26 dB, which is well above ambient noise level. Figure 7.54 shows the relationship between count rates, corrosion rates, and molarity level obtained during their test. Figure 7.55 shows a correlation between the AE count rate and corrosion rate. The test proves the following:

- There is a clear correlation between AE count rate and corrosion rate
- The ambient noise level is well below the threshold count rate

This test shows great promise for using AE in real time monitoring of corrosion process for infrastructure. But, significant further research is required to make it useful for structures such as bridges.

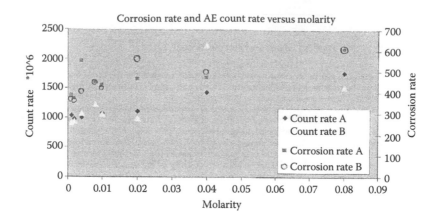

FIGURE 7.54   AE count, molarity, and corrosion rate. (Reprinted from ASNT Publication.)

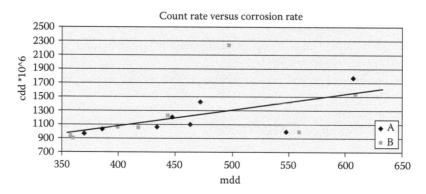

FIGURE 7.55   AE count versus corrosion rate. (Reprinted from ASNT Publication.)

7.5.6   PROOF LOADING (BURST AE)

A complete DMID involves identification of four damage attributes: occurrence, location, type, and size. The experiment reported by Carlos et al. (2006) showed how AE can be used to quantify all four DMID attributes in real time, *in-situ* situation. The experiment was performed on a full size storage vessel shown in Figure 7.56. The AE sensors were placed on a regular grid around the circumference of the vessel. As the vessel is pressurized in a fashion similar to a proof loading setup, the damages in the vessel shell started to emit signals in a burst mode similar to that in Figure 7.57. A special processing software was used (Figure 7.58). The software used signal processing techniques, such as wavelet transform (WT) analysis, and theoretical wave dispersion relations to detect the damage sources of the signals. A built-in discrimination logic that correlated type and size of damage sources with the detected AE signals, shown in Figure 7.59, was then used to completely identify damages on the vessel skin. The identified damages are shown in Figure 7.60.

7.5.7   QUALITY CONTROL (QC)

Another potentially important application of AE in the field of SHM/SHCE is QC and acceptance criteria. It was shown that the behavior of AE count rates can reveal differences in material composition, Pollock (2005). For example, when steel specimens with different degrees of annealment (underannealed, optimally annealed, and overannealed) are subjected to a monotonically increased loading, the AE energy rate behavior changes considerably, as shown in Figures 7.61, 7.62, and 7.63, respectively. This suggests that a carefully designed AE experiment can reveal the adequacy of in-service structural

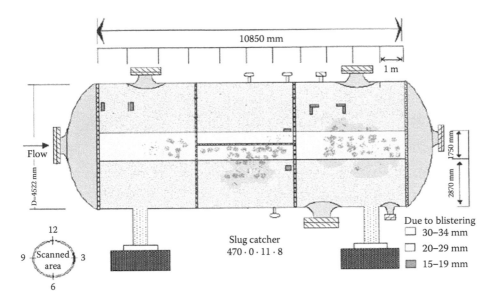

FIGURE 7.56 Pressure vessel with damage locations. (Courtesy of Physical Acoustics Corporation.)

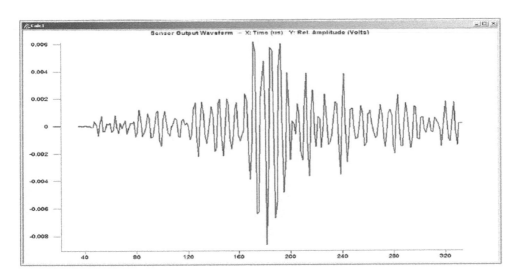

FIGURE 7.57 Typical AE signal. (Courtesy of Physical Acoustics Corporation.)

component. For example, the in-service behavior of gusset plates in a steel truss bridge can be examined without having to rely on numerical analysis. This becomes even more informative if the gusset plate is suspected to have a potentially brittle behavior that numerical analysis can't reveal.

Another potential QC application of AE would be in the emerging field of FRP bridge decks (Chapter 6 by Ettouney and Alampalli 2012). The body of experience of AE in the behavior of constructed FRP systems, such as pressure vessels, is vast. Utilizing this body of knowledge for load testing of FRP deck seems to be logical and thus further research is required in this area.

7.5.8 Incipient Failure Detection (IFD)

The real time capability of AE to detect, locate, and in sometimes size up, damage offer a unique potential application in the SHM/SHCE field; the potential of detecting incipient failure of global

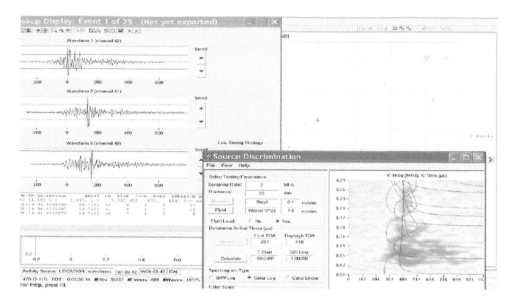

FIGURE 7.58 Processing software. (Courtesy of Physical Acoustics Corporation.)

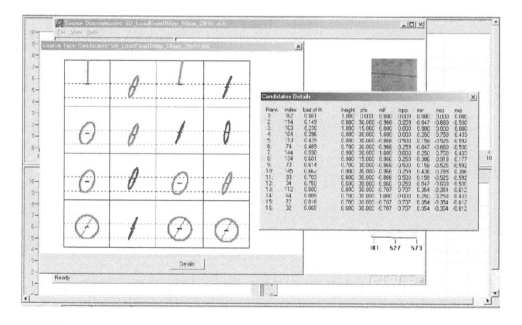

FIGURE 7.59 Damage discrimination screen. (Courtesy of Physical Acoustics Corporation.)

systems, or system components. This capability of AE was utilized in other fields. For example, by monitoring AE signals of pump discontinuities, signs of malfunction and ultimate failure were evident long before similar signs of failure were observed by monitoring lower frequency behavior, (see Figure 7.64).

The concept of IFD in civil infrastructure is an attractive one, and can save costs and improve safety. Recall that we explored monitoring the relationship between the stability condition of a truss compression member and its natural frequency. We showed that real time monitoring of compression members can reveal if there is an incipient loss of stability, which can lead to a catastrophic collapse of the truss. The concept can be generalized into even a higher frequency range; the frequency

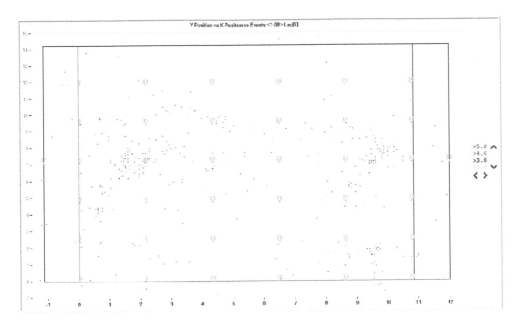

FIGURE 7.60 Damage locations during test. (Courtesy of Physical Acoustics Corporation.)

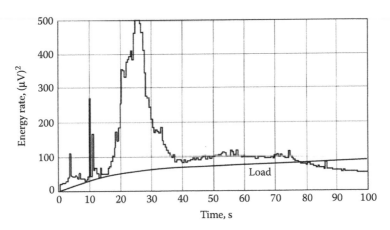

FIGURE 7.61 AE in QA of materials/construction: suboptimal composition. (Courtesy of Physical Acoustics Corporation.)

range of AE. Thus, the concept is, at least, logically feasible. This concept of monitoring AE signals at high frequencies in real time and trying to use potential early warning signs of incipient failure of either structural components, or whole structural systems needs to be researched fully: the potential rewards for such research, if successful are immense.

7.5.9 FAILURE OF HIGH STRENGTH WIRES

High strength wires are used extensively in civil infrastructure. They are used for prestressing/post-tensioning of reinforced concrete components in buildings, as well as bridges. High strength wires are also used in suspension and cable stayed bridges. Condition assessment of high strength wires has been of great interest to officials due to higher costs of retrofit and/or replacement. AE has been used successfully to monitor the event of breaking of wires. When a post-tensioned wire (or any high strength wire) breaks, the energy of the event generates burst-type AE signals that travel along

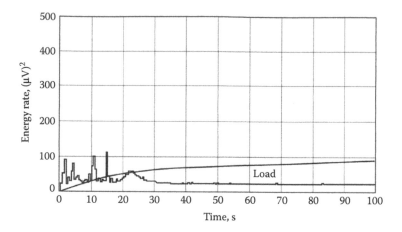

FIGURE 7.62 AE in QA of materials/construction: optimal composition. (Courtesy of Physical Acoustics Corporation.)

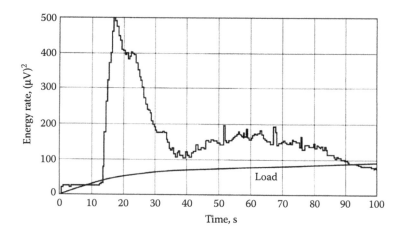

FIGURE 7.63 AE in QA of materials/construction: overoptimal composition. (Courtesy of Physical Acoustics Corporation.)

the wires for long distances due to the high intensity of the wire breaking. If AE sensors are placed along the wires, they can record the signals and through a simple TOF procedure (generally linear along the wire) the location of the break can be accurately computed.

Cullington et al. (2001) reported on a real time SHM project that utilized AE sensors to monitor post-tensioned wire breaking in real time. AE sensors were placed along the post-tensioned rail viaduct of Figure 7.65 as shown in Figure 7.66. When a wire breaks, the AE recorded signal is shown in Figure 7.67. The FT of the signal is shown in Figure 7.68. When a wire break is detected, it is easy for the official in charge to make appropriate decisions within reasonable time, thus ensuring safety of the system under consideration.

7.5.10 LOCAL AREA MONITORING (LAM)

Carlos et al. (2000) described the concept of LAM, and presented some of its practical applications. The concept of LAM with AE utilizes a limited number of AE channels and takes advantage of the structural loading resulting from normal traffic patterns. Applications suitable for LAM include

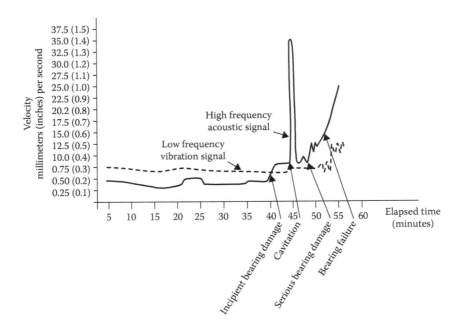

FIGURE 7.64 Incipient failure detection by AE: discontinuities of pump behavior. (From Green, A. et al., (1987). Acoustic emission applications in the petroleum and chemical industries, in *Nondestructive Testing Handbook: Volume Five, Acoustic Emission Testing*, ASNT, Columbus, OH. With permission. Reprinted from ASNT Publication.)

FIGURE 7.65 Railway viaduct, Huntington UK. (Courtesy of Pure Technologies US, Inc.)

- Determining if a preexisting or detected crack is active
- Monitoring of retrofits
- Monitoring of highly stressed areas to determine structural integrity
- Monitoring a suspected area that cannot be visually inspected

At first, AE was used on 15 bridges with "known defect" areas that were monitored to obtain more information on activity level of the defects. On the basis of this experience, an AE system for LAM was developed and described during their work.

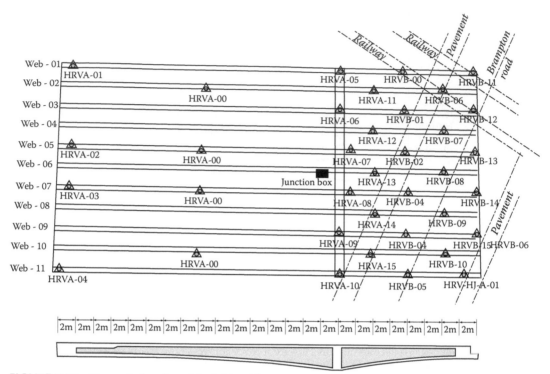

FIGURE 7.66   Plan and elevation of the bridge. (Courtesy of Pure Technologies US, Inc.)

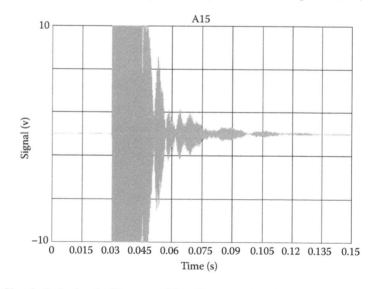

FIGURE 7.67   Signal of wire break. (Courtesy of Pure Technologies US, Inc.)

Local area monitoring system, shown in Figure 7.69, is rugged with shock and environmental protection, accommodates 8 channels (option for multiples of 2 up to 8), data can also be digitized and stored, can accommodate other than AE sensors, internal battery power if needed, and can be connected to PC through a modem. One possible sensor configuration is shown in Figure 7.70.

Field Test 1: In an SHM project for George P. Coleman Bridge in Yorktown, Virginia, in October 1996, sensors were mounted on the south, steel pivot box girder, and the remote operating features of the system were tested. A complete investigation of the north and south pivot box girder was performed and reported by the Virginia Transportation Research Council.

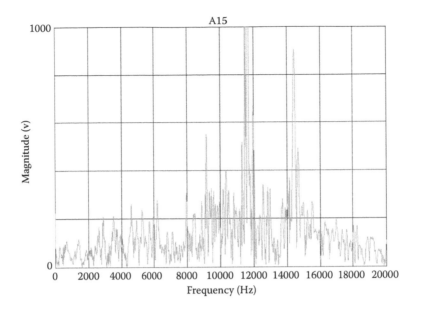

FIGURE 7.68 Fourier transform (FT) of wire break. (Courtesy of Pure Technologies US, Inc.)

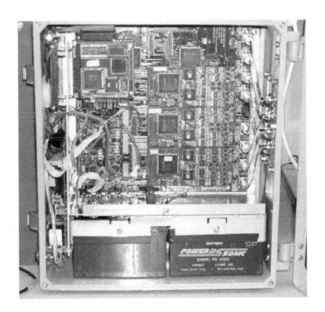

FIGURE 7.69 LAM instrument in NEMA 4 enclosure. (Courtesy of CRC Press.)

Second Field Test: In November 1998, on the Route 1 bridge south of Washington D.C., sensors were mounted near a weld crack in a longitudinal stiffener plate. The test was designed to determine if the crack had been arrested. The results indicated that it had been arrested.

7.5.11 ADVANTAGES AND DISADVANTAGES OF AE

Among the advantages of the AE method are the following:

- Can be local or global
- Passive: no need for external stimulus

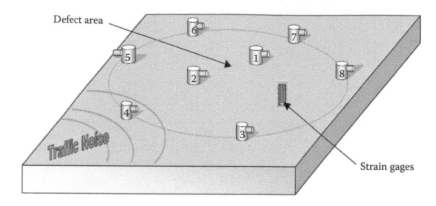

Defect area

Traffic Noise

Strain gages

FIGURE 7.70   Typical experimental setup. (Courtesy of CRC Press.)

- Has potential for proof testing, IFD, and QC
- Real time damage detection; damage is detected as it occurs
- Very amenable to long-term monitoring, that is, can span the leap from NDT to SHM easily

Disadvantages and limitations of AE are as follows

- Ambient noise can be a major issue, especially for continuous AE signals
- In many situations, AE can be mostly qualitative
- Direct contact with test object is needed
- More than one sensor is needed to locate the damage source

7.6   VIBRATION–BASED METHODS

7.6.1   GENERAL

This section addresses a popular DMID approach in civil infrastructure: the vibrations-based DMID. The basic premise of this approach is to use measured vibration properties of the system in the pristine and damaged states, then try to identify the damage through some analytical optimization process. In other words, the methods study the changes between baseline and damaged states. Generally speaking, these methods are based on modal, or parameter, identification and this relates them directly to the STRID field. We explore first the basic methodologies that utilize STRID method in vibration-based DMID. Next, we offer two examples of the use of these methods. Both examples point to the importance of being careful when choosing a vibration-based DMID technique in civil infrastructure.

7.6.2   OVERVIEW OF METHODS

7.6.2.1   Changes in Modal Properties

Since most methods are based on estimating effects of damage on modal properties, let us set some rules. Consider an SDOF with stiffness, and mass as k and m, respectively. The system will have a natural frequency of $\varpi = \sqrt{k/m}$. Rearranging we get

$$k = m\,\varpi^2 \tag{7.10}$$

For a baseline (no-damage) and damaged states, the change in the frequency is

$$\Delta\varpi = \varpi_1 - \varpi_0 = \frac{1}{\sqrt{m}}\left(\sqrt{k_1} - \sqrt{k_0}\right) \tag{7.11}$$

If the change in stiffness due to damage is

$$\Delta k = k_1 - k_0 \tag{7.12}$$

we can express the dimensionless change of frequency $\Delta\varpi/\varpi_0$ as

$$\frac{\Delta\varpi}{\varpi_0} = \left(\sqrt{1 + \left(\frac{\Delta k}{k_0} \right)} - 1 \right) \tag{7.13}$$

Figure 7.71 shows the relationship between dimensionless change of frequency $\Delta\varpi/\varpi_0$ and the damage as estimated by the dimensionless change of stiffness $\Delta k/k_0$.

To relate the frequency ranges with a simple beam situation, consider the frequency equation of simply supported beam:

$$\varpi_n = \frac{n^2}{\ell^2} \sqrt{\frac{\rho A}{EI}} \tag{7.14}$$

The mode number is n. The modulus of elasticity, the mass density, the cross-sectional area, and the moment of inertia are E, ρ, A, and I. The length of the beam is ℓ. Assuming a damage state as described in Table 7.26, the dimensionless change of frequency and dimensionless change of stiffness can be computed. For a simply supported steel WF 36X194, the damage states can be related to the changes in frequency and stiffness in Figure 7.72. For a dimensionless change in the range of 10% or more, 25% of the web, or more, need to be damaged for accurate vibration observations. Clearly, relying on frequency, or stiffness, changes alone might not be an efficient way of monitoring damage in this situation.

7.6.2.2 Utilization of Numerical Formulations FEM/BEM

Some DMID methods rely on numerical formulation to identify damage. Most of the methods use the changes in stiffness matrices ΔK for the DMID

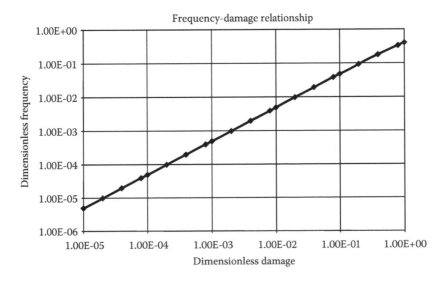

FIGURE 7.71 Frequency–damage relationship.

TABLE 7.26
Damage Levels of Steel Beam

Damage Level	Description
A	Loss of whole bottom half
B	Loss of bottom flange
C	25% of web
D	5% of web
E	1% of web
F	0% of web

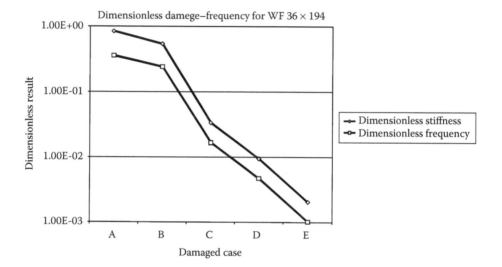

FIGURE 7.72 Dimensionless damage-frequency for steel beam.

The problem can be formulated as

$$[\Delta K] = [K_1] - [K_0] \tag{7.15}$$

The order of the matrices in Equation 7.15 is N, the total number of degrees of freedom. The matrix $[\Delta K]$ is an expression of damage, while $[K_0]$ and $[K_1]$ are the baseline and the damaged stiffness matrices, respectively.

Most of the techniques of this approach rely on finding $[\Delta K]$, subjected to some set of test results. This reminds us with parameter estimation methods of Chapter 6, only executed twice: for pristine and for damaged systems. The guidelines of using parameter estimation methods for DMID were discussed earlier in this chapter.

Detection of damage levels and location of the damage in the system are the main goals of most SHM projects. Several vibration-based damage detection schemes have been proposed in recent years. The next section investigates the accuracy of different damage detection techniques that utilize vibration-based methods. Two full-scale case studies are presented. Several popular DMID methods are applied for each case. We will find that all cases produced results with varying accuracy. When embarking on using vibrations-based DMID methods, the reader is advised to follow guidelines similar to that presented earlier in this chapter to ensure accurate results.

7.6.3   Case Study I: Theoretical Simulation of DMID Procedures

7.6.3.1   Introduction

Structural health monitoring is critical in several types of conventional civil engineering structures. For example, offshore oil platforms need continuous monitoring for structural damage that might occur below the waterline from extreme sea states and ship impact. Structural monitoring of bridges is necessary as the effects of corrosion, ship traffic, earthquakes, and, sadly, even terrorists continuously threaten the soundness of these vital structures. During the last two decades, attention has been focused on using the vibration characteristics of a structure as an indication of structural damage. See, for example, Pandey, A. K. and Biswas, M. (1994), Stubbs and Garcia (1996) and Faarar and Jaurengui (1996). The vibration characteristics of a structure may be defined by its natural modes of vibration. These depend only on the characteristics of the structure and not the excitation. The vibration characteristics of a structure may be determined from taking measurements at one or more points on the structure. Vibration signatures obtained before and after damage have been inflicted by excessive forces may be utilized to locate and estimate the severity of damage.

A structural member develops strain as it deforms due to an applied load. If the load is excessive, the yield strain (i.e., the strain above which the stress is no longer linearly proportional to the strain) of the member may be exceeded. Therefore, if the change in length of a structural member divided by its initial length is greater than the yield strain, the stiffness of a steel member will be reduced typically by a factor of 10 to 20. For a structural steel member, the yield strain is a function of the type of steel as well as the rate of loading. For typical commercial high strength low alloy steel with a static yield stress of 50,000 psi, the static yield strain is 0.0017.

The physics behind the detection of damage through the measurement of vibration signatures is based on the previously mentioned fact that if the strain in a structural member exceeds the yield strain, the stiffness of the member is reduced. If the stiffness of the member is reduced, then the characteristics of the vibration (natural frequency, mode shapes, and damping) of the member will change as well as those of the complete structure.

Two challenging problems present themselves. First, the number and location of structural monitoring measurements ideally should be kept to a minimum. Second, correlating the results of the measurements to a specific damage mechanism (e.g., hairline cracks, large localized strains, etc.) as well as to the location of the damage is difficult.

Ettouney et al. (1998) investigated the applicability and accuracy of different damage detection techniques to complex structures. A typical multi-jointed steel bridge will be considered. The damage will be simulated analytically in the structural models, and the damage detection algorithms will be applied to both the damaged and the undamaged structures.

The accuracy and efficiency of different damage detection schemes are investigated, including the capability of detecting damage location as well as the severity and extent of the damage. It is shown that the capability of the three algorithms to predict the location and damage levels depend on the number of modes considered. In general, good qualitative predictions were observed from all algorithms considered.

7.6.3.2   Vibration-Based DMID methods

7.6.3.2.1   General

Three health monitoring methods will be investigated in this study. The change in stiffness method, the change in flexibility method, and the change in damage index (DI) method. All these methods are based on the knowledge of the mode shapes of both the undamaged and the damaged structures $[\phi]$ and $[\phi^*]$, respectively. Also, the natural frequency diagonal matrices $[\Omega]$ and $[\Omega^*]$ for both damaged and undamaged structures are assumed to be known. The values of the mode shapes can be measured and computed for the structure of interest both in the undamaged and the damaged states. For the purpose of the current study, which is intended to investigate different analytical health monitoring methods, the values of $[\Omega]$, $[\Omega^*]$, $[\phi]$, and $[\phi^*]$ will be calculated numerically. In what follows, a summary of each of the three methods is presented. Note that the size of the mode shape matrices is $N \cdot M$, where M is the number of mode shapes considered in the problem, and N is the number of degrees of freedom of the structure. In the rest of this section, an * will refer to the damaged structure.

7.6.3.2.2   Change in Stiffness Method

Change of stiffness methods was investigated and discussed by several authors. Zimmerman and Kaouk (1994), Reese (1995), and Farrar and Jauregui (1996) discussed several aspects of the method. It is based on finding the change of the stiffness matrix of the structure $[\Delta K]$, where

$$[\Delta K] = [K] - [K^*] \tag{7.16}$$

The undamaged stiffness matrix $[K]$ and the damaged stiffness matrix $[K^*]$ can be calculated as

$$[K] = [\phi][\Omega][\phi]^T \tag{7.17}$$

$$[K^*] = [\phi^*][\Omega^*][\phi^*]^T \tag{7.18}$$

Note that the matrix $[\Delta K]$ contains stiffness terms, which include forces and bending moments. Since it is difficult to mix forces and moments, it is proposed to use a normalized matrix $[\overline{\Delta K}]$ such that

$$\left[\overline{\Delta K}\right] = \frac{[\Delta K]}{[K]} \tag{7.19}$$

Note that the above division is a scalar division, such that $\overline{\Delta K}_{ij} = \Delta K_{ij} / K_{ij}$.

To simplify the investigation of the matrix $[\overline{\Delta K}]$, the vector $\{DK\}$ will be considered. The vector $\{DK\}$ contains only the diagonal terms of the matrix $[\overline{\Delta K}]$. The relative values of the components of $\{DK\}$ should indicate the location of the damaged parts of the structure. The statistical approach used by Kim and Stubbs (1995a) to further isolate the damaged elements will be utilized here. The differential stiffness index (DSI) will be introduced as

$$DSI_i = \frac{DK_i - \overline{DK}}{\sigma_{DK}}, \tag{7.20}$$

where DK_i is the ith component of $\{DK\}$, $\overline{DK}$ is the mean of the vector $\{DK\}$ and σ_{DK} is the standard deviation of the vector $\{DK\}$. The subscript, i, has the range of $i = 1, 2, \ldots N$. The index DSI can be used to investigate the locations of the damage in the structure.

7.6.3.2.3   Change in Flexibility Method

Pandley and Biswas (1994) proposed the change of flexibility method. It is based on the computation of the differential flexibility matrix of the damaged and the undamaged structure. The flexibility matrix of the undamaged and the damaged structures can be computed as

$$[F] = [\phi][\Omega]^{-1}[\phi]^T \tag{7.21}$$

$$[F^*] = [\phi^*][\Omega^*]^{-1}[\phi^*]^T \tag{7.22}$$

From these expressions, the normalized differential flexibility matrix $[\overline{\Delta F}]$, can be computed using similar steps as in the differential stiffness matrix $[\overline{\Delta K}]$. The differential flexibility index (DFI) can be computed using the components of $[\overline{\Delta F}]$, following the procedure outlined in the previous section.

7.6.3.2.4 Damage Index (DI) Method

Kim and Stubbs (1995a) developed DI method. Unlike the change of stiffness and change of flexibility methods, which uses the structural nodal degrees of freedom as their basic variable, it is based on the structural elements as the basic variable. It was reported by Kim and Stubbs (1995b) that it has the advantage of requiring small number of structural modes to obtain an accurate damage detection results. The DI method results in a damage index DI_i, where subscript i has the range of i = 1, 2, … N_e, where N_e is the number of structural elements in the system. Excellent results using the DI method were reported in numerous reports. Stubbs et al. (1995) and Farrar and Jauregnui (1996) used it in an application to a highway bridge structure. Also, Garcia and Stubbs (1995) and Stubbs and Garcia (1996) reported on the successful use of the DI methodology. A comprehensive description of the method can be found in Kim (1995a and 1995b) and Stubbs and Kim (1994).

7.6.3.3 Results

7.6.3.3.1 General

The three damage detection algorithms, which were described above, were applied to a complex steel structure. Figure 7.73 shows a steel frame that is supported on five supports. The structure contains 241 structural elements, and 78 structural connections. All the connections of the structure are rigidly attached. The height of the structure is about 10 ft, and the longer dimension of the structure is about 70 ft. The structure was modeled using finite elements so as to have accurate enough modes up to a natural frequency of 250 Hz. The mode shapes $[\phi]$ and the natural frequencies $[\Omega]$, of the base line, undamaged structure were calculated. Two damaged states of the structure were then considered. A structure with only one damaged element, as shown in Figure 7.74 and a structure with 10 damaged elements, as shown in Figure 7.75. The damaged elements were simulated numerically by artificially specifying the modulus of elasticity of the damaged element E_d, as $E_d = \alpha E_u$. Where E_u is the modulus elasticity of the undamaged steel element. The magnitude of the factor $\alpha \leq 1.0$ represent the degree of damage in the element. Two damage levels were considered for this work, a moderate damage case, with $\alpha = 0.5$, and a severe damage case with $\alpha = 0.01$. An essential factor in

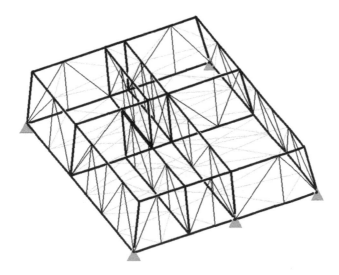

FIGURE 7.73 Model of steel structure.

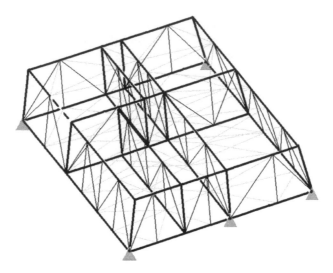

FIGURE 7.74   Model of steel structure, with one damaged member.

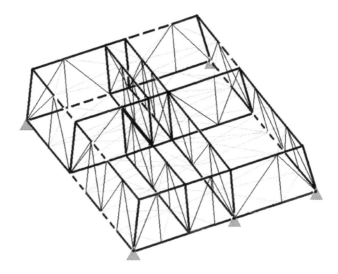

FIGURE 7.75   Model of steel structure, with ten damaged members.

any damage detection method is the number of structural modes M needed to accurately detect the structural damage. Two cases are studied in this work, namely $M = 14$ and $M = 140$. Note that the structural model has as many as 468 modes. It is desirable to limit the number of modes since this translates directly into cost savings of any long-term health monitoring scheme.

7.6.3.3.2   Change in Stiffness Method

The change of stiffness method is applied to the structures of Figures 7.73 through 7.75. The DSI were computed for the different conditions described above. For the case of one damaged element, Figure 7.76 shows the DSI for $\alpha = 0.5$ and both $M = 14$ and $M = 140$. Figure 7.77 shows similar results for $\alpha = 0.01$. Figures 7.78 and 7.79 show the same comparisons for the structure with 10 damaged elements. In Figures 7.76 through 7.79 the zones of nodal degrees of freedom that are connected to damage elements are highlighted. The DSI was fairly successful in detecting the location of the damaged one element case. It also predicted with a lesser degree of accuracy, the locations of the damaged 10 element case. Even with only 14 modes, the method seemed to have acceptable results.

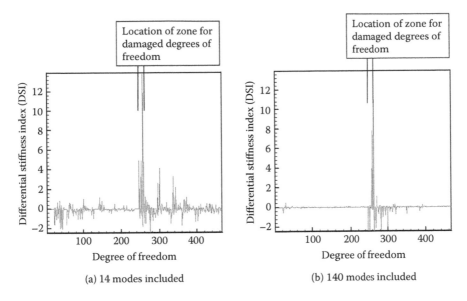

FIGURE 7.76   Change of stiffness method, 1 damaged element; $E_d/E_u = 0.50$ (a) 14 modes, and (b) 140 modes.

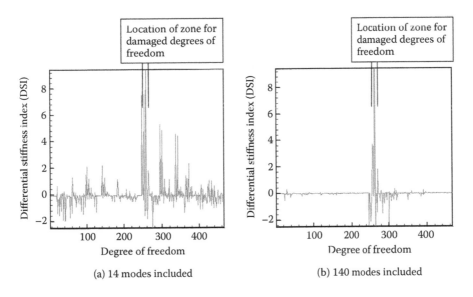

FIGURE 7.77   Change of stiffness method, 1 damaged element; $E_d/E_u = 0.01$ (a) 14 modes, and (b) 140 modes.

7.6.3.3.3   Change in Flexibility Method

Next, we apply the change of flexibility method to the structures of Figures 7.73 through 7.75. The DFI were computed. For the case of one damaged element, Figures 7.80 and 7.81 show the DFI for $\alpha = 0.5$ and $\alpha = 0.01$, each calculated for $M = 14$ and $M = 140$. Figures 7.82 and 7.83 show the same comparisons for the structure with 10 damaged elements. The results are similar to those of the change of stiffness method. In Figures 7.80 through 7.83, the zones of nodal degrees of freedom which are connected to damage elements are highlighted. The DFI was fairly successful in detecting the location of the damaged one element case. It also predicted with a lesser degree of accuracy the locations of the damaged 10 element case. Increase in the number of modes from 14 to 140

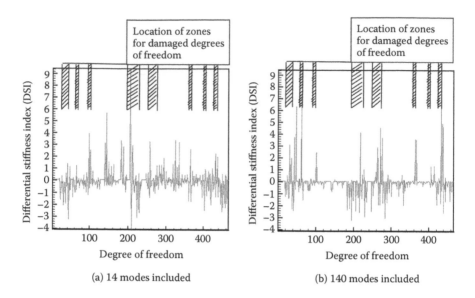

FIGURE 7.78   Change of stiffness method, 10 damaged elements; $E_d/E_u = 0.50$ (a) 14 modes, and (b) 140 modes.

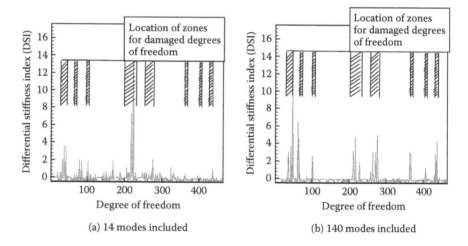

FIGURE 7.79   Change of stiffness method, 10 damaged elements; $E_d/E_u = 0.01$ (a) 14 modes, and (b) 140 modes.

improved the results, although the 14 modes case method still produced acceptable results. The DFI seems to predict the severity of the damage.

7.6.3.3.4   Damage Index Method

Finally, we investigate the DI method. The DI was computed for all the cases. Different results are shown in Figures 7.84 through 7.87. Note that the DI is a function of each structural element. Thus the DI would show the damage situation in each element, not a nodal degree of freedom. In Figures 7.84 through 7.87 the different elements that are damaged are highlighted. In general, the DI capabilities to predict the location and magnitude of damage are similar to those of the other two methods described earlier.

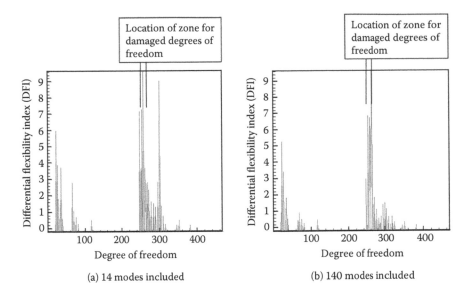

FIGURE 7.80 Change of flexibility method, 1 damaged element; $E_d/E_u = 0.50$ (a) 14 modes, and (b) 140 modes.

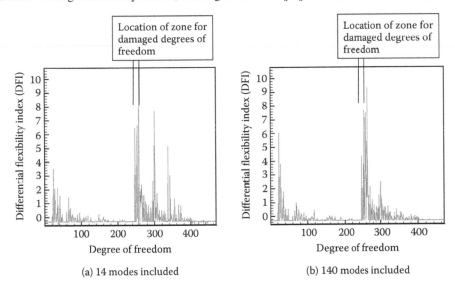

FIGURE 7.81 Change of flexibility method, 1 damaged element; $E_d/E_u = 0.01$ (a) 14 modes, and (b) 140 modes.

7.6.3.4 Conclusions

Three different damage detection algorithms were investigated in this section. A realistic complex steel structure was used for the evaluation. Different important parameters were studied. These parameters included number of structural modes, number of damaged elements and the extent of damage in each element. The three algorithms performed equally well in predicting both the locations of damaged elements and the extent of the damage.

Following these encouraging results, Ettouney et al. (1998) recommended to apply these algorithms to more complex structures, and to increase the parametric studies so as to include more important parameters, such as shape of damage and damages concentrated in structural joints. It is also of interest to investigate the performance of these algorithms when the mode shapes included are not consecutive. Experimental testing should eventually be used to verify the numerical testing.

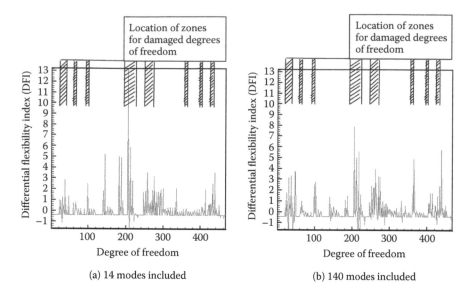

FIGURE 7.82   Change of flexibility method, 10 damaged elements; $E_d/E_u = 0.50$ (a) 14 modes, and (b) 140 modes.

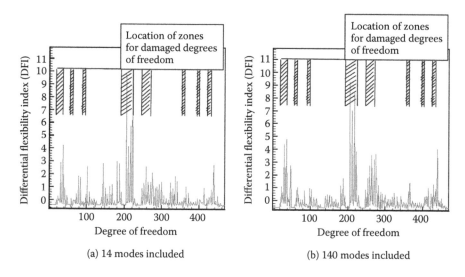

FIGURE 7.83   Change of flexibility method, 10 damaged elements; $E_d/E_u = 0.01$ (a) 14 modes, and (b) 140 modes.

7.6.4   Case Study II: I-40 Bridge Experiment

In a field experiment by Farrar et al. (1994) and Farrar and Jauregnui (1996), different DMID algorithms were applied to a bridge spanning highway I-40 in NM. The I-40 Bridge was a steel girder bridge supporting a reinforced concrete deck. An intentional damage was introduced to the steel girders to simulate fatigue cracks. Different vibration-based damage detection algorithms were then employed to see how well those algorithms fare in detecting the damage. Four damage scenarios were induced at the center of one of spans of the steel girder as shown schematically in Figure 7.88. After each damage state is introduced, the bridge was excited using a single hydraulic shaker that produced a 2000 pounds peak-force uniform random signal in the range of 2–12 Hz. Accelerations at different locations were measured. The test was designed to record a coarse set of accelerations

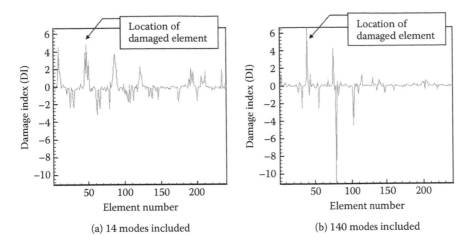

FIGURE 7.84   Damage index method, 1 damaged element; $E_d/E_u = 0.50$; (a) 14 modes, and (b) 140 modes.

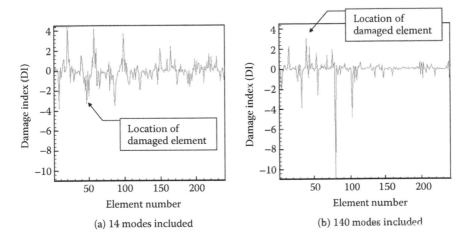

FIGURE 7.85   Damage index method, 1 damaged element; $E_d/E_u = 0.01$; (a) 14 modes, and (b) 140 modes.

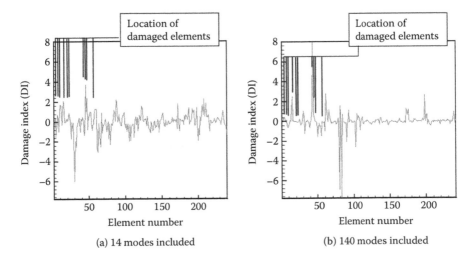

FIGURE 7.86   Damage index method, 10 damaged elements; $E_d/E_u = 0.50$; (a) 14 modes, and (b) 140 modes.

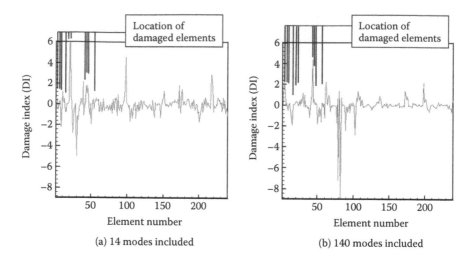

FIGURE 7.87   Damage index method, 10 damaged elements; $E_d/E_u = 0.01$; (a) 14 modes, and (b) 140 modes.

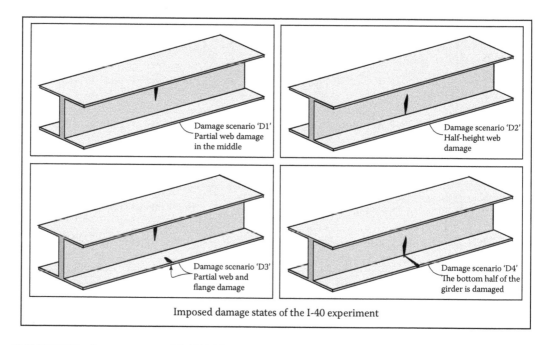

FIGURE 7.88   Damage states of I-40 Bridge.

at uniform stations along the whole bridge. Then a more refined set of accelerations were measured near the intentional damage location. A standard modal identification technique was then employed to identify the bridge vibration characteristics. As many as six modes were identified with their corresponding modes shapes, natural frequencies, and damping parameters. Those modes were identified for five bridge states: pristine state and the four intentional damaged states. The natural frequencies ranged from 2.3 Hz for the first mode to 4.69 Hz for the sixth mode. The damping ratio ranged from 0.62% to 1.6%. For more details on the experiment and its results see Farrar et al. (1994), Farrar and Jauregnui (1996), and Farrar and Doebling (1999).

What concerns us in this section is the application of several vibration-based damage detection algorithms to the I-40 results that were reported by Farrar and his colleagues. They employed five damage detection algorithms in trying to identify the known damages of Figure 7.88. We describe briefly the five algorithms, and then report on how those algorithms fared in detecting the I-40 experiment.

Damage Index Method: See description in previous section.

Mode Shape Curvature Method: This method is based on computing the analytic expression of beam curvature, $u''(x)$, in the form

$$u''(x) = \frac{M(x)}{EI} \tag{7.23}$$

The bending moment of the girder at station x is $M(x)$. The modulus of elasticity and the governing moment of inertia are E and I, respectively. Equation 7.23 is used to identify the damage, as reflected in the flexural stiffness EI from the measured beam curvature $u''(x)$. See Pandey et al. (1991) for more details on this method.

Change in Flexibility Method: See description in previous section.

Change in Uniform Load Surface Curvature: The method was suggested initially by Zhang and Aktan (1995). It is also based on computing curvature of the bridge girder with the implied assumption that it behaves as a beam. Simply stated, the method uses the test results to produce a deflected shape that is the resultant of a uniform loading on the bridge. It achieves that by adding all the columns in the flexibility matrix. When the experimental uniform deflected shape is computed, the curvature at each point can be computed using central difference approach. The difference between the pristine curvature and the damaged curvature is now computed. That curvature difference can be related to a quantified damage since we know the analytical expressions of the bending of beams.

Change in Stiffness Method: See description in previous section.

The results of applying the five vibrations-based damage detection methods to the four damage scenarios are shown for the coarse and refined accelerometers sets in Tables 7.27 and 7.28, respectively. It is clear that the accuracy of damage detection using different methods vary greatly. For in-depth discussions and conclusions of this informative and important experiment, see Farrar et al. (1994), Farrar and Jauregnui (1996), and Farrar and Doebling (1999).

The I-40 experiment was an important event in the evolution of SHM in civil infrastructure. It shed light on the different vibration-based DMID methods as stated above. Farrar and Doebling (1999) stated some of the lesson learned from that experiment: (a) Pretest visual inspection is important; (b) During the STRID phase, it is important to ensure that the results are both reciprocal and linear, since such assumptions are built-in most STRID methods; (c) Sensitivity of the modal identification method to environmental and testing procedures should be investigated; and (d) False-posivite analysis should be performed to ensure that the procedures will not identify pristine systems as damaged.

For our current purposes, we note that all the five methods were based on STRID modal identification techniques for both pristine and damage systems. After the STRID process, the DMID methods were applied. Earlier in this chapter we introduced a guide for using STRID as a DMID tool. We feel that this guide is applicable to all of the above five methods. It would be of interest to compute (quantify) the errors involved in damage using vibration-based approach for different types and sizes of damages. It should also be noted that the damages incorporated in the case studies are known before the testing. In many cases, the identified damages are relatively big.

7.7 SIGNAL PROCESSING AND DMID

7.7.1 TIME SIGNALS

The use of time signals (defined as a time-varying function) in the field of SHCE is extensive. Invariably, at some step of the process, the need to record and analyze time-dependent signals arises. The signal can be an input signal (earthquakes, blast pressures, fluid velocity, wind pressures,

TABLE 7.27
Results of Vibration-Based DMID Methods, Coarse Sensors Set

	DMID Method				
Damage Scenario	Damage Index Method	Mode Shape Curvature Method	Change in Flexibility Method	Change in Uniform Load Surface Curvature Method	Change in Stiffness Method
D1:	2	2	X	X	2
D2:	2	X	X	X	2
D3:	2	2	2	X	2
D4:	1	1	1	1	1

Note: 1—Damage located; 2—damage located using only two modes; 3—damage not located.
Source: Farrar, C. et al., Dynamic characterization and damage detection in the I-40 Bridge over the Rio Grande, *Los Alamos National Laboratory Report LA-12767-MS*, 1994.

etc.) or an output signal (measured responses such as accelerations, strains, displacements, etc.). Also, many of the NDT, STRID methods are based on measuring and analyzing signals. This shows that we need to have adequate knowledge of signal analysis for appropriate decision making in SHCE.

One of the basic requirements for any signal processing method is that the method must be compatible with the signal properties. An important signal property is whether it changes its characteristics (such as RMS) as time passes. Stationary signals do not change their characteristics with time. Nonstationary signals change their characteristics. Most of signals that affect civil infrastructure are nonstationary, such as earthquakes or blast pressures. Also, by definition, any nonlinear behavior would produce nonstationary signal. Sometimes the nonstationary requirement is dropped for the sake of simplicity; earthquake motions and responses are usually analyzed by FT, which is not suited to analyze nonstationary signals.

TABLE 7.28
Results of Vibration-Based DMID Methods, Refined Sensors Set

Damage Scenario	DMID Method				
	Damage Index Method	Mode Shape Curvature Method	Change in Flexibility Method	Change in Uniform Load Surface Curvature Method	Change in Stiffness Method
D1:	1	3	X	X	X
D2:	1	2	X	3	X
D3:	1	1	X	1	X
D4:	1	1	1	1	1

Note: 1—Damage located; 2—damage narrowed down to two locations; 3—damage narrowed down to three locations; X—damage not located.

Source: Farrar, C. et al., Dynamic characterization and damage detection in the I-40 Bridge over the Rio Grande, *Los Alamos National Laboratory Report LA-12767-MS*, 1994.

This section explores four popular signal processing methods: FTs, time-frequency transforms (TFT), wavelet transforms (WT), and Hilbert–Hwang transforms (HHT). The theoretical basis of each method is overviewed; the advantages/disadvantages and some examples of their use in STRID and DMID are presented.

7.7.2 Fourier-Based Transforms

7.7.2.1 Conventional Fourier Transforms

The frequency content of a continuous time signal $x(t)$ can provide valuable information during an SHM experiment. Perhaps the most popular tool for investigating the frequency content is its FTs. It can be expressed as

$$FT(f) = \int_{-\infty}^{+\infty} x(t)e^{-i2\pi ft}dt \tag{7.24}$$

The frequency spectrum $FT(f)$ is a complex illustration of the signal behavior at a given frequency f. Figure 7.89 illustrates the mathematical principles of $FT(f)$. One of the reasons of the popularity of FTs is that they exist also for discrete time signal $x(n)$ as

$$DFT(f) = \sum_{n=-\infty}^{\infty} x(n)e^{-in2\pi f} \tag{7.25}$$

assuming that

1. The discrete time signal is discretized with a time step of Δt
2. The total time steps is N

Then following Heisenberg uncertainty principle, Hubbard (1998), $DFT(f)$ is discretized with a frequency step of Δf with

$$\Delta f = \frac{1}{N\,\Delta t} \tag{7.26}$$

One of the advantages of the FT method is that the inverse transform exist. The continuous transform is expressed as

$$x(t) = \int_{-\infty}^{\infty} FT(f)e^{i2\pi ft}\,df \tag{7.27}$$

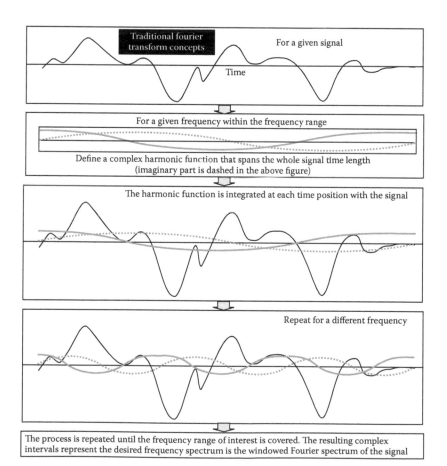

FIGURE 7.89   Principles of Fourier transform (FT).

The discrete transform is

$$x(f) = \sum_{n=-\infty}^{\infty} DFT(n)e^{-in2\pi n} \tag{7.28}$$

Thus, other advantages of the FT method are as follows:

- Can be used to analyze many applications
- Well understood and used by professionals
- Inverse transforms exist, as shown above

FT method is a very efficient analytical technique for evaluating Equation 7.25, for discrete FT, and Equation 7.28 for discrete inverse transform. The method is known as fast Fourier transform, FFT in short.

From Equation 7.26 it is clear that the Heisenberg principle puts severe limitations on the frequency or the time step of the signal. This has important implications during SHM experiments. For example, at low frequencies, the frequency step is relatively high, thus the resolution of the frequency spectrum is low. Another serious disadvantage of the Fourier spectrum is that it is based on harmonic functions only. The harmonic basis can have inaccurate effects on signals with short time duration, see Pendat and Piersol (2010). Perhaps the most limiting disadvantage is that the FT (or FFT) spectra do not reveal any information regarding the time dependency of frequency content of the signal. Thus, the FT method is not applicable for nonstationary signals (wind, flood, or earthquake signals) or nonlinear problems (such as analyzing failure potential of brittle or potentially unstable systems).

7.7.2.2 Time and Frequency Transforms

Before we continue to explore other methods of signal processing, we need to discuss further a very important derivative of FT, namely, the power spectral density (PSD). The PSD is defined as

$$G(f) = \frac{1}{2T}|FT|^2 \tag{7.29}$$

Where $G(f)$ is the PSD. Mathematically, it can be shown that for a given frequency width, say Δf, the area represented by $\Delta f \cdot |FT|^2$ is the mean square of the harmonic amplitude at such frequency (see Figure 7.90 for illustration). Thus, the PSD is a representative of the energy within a

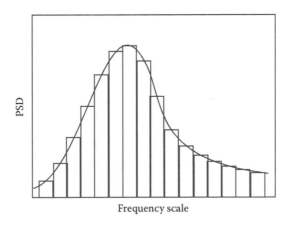

FIGURE 7.90 Physical illustration of PSD.

particular spectrum width. This is an important fact that can be used to understand and analyze time signals. Similar to the parent FT spectrum, we note a major limitation to the use of PSD in signal processing. The time dependency of the energy occurrence is lost. We only know the energy situation at a given frequency; we do not know *when* such energy occurred during the time signal.

There are several other derivatives of FT, such as correlation, coherence, and cross-correlation functions. For detailed description of those derivatives, the reader can refer to Pendat and Piersol (2010).

7.7.2.3 Time-Fourier Transforms (TFT)

To avoid the FT limitation of losing temporal information of frequency content, the TFT was introduced, see Boashash (1992). The time signal $x(\tau)$ is windowed by a window function $h(\tau)$. The windowed function is then centered about time t. We are now interested in the transform of $x(\tau)h(\tau - t)$ such that

$$TFT(f, t) = \int_{-\infty}^{+\infty} x(\tau)h(\tau - t)e^{-i2\pi f\tau}d\tau \qquad (7.30)$$

The frequency-time spectrum $TFT(f, t)$ is now an illustration of the signal behavior at a given frequency f and a given time t. Figure 7.91 illustrates the mathematical principles of $FT(f)$. Note that a discrete form of Equation 7.30 can be evaluated. We also note that Equation 7.26 still applies due to Heisenberg's principle. Similar to FT, the inverse transform of TFT exists.

The TFT resolves one of the limitations of the FT method. The TFT spectrum can now reveal useful information regarding the time dependency of frequency content of the signal. For example, time of excitation of particular vibration mode during a given experiment. Unfortunately, since TFT is still dependent on a fixed-length widowed harmonic function, TFT method is still not applicable for nonstationary signals or nonlinear problems. To summarize, the TFT method has the following advantages:

- Can be used to analyze many applications
- Well understood and used by professionals
- Inverse transforms exist, as shown above
- There are both continuous and discrete

Among the disadvantages of the TFT method are the following:

- Loses frequency resolution at low frequency (Heisenberg uncertainty principle)
- Loses time resolution at high frequency (Heisenberg uncertainty principle)
- Only harmonic functions
- Similar to the FT method, the TFT method is not applicable for nonstationary signals or nonlinear problems

7.7.3 Wavelets

7.7.3.1 Introduction

To resolve several of the FT or TFT limitations, another method of signal processing was developed in the past few decades: the WT, see Graps (1995) and Hubbard (1998). The continuous wavelet transform for a time signal $x(t)$ is

$$WT(s, \tau) = \int x(t)\,\psi(s, \tau, t)dt \qquad (7.31)$$

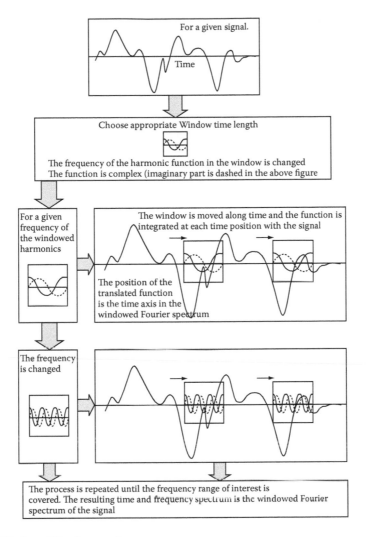

FIGURE 7.91   Windowed Fourier transform.

The basis function (also known as wavelet function, or mother wavelet) $\psi(s, \tau, t)$ include a scaling and translation variables s and τ, respectively; formally it can be expressed as

$$\psi(s, \tau, t) = \frac{1}{\sqrt{s}} \, \psi\left(\frac{t - \tau}{s}\right) \tag{7.32}$$

Upon inspection, it is clear that the scaling factor controls the amplitude of the wavelet function. The translation factor controls the location of the wavelet function on the time scale. As the scaling factor changes, the effective width of the wavelet function changes. The translation factor control the sweep of the wavelet function along the time scale. See Figure 7.92 for illustration of WT basic principles. There are several forms of wavelet functions will be discussed in Section 7.7.3.2. This is perhaps the most important distinction between WT and FT/TFT: the wavelet functions can be designed to accommodate the particular problem on hand. The FT/TFT are based on harmonic functions.

There is an inverse WT such that

$$x(t) = \int \int WT\left(s, \tau\right) \psi(s, \tau, t) ds \, d\tau \tag{7.33}$$

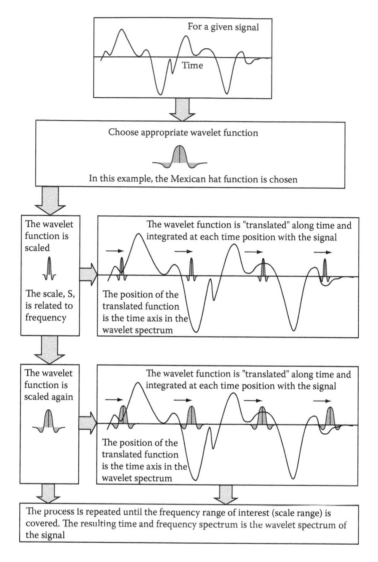

FIGURE 7.92   Basic principles of wavelet transforms (WT).

The frequency-time spectrum $WT(s, \tau)$ is now an illustration of the signal behavior at a given scale (frequency) s and a given translation (time) τ. Note that a discrete form of WT Equation 7.30, and the inverse WT, Equation 7.33, can be evaluated.

Some of the advantages of the WT method are as follows

- Can be used to analyze many applications
- Well understood and used by professionals
- Inverse transforms exist, as shown above
- There are both continuous and discrete forms
- Do not have the FT/TFT limits on frequency or time resolutions. Controlling the scaling and translation factors would result in a good frequency-time resolution compromise. This is a reasonable accommodation to the Heisenberg uncertainty principle, as discussed later.
- Can apply to nonstationary and nonlinear problems. Can also be used for pattern recognition and data compression

7.7.3.2 Wavelet Functions

One of the many differences between TFT and WT is the flexibility of the wavelet functions $h(\tau)$. Wavelet functions vary in shape as well as in time width. They can be chosen to better suite the signal under consideration. Examples of wavelet functions are the following

Morlet Wavelet: They are complex in shape and resemble a decaying harmonic wave form, They are used in many applications in SHM field such as wind and earthquake analysis (Figure 7.93).

Haar Wavelet: This wavelet is the simplest wavelet form (Figure 7.94). It is not a continuous function, yet its simplicity makes it a favorite for illustration as well as sensitivity studies.

Mexican Hat Wavelet: This wavelet (Figure 7.95) has simple analytical expression. It is used in several physics and engineering applications such as astronomy, ocean engineering, and seismic waves.

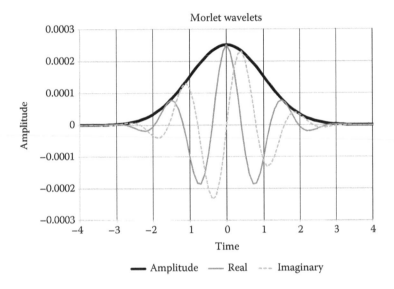

FIGURE 7.93 Morlet wavelets.

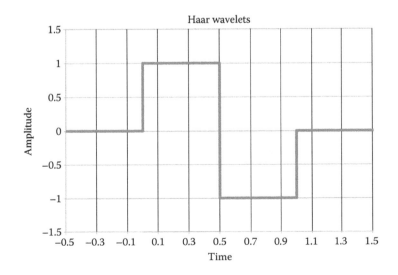

FIGURE 7.94 Haar wavelets.

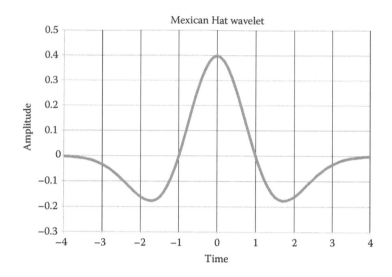

FIGURE 7.95   Mexican hat wavelets.

Daubechies Wavelets: One of the features of Daubechies wavelets is that they are produced numerically by iterations; they are not analytical functions. However, these types of waves are not finite in width, they cover the whole signal. Because of this, they do not have the truncation problems experienced by other types of wavelets.

7.7.3.3   Heisenberg Principle

In discussing the Heisenberg principle earlier, we showed that it relates time step, frequency step, and number of time samples for FT processes. A generalized form of Heisenberg principle is

$$\Delta t \cdot \Delta f = A \tag{7.34}$$

where A is a constant for the specific process under consideration; Figure 7.96 shows the implications of the Heisenberg principle for different signal processing techniques:

Time Signal: High temporal resolution, but no information about the frequency content of the signal.

FT: Can produce high frequency resolution, but temporal information is missing.

TFT: The equal time and frequency resolutions produces coarse information at the tail ends of both time and frequency scales.

WT: The relative time and frequency resolutions changes along the time and frequency scales. WT produces higher time resolution at lower frequency range. It also produces higher time resolution at higher frequency range. This behavior produces more accuracy than other processes.

7.7.3.4   Basic Example of DWT

Let us consider now a simple example of DWT. The signal

$$f(t) = f_1(t) + f_2(t) + f_3(t) \tag{7.35}$$

is sampled at time step of Δt with sample size of $N = 2^m = 2048$, which means $m = 11$. We assume

$$f_1(i\,\Delta t) = \sin\left(\frac{a_1\,\pi(i\,\Delta t)}{180}\right) \tag{7.36}$$

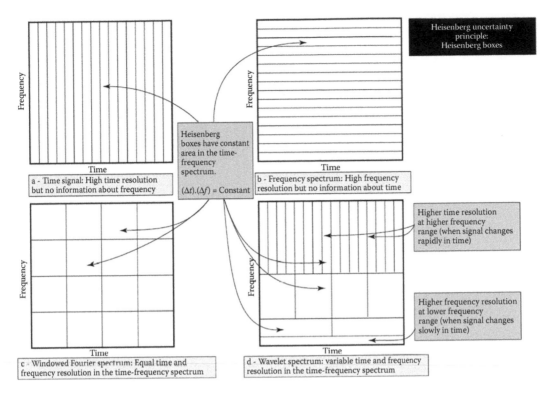

FIGURE 7.96 Heisenberg boxes for different signal transform methods.

$$
\left.
\begin{aligned}
f_2\left(i\,\Delta t\right)\big|_{i<512} &= 0 \\
f_2\left(i\,\Delta t\right)\big|_{i>512} &= \sin\!\left(\frac{a_2\,\pi\left(i\,\Delta t\right)}{180}\right)
\end{aligned}
\right\}
\tag{7.37}
$$

$$
\left.
\begin{aligned}
f_3\left(i\,\Delta t\right)\big|_{i<1024} &= 0 \\
f_3\left(i\,\Delta t\right)\big|_{i\geq 1024} &= \sin\!\left(\frac{a_3\,\pi\left(i\,\Delta t\right)}{180}\right)
\end{aligned}
\right\}
\tag{7.38}
$$

The sampled function $f(i\Delta t)$ is shown in Figure 7.97a. For our purpose, there are two main features of $f(i)$. First, it contains three distinct frequencies ($a_1\,\pi/180$), ($a_2\,\pi/180$) and ($a_3\,\pi/180$). Second feature is that the three signals f_1, f_2, and f_3, arrive at three distinct times $t = 0$, $t = 512$, and $t = 1024$, respectively.

Performing an FFT on $f(i)$ will produce the discrete Fourier spectrum of Figure 7.97b. The Fourier spectrum shows, as expected, the three distinct frequencies along the frequency scale. Also, as expected, the times of arrival of the three component signals are nowhere to be found in the FFT. Performing WT on the signal produced the spectra in Figure 7.98. The arrivals of the three sinusoids are clearly identified in an accurate manner, especially in the contour of Figure 7.98a. The head-on view of Figure 7.98c (the amplitude-scale view) resembles the FT spectrum of Figure 7.97b. It shows the three distinct frequencies, but the times of arrival of the signals are clear. The superior performance of WT as compared to the FT is evident.

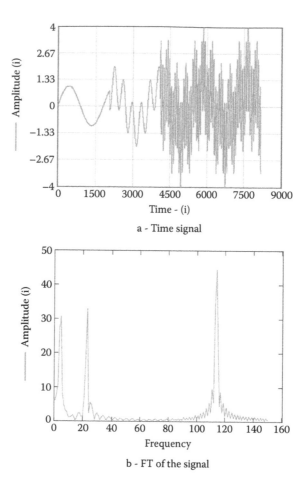

a - Time signal

b - FT of the signal

FIGURE 7.97   Three-sinusoids signal (a) time signal and (b) FT of the signal.

7.7.3.5   Applications of Wavelets in Damage Identification

The flexibility and ease of WT in extracting information from signals and measurements resulted in numerous applications by users and researchers. For example, Spanos and Failla (2004) developed a wavelets-based method to estimate the evolutionary power spectral density (EPSD) of nonstationary stochastic processes. The method can help in earthquake engineering studies by accommodating the nonstationary property of earthquake motions. Yan et al. (2004) used WT for STRID of bridge structures. They also considered potential uncertainties in their work while applying the WT technique. Another innovative utilization of WT was offered by Duan et al. (2004). They used WT as a teaching mechanism for neural networks.

Hamstad et al. (2002) showed a practical example on using WT for detecting damage in plates. The experiment was based on measuring AE signals from damages, then identifying damage information using WT. The experimental setup is shown in Figure 7.99. Seven AE sensors are placed on the plate. An induced damage was placed inside the plate under the middle sensor (#4). The identification process is based on identifying dispersion curves of the plate of interest. Figure 7.100 shows typical plate dispersion curves, Rose (2004). The WT and dispersion curves are utilized during the experiment as follows: (a) measure emission signals at different sensor locations, (b) use WT to produce time-frequency spectrum, (c) produce theoretical dispersion curves, and (d) match appropriate dispersion curve with the WT spectrum. Figure 7.101 shows a typical emission signal, its WT spectrum overlaid on a symmetric (S_0) and asymmetric (A_0) dispersion lines. A very good match between the experimental measurements and theoretical computations is shown. Since theoretical

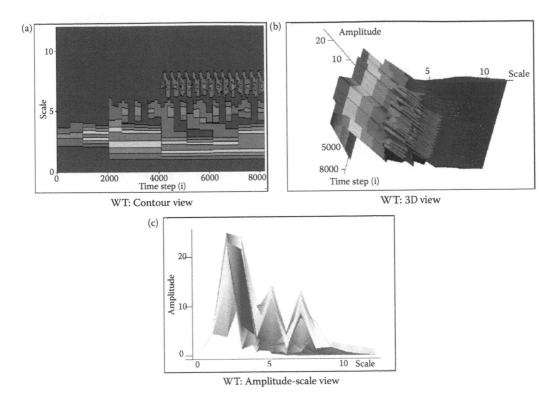

FIGURE 7.98 WT of three-sinusoid signal (a) contour view, (b) 3D view, and (c) 2D view.

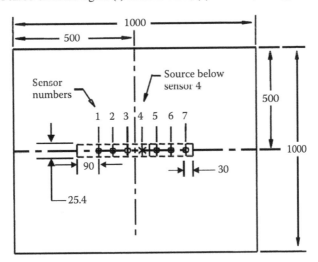

FIGURE 7.99 Testing specimen. All dimensions are in mm. The sensors are spaced 60 mm. (From Hamstad, M. A. et al., *J. Acoustic Emission*, 20, 39-61, 2002. With permission.)

curves are associated with specific stress wave with known wave velocity, and wave number, a simple computation can reveal the distance between the sensor location and the source of the signal, thus identifying the location of the damage. Figure 7.102 shows the conventional FT spectrum of the same signal. Clearly the FT does not provide any information regarding the damage location. Figure 7.103 shows the WT spectra as captured by different sensors. The effects of distance on the signals are evident. This shows that a triangulation scheme can pinpoint unknown damage location

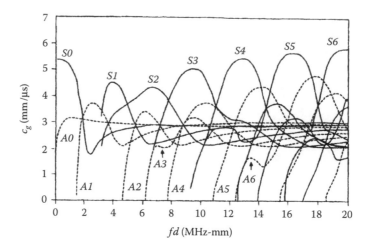

FIGURE 7.100 Dispersion curves (group velocities) of a plate. (From Rose, J., *Ultrasonic Waves in Solid Media,* Cambridge University Press, Cambridge, UK, 2004. Reprinted with the permission of Cambridge University Press.)

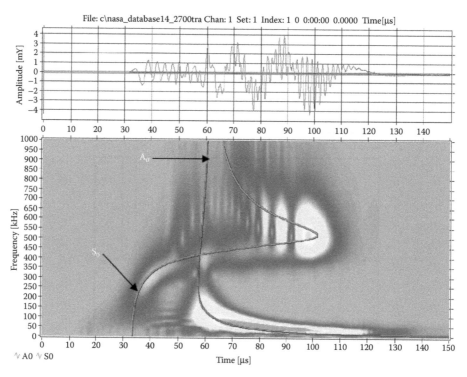

FIGURE 7.101 Recorded AE signal and its wavelet spectrum. Note the correlation with first mode dispersion curves A0 and S0. Wavelet scales are 0–150 μs, and 0–1.0 MHz. (From Hamstad, M. A. et al., *J. Acoustic Emission,* 20, 39-61, 2002. With permission.)

with accuracy. Figure 7.104 shows that the method is also applicable for detecting damage locations that are also deep within the plate thickness. Finally Figure 7.105 shows the equivalent FT of the measure signals. Again, the FT signals fail to provide information regarding the location of the damage. This experiment shows that careful integration of signal proceeding technique, theoretical analysis of wave motion (in a way, this can be considered STRID step), and NDT technique (AE) can provide accurate and informative damage detection methodology.

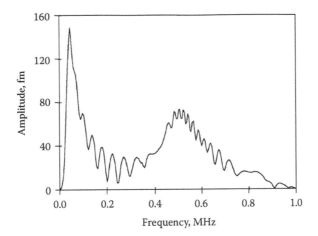

FIGURE 7.102 Fourier spectrum of the AE signal of Figure 7.101. (From Hamstad, M. A. et al., *J. Acoustic Emission*, 20, 39-61, 2002. With permission.)

7.7.4 HILBERT–HWANG TRANSFORM

7.7.4.1 General

Hilbert–Hwang transform is an increasingly popular method for signal analysis processes. It promises to have many successful applications in the field of SHM, especially when applied to civil infrastructure. The method was described in detail by Hwang et al. (1998). This section summarizes the essential theoretical basis of the method. Applications to SHM field are provided next.

7.7.4.2 Instantaneous Frequency

Hilbert transform $\hat{x}(t)$, of an arbitrary time series $x(t)$ can be defined as

$$\hat{x}(t) = \frac{1}{\pi} P \int_{-\infty}^{+\infty} \frac{x(\tau)}{t - \tau} d\tau \tag{7.39}$$

Where P is the Cauchy principal value. Note that $x(t)$ and $\hat{x}(t)$ form a complex conjugate pairs such that an analytic series can be defined as

$$z(t) = x(t) + i \, \hat{x}(t) \tag{7.40}$$

Defining

$$A(t) = \left(x^2(t) + \hat{x}^2(t) \right)^{0.5} \tag{7.41}$$

$$\theta(t) = \tan^{-1}\left(\frac{x(t)}{\hat{x}(t)} \right) \tag{7.42}$$

we end up with

$$z(t) = A(t) \, e^{i\,\theta(t)} \tag{7.43}$$

Hwang et al. (1998) showed that an instantaneous frequency, ω can be defined as

$$\omega = \frac{d\theta(t)}{dt} \tag{7.44}$$

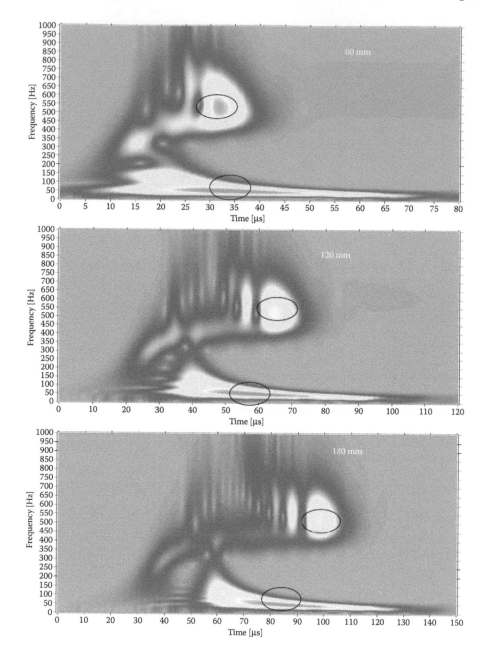

FIGURE 7.103 Effects of source distance on wavelet specturm. The time scales in the three spectra are 0–80 μs, 0–120 μs, and 0–150 μs, from top to bottom, respectively. The frequency scales are all the same 0–1.0 MHz. (From Hamstad, M. A. et al., *J. Acoustic Emission,* 20, 39–61, 2002. With permission.)

if the function $A(t)$ satisfy certain conditions:

1. The number of extreme values and the number of zero crossings must be either equal, or differ only by one at most.
2. The mean value of the envelopes that define the local maxima and the local minima at any point is zero.

A function that satisfies the above two conditions is defined as an intrinsic mode function (IMF). A simple method of decomposing the time series $x(t)$ into a set of IMF components is described next.

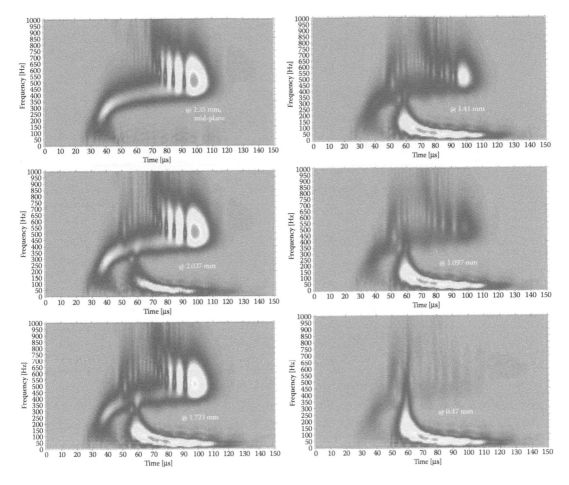

FIGURE 7.104 Effect of depth of AE source. The measurements were all at horizontal distance of 180 mm from AE source. (From Hamstad, M. A. et al., *J. Acoustic Emission,* 20, 39-61, 2002. With permission.)

7.7.4.3 Sifting Process

An ingenious process that is called *empirical mode decomposition* (EMD) was devised by Hwang et al. (1998) that identify IMF components from an arbitrary signal $x(t)$, as in Figure 7.106a. The steps are as follows:

1. Assign $i = 1$, where i is the counter of the IMF component of interest
2. Assign working signal, $W_{ij}(t) = x(t)$
3. Define a sifting counter, $j = 1$
4. Define $X_{MAX}(t)$ as a cubic spline function that connects all the local maxima in $W_{ij}(t)$. Note that $X_{MAX}(t)$ now represents the upper envelope of $W_{ij}(t)$ (see Figure 7.106b)
5. Define $X_{MIN}(t)$ as a cubic spline function that connects all the local minima in $W_{ij}(t)$. Note that $X_{MIN}(t)$ now represents the lower envelope of $W_{ij}(t)$ (see Figure 7.106b)
6. Define the average $\mu_{ij}(t) = ((X_{MAX}(t) - X_{MIN}(t))/2)$ (see Figure 7.106c)
7. Define an initial estimate for the ith IMF as $h_{ij} = W_{ij}(t) - \mu_{ij}(t)$ (see Figure 7.106d)
8. The estimate $h_{ij}(t)$ needs to be improved because: (a) riding waves exist, in the estimate, and they need to be removed, and (b) it needs to have smoother uneven amplitudes. This improvement can be done by increasing the sifting counter by 1 as $j = j + 1$.
9. Assign new working signal $W_{ij}(t) = h_{ij}(t)$

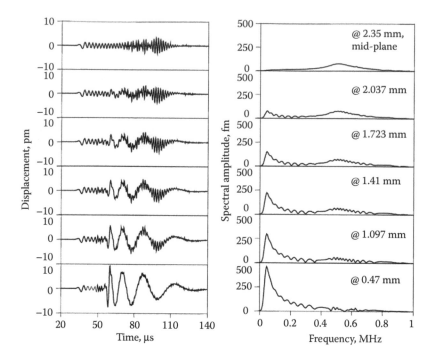

FIGURE 7.105 Measured signals and their Fourier spectra that correspond to Figure 7.104 plots. (From Hamstad, M. A. et al., *J. Acoustic Emission*, 20, 39-61, 2002. With permission.)

10. Repeat steps 4 through 9 until the standard deviation *SD* between two consecutive sifting signals as defined by $SD = \sum_t (|W_{i(j-1)}(t) - W_{ij}(t)|^2 / W_{i(j-i)}^2(t))$ is less than a preset limit. Such a limit can be in the range of 0.2–0.3. When that limit is reached, the ith IMF signal can be defined as $c_i = W_{ij}$
11. Assign $i = i + 1$
12. Repeat steps 3 through 10, until the new working signal $W_{ij}(t)$ becomes too small (less than a preset tolerance level), or when it becomes monotonic, where no more IMF can be extracted. In such a situation a residual, $r_n(t)$ can be defined as $r_n(t) = c_{n-1} - W_{nj}$. Note that $n = i$: the number of IMF components of the original signal $x(t)$

From the results of the above steps, the original signal can now be expressed as

$$x(t) = \sum_{i=1}^{i=n} c_i(t) + r_n(t) \tag{7.45}$$

The residue $r_n(t)$ can be a small number, or a trend. All empirical components $c_n(t)$ have the properties of IMF.

7.7.4.4 Hilbert Spectrum

The final step in evaluating the Hilbert–Hwang spectrum is to compute the Hilbert transform to each of the IMC components and computing the instantaneous frequency according to Equation 7.44. Thus the original signal can be expressed as

$$x(t) = \sum_{j=1}^{n} a_j(t) e^y \tag{7.46}$$

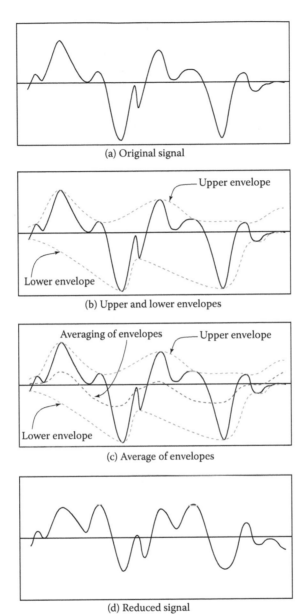

(a) Original signal

(b) Upper and lower envelopes

(c) Average of envelopes

(d) Reduced signal

FIGURE 7.106 Steps for generating IMF (a) original signal, (b) upper and lower envelopes, (c) averages of envelopes, and (d) reduced signal.

with

$$y = \sqrt{-1} \int \omega_j(t) dt \qquad (7.47)$$

Hwang et al. (1998) observed that the IMF expansion of $x(t)$ as shown in Equation 7.46 and Equation 7.47 is a generalized expression of the conventional discrete FT as expressed earlier in this section, which we repeat for convenience as

$$x(t) = \sum_{j=1}^{\infty} a_j e^{i\omega_j t} \qquad (7.48)$$

The differences between FT and IMF expansions are obvious:

- The FT produces constant amplitude for each component, while the IMF expansion produces time-dependent amplitude for each component.
- Similarly, the FT is based on constant frequency for each Fourier component, while the IMF expansion produced a time-dependent instantaneous frequency for each component.

The just-discussed time-frequency spectrum is called *Hilbert–Hwang spectrum*, for consistency we refer to it as HHT. It was shown to have the following advantages

- Can be used to analyze many applications
- Applicable for nonstationary signals or nonlinear problems
- The adaptivity of the HHT makes it problem dependent, which promises to produce more physically meaningful results than other more mathematically rigid transforms
- Adaptivity of the HHT makes it produce adequate frequency and time resolution to the problem on hand within the time and frequency ranges of interest

7.7.4.5 Use in SHM: Earthquake Applications

The use of HHT in studying earthquake time signals was illustrated by Hwang et al. (1998). Consider the El Centro earthquake acceleration time history of Figure 7.107. Using the procedures of IMF evaluations, the ten IMF components of the signal are shown in Figure 7.108. Note how the frequency content is changed for each IMF component. Figure 7.109 shows the time-frequency spectrum of the HHT of El Centro. For comparison with WT (using Morlet wavelets), Figure 7.110 shows the WT of the same time signal. The HHT of Figure 7.109 shows too many details as compared with the WT spectrum. When the HHT is smoothed as in Figure 7.111, the similarity of energy patterns of both WT and HHT becomes evident. However, there are pronounced differences between the two spectra that require additional study.

It is of interest to compare the FT behavior of El Centro with the marginal HHT along the frequency scale. This is important since the FT of El Centro have been studied extensively by numerous authors. Figure 7.112 shows the FT spectra (Figures 7.112a and 7.112b) and the HHT spectra (Figures 7.112c and 7.112d). The HHT shows higher frequency content at low frequencies that the

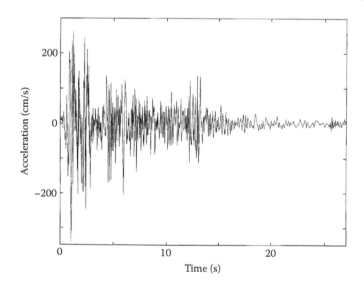

FIGURE 7.107 El Centro signal. (From Hwang, N. E. et al., The empirical mode decomposition and the Hilbert spectrum for nonlinear and non-stationary time series analysis, *Proceedings of the Royal Society*, London, 454, 903–995, 1998. Courtesy of Royal Society, London.)

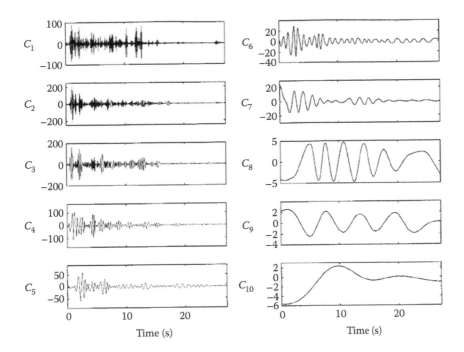

FIGURE 7.108 IMF Components of El Centro Signal. (From Hwang, N. E. et al., The empirical mode decomposition and the Hilbert spectrum for nonlinear and non-stationary time series analysis, *Proceedings of the Royal Society*, London, 454, 903–995, 1998. Courtesy of Royal Society, London.)

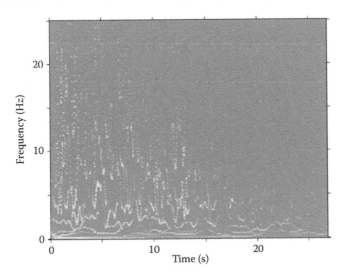

FIGURE 7.109 Hilbert spectrum of El Centro time signal. (From Hwang, N. E. et al., The empirical mode decomposition and the Hilbert spectrum for nonlinear and non-stationary time series analysis, *Proceedings of the Royal Society*, London, 454, 903–995, 1998. Courtesy of Royal Society, London.)

FT does not produce. This low frequency behavior can have a major effect on the seismic behavior on many important structures such as suspension and cable stayed bridges, as well as tall buildings.

7.7.4.6 WT versus HHT in Infrastructure Applications

Both Wavelet and EMD methods are capable of detecting time of changes in the properties of a given time signal. Vincent et al. (2000) investigated this capability in both techniques in a purely

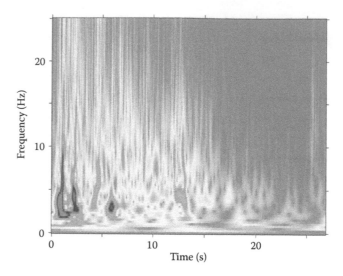

FIGURE 7.110   Wavelet transform (Morlet) of El Centro time signal. (From Hwang, N. E. et al., The empirical mode decomposition and the Hilbert spectrum for nonlinear and non-stationary time series analysis, *Proceedings of the Royal Society*, London, 454, 903–995, 1998. Courtesy of Royal Society, London.)

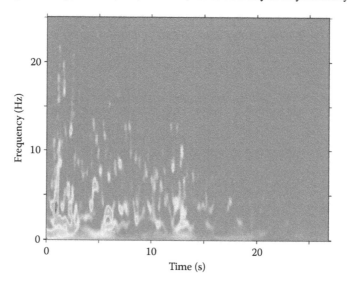

FIGURE 7.111   Smoother Hilbert transform of El Centro. (From Hwang, N. E. et al., The empirical mode decomposition and the Hilbert spectrum for nonlinear and non-stationary time series analysis, *Proceedings of the Royal Society*, London, 454, 903–995, 1998. Courtesy of Royal Society, London.)

numerical approach. They modeled a 3-DOF close coupled spring-mass system that is subjected to a sinusoidal base motion that lasts 20 seconds. After 10 seconds, they suddenly reduced the stiffness of one of the three springs in the system. The computed response of the third mass was then analyzed by both EMD and wavelet approach. The writers found that the sudden reduction in the spring (resulting in a softening of the whole system) was detected accurately by both methods, at 10 seconds. In addition, both methods gave some indication as to the severity of the damage, albeit in a different manner. The changes in the frequencies that correspond to vibration modes in the EMD approach would enable the estimation of damage level (loss of spring stiffness). The continuous wavelet approach might also give an indication of damage severity through its fringe patterns. In all, both methods proved to be worthy in detecting occurrence and level of the damage.

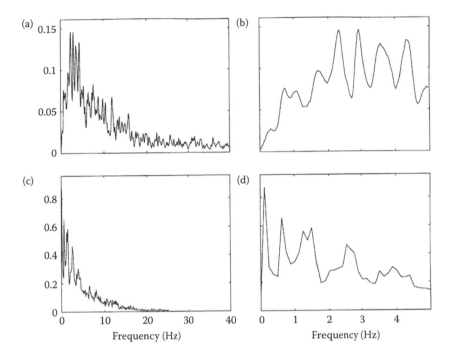

FIGURE 7.112 Fourier and marginal Hilbert spectra of El Centro time signal (a) Fourier spectrum in the range of 0-40 Hz; (b) detailed Fourier spectrum in the range of 0-5 Hz; (c) marginal HHT spectrum in the range of 0-20 Hz; and (d) detailed marginal HHT spectrum in the range of 0-5 Hz. (From Hwang, N. E. et al., The empirical mode decomposition and the Hilbert spectrum for nonlinear and non-stationary time series analysis, *Proceedings of the Royal Society*, London, 454, 903–995, 1998. Courtesy of Royal Society, London.)

In a similar study that addressed only the usability of wavelet method for DMID, Hou and Noori (2000) looked at an analytically simulated 3-DOF closed coupled system. Their investigations addressed the effects of level of damage and the noise contamination on the efficiency of the wavelet method in identifying the time and location of damage. They observed that the method is effective at higher damage levels (around 20% for their experiment) and/or low noise contamination. Conversely, they observed that, at low damage levels (around 5% for their experiment), the noise contamination might reduce the accuracy of prediction.

7.7.5 Signal Processing as an Integrator of DMID Methods

7.7.5.1 Use of Correlation and Operational Modal Analysis

As an example of an integral use of SHM different techniques in increasing safety, consider the SHM project by Lin et al. (2005). The principals of that project wanted to (a) investigate numerical modeling assumptions of elastomeric bearings in bridges, and (b) the temperature effects on such a behavior. It is noted that during modeling a bridge that is supported by elastomeric bearing, those bearings are assumed to be either hinges, or fixed boundary conditions. Such assumptions were shown to be inaccurate (see Sanayei et al. (1999). The sensitivity of analysis and design of bridges to such assumptions made the SHM project important. The principals of the project decided to use dynamic modes of the bridge as the metrics in their project. However, since it is expected that temperature can affect modal behavior of bridges, it was important to include sensitivity to temperature variations in the project activities.

A simple span, skew composite steel bridge in Amherest, NY was used for the project (Figure 7.113). There were ten steel girders that were about 10 ft apart. The girders were connected by a set of cross bracings. The skew angle was about 40°–45°. Each end of the girder is supported

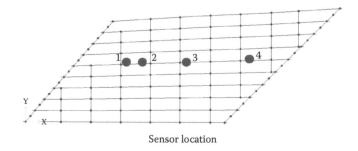

Sensor location

FIGURE 7.113   Plan of the bridge, and location of sensors. (Courtesy of Jerome O'Connor.)

by an elastomeric bearing that is 2.5-in thick, and in-plane dimensions of 30 in × 12 in. Four accelerometers were used. The accelerometers have a sampling range of 0–300 Hz; fairly adequate to process the project frequency range of 0–20 Hz. The plan of the bridge and the locations of the accelerometers are shown in Figure 7.113. As much as 168 hours of measurements were recorded. The recordings were made at 2 hours intervals. The measurements were made at different ambient temperatures, ranging from 80° to 32°F.

One of the challenges of that project was the method to excite some of the bridge dynamic modes. The bridge under consideration is a heavy bridge, and to use forced vibrations to excite the needed modes would be costly. Because of this it was decided to use the ambient vibration measurements to evaluate the modal behavior. The measured vibrations would then be analyzed using signal processing techniques to estimate the needed information. Measured ambient vibrations contain both the bridge natural modal behavior (forced and free vibrations) as well as unavoidable noise. Using signal processing techniques if it was possible to eliminate, or reduce the noise, then it would be possible to estimate accurately the bridge behavior. Two signal processing techniques were used.

Cross-Correlation Analysis: As was mentioned earlier, cross-correlation and the cross power density function of two time signals x_1 and x_2 can be described as

$$R_{12} = \int_{-\infty}^{+\infty} x_1(t) x_2(t+\tau)\, dt$$

(7.49)

$$P_{12} = FT(R_{12})$$

(7.50)

Note that the effects of the noise will be reduced in P_{12}, leaving the bridge modal effects in the signal. To improve accuracy, the authors used an average of several cross power spectra for their estimation. Figures 7.114 and 7.115 show the results of the cross-correlation analysis at 40° and 80°F, respectively. The first two natural frequencies of the bridge were insensitive to temperature at about 2 Hz and 4 Hz. The third natural frequency of the bridge was estimated as 6.3 Hz at 40° and 5.7 Hz at 80°.

Random Decrement (Operational Modal Analysis): Random decrement $y(t)$ of a function $x(t)$ of length n is defined as

$$y(\tau) = \frac{1}{N} \sum_{i=1}^{N} x(\tau + t_i)$$

(7.51)

The Fourier spectra of $y(t)$ will also reduce noise effects and enhance modal behavior of the bridge. It was found that the first two natural frequencies of the bridge were similar to the correlation analysis at 2 Hz and 4 Hz, respectively. The third frequency was estimated at 5.8 Hz, fairly similar to the correlation results.

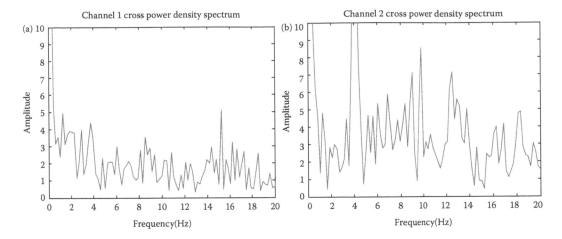

FIGURE 7.114   Cross power spectral accelerations (40°) (a) channel 1, and (b) channel 2. (Courtesy of Jerome O'Connor.)

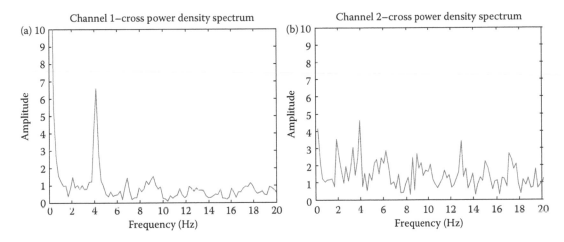

FIGURE 7.115   Cross power spectral accelerations (80°) (a) channel 1, and (b) channel 2. (Courtesy of Jerome O'Connor.)

The above estimations showed some sensitivity of the bridge behavior (third mode) to temperature effect. To further investigate the modeling effects of the elastomeric bearings, the authors used a finite element modeling technique. They showed that by accurately modeling the bearings, it was possible to produce natural frequencies of the bridge similar to the measurements. They found that only vertical stiffness of the bearings affect bridge natural frequencies. They concluded that in numerical bridge modeling, the vertical stiffness of elastomeric bearings must be assigned with care to ensure accurate modeling results.

7.7.5.2   Filtering Multihazards Effects

Connor et al. (2002) used a specifically designed signal processing techniques that meets the goals of the SHM experiment. They utilized high and low-pass filters to isolate signals from different sources. The bridge was a four span continuous haunched steel deck truss. The deck was designed to be fully composite with the main upper chord of the steel truss as well as the longitudinal stringers and transverse floor beams. Thus, due to this unique structural characteristic, it was decided to instrument and monitor the structure to better understand its short- and long-term behavior. Items of interest were: live load response, quantifying the effects of long-term creep and shrinkage of the deck, and degree of local and global composite action.

Due to the interest in long-term response from construction to in service, vibrating wire strain gages wee used for static type of measurements and regular strain gages for short duration dynamic measurements (load tests, etc.). Some gages were also installed inside the deck on the rebars of interest. Special attention was paid to temperature compensation effects. A wireless modem installed at the site was used to communicate the data from the field to a remote office location. Nine 12-volt batteries were used to power the system with a solar panel. A methodology was adopted to conserve power based on observed activity.

Strain data was processed in three steps. First, a digital low-pass filter was used to remove the noise spikes from the data stream caused by phenomena such as quickly changing magnetic fields and radio transmissions.

A copy of filtered data (low pass) was passed through a second high-pass filter to remove the strain effects from thermal effects. Thus, one file contains mechanical and thermal and the other thermal alone. Subtracting the later from the former, gives mechanical signals alone.

The response was as expected and agreed with designers' estimates. Truss acted compositely with the entire deck system. Hence, live load stresses decreased and a more redundant structure system is attained overall.

7.7.6 RULES FOR USING SIGNAL PROCESSING TOOLS

The properties of the four popular signal processing methods are now compared in Table 7.29. Clearly, the methods can be helpful in SHM applications depending on their properties. Table 7.30 shows some SHM situations where these four methods can be used. We conclude this section with some rules for applying those different signal processing methods in general SHM situations.

TABLE 7.29
Comparison of Signal Processing Methods

Issue	FT	TFT	WT	HHT
Scales	Frequency	Frequency and time	Frequency and time	Frequency and time
Discrete/Continuous	Yes/yes	Yes/yes	Yes/yes	NA
Stationary	No	No	Yes	Yes
Nonlinear systems	No	No	Yes	Yes
Adaptive	No	No	No	Yes

TABLE 7.30
Potential SHM Uses of Different Signal Processing Methods

Method	Possible use in SHM	Comment
FT	Earthquake engineering	FT has been used extensively in earthquake engineering, even though earthquake signals are not stationary. The potential nonlinear response of the structures to earthquakes might limit the accuracy of FT in studying earthquake responses and damages
TFT	Strain history due to truck motions	Sine response of bridges to truck motions can be mostly linear, TFT might provide accurate tool for processing the strain history. The time-varying frequency response can be captured by TFT (WT and HHT can also be used for this task)
WT	Flood and scour	Potential nonlinearities and nonstationary response of bridges to floods and scour are reasons for considering WT (and HHT) to analysis in this situation
HHT	Flood and scour	Same as WT, above

- **Plan Ahead:** The frequency and time steps, as well as the frequency and time ranges should be well know in advance of any SHM experiment.
- **Structural Identification:** Ensure that the above steps and ranges are consistent with the structural properties of interest. For example, the frequency range should cover all structural modes of interest in a STRID situation.
- **Damage Scales:** Justify in advance how signal processing can aid in identifying the damage of interest. Ensure that the frequency range and steps are consistent with the expected damage type, location, and size.
- **Uncertainties:** Make allowance for any uncertainty or variability in the experiment. This includes, but not limited to, setting up potential probabilistic methods that are consistent with the signal processing method.

7.8 DAMAGE IDENTIFICATION IN SHM

7.8.1 GENERAL

The previous sections of this chapter explored several methods of DMID. It should be remembered that the choice of a particular method for DMID should be consistent with the overall goals and objectives of the project. They should also be consistent with the damage parameters as explained earlier in this chapter. This section will present some important considerations that might be of help to the professional while choosing a DMID method. The presence of DMID as only one component within the overall SHM project means that the interactions of DMID with other components must be considered.

7.8.2 IMPORTANT CONSIDERATIONS

Quality: Quality of results of DMID method needs to be considered carefully. How capable the method is in identifying the damage of interest? Some methods can produce qualitative damage description, such as "corrosion is extensive in near the support" or "the connection has an unacceptable deformation." Other methods would produce a more quantitative damage estimate such as "a 0.25 inch crack is detected at a depth of 0.12 inch below the surface." Both qualitative and quantitative damage estimates are useful. The professional will need to ensure that the quality of the damage estimate is consistent with the objectives of the SHM project on hand.

Resolution: DMID methods must be capable of addressing time (frequency) and spatial resolution of the damage of interest.

Reliability: Uncertainties need to be considered as a part of the DMID experiment. This can be accomplished by taking sufficient sample points, repeating the experiment, or any other activity that would accommodate uncertainties or unexpected breakdowns in parts of the experiment.

Fiscal: The cost of the DMID method must be reasonable and consistent with the potential value of information that is expected from the activity. There are numerous cost–benefit examples in this volume that can be used for such an evaluation.

Others: There are many other considerations for the use of a DMID method in an SHM project. Some of those issues are environmental considerations, traffic conditions, experience of the professionals, and prevailing rules and regulations.

7.8.3 INTERACTIONS WITH OTHER SHCE COMPONENTS

Damage Identification processes can only be one part of a whole SHCE process. It needs to be preceded by, or performed in conjunction with, sensing or measurement phase. DMID process is also usually followed by a decision making process. DMID can be executed alone, or in conjunction with STRID activity. Table 7.31 explores some of these interrelationships between DMID for different type of damages/hazards.

7.8.4 Common Bridge Damages and SHM Applications

Damages affect bridges in many forms. They depend on the type of bridge, the bridge component as well as the cause of the damage. Table 7.32 shows some types of damages that can occur in reinforced concrete bridges (Figures 7.116 and 7.117). Similarly, Tables 7.33 and 7.34 show types of damages that can occur in steel truss and steel girder bridges, respectively (Figures 7.118–7.123). Finally, Table 7.35 shows some types of damages that might occur in bridge pedestal seats. Other damage situations will be explored in depth by Ettouney and Alampalli (2012) (Figures 7.124–7.126).

TABLE 7.31
Interactions of DMID with Other SHCE Phases

| SHCE Phase | Hazard | | | |
	Fatigue	Corrosion	Earthquakes	Scour
Sensing	Real time or intermittent sensing is suitable		Pre-event sensing would be different from Post-event sensing	Real time or intermittent sensing is suitable
STRID	Effects of monitored fatigue damage on structural performance needs to be investigated. Special considerations need to be taken in the analysis phase due to the localized nature of fatigue damage	Accommodating corrosion effects in STRID efforts might be difficult. Loss of section can be modeled, however, accurate information is needed	Nonlinear analysis is needed for this type of STRID. For both pre- and afterevent analysis	Soil-foundation-water interactions need to be simulated. Direct modeling might be difficult. Informal analysis techniques such as neural networks might prove to be convenient

TABLE 7.32
Types of Damages and Reinforced Concrete Bridge Types

Bridge Type	Damages
Cast-In-Place slabs Tee beams Concrete girders Concrete channel beams Concrete rigid frames	Cracking, scaling, delamination, spalling, efflorescence, honeycombs, pop-outs, wear, collision damage, abrasion, overload damage, reinforced steel corrosion (Figure 7.116)
Concrete arches Precast and prestressed slabs	In addition to the above, stress corrosion
Prestressed double tees (Figure 7.117)	Cracking, scaling, delamination, spalling, efflorescence, honeycombs, pop-outs, wear, collision damage, abrasion, overload damage, reinforcing steel corrosion, stress corrosion of prestressing strands
Prestressed I-beams Concrete box girders including segmental	Cracking, Delamination, Spalling, Collision damage, Overload damage, Reinforcing/prestressing steel corrosion, stress corrosion, Efflorescence, Pop-outs
Concrete box culverts	Cracking, scaling, delamination, spalling, efflorescence, honeycombs, pop-outs, wear, abrasion, overload damage, Section loss of exposed reinforcing bars, embankment scour at culvert inlet and outlet, roadway settlement

FIGURE 7.116   Delamination of reinforced concrete beam. (Courtesy of New York State Department of Transportation.)

FIGURE 7.117   Spalling of concrete and corroded rebars. (Courtesy of New York State Department of Transportation.)

7.9   APPENDIX: LAMB WAVES

Lamb waves (also known as guided waves, or plate waves) are linear elastic waves that propagate in solids at UT frequencies. They can travel for relatively long distances. They are guided waves that travel in plates or shell systems, that is, systems with small thicknesses. The Lamb wave velocity is proportional to both the plate thickness and the frequency of vibration. It has two main modes of propagation: symmetric and asymmetric. In civil infrastructure, they have been used successfully in detecting damages in several types of system components such as pavements and truss gusset plates. One major limitation of utilization of Lamb waves in civil infrastructure damage detection is that they are fairly limited to localized damage detection, as opposed to global damage detection. For detailed discussions on Lamb waves, see Rose (1999).

7.10   APPENDIX: DISPERSION CURVES

The 3D elasticity equations in Cartesian coordinates can be expressed as

$$\sigma_{ij,j} + \rho\, f_i = \rho \ddot{u}_i \tag{7.52}$$

$$\varepsilon_{ij} = 0.5\left(u_{i,j} + u_{j,i} \right) \tag{7.53}$$

TABLE 7.33
Type of Damages in a Steel Truss Bridge

Truss Components	Damages
Truss members	Section losses due to corrosion (Figure 7.118). Corrosion between plates of built-up members. Effects of de-icing salts
	Damage due to direct collision impact resulting in cracks or large deformations. Imperfections in welds, especially plug-welds. Damages in bolted or welded retrofits due to drilling, flame cutting, or punching
	Looseness in truss rods should be considered a type of damage, since they are not carrying the intended tensile force
Alignment of individual truss members	Truss member misalignment should be considered as a type of damage. Such misalignment can produce force redistribution, which is inconsistent with design condition that might lead to undesirable overstressing. DMID of alignment can be achieved by tilt meters, laser sensing, etc.
Member overstressing	Lateral or local buckling are signs of overstressing in compression members (Figure 7.119). Necking down of cross-sections is a sign of overstressing in tension members. Special attention should be given to less ductile steels such as high strength steel, or older cast iron construction.
Connections	Lost rivets or welds can result in a higher undesired stresses. Acoustic monitoring can help in detecting noises from loose connections (the limit is an AE monitoring). Visual inspections of corroded connections or supports can be of help (Figure 7.120)

Source: NYSDOT, *Bridge Inspection Manual,* New York State Department of Transportation, Albany, NY, 1997.

TABLE 7.34
Type of Damages in a Steel Girder Bridge

Steel Girder Bridge Components	Damages
Alignment of individual girders and/or the whole girder assembly	Misalignment should be considered as a type of damage. Such misalignment can produce force redistribution, which is inconsistent with design condition that might lead to undesirable overstressing.
Web crippling/section losses	This can happen in sections with large shearing forces, such as those near supports. Crippling indicates excessive stresses. Section losses occur due to corrosion. DMID methods for detecting these damages are not the same and should be considered separately (Figure 7.121)
Corrosion	Corrosion would cause section losses in some hidden areas such as under the deck, behind sidewalks, and above supports. Careful choice of DMID methods is essential so as to ensure accuracy in detection (Figure 7.122)
Riveted or built-up sections	These components are particularly susceptible to crevice corrosion (corrosion in hidden surfaces). Careful choice of DMID methods is essential so as to ensure accuracy in detection

Source: NYSDOT, *Bridge Inspection Manual,* New York State Department of Transportation, Albany, NY, 1997.

$$\sigma_{ij} = \lambda \varepsilon_{ij} \delta_{ij} + 2\mu \varepsilon_{ij} \tag{7.54}$$

Eliminating stresses and strains from Equations 7.52, 7.53, and 7.54 we get

$$\mu u_{i,jj} + (\lambda + \mu) u_{j,ji} + \rho f_i = \rho \ddot{u}_i \tag{7.55}$$

FIGURE 7.118   Corrosion loss in a steel member. (Courtesy of New York State Department of Transportation.)

FIGURE 7.119   Local buckling in a steel member. (Courtesy of New York State Department of Transportation.)

FIGURE 7.120   Heavily corroded steel bearing. (Courtesy of New York State Department of Transportation.)

FIGURE 7.121   Section loss in steel girders. (Courtesy of New York State Department of Transportation.)

FIGURE 7.122   Corrosion of flange near sidewalk due to de-icing. (Courtesy of New York State Department of Transportation.)

FIGURE 7.123   Steel girder damage near supports. (Courtesy of New York State Department of Transportation.)

TABLE 7.35
Type of Damages in a Bridge Seats and Pedestals

Type of Bridge Pedestal	Damages
Concrete	Damages include cracking, scaling, spalling, delaminating, or leaching. See Figures 7.124 and 7.125. Figure 7.126 shows a heavily damaged concrete bearing seat. A temporary bridge seat is used to support the bridge girder (shown in the foreground of Figure 7.126)
Masonry	Damages include cracks and deterioration of mortar joints or stone bricks. Loss of masonry might occur as the pedestal deteriorates
Steel	Damages include loss of section, bowing, or buckling

Source: NYSDOT, *Bridge Inspection Manual,* New York State Department of Transportation, Albany, NY, 1997.

FIGURE 7.124   Bridge pedestal with minor damages. (Courtesy of New York State Department of Transportation.)

FIGURE 7.125   Bridge pedestal with moderate damages. (Courtesy of New York State Department of Transportation.)

FIGURE 7.126   The inspector shows a heavily damaged bridge bearing concrete seat, where the reinforcement rebars are exposed.

For plate with thickness h in the x_3 direction that is free from tractions at top and bottom surfaces ($x_3 = -0.5h$ and $x_3 = 0.5h$), Equation 7.55 can be decomposed to two equations as follows

$$\frac{\partial^2 \phi}{\partial x_1^2} + \frac{\partial^2 \phi}{\partial x_3^2} = \frac{1}{c_L^2} \frac{\partial^2 \phi}{\partial t^2} \tag{7.56}$$

$$\frac{\partial^2 \psi}{\partial x_1^2} + \frac{\partial^2 \psi}{\partial x_3^2} = \frac{1}{c_T^2} \frac{\partial^2 \psi}{\partial t^2} \tag{7.57}$$

Equation 7.56 governs longitudinal wave motion while Equation 7.57 governs the shear wave motion. Assuming plane strain motions in the plate, the displacements are related to ϕ and ψ by

$$u_1 = \frac{\partial \phi}{\partial x_1} + \frac{\partial \psi}{\partial x_3} \tag{7.58}$$

$$u_2 = 0 \tag{7.59}$$

$$u_3 = \frac{\partial \phi}{\partial x_3} - \frac{\partial \psi}{\partial x_1} \tag{7.60}$$

By separating the time and space variables, we assume the wave form solutions for Equations 7.58 and 7.60 as

$$\phi = \Phi(x_3) e^{i(kx_1 - \omega t)} \tag{7.61}$$

$$\psi = \Psi(x_3) e^{i(kx_1 - \omega t)} \tag{7.62}$$

Defining the wave number as k, the driving frequency as ω, the wave phase velocity as c_p and the wave length as λ we have the following relationships:

$$k = \frac{\omega}{c_p} \tag{7.63}$$

and

$$c_p = \frac{\lambda \omega}{2\pi} \tag{7.64}$$

We can further introduce the following variables:

$$p^2 = \frac{\omega^2}{c_L^2} - k^2 \tag{7.65}$$

and

$$q^2 = \frac{\omega^2}{c_T^2} - k^2 \tag{7.66}$$

Substituting Equations 7.63 through 7.66 into Equations 7.61, and 7.62, we obtain

$$\frac{\tan(qh)}{q} + \frac{4k^2 p \tan(ph)}{\left(q^2 - k^2\right)^2} = 0 \tag{7.67}$$

$$q \tan(qh) + \frac{\left(q^2 - k^2\right)^2 \tan(ph)}{4k^2 p} = 0 \tag{7.68}$$

Equation 7.67 governs the symmetric wave motions while Equation 7.68 governs antisymmetric wave motions along x_1. For detailed discussions on dispersion curves, see Rose (1999).

7.11 APPENDIX: HELMHOLTZ EQUATION

Let us define, as usual, the wave number k, as

$$k = \frac{\omega}{c} \tag{7.69}$$

Then Helmholtz equation can be written as

$$\left(\nabla^2 + k^2\right) A = 0 \tag{7.70}$$

$$A = A(x_1, x_2, x_3) \tag{7.71}$$

For more details, see Rose (1999).

7.12   APPENDIX: ANGULAR SPECTRUM METHOD

By inspecting Equation 7.71, for a constant x_3, we can perform the FT in the (x_2, x_3) plane as

$$A = \iint_{-\infty}^{+\infty} \hat{A}\, e^{i\left(x_2 \hat{x}_1 + x_2 \hat{x}_2\right)}\, d\hat{x}_1\, d\hat{x}_2 \qquad (7.72)$$

The inverse transform of this equation is

$$\hat{A} = \frac{1}{4\pi^2} \iint_{-\infty}^{+\infty} A\, e^{-i\left(x_2 \hat{x}_1 + x_2 \hat{x}_2\right)}\, dx_1\, dx_2 \qquad (7.73)$$

Note that

$$\hat{A} = \hat{A}\left(\hat{x}_1, \hat{x}_2, x_3\right) \qquad (7.74)$$

Substituting Equation 7.74 in Equation 7.70 we get the transformed Helmholtz equation as

$$\frac{\partial^2 \hat{A}}{\partial x_3^2} + \omega^2\, \hat{A} = 0 \qquad (7.75)$$

With

$$\omega^2 = k^2 - x_1^2 - x_2^2 \qquad (7.76)$$

The solution to Equation 7.75 is

$$\hat{A} = \hat{A}_1\, e^{i\omega x_3} + \hat{A}_2\, e^{-i\omega x_3} \qquad (7.77)$$

With the separation of variables we have $\hat{A}_1 = \hat{A}_1(\hat{x}_1, \hat{x}_2)$ and $\hat{A}_2 = \hat{A}_2(\hat{x}_1, \hat{x}_2)$. We can now express the general solution of Helmholtz equation as

$$A = \iint_{-\infty}^{+\infty} \hat{A}_1\, e^{i\left(x_2 \hat{x}_1 + x_2 \hat{x}_2 + \omega x_3\right)}\, d\hat{x}_1\, d\hat{x}_2 + \iint_{-\infty}^{+\infty} \hat{A}_2\, e^{i\left(x_2 \hat{x}_1 + x_2 \hat{x}_2 - \omega x_3\right)}\, d\hat{x}_1\, d\hat{x}_2 \qquad (7.78)$$

The two parts of Equation 7.78 represent incoming waves and outgoing waves for $e^{+i\omega x_3}$ and $e^{-i\omega x_3}$, respectively. Each incoming or outgoing wave can either be propagating wave, or exponentially decaying wave in the x_3 direction when $\omega^2 \geq 0$ or $\omega^2 < 0$, respectively. The four types of waves can represent any wave field in a plate. Equation 7.78 have thus reduced the 3D representation of complex wave field into a set of planar waves in the $(\hat{x}_2, \hat{x}_3)$ plane. The spectrum represented by $\hat{A}_1 = \hat{A}_1(\hat{x}_1, \hat{x}_2)$ and $\hat{A}_2 = \hat{A}_2(\hat{x}_1, \hat{x}_2)$ is called the *angular spectrum* since the coordinates $\hat{x}_2$ and $\hat{x}_3$ can be related to the angle between the original (x_2, x_3) plane and the transformed $(\hat{x}_2, \hat{x}_3)$ plane.

REFERENCES

Alkhalifah, T. (2005). "Nondestructive Testing Resolution: Electromagnetic Waves versus Seismic Ones," 3rd Middle East Nondestructive Testing Conference & Exhibit, Manama, Bahrain.

ASNT. (1995). "Laser Based Nondestructive Testing Methods," in *Nondestructive Testing Handbook, 2nd Edition, Vol. 9, Special Nondestructive Testing Methods*. ASNT, Columbus, OH.

ASNT. (1999). *Choosing NDT: Applications, Costs, and Benefits of Nondestructive Testing in Your Quality Assurance Program*, American Society for Nondestructive Testing, Inc., Columbus, OH.

Baron, J. and Ying, S. (1987). "Acoustic Emission Source Location," in *Nondestructive Testing Handbook: Vol. 5, Acoustic Emission Testing*, ASNT, Columbus, OH.

Beissner, R. (2005). "Magnetic Field Testing," in *ASM Handbook: Non Destructive Evaluation and Quality Control, Vol. 17*, ASM, Materials Park, OH

Beissner, R. and McGinnis, C. (1984). "Laboratory Test of Magnetic Field Disturbance (MFD) System for Detection of Flaws in Reinforcing Steel," Report No. DTFH 61-80-C-00002, FHWA, Washington, DC.

Bendat, J. and Piersol, A. (1971). *Random Data: Analysis and Measurement Procedures*, Wiley-Interscience, New York.

Birks, A., Golis, M., and Green, R. (2007). "Introduction to Ultrasonic Testing," in *Nondestructive Testing Handbook: Vol. 10*, ANST, Columbus, OH.

Boashash, B. (1992). *Time-Frequency Signal Analysis*, Wiley, New York.

Bossi, R., Mengers, P., and Qien C. (2007). "Radioscopy and Tomography," in *Nondestructive Testing Handbook: Vol. 10*, ASNT, Columbus, OH.

Botsco, R. and Jones, T. (2007). "Thermography and Other Special Methods," in *Nondestructive Testing Handbook: Vol. 10*, ASNT, Columbus, OH.

Botsko, R. (2007). "Introduction to Ultrasonic Testing," Section 13, Part 2 in *Nondestructive Testing Handbook: Vol. 10*, ASNT, Columbus, OH.

Butt, S. and Limaye, V. (2002). "Damage Detection in Concrete Bridge Deck Slabs Using Acoustic Attenuation," Proceedings of 1st International Workshop on Structural Health Monitoring of Innovative Civil Engineering Structures, ISIS Canada Corporation, Manitoba, Canada.

Carlos, M. F., Miller, R. K., and Tamutus, T. A. (2000). "Acoustic Emission Local Area Monitoring System," *Structural Materials Technology: an NDT Conference*, ASNT, Atlantic City, NJ.

Carlos, M. F., Wang, D., Vahaviolos, S., Anastasopoulos, A. (2006). "Advanced Acoustic Emission for On-stream Inspection of Petrochemical Vessels," Proceedings of the 3rd International Conference on Emerging Technologies in NDT, A. A. Balkema, Netherlands 2004, ISBN 90 5809 645 9 (Volume)-. 90 5809 645 7(CD), pp. 167–172.

Carlyle, J. H. (1993). "Acoustic Emission Monitoring of the I-10 Mississippi River Bridge," *Phase Report No. R90-259*, Physical Acoustics Corporation, Lawrenceville, NJ.

Carlyle, J. H. and Ely, T. M. (1992). "Acoustic Emission Monitoring of the I-95 Woodrow Wilson Bridge," *Phase Report No. R90-259*, Physical Acoustics Corporation, Lawrenceville, NJ.

Carlyle, J. H. and Leaird, J. D. (1992). "Acoustic Emission Monitoring of the I-80 Bryte Bend Bridge," *Phase Report No. R90-259*, Physical Acoustics Corporation, Lawrenceville, NJ.

Colla, C. and Wiggenhauser, H. (2000). "Developments in Impact-Echo for the investigation of Concrete Structures," *Structural Materials Technology: an NDT Conference*, ASNT, Atlantic City, NJ.

Connor, J. (1976). *Analysis of Structural Member Systems* The Ronald Press Company, New York.

Connor, R. J., Santousuosso, B., and Pessiki, S. P. (2002). "Long-Term Wireless Remote Monitoring of the Lehigh River Bridge," *Proceedings, NDE Conference on Civil Engineering*, ASNT, Cincinnati, OH.

Cullington, D. W., MacNeil, D., Paulson, P., and Elliot, J. (2001). "Continuous acoustic monitoring of grouted post-tensioned concrete bridges," *NDT & E International, Vol. 34, No 2*.

De Witt, D. (1988). *Theory and Practice of Radiation Thermometry*, John Wiley, New York.

Demma, A., Alleyne, B., and Pavlakovie, B. (2005). "Testing of Buried Pipelines using Guided Waves," Proceedings of the 2005 ASNT Fall Conference, Columbus, OH.

Dennis, M. (2005). "Industrial Computed Tomography," in *ASM Handbook: Non Destructive Evaluation and Quality Control, Vol. 17*, ASM, Materials Park, OH

Doebling, S.W., Farrar, C.R., Prime, M.B., and Shevitz, D. W. (1996). "Damage Identification and Health of Structural and Mechanical Systems from Changes in their Vibrational Characteristics: A Literature Review," *Los Alamos National Laboratory Report, LA-13070-MS*.

Duan, Z., Yan, G., and Ou, J. (2004). "Structural Damage Detection in Ambient Vibration Using Wavelet Packet Transform and Probabilistic Neural Network," Proceedings of 2nd International Workshop on Structural Health Monitoring of Innovative Civil Engineering Structures, ISIS Canada Corporation, Manitoba, Canada.

Duke, J. C. and Horne, M. R. (2006). "Viability of Monitoring Acoustic Emission from Crack Growth in Bridges," *NDE Conference on Civil Engineering*, ASNT, St. Louis, MO.

Dunegan, H., Harris, D., and Tetelman, A. (1970). "Detection of Fatigue Crack Growth by Acoustic Emission Techniques," *Materials Evaluation, Vol. 28, No. 10*, Columbus, OH.

Emigh, C., Iddings, F., and Willenberg, J. (2007). "Radiation Principles and Sources," in *Nondestructive Testing Handbook*: *Vol. 10*, ASNT, Columbus, OH.

Endo, K. (2000). "A study on the Application of the Acoustic Emission Method for Steel Bridges," Masters of Science in Engineering Thesis, The University of Texas at Austin, Austin, TX.

Ettouney, M. and Alampalli, S. (2000). "Engineering Structural Health." *ASCE Structures Congress 2000*, Philadelphia, PA.

Ettouney, M. and Alampalli, S. (2012). *Infrastructure Health in Civil Engineering, Applications and Management*, CRC Press, Boca Raton, FL.

Ettouney, M., Daddazio, R., Hapij, A., and Aly, A., (1998), "Health Monitoring of Complex Structures," Proceedings, Smart Structures and Materials 1998: Industrial and Commercial Applications of Smart Structures Technologies, SPIE, Vol. 3326.

Farrar, C. and Doebling, S. (1999). "Damage Detection and Evaluation II," in *Modal Analysis and Testing*, Editors: Silva, J. and Maia, N. NATO Science Series, Kluwer Academic Publishers, London, UK.

Farrar, C. and Jauregui, D. (1996). "Damage Detection Algorithms Applied to Experimental and Numerical Modal Data from the I-40 Bridge," *Los Alamos National Laboratory Report No. LA-13074-MS*, Los Alamos, NM.

Farrar, C., Baker, W., Bell, T., Cone, K., Darling, T., Duffey, T., Eklund, A., and Migliori, A. (1994). Dynamic Characterization and Damage Detection in the I-40 Bridge over the Rio Grande, *Los Alamos National Laboratory Report LA-12767-MS*.

Fisk, P. (2008). "Sonic / Ultrasonic Testing for Bridge Inspections," *NDE/NDT for Highway and Bridges: Structural Materials Technology (SMT)*, ASNT, Oakland, CA.

Fu, Y. and Dewolf, J. (2001). "Monitoring and Analysis of a Bridge with Partially Restraind Bearings," *ASCE Journal of Bridge Engineering,* 6(1).

Garcia, G. and Stubbs, N. (1995). "Effect of damage size and location on the stiffness of a rectangular beam," *Smart Structures and Materials*, San Diego, CA.

Graps, A. (1995). Introduction to Wavelets, *IEEE Computational Science and Engineering, IEEE*, Los Alamitos, CA.

Green, A., Blackburn, P., Craig, B., Cross, N., Ferdinand, M., Fowler, T., and Robinson, D. (1987). "Acoustic Emission Applications in the Petroleum and Chemical Industries," in *Nondestructive Testing Handbook: Volume Five, Acoustic Emission Testing*, ASNT, Columbus, OH.

Greene, A. (2005). "Radiographic Inspection," in *ASM Handbook: Non Destructive Evaluation and Quality Control, Vol. 17*, ASM, Materials Park, OH

Gucunski, N., Rascoe, C., Wang, Z., Chen, Y., and Fang, T. (2008). "Automated Interpretation of Bridge Deck Impact Echo Data Using Three - Dimensional Visualization," *NDE/NDT for Highways and Bridges: Structural Materials Technology (SMT)*, ASNT, Oakland, CA.

Hamstad, M. A., O'Gallagher, A., and Gary, J. (2002). "A Wavelet Transform Applied To Acoustic Emission Signals: Part 1: Source Identification," *Journal of Acoustic Emission*, 20, 39–61.

Hardy, G. and Bolen, J. (2005). "Thermal Inspection," in *ASM Handbook*, ASM International, Materials Park, OH.

He, J. and Fu, Z. (2001). *Modal Analysis*, Butterworth Heinemann, Boston.

Hou, Z. and Noori, M. (2000). "Application of Wavelet Analysis for Structural Health Monitoring," Proceedings of 2nd International Workshop on Structural Health Monitoring, Stanford University, Stanford, CA.

Housner, G., and Masri, S. (1996). "Structural Control Issues Arising from the Northridge and Kobe Earthquakes," Proceedings of 11th World Conference on Earthquake Engineering, Mexico, Paper No. 2009.

Hwang, N. E., Shen, Z., Long, S. R., Wu, M. C., Shih, H., Zheng, Q., Yen, N. C., Tung, C. C., and Liu, H. H. (1998). "The Empirical Mode Decomposition and the Hilbert Spectrum for Nonlinear and Non-Stationary Time Series Analysis," *Proceedings of the Royal Society*, London, 454, 903–995.

Hubbard, B. (1998). *The World According to WAVELETS*, A. K. Peters, Natick, MA.

Jiles, D. C. (1986). "Evaluation of the Properties and Treatment of Ferromagnetic Steels Using Magnetic Measurements," Proceedings of the Nondestructive Evaluation of Ferromagnetic Materials, Houston, TX.

Karbhari, V., Kaiser, H., Navada, R., Ghosh, K., and Lee, L. (2005). "Methods for Detecting Defects in Composite Rehabilitated Concrete Structures," Submitted to Oregon Department of Transportation, and Federal Highway Administration, Report No. SPR 336.

Kim, J. and Stubbs, N. (1995a). "Damage Detection in Offshore Jacket Structures from Limited Modal Information," *International Journal of Offshore and Polar Engineering*, 5(1), 58–66.

Kim, J. and Stubbs, N. (1995b). "Modal-Uncertainty Impact and Damage-Detection Accuracy in Plate Girders," *Journal of Structural Engineering*, ASCE Publication, 121(10).

Klausenberger, F. and Barton, J. (1977). "Detection of Flaws in Reinforcing Steel in Prestressed Concrete Bridge Members," Report on Contract No. DOTFH-11-8999, FHWA, Washington, DC.

Koerner, R. and Lord, A. (1982). *Use of Acoustic Emissions to Predict Ground Stability* Federal Highway Administration Report No. FHWA / RD-82/052, Washington, DC.

Koerner, R., McCabe, W., and Lord, A. (1981). "Acoustic Emission Behavior and Monitoring of Soils," *Acoustic Emissions in Geotechnical Engineering Practice*. American Society for Testing and Material. ASTM STP 750. Philadelphia, PA

Lin, L., Lee, G., Liang, Z., and O'Connor, J. (2005). "Health Monitoring of Bridge Bearings," Proceedings of the 2005 ASNT Fall Conference, Columbus, OH.

Maclachlan-Spicer, J., Kerns, W., Aamodt, L., and Murphy, J. (1991). "Time Resolved Infrared Radiometry of Multilayer Organic Coatings Using Surface and Subsurface Heating," Thermosense XIII, Editor: Baird, G. Proceedings of Society of Photo-Opt. Instrumentation Engineering (SPIE), *Vol. 1467.*

Maldague, X. (2002). "Introduction to NDT by Active Infrared Thermography," *Materials Evaluation, ASNT, Vol. 60*, No. 9, Columbus, OH.

Miller, R. (2007). "Acoustic Emission Testing," in *Nondestructive Testing Handbook: Vol. 10, Nondestructive Testing Overview*, ANST, Columbus, OH.

Moore, P. (2001). *NDT Handbook on Infrared Technology*, Editor: Maldague, X. ASNT Handbook Series, ASNT Press, Columbus, OH.

NYSDOT. (1997). Bridge Inspection Manual. New York State Department of Transportation, Albany, NY.

Owens, S., Rose, J., Mu, J., and Zhang, L. (2005). "A new High Frequency Long Range Ultrasonic Guided Wave Inspection System for Pipes," Proceedings of the 2005 ASNT Fall Conference, Columbus, OH.

Pandey, A. K. and Biswas, M. (1994). "Damage Detection in Structures using Changes in Flexibility," *Journal of Sound and Vibration*, 169(1), 3–17.

Pandey, A. K., Biswas, M., and Samman, M. (1991). "Damage Detection in from changes in curvature mode shapes," *Journal of Sound and Vibration*, 145(9), 312–332.

Pendat, J. and Piersol, A. (2010). *Random Data: Analysis and Measurement Procedures*, Wiley, New York.

Pollock, A. (2005). "Acoustic Emission Inspection," in *ASM Handbook: Non Destructive Evaluation and Quality Control, Vol. 17,* ASM, Materials Park, OH

Post, D., Han, B., and Ifju, P. (1994). *High Sensitivity Moiré: Experimental Analysis for Mechanics and Materials*, Springer-Verlag, New York.

Reese, G. J. (1995) "The Effects of Finite Element Grid Density on Model Correlation and Damage Detection of a Bridge," Proceedings of the 36th AIAA/ASME/ASCE/AHS/ASC Structures, Structural Dynamics and Materials Conference.

Riahi, M. and Khosrowzadeh, B. (2005). "Development of Acoustic Emission AE) Method for the Diagnosis of Corrosion in Steel Pipes," Proceedings of the 2005 ASNT Fall Conference, Columbus, OH.

Rollwitz, W. (2005a). "Magabsorption NDE," in *ASM Handbook: Non Destructive Evaluation and Quality Control, Vol. 17,* ASM, Materials Park, OH.

Rollwitz, W. (2005b). "Microwave inspection," in *ASM Handbook: Non Destructive Evaluation and Quality Control, Vol. 17,* ASM, Materials Park, OH.

Rose, J. (2004). *Ultrasonic Waves in Solid Media*, Cambridge University Press, Cambridge, UK.

Sanayei, M. and Saletnik, M. J. (1996a). "Parameter Estimation of Structures From Static Strain Measurements, Part I: Formulation," *Journal of Structural Engineering, ASCE,* 122(5), 555–562.

Sanayei, M. and Saletnik, M. J. (1996b). "Parameter Estimation of Structures From Static Strain Measurements, Part II: Error Sensitivity Analysis," *Journal of Structural Engineering, ASCE,* 122(5), 563–572.

Sanayei, M., McClain, J., Wadia-Fascetti, S., and Santini, E. (1999). "Parameter Estimation Incorporating Modal Data and Boundary Conditions," *Journal of Structural Engineering, ASCE,* 125(9), 1048–1055.

Shepard, S., Lhota, J., Ahmed, T., Rubadeux, B., and Wang, D. (2001). "Qualification and Automation of Pulse Thermographic NDE," in *Nondestructive Evaluation of Materials and Composites*, Editor: Baaklini, et al. Proceedings of Society of Photo-Opt. Instrumentation Engineering (SPIE), Vol. 4336.

Spanos, P. D. and Failla, G. (2004). "Evolutionary Spectra Estimation Using Wavelets," *Journal of Engineering Mechanics, ASCE*, Reston, VA, 130(8), 952–960.

Stubbs, N. and Garcia, G. (1996). "Application of Pattern Recognition to Damage Localization," *Microcomputers in Engineering, Vol. 11.* Computer-Aided Civil and Infrastructure Engineering, pp 395–409.

Stubbs, N. and Kim, J. T. (1994). "Field Verification of a Nondestructive Damage Localization and Severity Estimation Algorithm," Report prepared for New Mexico State University, Las Cruces, NM.

Stubbs, N., Kim, J. T., and Farrar, C. R. (1995). "Field Verification of a nondestructive damage localization and severity estimation algorithm," 13th International Model Analysis Conference, Nashville, TN.

Tracy, N. (2007). "Liquid Penetrant Testing," in *Nondestructive Testing Handbook*: *Vol. 10*, ANST, Columbus, OH.

Vincent, H., Hu, J. and Hou, Z. (2000). "Damage Detection Using Empirical Mode Decomposition Method and a Comparison with Wavelet Analysis," Proceedings of 2nd International Workshop on Structural Health Monitoring, Stanford University, Stanford, CA.

Yan, B., Goto, S., and Miyamoto, A. (2004). "Time-Frequency Analysis Based Methods for Modal Parameter Identification of Bridge Structure Considering Uncertainty," Proceedings of 2nd International Workshop on Structural Health Monitoring of Innovative Civil Engineering Structures, ISIS Canada Corporation, Manitoba, Canada.

Zhang, Z. and Aktan, E. (1995). The Damage Indices for the Constructed Facilities, Proceedings of the 13th International Modal Analysis Conference, *Vol. 2*, Nashville, TN.

Zimmerman, D. C. and Kaouk, M. (1994). "Structural Damage Detection using a Minimum Rank Update Theory," *Journal of Vibration Acoustics*, (2), 222–231.

8 Decision Making in IHCE

8.1 INTRODUCTION

There should be a reason for any engineering activity. This applies well to any structural health monitoring (SHM) project. The ultimate reasons for SHM activities should be either reducing costs or increasing safety. The way to ascertain that the SHM activities for a particular project would ultimately achieve the cost reduction and/or increasing safety is through the decision making tool chest. We define the decision making tool chest as a collection of quantitative or semiquantitative analytical methods and techniques that help the professional in analyzing different aspects of the SHM project and making appropriate decisions that would help to reduce costs or increase safety. This relationship is shown in Figure 8.1. Note that the decision making tool chest can be utilized to design the SHM experiment. It can also be used to assess the results of the SHM experiment. The utilization of the decision making tool chest depends on the type of SHM project, as well as the ultimate goals of embarking on the SHM project to start with.

Decision making processes involve all aspects of SHM. As defined earlier, Decision making is the fourth step in any structural health engineering effort. We aim to generalize this concept. Decision making is a part of every step in the first three steps, generally referred to as SHM, as well as a separate part by itself. Figure 8.1 shows these integral interactions between decision making (DM) and the other SHM steps. As Figure 8.1 shows, any SHM effort would have one (or both) of the goals of (a) saving costs, and (b) improving safety. DM should always be the gateway to any/all of the three components of SHM, it also should be present during the execution of any/all of those components, and finally, DM should be the last step to be executed in the effort. The final step in any effort should be an attempt to answer the question "how much additional safety, or reduced cost, or both, have been achieved by this effort?" Lack of appropriate and timely execution of DM might result in less efficient project, which can result in costlier, and perhaps less-safe results.

DM tools that can be utilized before, during, and after the execution of any/all of the SHM project component(s) are immense. To streamline and formalize the discussion of the interrelationship between the DM tools and the SHM components (and subcomponents) we first introduce the concept of decision making tool box (DMTB). The DMTB will enumerate the main subcomponents of DM tools, as shown in Table 8.1.

Decision making components, as shown in the DMTB (Figure 8.3) have been presented and studied by numerous authors in many fields. We note that one, or more, of the techniques employed by those tools will play a role during an SHM project. As shown earlier, SHM includes measurements, structural identifications, and damage identifications (DMIDs) components. As such, each of the decision making tools mentioned above can be used in any of these components. Finally, since SHM projects are usually designed to assess one or more hazards (corrosion, scour, earthquakes, impact, normal wear and tear, etc.); it is also reasonable to conclude that the particular decision making technique to be utilized in the project will be greatly affected by the hazard (or hazards) under consideration.

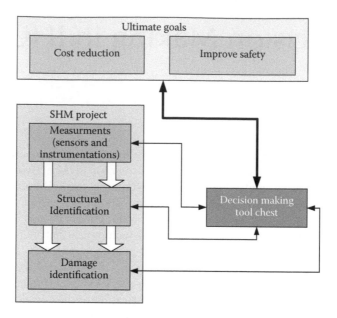

FIGURE 8.1   Decision making and SHM effort.

TABLE 8.1
Decision Making Tool Box (DMTB)

Main components of DMTB

Risk management

Risk assessment

Decision and utility theorems (Figure 8.2)

Cost–benefit analysis

Life cycle cost analysis

Statistical analysis

Probabilistic methods and techniques

Stochastic processes, including Markov Processes

Basic economics

8.2   DECISION MAKING PROCESS AND STRUCTURAL HEALTH COMPONENTS

8.2.1   INTRODUCTION

Decision making tools should be employed at every phase of any SHM project. Many of the decisions that involve the needs, goals, and methodologies of the project should be established *before* embarking on an SHM project, not after the project is finished. Of these, the need for the project is perhaps the most important. The decision maker should embark on a process that established the need for such an effort. In performing such an effort, the decision maker will invariably use one, or more, of the techniques of this chapter.

After the need for an SHM project is established, all of the four phases of the structural health in civil engineering (SHCE) project as described in Chapter 2, will require the use of ensemble of decision making tools, as shown in Figure 8.4. In the next four sections we discuss the utilizations of the DMTB in each of the four phases of the SHCE field. We then discuss how the use of ensemble of the DMTB techniques can produce a successful SHCE project.

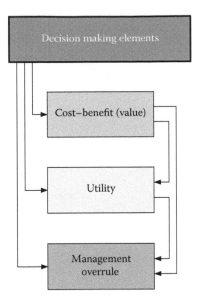

FIGURE 8.2   Elements of decision making.

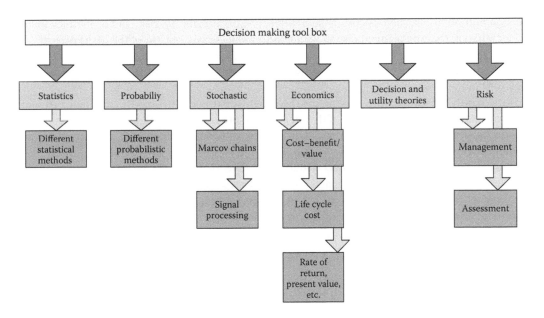

FIGURE 8.3   Decision making tool box (DMTB).

8.2.2   SENSORS AND MEASUREMENTS: DATA COLLECTION

Sensing and measurements are the first part of any SHM project. Different decision making tools and processes will be needed to ensure successful execution of this phase. Among the interrelationships between the sensing/measurements efforts and decision making tools are the following:

Planning of the Measurement and Sensing Phase: Optimizing costs of the SHM experiment and the costs of measurements must be one of the goals of the SHM project. Chapter 2 offered "General Theory of Experiments" for optimizing cost of experiment. "Special Theory of Experiments" for optimizing cost of measurements was also presented in Chapter 2. The utilization of different

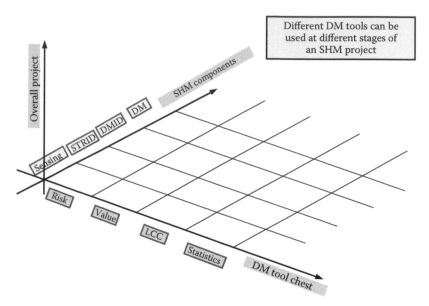

FIGURE 8.4   Decision making tool box and elements of structural health.

decision making tools, such as cost–benefit/value, utility, probability, risk, and economics theories and techniques must be planned during the measurements/sensing phase of the SHM project.

Data Organization: Organizing data in efficient and simple databases is not an easy task, especially for large and long SHM projects. Signal processing and statistical techniques needs to be employed in this phase.

Data Analysis: Generating histograms, probability density functions (PDFs) and cumulative distribution functions (CDFs) is part of this step. Signal processing techniques (see Chapter 7) might also be needed. Efficient archiving and data sorting retrieval methods and software need to be planned and employed.

Consistency with Overall Project Goals: The lore of advanced technologies can have an undesirable effect on achieving the goals of the SHM project. Use of advanced technology must be consistent with the overall goals of the project. All parts of the experiment: sensors, data processing and instrumentation must be well integrated, and in full conformity with the overall project objectives.

8.2.3   STRUCTURAL IDENTIFICATION: VULNERABILITIES AND MITIGATION

Decision making process, including the many methods of the process relies heavily on STRID component of SHM. Such interaction between the two subjects should be accommodated in any decision making process (see Figure 8.5). Some of the ways in which the two fields interact together are as follows:

Theories of STRID: Some STRID methods are based on DM Methods. For example, artificial neural network (ANN) was presented as an STRID method in Chapter 6. The ANN method can also be used to aid in project prioritization and cost–benefit analysis. We note that structural reliability method that is considered a decision making technique in this document (it aids the decision maker to decide on the safety of the structure) can also be considered a structural identification technique, since it explore structural behavior at its very core.

Risk: Hazards, Vulnerabilities, and Consequences: The concept of risk, as described later in Section 8.6, deals with all three aspects of structural health: hazards, vulnerabilities, and consequences. All structural identification methods explore the relationships between these three aspects from structural behavior view point. All STRID methods deal with inputs (hazards) structural

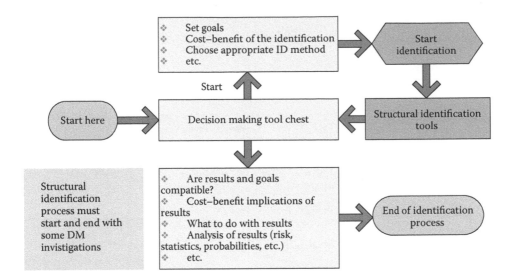

FIGURE 8.5 Structural identification and decision making process

properties (vulnerabilities) and consequences (structural response). The methods can vary from simplistic to fairly complex, but the process is still the same. Because of this, the use of risk methods as a decision making tool must utilize STRID process in one form or another.

Decision Making Pre-STRID: In many situations, the STRID process itself must be preceded by decision-making analysis. For example, in a situation where the professional needs to evaluate maintenance budget allocation, an STRID condition assessment investigation of some of the bridge network might be needed to aid in prioritizing available funding.

Decision Making Post-STRID: The reverse to the previous situation is also possible in some cases. For example, in a large network of bridges, it might be necessary to decide *a priori* which bridges might require further STRID condition assessment. In such situations, a simple risk-based assessment can help the decision maker to target fewer bridges for further STRID condition assessment.

Capacity/Demand (Reliability): As was discussed earlier, structural reliability methods can be considered both a DM tool and a STRID tool. Efficient and accurate reliability studies can be successful only if this dual identity is recognized. This recognition means, among other things, that a reliability analysis should pay equal attention to different structural analysis rules and needs and the uncertainties of parameters rules and needs.

Probabilistic Structural Methods: Probabilistic structural methods are well known. Some of these methods are component based, such as probabilistic behavior of beams or axial members, see, for example, Elishakoff (1983), Neubeck (2004), or Melchers (2002). Other methods can analyze the uncertainties in the structure on a system level, such as probabilistic finite element methods and probabilistic boundary element method, Ettouney et al. (1989a and 1989b). These types of methods can help the decision maker to accommodate structural uncertainties during any decision making analysis that deals with uncertainties of structures. Section 8.8 of this chapter is devoted to this subject.

8.2.4 DAMAGE IDENTIFICATION

Damage identification process is interrelated to decision making process in many ways. Decision-making efforts must precede DMID efforts; they must also follow DMID effort. We describe these relationships and others next (Figure 8.6).

Decision Making before DMID Efforts: Before embarking on any DMID, the decision maker should decide if a simple nondestructive testing (NDT) effort (including visual inspection) can be

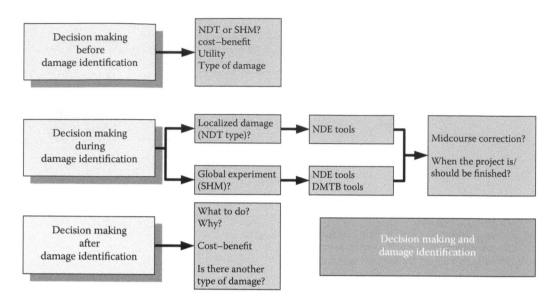

FIGURE 8.6 Decision making and damage identification process.

used instead of the more costly SHM? The advantages and limitations of both NDT and SHM are discussed in several other parts of this document (see, for example, Chapters 2 and 7). Another important decision to make is to evaluate the utility, or cost–benefit (value) of DMID. In particular: the decision maker should evaluate, before embarking on a DMID process, the benefits, or utilities that will be realized from the process.

Another less obvious decision that needs to be considered in the DMID process is the type of damage that needs to be identified. As discussed in Chapter 7, there are numerous types of damages. The tools to identify those different damages are different in technology, cost, and types of results. A careful analysis is needed to ensure the most efficient DMID process.

Decision Making during DMID Efforts: Decision making tools during the DMID effort will depend largely on whether the DMID effort is localized, NDT type effort, or a more general SHM effort. For NDT type effort, the decision making effort is mostly technical. In such a situation, analytical NDE tools can be used. For the more general SHM projects (which cover larger areas of the structure, and are usually done on a continuous or intermittent time spans), both NDE analytical tools as well as decision making effort that is specific to the SHM project of interest must be performed. For example: the quality and applicability of the data that is being gathered by the SHM project must be evaluated often to ensure that they are consistent with the overall project goals. Also, the data must be evaluated to see if there is a need for midcourse correction, or changes. See Section 8.4.5 on sequential decision making for more on this subject. Another decision that needs to be addressed during the DMID effort is if/when to finish/stop the SHM project.

Decision Making after DMID Efforts: After the DMID process ends, the implications of finding/not finding the sought after damage must be studied. A plan describing what to do, and why, if any damage is identified must be part of the original project plan. Finally, the decision maker must be ready for the *serendipity principle,* which was presented in Chapter 1. The serendipity principle in the DMID situation is whether a damage that is different from the originally sought after damage is discovered.

8.2.5 DECISION MAKING

Decision making process is involved with all components of SHM, as we have discussed. However, after all information from sensing/measurements, structural identification, DMID has been gathered, the decision making now becomes the central subject: what to do with all of

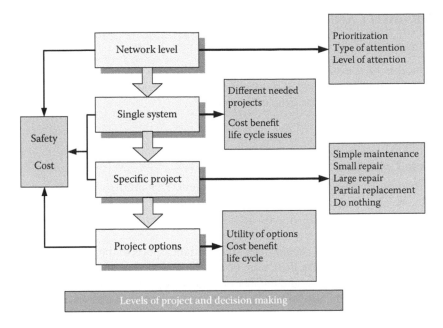

FIGURE 8.7 Levels of projects and decision making issues.

this information? The answer of course depends on the needs of the decision maker. There are many possibilities at graduating level of detail, as shown in Figure 8.7. Depending of the level of interest, the type of decision making process differs. Actually, it is paramount that such level and needs established *before* embarking on an SHM project, not after the project is finished. Different levels and types of decision making issues for each level are discussed next. Bridge structures will be used in this discussion; however, the discussion is applicable to any type of structural systems.

Network Asset Level: DM tools can identify and prioritize systems (bridges) that need attention. For example, simple DM vulnerability assessment tools can help in identifying bridges that are vulnerable to scour or seismic damage, and the degree of that vulnerability. In addition, the type and level of attention can also result from such network-level investigation.

Single System (Bridge) Level: DM tools can prioritize between needed projects. For example: painting, seismic retrofit, or diagnostic testing. This level can use STRID, DMID, or sensing type projects. The varied nature of types of projects makes this particular process difficult, yet, with the choice of appropriate DM tool, or tools, adequate decisions can be made.

A specific Project Level: After a specific project has been identified, the decision maker needs to decide on what type and level of effort is needed: (a) simple maintenance, (b) small repair, (c) large repair effort, (d) partial replacement effort, or (e) do nothing. This level of decision making might require several decision making tools that range from simple cost–benefit analysis to complex risk analysis, depending on the particular circumstances of the project.

A Given Project Option: When the professional decides upon the type of effort for a particular project, the type of action needs to be explored. For example, after deciding to retrofit and strengthen a group of reinforced concrete columns, the decision maker needs to decide on the most beneficial type of retrofit. Among potential retrofit option might be to use FRP (fiber reinforced polymers) wrapping, replace steel rebars, or add steel jacketing. There are several issues that need to be considered: useful life of retrofit, initial cost of different retrofit options, and life cycle costs.

At All Decision Making Levels: The main objective in SHCE is to ensure safety at reasonable costs. As such, it is important to explore safety, costs, benefits, utility during all decision making processes, if pertinent.

8.3   PROBABILITY AND STATISTICS

8.3.1   Introduction

Different concepts of statistics, probability, regression, and correlation will be discussed in this section. Whenever possible, the use of those concepts in the SHM/SHCE field is highlighted. These concepts are used in examples throughout this volume.

8.3.2   Statistics

8.3.2.1   Data Collection

Almost all SHM project collect information about system. This information is then grouped in data sets or databases. One of the most important steps in an SHM project is the details of these databases and how they are organized. A good data structure would have the following attributes:

- Complete/encompassing: it should cover all important aspects of the project. For example: in a modal identification project, there must be enough sensors that are located at adequate locations for producing the minimum number of mode shapes that are required by the project.
- Simple: The simpler the data structure, the easier it would be to extract information from.
- Expandable: Can be added upon at later times without major difficulties.
- Suitable for analysis: The analysis that would be performed on the collected data must be kept in mind while deciding upon the data and the data structure itself. Mismatch between the data, the data structure, and the analysis type can lead to a failed project.

8.3.2.2   Frequency and Histograms

A group of data points x_i that is collected in a vector $X = \{x_i\}$, can be grouped in ranges, where the number of occurrences of x_i within a predefined ranges are computed into another array Y. Note that $i = 1, 2, ..., N_i$ and N_i is the size of X. This array is the frequency array. Thus, by definition

$$Y = \left\{ y_j \right\} \tag{8.1}$$

$$y_j = \text{Number of occurrences which } x_i \text{ satifies the condition } R_j \leq x_i < R_{j+1} \tag{8.2}$$

Where R_j describes the range limits, with N_j describes the total number of ranges.

Identifying the frequency of a data set is used extensively in SHM projects to identify statistical properties of the data. Consider, for example, the partial data set in Table 8.2. The complete set contains 1000 records of sampling savings (units of $1,000) of a particular project as estimated by an official. The maximum and the minimum records of this data set are $5,402.29 and $4,518.37, respectively. The official decided that ten range limits are adequate to study the frequency of occurrences of this data set. The ranges and the frequency of occurrences are shown in Table 8.3. Note that the total number of occurrences is 1000, which is the same as the total number of records in the data set. A graphical display of the frequency is called the *histogram* is shown in Figure 8.8. The histogram is also used extensively in analyzing data during SHM projects.

8.3.2.3   Random Variables

There are numerous ways to define a random variable, see, for example, Benjamin and Cornell (1970) or Schlaifer, R (1969). These definitions range from mathematically complex to the fairly qualitative toss of coin experiment. For our purposes, we observe that almost all parameters that pertains to SHM are random in nature, that is, they do not have a constant value. Random variables can be defined using their statistical properties (mean, variance, etc.) In some instances, the changes in the parameter are small enough to permit the professional to deem it constant, or deterministic at a given time and/or space. Deterministic variables can be defined completely by a single valued number.

TABLE 8.2
Example of a Data Set

Sampling of Present Value of Savings of an SHM Project (Only first 50 records of 1000 records are shown)

Record #	Value	Record #	Value	Record #	Value	Record #	Value	Record #	Value
1	5111.05	11	5048.01	21	4834.12	31	4936.97	41	4938.61
2	5066.48	12	4899.32	22	4932.45	32	4906.50	42	4948.48
3	5007.72	13	5133.57	23	5032.43	33	5156.15	43	5057.90
4	4929.55	14	4764.81	24	4897.59	34	5150.22	44	4846.75
5	4675.08	15	4944.34	25	5039.52	35	5167.41	45	4890.13
6	4906.10	16	4891.67	26	4802.79	36	4941.53	46	5084.11
7	4751.34	17	4917.50	27	5075.22	37	5044.49	47	5059.83
8	5038.29	18	5059.30	28	4954.99	38	5001.62	48	4916.91
9	4889.03	19	4914.10	29	5244.73	39	4931.59	49	4855.30
10	5010.23	20	4886.22	30	4902.96	40	5058.99	50	4813.91

Note: The maximum and minimum of the complete data set are 5402.29 and 4518.37, respectively.

TABLE 8.3
Ranges and Frequencies of Occurrences

Range	Frequency
4500	0
4600	1
4700	11
4800	55
4900	171
5000	274
5100	271
5200	153
5300	57
5400	6
5500	1
5600	0

Random variables can be either discrete or continuous. Discrete random variables as the name implies can take only finite set of values: they can be counted. Continues random variables have infinite set of values. In SHM, both types of random variables can be encountered. Mathematical tools exist for both discrete and continuous versions of the random variables. We will encounter both types throughout this document.

A discrete random variable is then collected in the form x_i, with $i = 1,2, ...N_i$ and N_i is the size of data set (sample). A continuous random variable is expressed simply as x, with $-\infty<x<\infty$.

8.3.3   PROBABILITY DISTRIBUTION

8.3.3.1   Discrete versus Continuous Distributions

The CDF of a continuous random variable X is defined as

$$F(x) = P(X \leq x) \tag{8.3}$$

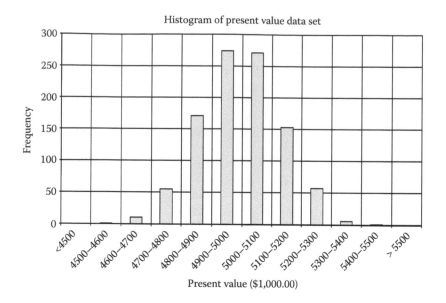

FIGURE 8.8   Histogram example: present value data set.

A PDF of a continuous random variable is

$$f(x) = \frac{dF(x)}{dx} \tag{8.4}$$

Similar expressions exist for discrete random variables. We note that the PDF of a discrete random variable can easily be generated from the frequency data set that was described earlier. See Benjamin and Cornell (1970) for more details.

8.3.3.2   Expectations and Moments of Random Variables

A general expression for the nth expectation (moment) of a random variable x is

$$E(x^n) = \int_{-\infty}^{+\infty} x^n f(x) dx \tag{8.5}$$

Where $f(x)$ is the PDF of x.

For $n = 1$, the expectation is commonly called *expected value*, *mean*, or *average* of the random variable. It is expressed as

$$\mu = E(x) = \int_{-\infty}^{+\infty} x f(x) dx \tag{8.6}$$

The variance V and the standard deviation σ^2 are defined as

$$V = \sigma^2 = E\left[x - E(x)\right]^2 \tag{8.7}$$

We get the useful relation

$$V = \sigma^2 = E(x^2) - E[x]^2 \tag{8.8}$$

$$V = \sigma^2 = \int_{-\infty}^{+\infty} (x - \mu)^2 f(x) dx \tag{8.9}$$

Similar expressions exist for discrete random variables.

Higher moments $n > 2$ are used in statistical analysis applications.

8.3.3.3 Uniform Distribution

Uniform PDF is expressed as

$$f(x) = \frac{1}{b-a} \qquad \text{for} \qquad a < x < b \tag{8.10}$$

$$f(x) = 0 \quad \text{elsewhere} \tag{8.11}$$

The mean and variance are

$$\mu = \frac{a+b}{2} \tag{8.12}$$

$$V = \frac{(b-a)^2}{12} \tag{8.13}$$

Because of its simplicity, it is used frequently in SHM projects (see Figure 8.9).

8.3.3.4 Normal and Truncated Normal Distribution

Normal distribution is expressed as

$$f(x) = \frac{1}{\sigma\sqrt{2\pi}} \exp\left(-\frac{(x-\mu)^2}{2\sigma^2}\right) \tag{8.14}$$

The mean is μ and the variance is $V = \sigma^2$. Equation 8.14 has infinite tails on both sides of the scale. In many physical applications, it is required to truncate these tails to reasonable physical limits. For a truncated normal distribution, Equation 8.14 mentioned above, can be rewritten as (Elishakoff 1983)

$$f(x) = \frac{A}{\sigma\sqrt{2\pi}} \exp\left(-\frac{(x-\mu)^2}{2\sigma^2}\right) \quad \text{for} \quad a \le x \le b \tag{8.15}$$

$$f(x) = 0 \quad \text{for } a \le x \le b \tag{8.16}$$

$$A = 12\,erf(k) \tag{8.17}$$

$$erf(k) = \frac{1}{\sqrt{2\pi}} \int_0^k \exp\left(-\frac{y^2}{2}\right) dy \tag{8.18}$$

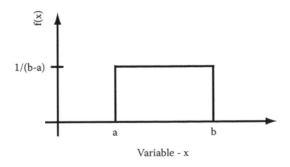

FIGURE 8.9 Uniform PDF.

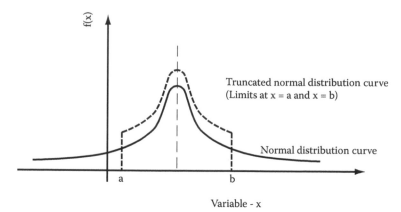

FIGURE 8.10   Uniform and truncated uniform PDF.

See Figure 8.10 for illustration of normal and truncated normal distributions. Uniform distribution is used in many SHM situations. For example, it can be used to represent the random behavior of strength of large number of concrete test samples.

8.3.3.5   Poisson's Distribution

$$f(x) = \frac{\mu^x e^{-\mu}}{x!} \tag{8.19}$$

The mean and the variance of Poisson distribution is μ. Poisson distribution is used in numerous physical applications. For example, it can be used to represent the number of arriving trucks to a highway bridge during a given interval of time.

8.3.3.6   Marginal Distributions

In many SHM applications we are confronted with more than one random variable. Consider a situation with two continuous random variables x and y. A joint CDF can be define as

$$F(x,y) = P(X \le x, Y \le y) \tag{8.20}$$

The joint PDF is

$$f(x,y) = \frac{\partial^2 F(x,y)}{\partial x\, \partial y} \tag{8.21}$$

The marginal CDF is

$$F(x) = \int_{-\infty}^{x} \int_{-\infty}^{+\infty} f(x_1, x_2) dx_1\, dx_2 \tag{8.22}$$

And the marginal PDF is

$$f(x) = \int_{-\infty}^{+\infty} f(x, x_1) dx_1 \tag{8.23}$$

8.3.4   Conditional Probabilities: Bayes Theorem

For two random variables the conditional probability that relate them is

$$f(y|x) = \frac{f(x,y)}{f(x)} \tag{8.24}$$

The conditional probability expression 8.24 is generalized to Bayes theorem which states that

$$P\left(\theta_i \mid A_j\right) = \frac{P\left(A_j \mid \theta_i\right) P\left(\theta_i\right)}{\sum\limits_{k=1}^{k=N} P\left(A_j \mid \theta_k\right) P\left(\theta_k\right)} \tag{8.25}$$

Bayes theorem expresses the probability of a state θ_i, given another state A_j in terms of

- The marginal probability of that state θ_i
- The conditional probabilities of A_j given all other possible states of θ
- All marginal probabilities of θ

Symbolically, Benjamin and Cornell (1970) presented Equation 8.26 as

$$P\left(\text{State} \mid \text{Sample}\right) = \frac{P\left(\text{Sample} \mid \text{State}\right) P\left(\text{State}\right)}{\sum\limits_{\text{All States}} P\left(\text{Sample} \mid \text{State}\right) P\left(\text{State}\right)} \tag{8.26}$$

Bayes theorem has many applications in the field of SHM. There are many examples that show its use in many parts of this volume.

8.3.5 CONFIDENCE INTERVALS AND CHEBYSHEV INEQUALITY

For any random variable x the knowledge of the mean μ and standard deviation σ can be used in producing the probability of that variable within any given range, as shown in the following equations

$$P\left(x_1 \leq x \leq x_2\right) = 1 - \frac{1}{a^2} \tag{8.27}$$

$$x_1 = \mu - a\,\sigma \tag{8.28}$$

$$x_2 = \mu + a\,\sigma \tag{8.29}$$

The above equation is known as Chebyshev inequality. It was described in full by Benjamin and Cornell (1970). Daddazio and Ettouney (1990 and 1991) utilized the method in acoustic radiation problems, while Ettouney et al. (1993) used it to solve soil problems, and Daddazio and Ettouney (1987) used it to address uncertainties in boundry conditions for impulsive loads on structural systems, respectively. The expressions for one sided inequalities are

$$P\left(x_1 \leq x\right) = \frac{a^2}{1+a^2} \tag{8.30}$$

$$P\left(x \leq x_2\right) = \frac{a^2}{1+a^2} \tag{8.31}$$

8.3.6 FUNCTIONS OF N RANDOM VARIABLES

8.3.6.1 General

It is easy to express the expectations of simple functions in an analytical form as shown in Table 8.4. Note that Y is a function of the random variables x, x_1, or x_2.

8.3.6.2 Taylor Series

There are numerous situations when the known functional dependence of several random variables is too complex. In such situations the use of the simple expressions of Table 8.4 is not possible. A simple, yet not too widely know method of finding the expectations of such complex function is Taylor series approach. Consider the functional relation

$$Y = Y\left(x_1, x_2, \cdots x_i, \cdots, x_N\right) \tag{8.32}$$

Where x_i is a set of independent random variables and $i = 1, 2, 3, \ldots, N$. It is assumed that the mathematical expectations of x_i are known. Recall that the expectations of order n are related to the PDFs $f(x_i)$ as

$$E\left(x_i^n\right) = \int_{-\infty}^{+\infty} x_i^n \, f\left(x_i\right) dx \tag{8.33}$$

In practice, the first two moments are used, for example, the mean (expected value) $E(x_i)$ and the mean square $E\left(x_i^2\right)$.

For simplicity, we define the mean of x_i as

$$\bar{x}_i = E\left(x_i\right) \tag{8.34}$$

We need to evaluate the mathematical expectations of Y. Specifically, we need to evaluate the mean (expected value) $E(Y)$ and the mean square $E(Y^2)$.

For complex functional relationships, it is not easy to express $E(Y)$ and $E(Y^2)$ in simple expressions of $E(x_i)$ and $E\left(x_i^2\right)$. A general solution that can be used is by utilizing the Taylor serious expansion of Equation 8.32. The expansion around $\bar{x}_i$ can be expressed as

$$Y = \bar{Y} + \sum_i^N \varepsilon_i \, \bar{Y}_{,i} + \frac{1}{2}\sum_i^N \sum_j^N \varepsilon_i \, \varepsilon_j \, \bar{Y}_{,ij} + \cdots \tag{8.35}$$

Higher order terms $O\left(\varepsilon_i^3\right)$ and higher, are assumed to be of negligible value. Note that

$$\varepsilon_i = x_i - \bar{x}_i \tag{8.36}$$

$$\bar{Y} = Y\big|_{x_i = \bar{x}_i} \tag{8.37}$$

$$\bar{Y}_{,i} = \frac{\partial Y}{\partial x_i}\bigg|_{x_i = \bar{x}_i} \tag{8.38}$$

and

$$\bar{Y}_{,ij} = \frac{\partial^2 Y}{\partial x_i x_j}\bigg|_{x_i = \bar{x}_i, \, x_j = \bar{x}_j} \tag{8.39}$$

TABLE 8.4
Expectations of Simple Functions

Function	E(Y)
$Y = cx$	$cE(x)$
$Y = x_1 + x_2$	$E(x_1) + E(x_2)$
$Y = x_1 x_2$	$E(x_1)\,E(x_2)$: assuming that x_1 and x_2 are independent

---

Taylor series expansion Equation 8.35 is now in a form that can easily define $E(Y)$ and $E(Y^2)$. From Equation 8.36, by definition:

$$E(\varepsilon_i) = 0 \tag{8.40}$$

and

$$E(\varepsilon_i^2) = E(x_i^2) = V_i \tag{8.41}$$

Given that the random variables x_i are also independent, then

$$E(\varepsilon_i \varepsilon_j) = 0 \tag{8.42}$$

Taking the expectations of Equation 8.35 we get

$$E(Y) = \bar{Y} + \frac{1}{2}\left[\sum_i^N E(\varepsilon_i^2)\bar{Y}_{,ii}\right] \tag{8.43}$$

and

$$E(Y^2) = \bar{Y}^2 + \left[\sum_i^N E(\varepsilon_i^2)\left(\bar{Y}\bar{Y}_{,ii} + (\bar{Y}_{,i})^2\right)\right] \tag{8.44}$$

If the random variables x_i are dependent, that is, $E(\varepsilon_i \varepsilon_j) \neq 0$, the expectations for $N = 2$ case are defined as

$$E(Y) = \bar{Y} + \frac{1}{2}\left[E(\varepsilon_1^2)\bar{Y}_{,11} + E(\varepsilon_2^2)\bar{Y}_{,22}\right] + E(\varepsilon_1\varepsilon_2)\bar{Y}_{12} \tag{8.45}$$

and

$$E(Y^2) = \bar{Y}^2 + \left[E(\varepsilon_1^2)\left(\bar{Y}\bar{Y}_{,11} + (\bar{Y}_{,1})^2\right) + E(\varepsilon_2^2)\left(\bar{Y}\bar{Y}_{,22} + (\bar{Y}_{,2})^2\right)\right]$$
$$+ 2E(\varepsilon_1\varepsilon_2)\left(\bar{Y}\bar{Y}_{,12} + \bar{Y}_{,1}\bar{Y}_{,2}\right) \tag{8.46}$$

The expectations for $N = 3$ are

$$E(Y) = \bar{Y} + \frac{1}{2}\left[E(\varepsilon_1^2)\bar{Y}_{,11} + E(\varepsilon_2^2)\bar{Y}_{,22} + E(\varepsilon_3^2)\bar{Y}_{,33}\right]$$
$$+ E(\varepsilon_1\varepsilon_2)\bar{Y}_{12} + E(\varepsilon_1\varepsilon_3)\bar{Y}_{13} + E(\varepsilon_2\varepsilon_3)\bar{Y}_{23} \tag{8.47}$$

and

$$E(Y^2) = \bar{Y}^2 + \left[E(\varepsilon_1^2)\left(\bar{Y}\bar{Y}_{,11} + (\bar{Y}_{,1})^2\right) + E(\varepsilon_2^2)\left(\bar{Y}\bar{Y}_{,22} + (\bar{Y}_{,2})^2\right) + E(\varepsilon_3^2)\left(\bar{Y}\bar{Y}_{,33} + (\bar{Y}_{,3})^2\right)\right]$$
$$+ 2E(\varepsilon_1\varepsilon_2)\left(\bar{Y}\bar{Y}_{,12} + \bar{Y}_{,1}\bar{Y}_{,2}\right) + 2E(\varepsilon_1\varepsilon_3)\left(\bar{Y}\bar{Y}_{,13} + \bar{Y}_{,1}\bar{Y}_{,3}\right)$$
$$+ 2E(\varepsilon_2\varepsilon_3)\left(\bar{Y}\bar{Y}_{,23} + \bar{Y}_{,2}\bar{Y}_{,3}\right) \tag{8.48}$$

Once the expectations of Y are computed, the variance V, and standard deviation σ are computed using

$$V = E(Y^2) - E^2(Y) \tag{8.49}$$

and

$$\sigma = \sqrt{V} \qquad (8.50)$$

respectively.

Taylor series approach is useful when the functional form of the random variables is known. For example, it can be used in evaluating the probabilistic behavior of load rating of bridges or cost estimates of failure. Also, the formal expressions of scour hazards are good applications for the Taylor series approach.

8.3.6.3   Numerical Integration

Another method of obtaining the expectations of a function such as Y in Equation 8.35 is to utilize numerical integration scheme. Such an approach would offer less insight to the problem than the Taylor series approach. In such situations, the reader is advised to consider the Monte Carlo simulation method as an alternative. Monte Carlo method offer more general solution, with similar level of computational demands.

8.3.7   REGRESSION AND CORRELATION ANALYSIS

Consider an SHM experiment that measure the humidity levels x at a reinforced concrete bridge site and the rate of corrosion y in the bridge. After N readings, the professional wanted to use the findings to estimate the relationships of corrosion rate at other locations that are similar to the SHM testing location. Also, the professional wants to study if there is any correlation between the humidity level and the rate of corrosion. Formally, for a given set of data pairs, x_i and y_i, where $i = 1, 2, ..., N$, find a relationship

$$y(x) = ax + b \qquad (8.51)$$

where a and b are constants. Find also a factor r that indicates the degree of correlation between humidity levels and rate of corrosion. We discuss both methods next.

8.3.7.1   Linear Regression

The constants a and b can be estimated using a minimum least square approach. It aims at reducing the least square of the differences between the line and the data pairs x_i and y_i. It can be shown that

$$b = \frac{\sum_{i=1}^{i=N}(x_i - \mu_x)(y_i - \mu_y)}{\sum_{i=1}^{i=N}(x_i - \mu_x)^2} \qquad (8.52)$$

and

$$a = \mu_y - b\mu_x \qquad (8.53)$$

Note that μ_x and μ_y are the averages of the data sets x_i and y_i, respectively.

Equations 8.51 through 8.53 define the linear regression method. The parameter a is the intercept of regression line, and b is the slope of the regression line. Studying the regression line can give insights to data trends as well as the relationships between the two sets of parameters under consideration. Figure 8.11 illustrates the concept of regression lines.

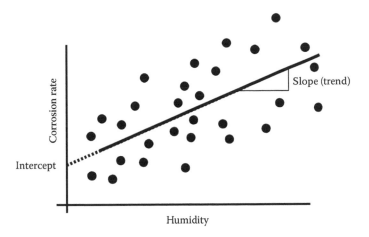

FIGURE 8.11 Regression analysis concepts.

Note that regression analysis can be done with nonlinear equations. It also can be used to identify moving averages for stochastic (time dependent) problems. It offers many powerful tools that can have many applications in the field of SHM/SHCE.

One final comment on regression analysis: we note that the method of ANN (see Chapter 6) offers a stronger tool that relates data sets with complex interrelationships. It lacks the simplicity of regression analysis, but it offers in return a far more powerful technique that can be used in many SHM situations.

8.3.7.2 Correlation Analysis

To obtain the correlation between x_i and y_i we use the equation

$$r = \frac{(1/N-1)\sum_{i=1}^{i=N}(x_i - \mu_x)(y_i - \mu_y)}{\sigma_x \sigma_y} \tag{8.54}$$

where σ_x and σ_y are the standard deviations of x_i and y_i. The correlation coefficient $-1 \leq r \leq +1$ indicates how closely the two parameters correlate. For $r = 0$, there is no correlation between x and y, and they are completely independent. For $r = +1$, there is strong positive correlation, that is, when one variable changes, the other variable changes in same direction. For $r = -1$, there is strong negative correlation, that is, when one variable changes, the other variable changes in opposite direction.

Correlation coefficient of Equation 8.54 can be generalized to form correlation matrix between more than two parameters. In such a situation, the correlation coefficients can form a correlation matrix that can be of benefit in the SHM field, see Pendat and Piersol (2010).

8.4 TRADITIONAL THEORIES OF DECISION MAKING

8.4.1 ELEMENTS OF TRADITIONAL DECISION MAKING

8.4.1.1 Value versus Utility Methods

Perhaps the most obvious decision making metric is the monetary metric. If there are several options that are available to a decision maker, the optimal choice would be the one that is the least expensive, or the one that provide the most savings, and so on. These have produced the traditional monetary cost–benefit decision making methods (see, for example, Treuman 1977). The concept of value can be linked to the cost–benefit in a mathematical form, as a ratio or subtraction. Because of this, value will be used in this volume interchangeably with cost–benefit.

Traditionally, it was argued that the two main disadvantages of the highly quantitative cost–benefit approach are that

- It requires too much information that might not be available to the decision maker
- It lacks the subjective feel that many decision makers have; such subjective feel is extremely difficult to enumerate in a true cost–benefit analysis

To avoid such limitations, Von Neumann and Morgenstern (1944) developed a new decision making process that is based on developing utility functions. The theory of utility has been used since then as an alternate method for decision making processes. Table 8.5 shows the main differences between traditional cost–benefit/value and utility decision making methods.

Before we explore the two methods in more detail, we would like to note that we propose to generalize the concept of cost–benefit/value beyond monetary or quantitative methods to include more subjective issues. We already mixed objective and subjective parameters in Chapters 4 (when evaluating values of experiments) and 5 (when evaluating suitability of sensors to experiments). As such, we believe that our generalized utilization of the cost–benefit/value actually encompasses both attributes of traditional cost–benefit and utility methods.

8.4.2 Utility Method

8.4.2.1 Theory of Utility

As was mentioned earlier, the utility method has several advantages to the traditional cost–benefit (value) method in decision making. We present the utility theory, its mathematical basis, and its use in the field of SHM in this section. We start by presenting a definition of utility by Trueman (1977) as, "Utility is the subjective numerical measure of the value of an act to a decision maker when a particular event occurs." In other words, a utility is a subjective definition of value.

There are as many types of utility metrics as there are decision makers. Every decision maker can establish the type of utility as it suites the project on hand. Of course, care must be taken when deciding upon the type of utility used. Some good attributes of utility metrics are as follows:

- **Accuracy:** The utility used must be accurate in describing the type of project. At least it must be consistent with the *spirit* of the project. For example, monetary or cost utility is suited for projects where costs are important factor in the decision making process. Safety as a metric might be a good utility for bridge security measure. Availability is a good metric for research oriented SHM project.
- **Simplicity:** The utility metric should be simple to sort out. For example: monetary utility is fairly simple to understand and apply. In some risk-related decision making processes, utilities such as vulnerability, asset value, threat, or consequences are used. They are fairly simple to apply in a qualitative sense.

TABLE 8.5
Comparison between Value and Utility Methods in Decision Making

Value	Utility
Mostly quantitative foundations	Mostly qualitative
Needs mostly objective/mathematical descriptions	Can be adjusted to account for subjective situations
Analytically rigorous	Less analytically rigorous
Due to its rigor it does not need as much care in formulating the problem	Needs more care in formulating the problem
Needs less care in interpreting the results	Needs care in interpreting the results

- **Efficiency:** The utility metric should be efficient in describing as many aspects of the project as possible. Short of this, multiple utilities might be needed to accomplish this objective. Relative risk analysis is a good example to this need. Three utility metrics are used to describe risk levels in many situations: hazard, vulnerability, and consequences. Failure mode and effect analysis can be used to address this issue. See Johnson and Simon (1997) and Johnson and Dock (1998) for in-depth description of this topic.

After deciding upon the type of utility X, we need to define the utility function $U(X)$. To accomplish this, we need to establish few points that relates X to $U(X)$. Some examples will be provided later.

The choice of the utility value depends on the level of experience of the analyst. Thus, in using utility approach, it is prudent for the analyst to be as familiar with the subject as possible. SHM projects can also help the analyst to produce accurate utility values.

8.4.2.2 Types of Utility

8.4.2.2.1 Monetary/Cost Utility

The monetary utility is a popular utility that is commonly used in financial decision. Well-known qualitative illustrations of different monetary utility function are shown in Figure 8.12.

Cost utility describes cost attributes of different aspects of the SHM project. It is especially beneficial when exact costs are not well known to the analyst. For example, if the decision maker needs to establish a cost-utility function for cost of executing and maintaining an SHM project over the course of 10 years. There is no available information that can help the decision maker to establish an objective cost–benefit analysis. The utility approach is then chosen. The utility function bounds are established first: what are the maximum and minimum expected costs for the project? The decision maker estimates that the maximum cost is $500,000.00 and the minimum cost is $0 (since this would be the case of no SHM project to start with). The utility function is assigned values of 1.0 and 0.0 for these bounds. The midpoint $U(X) = 0.5$, is established next. If the decision maker is risk neutral, then the midpoint would be $X = 0.5(0+500,000) = 250,000$. For a more aggressive decision maker, the midpoint would be higher than 250,000, say $X = 350,000$. For a conservative decision maker, the midpoint would be less than 250,000, say $X = 150,000$.

The three-point utility function is now formed, as shown in Table 8.6, and Figure 8.13. The advantage of the utility approach is now clear: (a) The lack of concise information did not prevent the decision maker from establishing a utility function using previous experiences, and (b) The attitude of the decision maker (conservative, neutral or aggressive) is clearly reflected in the utility function.

A final comment on the construction of utility functions: we used three points for the above example. It is recommended to use five points to establish these functions for reasonable accuracy, the five-points approach can be done with similar logic to the above.

8.4.2.2.2 Benefit Utility

We propose another possible utility in the field of SHM: the benefit utility. The benefit utility can be established in a manner that is similar to the cost utility, above. Similar to cost utility, the benefit

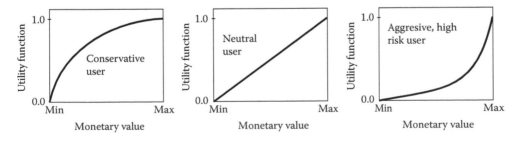

FIGURE 8.12 Monetary utility functions.

TABLE 8.6
Cost Utility Functions

| Cost Utility | Cost of SHM Project ($ Thousands) | | |
	Conservative	Neutral	Aggressive
0.00	0	0	0
0.50	150	250	350
1.00	500	500	500

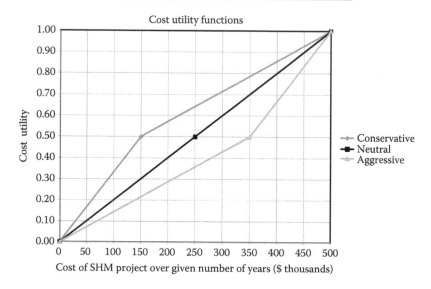

FIGURE 8.13   Cost utility functions.

TABLE 8.7
Benefit Utility Functions

| Benefit Utility1 | Benefit of SHM Project ($ Thousands) | | |
	Conservative	Neutral	Aggressive
0.00	0	0	0
0.50	175	300	425
1.00	600	600	600

utility for both conservative and aggressive decision makers will be similar in shape. Aggressive decision maker expects more benefit for the same cost than the conservative decision maker and the reverse is true. For example, Table 8.7 and Figure 8.14 show the benefit utility for the above-mentioned SHM project.

8.4.2.2.3   Utility of Experimentations

Chapter 4 presented several methods that can help the decision maker in identifying the value of the experiments in a rigorous quantitative manner. We introduce here another method that is based on the utility theory to evaluate experiments: utility of experimentation. The utility function can be formed in same fashion as before. The professional would estimate the maximum and minimum values that are to be obtained from performing the experiment. A maximum and minimum utility of 1.0 and 0.0 is assigned to the function. The midpoint is then estimated following similar procedures as before.

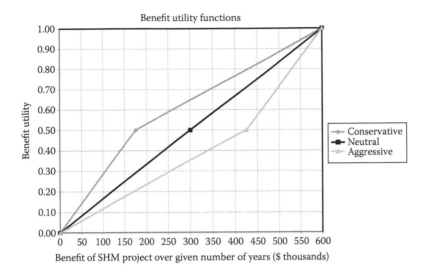

FIGURE 8.14 Benefit utility functions.

8.4.2.2.4 Other Types of Utility

There are many other potential utilities to be used in the SHM/SHCE field. For example, utilities of consequences, vulnerability, damage, and so on can all be established to help the professional in making decisions.

8.4.2.3 Multidimensional Utilities

Utilities can also be functions of other utilities. Such as the following:

$$U(X_1, X_2) = \alpha_1 U(X_1) + \alpha_2 U(X_2) \tag{8.55}$$

Where $U(X_1, X_2)$ is a utility function of the addition of two utilities The factors α_1 and α_2 are constants. Also the expression:

$$U(X_1, X_2) = \alpha_1 U(X_1) U(X_2) \tag{8.56}$$

represents another multidimensional utility.

A savings utility function can be established with X_1 representing the SHM project costs in the above example, and X_2 representing the benefits of the SHM project. The SHM savings utility function over the next 10 years is then

$$U(X_1, X_2) = U(X_2) - U(X_1) \tag{8.57}$$

Using Tables 8.6 and 8.7 the savings utility function can be computed as in Table 8.8 and Figure 8.15.

8.4.2.4 Expected Utilities

After forming the utility functions, they can be used to compute the expected utility. For example, suppose that the decision maker has two options to proceed in the SHM project:

- Option A: Have a 25% probability of costing $100,000 and 75% probability of costing $300,000.
- Option B: Have a 25% probability of costing $150,000 and 75% probability of costing $250,000.

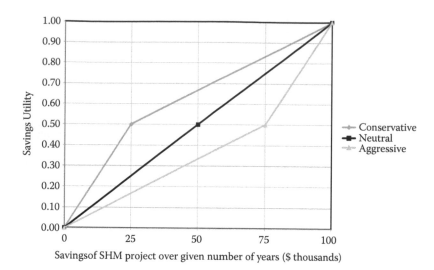

FIGURE 8.15   Savings utility function.

TABLE 8.8
Savings Utility Function

Savings Utility	Savings of SHM Project ($ Thousands)		
	Conservative	Neutral	Aggressive
0.00	0	0	0
0.50	25	50	75
1.00	100	100	100

TABLE 8.9
Expected Cost Utility Comparison

Option	P	X_1 ($)	(1 – P)	X_2 ($)	$U(X_1)$	$U(X_2)$	E[U]
A	0.25	100,000	0.75	300,000	0.33	0.72	0.62
B	0.25	150,000	0.75	250,000	0.41	0.65	0.59

Assuming that the decision maker is using the conservative utility function curve of Figure 8.13, the expected utility can be estimated using

$$E[U] = p\,U\!\left(X_1\right) + \left(1-p\right)U\!\left(X_2\right) \tag{8.58}$$

Table 8.9 shows the details of the computations. The expected utility of option A is higher, thus it is the better choice in this case.

The process can continue to accommodate the benefit and savings utilities. Such an exercise is more involved computationally, but it is similar to the cost-only exercise; we will leave resolving it to the reader.

8.4.3   Cost–Benefit/Value (Monetary) Method

The cost–benefit/value (some authors refer to it as the monetary) method in decision making is based on constructing a payoff table similar to Table 8.10. The decision maker is faced with several

TABLE 8.10
Example of Payoff Table in Decision Making

State of Nature/Events	Probability	Decisions/Actions			
		V_1	V_2	V_j	V_N
θ_1	p_1	u_{11}	u_{12}	u_{1j}	u_{1N}
θ_2	p_2	u_{21}	u_{22}	u_{2j}	u_{2N}
θ_i	p_i	u_{i1}	u_{i2}	u_{ij}	u_{iN}
θ_M	p_M	u_{M1}	u_{M2}	u_{Mj}	u_{MN}

TABLE 8.11
Payoff Table for Severe Flood Mitigation

Event	Probability	Severe Flood Mitigation Costs ($1,000)			
		Do Noting	Method – 1	Method – 2	Method – 3
0–1 years	0.10	1000	100	150	200
1–10 years	0.30	500	75	40	20
10–50 years	0.60	100	10	5	3
Total	1.0	310.0	38.5	30.0	27.8

states of nature (events) that are placed in the rows of Table 8.10. They are designated with θ_i where $i = 1,2,...M$, with M is the total number of possible events. The payoff table will also include a set of decisions/actions that the decision maker would choose from. They are designated V_j with $j = 1,2, ..., N$, with N is the total number of available choices/actions. The body of the payoff table is filled with the cost that will be incurred if event θ_i occurred and the action V_j is taken. These costs are designated u_{ij}. Finally, the knowledge of the probability of occurrence of the events θ_i is needed. They are designated p_i. Note that, by definition $\Sigma_1^M p_i = 1.0$.

After the payoff table is fully formed, the cost of each decision is computed as follows:

$$C_j = \sum_{i=1}^{i=M} p_i u_{ij} \tag{8.59}$$

The preferred decision is the decision with the minimum C_j.

As an example of the payoff table, consider Table 8.11, which explores severe flood conditions and different mitigation options. Three potential state of nature can happen: the flood might occur in 0–1 years, 1–10 years, or 10–50 years. Four decisions are available. Three mitigation options plus a do-nothing option. The costs of different states for different actions fill the body of the payoff table. Also the probabilities that a severe flood occur for different events are placed in the payoff table. The total cost of each action is then computed using Equation 8.59. Obviously the do-nothing option is the least cost-effective option. The three options are fairly close to each in cost, with mitigation option 3 costing the least.

8.4.4 More on Traditional Decision Making Methods

The payoff and utility methods as presented above can have many extensions that make them much more useful for decision making process, especially when used in conjunction with an SHM project. We summarize some of those techniques next.

Posterior Analysis: Note that both utility and monetary methods utilize probability of different events θ_i. If there are additional information that are available that relate the events to the different choices in form of conditional probabilities, this would enable the decision maker to use Bayes theorem in improving the probabilities used in the payoff table. Such improvements will ensure more accurate decision making process.

Preposterior Analysis: In addition to the conditional probabilities in the posterior analysis, if specific information regarding the problem on hand is available, it can be used to improve the decision making process even more. Such information can be in form of additional testing, or additional data sets.

Sequential Decision Making: During the decision making process if additional information becomes available it can change the course of the process. Such information can be obtained from additional testing that is performed by the decision maker to enhance the accuracy of the decision.

Probabilistic versus Deterministic: Even though the methods of this section are all based on available probabilities of different parameters, all of these methods produce single valued results. Because of this, they sometimes are referred to as deterministic methods. Those averages will influence the decision. In many situations there is a need to also know the variances of those results. To produce both averages and variances, probabilistic decision making techniques are needed. Some of those probabilistic techniques are discussed in Section 8.6.

There are several examples of those three methods throughout this manuscript. All the examples are presented in relation to the SHM/SHCE field.

We also note that the above-mentioned three forms of analysis are particularly suited to SHM projects, since SHM projects are meant for that: to provide information that can help decision makers to improve their decisions. The only requirement is that such an interaction between SHM results and decision making needs must be recognized before embarking on either (SHM project and decision making process).

8.4.5 USE IN SHM

Decision making method must be performed in every step of an SHM project. This is especially needed for such an emerging new field. Some of the benefits of decision making methods can be used in prioritizations of decisions, including (a) comparing costs of SHM options, (b) comparing benefits, costs, and/or, savings, and (c) helping decision maker in arriving to a subjective decision as to the best course of action.

SHM can also be used to increase accuracy of utility methods to (a) create measurements (datasets) that can be used to compute different probabilities of events p, (b) create data sets that can help in establishing accurate utility functions, (c) estimate payoff table entries, (d) produce conditional probabilities, and (e) use measurements results in accurate probability estimates.

8.4.6 DECISION TREE

Decision tree method is a graphical technique that can be used with any of the methods that were discussed in this section. It is particularly useful for the more complex decision making techniques such as posterior, preposterior or sequential analysis. It has the additional advantage of showing a global overview of the whole problem, rather than splitting the analysis into series of tables. Several SHM/SHCE examples for the use of decision trees are presented elsewhere in this chapter, and throughout this volume.

8.4.7 CLOSING REMARKS

We explored many traditional methods for decision making in this section. Several examples were given, and many more examples are given throughout this manuscript. The reader is encouraged to refer to these examples. There are two more points we offer regarding traditional decision making methods. First, those methods were largely developed before the era of modern computing, as such they are mostly visual, and with simple computational demands. With the availability of fast and efficient computing, the reliance on computationally demanding simulation methods paved the way to other techniques that will be explored in other sections in this chapter.

Another attribute of traditional methods is that they seem to lump together hazards, vulnerabilities, and consequences without a formal way to study these issues and their individual and combined effects on decision making process. Risk and reliability techniques, as will be presented later, seem to provide a more formal fashion to address those different issues.

This does not preclude the usefulness of the traditional methods in any manner. It simply adds to the wealth of methods that are available to the decision maker. We recommend that decision makers utilize all pertinent decision making methods; this will ensure accurate decisions for problems they face.

8.5 RELIABILITY

8.5.1 Introduction: Reliability and SHM

We discuss in this section the concept of structural reliability: how it can benefit from SHM practices, and how, in turn, it can aid decision making process. Structural reliability has been studied extensively by many authors, see Rausand and Royland (2004). We will first summarize essential concepts of structural reliability: capacity demand, reliability (safety index), and probability of failure. For the purpose of this document we will be defining structural reliability as simply the probability of failure. We next discuss general structural reliability evaluation methods, with discussions of the utilization of these methods for and by SHM technologies. The remainder of the section will be devoted to detailed examples of utilizing reliability techniques in the fields of SHM.

8.5.1.1 Capacity, Demand, and SHM

Conventional structural design methods included the evaluation of structural capacity C and the demands from the loadings D. Structural capacity includes ultimate strains, yield stresses, allowable displacements, allowable accelerations, and so on. Demands from loadings include gravity loads, live loads, seismic motions, wind pressures, and so on. In general, structural design require that

$$C \geq D \tag{8.60}$$

Since both C and D are highly uncertain parameters, these uncertainties are accounted for by several methods. For example, if the PDFs of the uncertain capacities and demands are defined as $C(x)$ and $D(x)$, where x is a random variable, then probability of failure:

$$p_f = P(C \geq D) \tag{8.61}$$

Also

$$p_f = P((C-D) \geq 0) \tag{8.62}$$

From Equation 8.61 and given $C(x)$, and $D(x)$, it can be shown that

$$p_f = \int_0^{+\infty} (C(x) - D(x)) \, dx \tag{8.63}$$

The probability of failure p_f is the area under the failure density function. The process is shown in Figure 8.16.

We reach finally a definition of reliability R as the probability that the structure does not fail, that is,

$$R = 1 - p_f \tag{8.64}$$

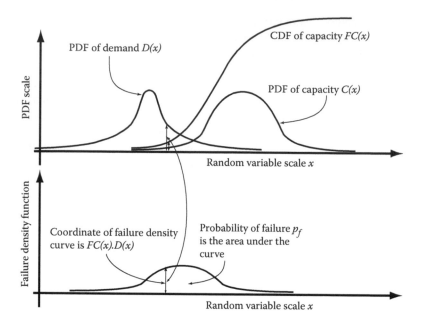

FIGURE 8.16   Concepts of probability of failure.

8.5.1.2   Reliability (Safety) Index and SHM

If $C(x)$ and $D(x)$ are both normally distributed, then for

$$Y = C - D \tag{8.65}$$

Then the mean and variance of Y are

$$\mu_Y = \mu_C - \mu_D \tag{8.66}$$

$$V_Y = V_C + V_D \tag{8.67}$$

respectively.

The probability of failure is then

$$p_f = P(Y \le 0) = \Phi\left(\frac{-\mu_Y}{\sigma_Y}\right) \tag{8.68}$$

With

$$\sigma_Y = \sqrt{V_Y} \tag{8.69}$$

Note that $\Omega(\phi)$ is the standard normal distribution function. From Equations 8.66, 8.67, and 8.68

$$p_f = \Phi\left(\frac{(-\mu_C + \mu_D)}{\sqrt{\sigma_C^2 + \sigma_D^2}}\right) \tag{8.70}$$

Defining

$$\beta = \frac{(\mu_C - \mu_D)}{\sqrt{\sigma_C^2 + \sigma_D^2}} \tag{8.71}$$

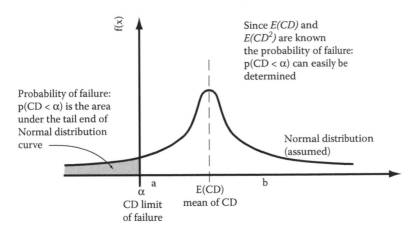

Random variable: capacity demand ratio (CD)

FIGURE 8.17 Probability of failure in a capacity/demand paradigm.

We end up with the well-known expression for probability of failure

$$p_f = \Phi(-\beta) \tag{8.72}$$

The function β is known as the reliability, or safety, index. It is the shaded area to the left of the normal distribution curve of Figure 8.17.

Reliability index is used in SHM and decision making situations extensively. It is discussed in Chapters 9 and 10 by Ettouney and Alampalli (2012).

8.5.2 Reliability Methods as Applied to SHM

8.5.2.1 Structural Components

Reliability of structural *components* have been explored by many authors, and well documented. See, for example, Neubeck (2004) or Melchers (2002). We can relate capacity and demand of components in as in Equation 8.65 and evaluate probability of failure as in Equation 8.72. If probabilistic properties of $C(x)$ and $D(x)$ are known p_f can be computed analytically. If the expressions are complex p_f can be computed using numerical methods such as Taylor series, or general simulation methods such as Monte Carlo method.

Monitoring techniques can be used to monitor and compute statistical properties of important variables of Equation 8.72. This will ensure more accurate estimations of p_f. An example of computing p_f of bridge columns that loses support due to scour is given in Chapter 1 by Ettouney and Alampalli (2012).

8.5.2.2 Structural Systems

Even the simplest structural system is composed of several components. Because of this, simple solutions for probability of failure of the system p_{fs} are not possible. In some cases, however, structural systems can be categorized into systems with simple behavioral attributes. In such situations, computing p_{fs} can still be done in a reasonable simple fashion.

8.5.2.2.1 Parallel Systems

Parallel structural systems are shown schematically in Figure 8.18. In the literature, the parallel systems are of interest since the p_{fs} can be computed in a simple fashion (see Neubeck 2004). For example, for a system with two parallel components the system probability of failure is

$$p_{fs} = 1 - (R_1 + R_2 - R_1 R_2) \tag{8.73}$$

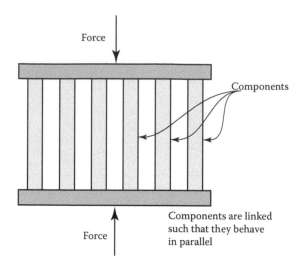

FIGURE 8.18 Parallel systems.

with

$$p_{fi} = 1 - R_i \tag{8.74}$$

For three parallel components

$$p_{fs} = 1 - \left(R_1 + R_2 + R_3 - R_1 R_2 - R_1 R_3 - R_2 R_3 + R_1 R_2 R_3 \right) \tag{8.75}$$

Note that p_{fi} and R_i are the probability of failure and reliability of the ith component. Equations 8.73 and 8.75 can be generalized for systems with higher number of components.

Parallel systems are simple in their structural behavior, so, complex building or bridge super structures are definitely not even remotely resemble parallel systems. However, on closer look, it appears that bridge bents are good example of parallel systems. This is especially true when the beams at the top of the bents are more rigid than the columns; this is the situation for relatively longer column spans as compared to beam spans. This means that it is fairly simple to formulate an algorithm to monitor the Pf of bridge bents due to most hazards, using 8.73, above, without the need of complex structural identification system. We give an example of using SHM in monitoring p_{fs} of bridge bents in case of seismic hazards in Chapter 2 by Ettouney and Alampalli (2012). Similar SHM setups are possible for other hazards.

8.5.2.2.2 Serial Structural Systems

Serial structural systems are shown schematically in Figure 8.19. Similar to parallel systems, the serial systems are of interest since its failure probability p_{fs} can be easily computed. The serial system will fail if the weakest component of the system fails. Thus, the system p_{fs} is

$$p_{fs} = 1 - \prod_{i=1}^{i=N} R_i \tag{8.76}$$

Note that N is the number of serial components and R_i is the reliability of the ith component.

In-series systems can be recognized, after some reflection, as nonredundant or very-low redundant structural systems. An example of SHM use in case of nonredundant steel truss system is given later in this section.

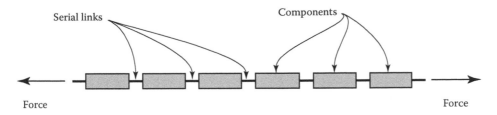

FIGURE 8.19 Serial systems.

8.5.2.2.3 General Systems

General structural systems that do not conform to the parallel or in-series categorization are much more difficult to evaluate their reliability. There are several general methodologies that can be used for this purpose:

Failure Modes and Effects Analysis (FMEA): FMEA method of prioritization and assessment of system failure have been considered a reliability method. We discuss this method in the risk section, since we consider it a risk, not reliability method, since it includes as one of its parameters the consequences of failure.

Failure Modes, Effects, and Criticality Analysis (FMECA): FMECA method of prioritization and assessment of system failure is similar to FMEA except that it also adds the criticality of failure modes in the analysis. It is also based on tabular format in its description of failure modes.

Fault Tree Analysis: Fault tree analysis is a graphical-based method that utilizes the information that results from FMECA analysis; it identifies the critical failure modes at any level. It relies on logical relationships such as "AND" and "OR" to relate logical events. During fault tree analysis, primary and secondary failure modes are identified by the analyst.

Monte Carlo Simulation: This is a general purpose simulation technique that is widely used in many reliability situations. It is used extensively in this document.

For more extensive discussions about reliability methods of structural systems, see Melchers (2002) or Neubeck (2004).

8.5.2.2.4 SHM Techniques and Structural Reliability

Structural health monitoring (SHM) techniques can help in evaluating structural reliability in many ways. For example:

Capacity Monitoring: This type of monitoring aims to monitor and record capacity of systems. For example, by measuring strains, displacements, or motions, the different capacities of the structure can be inferred. The different probabilistic properties of capacities can be used in any desired reliability investigation.

Demands Monitoring: Examples of demands are magnitude of loads, frequency of load applications (such as truck traffic), hazards properties (such as earthquakes, floods, wind, etc.), or changes in gravity loads over the years. Monitoring of these demands can produce different probabilistic properties of the demands that can be used in any desired reliability investigation.

Separate and Integrated Capacity—Demand Monitoring: Separate capacity—demand monitoring involves measuring metrics of capacity and demands simultaneously, then correlating them to produce appropriate reliability estimates. Integrated capacity—demand monitoring involves using the virtual sensing (VS) concept: trying to find a metric to monitor that can infer failure of a component of a system.

Examples of different types of usage of SHM to estimate reliability of structures are given next.

8.5.2.3 NDT versus SHM Reliability

The probability of failure of a whole system considers all potential failure modes of the system and their interaction. This is in contrast with the failure of components and their probabilities. Recall that one of the differences between NDT and SHM is that NDT presumes the knowledge of the approximate location of the damage. As such, NDT is more suited to studying the probability of failure of structural components than the probability of failure of the whole structural system. SHM, on the other hand looks for DMID of the whole structure, or DMID of several components assembly. Because of this, SHM can be used for more accurate estimations of structural failure probabilities.

An example of this distinction between NDT and SHM is in the use of global vibration modes in identifying potential loss of flexural strength as reported by Farrar and Doebling (1998). SHM experiment helped in identifying the bridge flexural modes.

8.5.3 Capacity Monitoring: Avoiding Catastrophic Failure

The sudden and catastrophic failure of the I-35 Bridge in Minneapolis in 2007 brought into attention the dangers of sudden bridge failures. Sudden failures of bridges, or any other type of infrastructure should be avoided due to the tragic loss of lives, and immense social costs that will follow. One possible way to address this problem is through the utilization of SHM technologies.

A potential possibility of sudden failures has always been failure of critical compression members in truss bridges. Note that tension members do not usually fail suddenly since they will experience noticeable elongation (ductile) behavior. Let us consider the truss of Figure 8.20. It is a typical truss system that is used in many bridges. Note that none of compression members in the truss is redundant. Thus: if any of them fails suddenly due to local buckling of the member, the whole truss might fail suddenly. Before we continue further, we need to differentiate between the global instability of the truss, and the local instability of a compression member. The global instability of the truss involves the whole system instability, its theory and analytical methods can be found

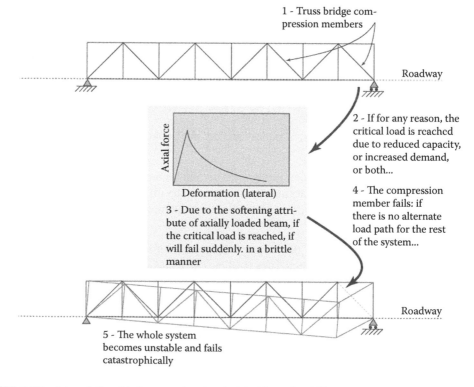

FIGURE 8.20 Interrelationship between local and global loss of stability.

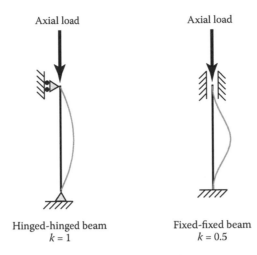

Axial load   Axial load

Hinged-hinged beam   Fixed-fixed beam
$k = 1$ $k = 0.5$

If boundary conditions change,
the k factor will increase, thus
the critical load capacity will
decrease

FIGURE 8.21 Effects of boundary conditions on beam critical load capacity.

elsewhere, Timoshenko and Gere (1961). Our current interest is in the local instability (buckling) of a truss member (see Figure 8.21). The decision maker wanted to monitor the potential local buckling of a compression member in the truss. The first step in the project is to identify the reasons for local buckling of the member. This can be done by studying the critical (buckling) axial force in the member, which can be expressed as

$$P_{cr} = \frac{\pi^2 EI}{(kL)^2} \tag{8.77}$$

Thus the buckling load P_{cr} is dependent on the flexural stiffness of the member EI. (The modulus of elasticity is E and the moment of inertia is I.) It is also dependent on the length of the member L. The effective length factor k is a function of the boundary conditions of the member. Assuming that the two connections are not experiencing a relative lateral displacement: for a pin-pin connection $k = 1.0$, and for fixed-fixed member $k = 0.5$. If the relative lateral displacements between the two connections become larger than zero, then k increases. In the limit, as $k \to \infty$, the critical load $P_{cr} \to 0$, and the member becomes unstable, and fail (see Figure 8.20).

Thus the potential causes of local instability become clear.

- One potential reason is the reduction of flexural stiffness EI due to erosion of material, for example.
- A change in the factor k. Such a change can occur from changes in the boundary conditions of the member. Such changes can be caused by deterioration of the conditions of one, or both, of the member connections. Such a deterioration would cause k to increase, thus reducing P_{cr}.
- An unexpected large relative displacement between the two ends of the member. Such a deterioration would cause k to increase, thus reducing P_{cr}.

All of the above changes are complex changes, and they are mostly not observable by visual inspection. Moreover, due to the potential sudden nature of the instability failure visual observations might not be of value in this situation. Another complication would become apparent by examining

Equation 8.77. Note that the critical load is independent of the applied load. Because of this, using strain sensors is not useful for monitoring local instability of truss members. A novel monitoring solution is needed.

Before continuing the discussion, we note that this local failure-to-global failure situation occurs mainly because of lack of structural redundancy in the system. Such a problem is also known as the progressive collapse problem in building community. Furthermore, we also note that this situation can be considered as the in-series reliability problem, which was discussed earlier: if one component failed, the rest of the system would fail, thus the probability of failure of the components is the same as the probability of failure of the whole system.

The concept of VS can be useful to resolve this problem. The decision maker recalls that the natural frequency of a beam that is subjected to an axial force P is

$$f_P = \sqrt{\frac{k}{m}\left(1 - \frac{P}{P_{cr}}\right)} \tag{8.78}$$

The governing flexural stiffness of the beam is k and its governing mass is m. If we denote f_0 as the natural frequency of a beam without the effects of the axial force we have

$$f_0 = \sqrt{\frac{k}{m}} \tag{8.79}$$

Thus

$$f = f_0\sqrt{1 - \frac{P}{P_{cr}}} \tag{8.80}$$

Or

$$\bar{f} = \sqrt{1 - \bar{P}} \tag{8.81}$$

Where $\bar{f} = f/f_0$ and $\bar{P} = P/P_{cr}$ are the dimensionless frequency and axial loads, respectively. Figure 8.22 shows the relationship of Equation (8.81).

This means that for a dimensionless axial load of about 10%, the dimensionless flexural frequency is about 95%. As the dimensionless axial load increases to 50%, the dimensionless axial frequency decreases to about 70%. In more specific terms, let us assume that a fixed-fixed beam as in Figure 8.21 is designed for $\bar{P} = 0.3$, which is a reasonable design range for a compression member in a truss. The corresponding dimensionless flexural frequency is $\bar{f} \approx 0.84$. Let us assume that the beam loses its fixity in one of its ends so that it becomes a fixed-hinged beam, the effective length k factor can be shown to increase from $k = 0.5$ to $k = 0.699$. Using Equation 8.77 the critical load is decreased by 51%. Thus the new dimensionless axial load is $\bar{P} = 0.3/.51 = 0.59$. The dimensionless flexural frequency is reduced to $\bar{f} \approx 0.63$. This is a 25% reduction in the natural frequency, which can easily be detected by a reasonable monitoring scheme. If both sides of the beam lost their fixity, the changes become more severe. The new effective length factor becomes $k = 1.0$, and the new dimensionless axial load becomes $\bar{P} = 0.3/0.25 = 1.20$. The beam becomes unstable. If the flexural frequency was monitored, it would have observed that it is approaching zero (the instability threshold), long before the beam becomes actually unstable.

We thus have established a simple way to monitor reduction of axial-flexural capacity of beams by monitoring its flexural natural frequency. There is one major hurdle that remains. Beams flexural frequencies are much higher than global system natural frequencies. To resolve such a hurdle, we need to explore the scales of both global and local frequencies.

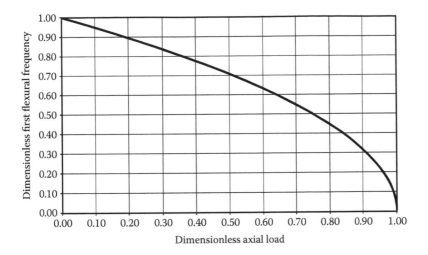

FIGURE 8.22   Axial load effects on beam flexural frequency.

Global Truss Frequency Scale: For the truss in Figure 8.20, we can estimate a simple expression for its first natural frequency in the following steps. First, we assume that all truss members have equal cross-sectional area. An approximate global moment of inertia of the whole truss is

$$I_{eq} \approx \frac{A\ell^2}{2} \tag{8.82}$$

Where A is the cross-sectional area of the truss member. If ρ is the mass density of the truss material, the total mass of the truss is

$$M = \rho A\left(2n\ell + (n+1)\ell + n\ell\sqrt{2}\right) \tag{8.83}$$

Assuming that $n \gg 1$ we get

$$M \simeq 4.14 n\rho\ell \tag{8.84}$$

The vertical stiffness of the truss can be approximated by

$$k \simeq \frac{24 EI_{eq}}{L^3} \tag{8.85}$$

Given that

$$L = n\ell \tag{8.86}$$

And using Equation 8.79 we get the first natural frequency of the truss as

$$f_G = \frac{1}{2\pi}\sqrt{\frac{24EI_{eq}}{L^3}\frac{1}{\alpha_G(4.14n\rho A\,\ell)}} \tag{8.87}$$

The equivalent mass in Equation 8.87 is $\alpha_G M$.

Equation 8.87 can be simplified to

$$f_G = \frac{0.46}{(\alpha_G L)^2}\sqrt{\frac{EI_{eq}}{m_{eq}}} \tag{8.88}$$

With

$$m_{eq} = \rho A \tag{8.89}$$

Substituting Equations 8.82 into 8.88 we get

$$f_G = \frac{0.46}{\alpha_G^2 \sqrt{2}} \frac{1}{\ell n^2} \sqrt{\frac{E}{\rho}} \tag{8.90}$$

Taking $\alpha_G = 0.5$ (Biggs 1964) the simplified global truss first natural frequency is

$$f_G = \frac{1.3}{\ell n^2} \sqrt{\frac{E}{\rho}} \tag{8.91}$$

Local Member Natural Frequency Scale: The diagonal member natural frequency scale can be estimated using similar steps as before. The local stiffness k_{Local} is

$$k_{Local} = \frac{24EI}{\left(\sqrt{2}\ell\right)^3} \tag{8.92}$$

The governing mass of the diagonal is

$$m_{Local} = \alpha_G \sqrt{2}\, \rho A \ell \tag{8.93}$$

The first local natural frequency of the diagonal can then be estimated as

$$f_{Local} = \frac{1}{2\pi} \sqrt{\frac{24EI}{\left(\sqrt{2}\,\ell\right)^3} \frac{1}{\alpha_G \sqrt{2}\, \rho A\, \ell}} \tag{8.94}$$

Simplifying

$$f_{Local} = \frac{\sqrt{2}}{\pi} \sqrt{\frac{EI}{\rho A\, \ell^4}} \tag{8.95}$$

Resolving Monitoring Scaling Hurdle: From Equations 8.91 and 8.95 the relative local to global frequencies ratio is

$$f_{LG} = \frac{1.3\sqrt{2}}{\pi} \frac{r n^2}{\ell} \tag{8.96}$$

Finally

$$f_{LG} = \frac{0.58\, r n^2}{\ell} \tag{8.97}$$

To get a practical sense for the range of f_{LG}, let us explore a steel truss with n = 20, r = 20 in, and height l = 15 ft; the f_{LG} ratio is about 26. This means that the first flexural frequency of the compression member is about 26 time the first global truss frequency. This large scale necessitates special attention during an SHM project to monitor f_local. This includes the following:

* Ensure that the sampling rate during SHM measurements is high enough to measure f_local accurately

- Utilize adequate signal processing schemes to filter away lower global frequencies
- Validate measurements using adequate structural identification methods

The final step in this project is to assign a threshold alarm level at which mitigation effort might be needed. Such a threshold might be an upper limit to P_bar, say 0.75. Given the high uncertainty of most of the parameters in this problem, and the high consequences of failure, it might be advisable to treat P_bar as a random variable, instead of a deterministic variable. It should also be noted that this may not be practical for most structures due to the nature of complexity involved, cost, and limitations of the available monitoring systems.

8.5.4 INTEGRATED CAPACITY—DEMAND MONITORING: VIRTUAL SENSING

Concept of VS (see Chapter 5 of Ettouney and Alampalli 2012) can be sued to monitor directly the probability of capacity-demand ratio, that is, probability of failure. An obvious application is the monitoring of strains at any location. If there is a preset failure strain level, then by estimating the probability of situations when the measured strain will exceed such a value, we obtain the need failure probability. Since measured strains represent demand, and preset failure strains represent capacity, what we have is a VS technique to monitor probability of failure. We illustrate this concept here using a reported experiment by Chakraborty and DeWolf (2006). We use the monitored location of the neutral axis of composite bridge girder to estimate probability of failure of the cross-section. The approach is as follows:

The SHM project by Chakraborty and DeWolf (2006) was installed on a multi-span bridge with steel girders and reinforced concrete deck. The steel girders and concrete deck were designed to perform in composite action. The bridge elevation, plan, and cross-section are shown in Figures 8.23, 8.24, and 8.25, respectively. The SHM system included 20 uniaxial strain gauges as shown in Figure 8.24. The strain gauges were arranged in pairs. The gauges in each pair were placed 2 in below the top flange and 2 in above the bottom flange at the same vertical cross-sectional plane, respectively. Eight of the pairs were placed at the center of the eight girders at the longer span. The other two pairs were placed at quarter of the span on two of the eight girders.

The SHM project measured strains at the 20 locations. The measured strains were then used to study different aspects of bridge behavior as it responded to normal traffic on the bridge over extended period of time. Among those behavioral aspects were (a) how the neutral axes of the composite cross-section vary during a typical weekday of traffic, (b) distribution of traffic load to the eight girders, (c) comparison of measured and AASHTO design requirements, (d) validation of analytical techniques (finite elements method) and measured results, and (e) effects of slab continuity. The results of that SHM project provided great insights to the behavior of bridges of the type they considered. For example, the measured neutral axes during typical weekday showed that the neutral axis actually changes its position during a typical weekday of traffic. The authors produced a histogram showing the relationship between N and X, where X is the location of the neutral axes and N is the number of

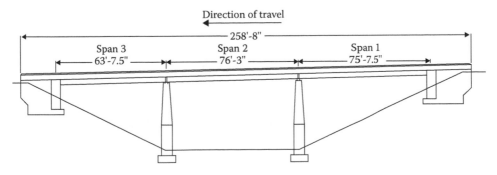

FIGURE 8.23 Elevation view of bridge. (From Chakraborty, S. and DeWolf, J. *J. Bridge Eng.*, ASCE, 11(6), 753–762, 2006. With permission from ASCE.)

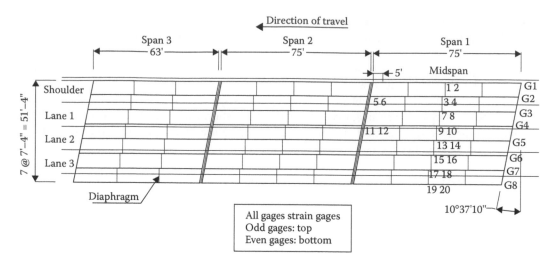

FIGURE 8.24 Plan view of bridge. (From Chakraborty, S. and DeWolf, J., *J. Bridge Eng.*, ASCE, 11(6), 753–762, 2006. With permission from ASCE.)

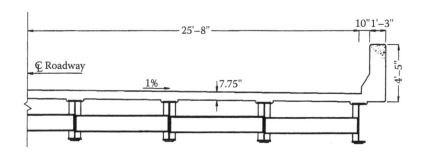

FIGURE 8.25 Typical cross-section. (From Chakraborty, S. and DeWolf, J., *J. Bridge Eng.*, ASCE, 11(6), 753–762, 2006. With permission from ASCE.)

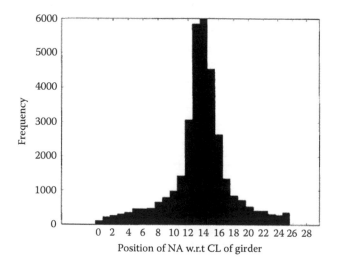

FIGURE 8.26 Histogram of neutral axis during a typical weekday of traffic. (From Chakraborty, S. and DeWolf, J., *J. Bridge Eng.*, ASCE, 11(6), 753–762, 2006. With permission from ASCE.)

occurrences of X per typical week day. Figure 8.26 shows an illustration of a typical histogram. The authors observed that the histogram is narrowly spread, and it is centered about the design value of the composite girder (X = 13.5 in). In addition, the measured location of the neutral axis does not extend to the tensile (bottom) part of the concrete slab (Figure 8.25). Because of the above, they concluded that the SHM experiment showed that the composite behavior of the subject concrete bridge is adequate.

Since composite behavior of steel and reinforced concrete deck is one of the most basic design assumptions in modern bridges, we would like to discuss the SHM experiment, and try to formalize it to gain the maximum value from such an experiment. First, we observe that the fact that the histogram of Figure 8.26 has a spread, it would indicate a nonlinear behavior of cross-section, which is not particularly surprising: after all, the composite steel-concrete design is a nonlinear condition that depends on the level of loading. This nonlinearity makes the knowledge of the histograms of the neutral axis during an SHM experiment extremely valuable. To study such a behavior further we need to look at the SHM-produced histogram in a structural reliably sense. To formalize the process, note that in the following discussion, we will be using hypothetical mathematical formulations, which have no relationship with the above-mentioned SHM project.

Let us assume that a PDF of the location of the composite neutral axis for a particular bridge is generated from a measured histogram during an SHM project. See Section 8.3 on PDF and histograms for more details. Keeping the reasonable assumption that the composite section deform in a linear fashion, as displayed in Figure 8.27, we can, after a bit of reflection, observe that as the load on the composite system increases, the location of the neutral axis X decreases and as the load on the composite section decreases, the value of X increases. There is actually an inverse nonlinear relationship between the load level on the composite section and X.

Proceeding further, let us assume that the governing design condition of the composite section results in $x = x_{DESIGN}$. Let us associate such a design condition with a load level of L_{DESIGN}, using Figure 8.17, we can compute the probability that the actual load, in a typical weekday, would exceed the design load condition $P(L \geq L_{DESIGN})$ to be

$$P\left(L \geq L_{DESIGN}\right) = A_1 \tag{8.98}$$

Where A_1 is the area under the PDF to the left of $x = x_{DESIGN}$.

The question now is how to relate the design loading condition L_{DESIGN} to the failure loading condition L_{FAILUR}. Obviously, this is a case-dependent situation; so, let us relate the two values as

$$L_{FAILURE} = \alpha L_{DESIGN} \tag{8.99}$$

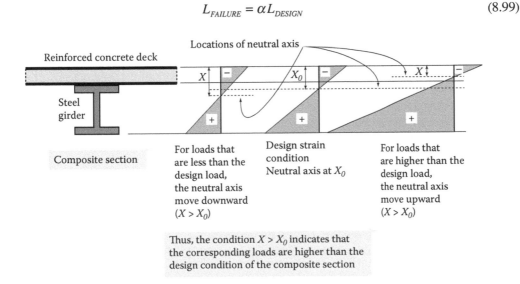

FIGURE 8.27   Behavior of neutral axis in a composite section.

With

$$\alpha \geq 1.0 \qquad (8.100)$$

Upon defining an adequate α, we can define the probability of failure (reliability of the composite system) as

$$P_f = P\left(L \geq L_{FAILURE}\right) = A_2 \qquad (8.101)$$

Where A_2 is the area under the PDF to the left of $x = x_{FAILURE}$. $x = x_{FAILURE}$ can easily be computed using the concepts of Figure 8.17 and basic composite design methods.

The above example shows how to compute reliability of composite systems using the result of simple SHM project that is based only on strain measurements. In cases of continuous or intermittent monitoring the official can detect the actual probability of failure of the system in real, or near real, time.

We end this example by noting that it is based on assuming full composite behavior between steel girder and reinforced concrete deck. Other SHM or NDT techniques can be utilized to validate such an important assumption, for example, by measuring relative strains between steel and concrete. If the full composite action assumption is not accurate, modifications of the equations of this section are needed.

8.5.5 Separate Capacity—Demand Monitoring

8.5.5.1 General

There are several situations where the structural reliability is computed by separate considerations of capacity and demands. This is accomplished as follows:

1. Compute statistical properties of demands (e.g., truck weights and frequency) as measured by the SHM system
2. Compute statistics of capacity (e.g., strain levels)
3. Compute reliability index using appropriate method (e.g., Taylor series or Monte Carlo simulation)

Two popular applications for this technique of reliability computations are proof load testing and load and resistance factor design (LRFD)/load and resistance factor rating (LRFR) fields.

8.5.5.2 Proof Load Testing

Proof load tests essentially apply loads to the structure gradually until the load reaches a prespecified level, or the structure displays signs of distress. This type of load testing is valuable since it gives information about the capacity of the structure to meet high loading demands. Careful analysis of both capacity and demand of the structure must be performed before, during, and after the proof load test to ensure successful testing. These analyses include both numerical and monitoring efforts.

When designing a proof load test, the professional must consider carefully the level of loading to be used in the test. This is demonstrated in Figure 8.28 (from Melchers 2002). Low-level proof loads (low-level q^*/L) will entail high uncertainty (as reflected in the higher testing variances V_R) in the probability of failure that are detected by the test. On the other hand, as the proof load level increases, the failure load probability ($P(H > 0)$) increases.

It should be noted that there are more limitations to proof load tests:

- There is a possibility of failure if the loads exceed the capacity of the structure.
- There might be some permanent damage due to exceedance of elastic limit states.

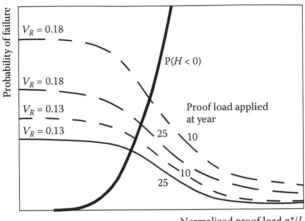

FIGURE 8.28 Reliability of proof test. (From Melcher, R. *Structural Reliability Analysis and Prediction.* 2002. Copyright Wiley-VCII Verlag GmbH & Co. KGaΛ. Reproduced with permission.)

- The loading pattern might not excite an important limit state, thus the test will not show the limiting capacity of that state. For example, gravity loading of a bridge will not be capable of exploring limit states of wind, or seismic loads.

Of course, proof load testing is still important and useful since it gives insight to behavior of structures under increasingly demanding loading conditions.

8.5.5.3 LRFD and LRFR

LRFD: LRFD method is a popular design method for structures; it is used for designing buildings, bridges, and all other types of structures, see Ravindra and Galambos (1978) and Barker and Puckett (2007). The LRFD method require for design metrics (such as beams, columns, connections, etc.):

$$C_i \geq D_i \tag{8.102}$$

$$C_i = \phi_i R_{in} \tag{8.103}$$

$$D_i = \sum_{j=1}^{j=N} \gamma_{ij} L_{ijn} \tag{8.104}$$

The capacity and demands of the ith design metric is C_i and D_i, respectively. The nominal resistance of the design metric is R_{in}. The nominal j^{th} load effect on the design metric is L_{ijn}. There are N load effects on the design metric under consideration. The resistance factor ϕ_i and the load factors γ_{ij} are factors that are obtained through probabilistic analysis, and extensive testing and monitoring. In important or sensitive structures, those factors might be improved by performing case-specific SHM project.

   LRFR: LRFR method is used in bridges to estimate their adequacy to resist required dead and live loads. The LRFR method is also based on evaluating factors through probabilistic analysis, and extensive testing and monitoring, see Moses (2001) and Murdock (2009). SHM projects can also improve the accuracy of the generic LRFR factors. See Chapter 8 of Ettouney and Alampalli (2012) for more on SHM use to improve LRFR predictions.

8.6 RISK

8.6.1 OVERVIEW

There are several definitions of risk as applied to civil infrastructure, see FEMA (2009), Aven (2003), and Koller (2000). For our purpose, we define the risk as a measure of severity of the outcome of uncertain events when applied to systems with a degree of uncertainties. This definition accommodates all recognized important parameters in traditional risk theories:

- Uncertainties of events/hazards
- Uncertainties within the system itself (vulnerabilities)
- The outcome (consequences)

We would like to differentiate between the concept of risk and reliabilities. Reliability concerns itself with the interrelationship between capacity and demand of systems. In other words, it concerns itself with hazards and vulnerabilities. The only outcome reliability is concerned with is the system behavior, for example, whether the system has failed, or damaged. Risk on the other hand, concerns itself with consequences of all types: it could be system failure, or consequences beyond system failure such as economic, social, or political consequences.

There are several quantitative ways to compute risk: linear programming, decision tree, and Monte Carlo techniques (see Table 8.12). Also, there are several qualitative methods for computing risk.

In this section we explore risk management and its underlying different type of risk. The role of SHM in each of those types of risk is discussed briefly. The following subsections include discussions of relative versus absolute risk, deterministic versus probabilistic risk, and the general approaches of evaluating different risk components. Four specific examples of the utilization of SHM in risk management are presented. Each example is fairly different, but the four examples as a whole show the rich and varied ways of interaction of risk techniques and SHM technologies: such interaction would benefit decision makers in arriving to accurate, safe, and cost-effective decisions.

8.6.2 COMPONENTS OF RISK MANAGEMENT

For the purpose of this document we define risk management as the main risk module that encompasses several types of risk. Thus, risk manager needs to handle several types of risk (see Figure 8.29). In addition, the role and utilization of SHM vary depending on the type of risk. In what follows, we discuss briefly those different types of risk and how SHM is related to each.

8.6.2.1 Risk Acceptance

The first step in risk management is to establish the presence of a risky situation. For our purposes, we use a generalized definition of risky situation as any situation that would affect the health and

TABLE 8.12
Techniques for Computing Risk

Method	Advantage	Disadvantage
Linear programming, not covered in this manuscript, see Trueman (1977) for more details	Formal	Mostly Deterministic some probabilistic applications, Elsayed and Ettouney (1994). Difficult to relate linkages and dependencies between variables
Decision tree/fault tree	Visual, some linkages between variables	Can be limiting for complex cases. Probabilistic decision tree models are not as widely used as deterministic decision tree models
Monte Carlo	General and flexible. Easy to program in computer codes	Can be computationally demanding. Difficult to "feel" the physical meaning of the results

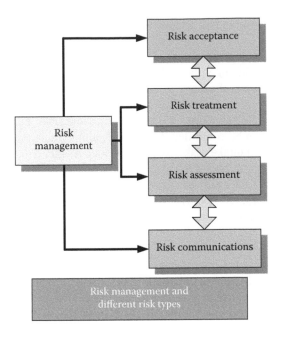

FIGURE 8.29 Risk management ant its components.

well-being of the infrastructure under consideration. Following such a general definition, risky situations include the smallest of effects such as painting boils and catastrophic effects as general bridge collapse. Identifying the presence of risk and accepting it can be done qualitatively using the experience of bridge officials, or quantitatively using different components of SHM.

8.6.2.2 Risk Assessment

After the decision of accepting a particular risk, assessment of that risk follows. Risk assessment results can be used in a decision making process in one of the three forms:

- Direct utilization
- Cost–benefit process
- Life cycle analysis (LCA) process

Risk values can be used directly in several manners, as follows:

1. Studying the components of risk (threats, vulnerabilities, and consequences) can give an insight to causes of higher risk issues. The decision maker can then address these issues to reduce overall risk.
2. Prioritize different projects based on risk assessment results. For example, let us assume that there are three available projects that can reduce risk of a particular situation from R0 to R1, R2, and R3. If R1 is the least of the three, then the first project is the preferred project since it produces the least risk.

8.6.2.3 Risk Treatment

Soon after a type of risk assessment is performed, risk treatment phase starts. Risk treatment (sometime also referred to as risk response) shows how the risk assessment results are handled. There are several types of risk treatment such as

- **Accept Risk**: the identified risk is within the previously identified acceptable level. Levels of acceptable risk are usually identified via cost–benefit analysis, LCA, or similar type of quantifiable decision making techniques.

Many of the methods of assessing risk are explored later in this section.

- **Apply Appropriate Risk Reduction Solutions**: Risk reduction or mitigation solutions can be accomplished by reducing vulnerabilities, minimizing threats, reducing consequences, or a combination of some of each. For example, a painting project would mitigate risk of corrosion hazard.
- **Risk Avoidance**: In some situations it is possible to avoid risks, but only if the actions that may cause the risks can be controlled. In most situations where the risk is due to a natural hazard, such as an earthquake, risk avoidance is not possible.
- **Risk Transfer**: Risk can be transferred to other organizations in some situations. The typical risk transfer is an insurance against the risk. Since insurance involves monetary compensation for losses, it is not always an appropriate way of risk treatment, especially if losses may involve casualties.

8.6.2.4   Risk Communication

Risk communications establishes interchange between civil infrastructure authorities (owners, managers, professionals) and the users of those infrastructure. Such an exchange of information should be two ways and continuous. It also should include as many stakeholders from the two sides as possible. The goals of risk communications are (a) Convey that no harm is going to happen. This will ease public concerns, (b) Give some guidance on how the public should respond to the situation, and (c) Reduce anxieties regarding the situation.

Risk communication occurs during one of three phases that relate to hazardous events (severe floods, earthquakes, etc.) that might affect the infrastructure:

- **Preevent Risk Communications**: involves communicating risk assessment results, the values of any needed mitigation solutions, training activities, and so on.
- **During Event Risk Communications**: On-going risks need to be communicated to the public efficiently and in a timely fashion. Examples are prolonged hurricane events.
- **After Event Risk Communications**: Communicating information on the causes of events and future mitigating solutions is essential.

Other reasons for the paramount importance of risk communications are due to the fact that it helps the public to know and understand the relevant facts that relate to risks (before, during, and after events). This way, the public can take appropriate action, such as orderly evacuations during a severe wind storm. Note that risk communication involves more than just relaying information. Risk communication should also accommodate personal factors and emotions such as unreasonable fears, trust in authorities, and uncertainties. Generally speaking, there are three modes of risk communications, namely (a) written (e.g., newspaper, bulletins, flyers), (b) verbal (personal communications), and (c) visual (TV messages).

8.6.3   Other Forms of Risk

8.6.3.1   Relative versus Absolute Risk

Risk can also be classified into two types: relative and absolute. Relative risk is generic, requires smaller amounts of input data and forecasting and is easier to formulate. Absolute risk is more specific, requires more input data and can be more difficult to formulate. The two types of risks can be classified, arguably, as qualitative and quantitative for relative and absolute risks, respectively.

8.6.3.2   Deterministic versus Probabilistic Risk

At first glance, the title of this section seems to be odd. The reader might ask: isn't risk based on uncertainties to begin with? If so, then how can it be deterministic? In a way, the question is valid.

However, we need to step back and try to observe the methods of infrastructure that are supposed to be deterministic. Recall that all inputs in deterministic methods are based on best guesses, or averages. In additions, many parameters in deterministic methods, such as load factors in LRFD design approaches, are based on probabilistic evaluations. On the basis of this, we identify deterministic risk method as a method that would produce a single valued result: the expected risk value. We identify the probabilistic risk method as the method that produces expectations (at lest two lowest expectations) or a complete histogram of the risk as a random event.

8.6.4 EVALUATION OF RISK AND ITS COMPONENTS

8.6.4.1 General Risk Expression

We use in this document a general form of risk, R as

$$R = \int HVC \tag{8.105}$$

In Equation 8.105, H, V, and C are the hazard, vulnerability, and consequence. The integral in Equation 8.105 is over the hazard space. A discrete form of Equation 8.105 is

$$R = \sum HVC \tag{8.106}$$

Again, the sum in Equation 8.106 is over all possible hazard levels. A simple form of Equation 8.105 is

$$R = HVC \tag{8.107}$$

This simple form presumes that the hazard is single valued. Note that both Equations 8.105 and 8.107 imply a dependence between the hazard level, or levels, and both vulnerabilities and consequences.

Structural health monitoring can help in estimating different risk components. We explore different risk components-SHM interrelationships next.

8.6.4.2 Hazards

Different methods of risk evaluation process different hazards differently. The use of SHM to aid in evaluating hazards, or the way SHM handles hazards in a risk-related activity are interrelated. Table 8.14 shows these different interrelationships.

8.6.4.3 Vulnerability

Measures of vulnerability, as used by risk professionals, vary depending on the type of risk analysis being performed. When used in relationship with SHM, the SHM project is also affected, as shown in Table 8.15.

8.6.4.4 Cost—Consequences

Evaluating cost or consequences is perhaps the most difficult component in evaluating risk. This is simply because it is not well defined. For example:

- Is consequence just the direct cost of damage retrofit?
- Is it cost of damage, or failure?
- Does it include management costs?
- Does it include loss of business, or business interruption?
- Does it include other indirect costs?
- Does it include social or political costs?

A simple cost of bridge failure expression $C_{Failure}$, measured in \$, was developed by Stein and Sedmera (2006) as

$$C_{Failure} = C_1 e WL + \frac{TDAd}{100}\left\{\left[C_2\left(\frac{100}{T}-1\right)+C_3\right]+\frac{1}{S}\left[C_4 O\left(\frac{100}{T}-1\right)+C_5\right]\right\}+C_6 X \quad (8.108)$$

Where

C_1 = Unit rebuilding costs (Table 8.52) unit is in \$

e = Cost multiplier (Table 8.53) unit is in \$

W = Bridge width, can be obtained from National Bridge Inventory (NBI), field 52 (unit is in ft.)

L = Bridge length, can be obtained from NBI, field 49 (unit is in ft.)

C_2 = Cost of running cars (Table 8.54)

C_3 = Cost of running trucks (Table 8.54)

D = Detour length, can be obtained from NBI, field 19 (unit is in Miles)

A = Average daily traffic on bridge (ADT), can be obtained from NBI, field 29

d = Duration of detour (Table 8.55) unit is in days

C_4 = Value of time per passenger (Table 8.56)

O = Average occupancy rate per car (Table 8.57)

T = Average daily truck traffic on bridge (ADTT), can be obtained from NBI, field 109 (note that it is a % of ADT)

C_5 = Value of time per truck (Table 8.58 or Table 8.59)

S = Average detour speed, units are in mph

X = Cost per life lost

C_6 = Number of deaths from failure (Table 8.60)

TABLE 8.13
Typical Values for Some
Parameters of Equation 8.108

Parameter	Typical Value	Units
O	1.63	Scalar
S	40	mph
X	500,000	\$

TABLE 8.14
Hazard Estimation and SHM Methods

Hazard	Quantitative (Absolute Risk)	Qualitative (Relative Risk)
Magnitude	Measure earthquake motions, flood water velocities and pressures, traffic loads, etc.	
Recurrence	Example: traffic patterns and frequency	Use available records, e.g., flood maps, seismic maps, etc.
Large-scale: earthquakes	Measure seismic motions at different regions. Also, estimate effects of local soil conditions	
Small-scale: corrosion	Measure actual loss of area, or rusting volume, see Chapters 3 and 4 in Ettouney and Alampalli (2012)	Measure corrosion rates, see Chapters 3 and 4 in Ettouney and Alampalli (2012)
Small scale: fatigue	Measure fatigue cracks using different NDT methods (e.g., ultrasound or acoustic emission)	Estimate remaining fatigue life. See Chapter 5 in Ettouney and Alampalli (2012)

TABLE 8.15
Vulnerability and SHM Methods

Hazard	Sensing	STRID	DMID
Earthquakes	Use accelerometers or strain sensors for modal identification and nonlinear behavior	Before event: use modal identification methods, Chapter 6	After event: use different NDT methods to estimate earthquake damage
Scour	See Chapter 1 of Ettouney and Alampalli (2012)		
Fatigue	See Chapter 5 of Ettouney and Alampalli (2012)		
Gravity loads	Use vibration and strain sensors to estimate responses to gravity loads	Use parameter or modal identification to estimate structural properties during exposure to gravity loads	Use different NDT methods to estimate deterioration of bridge components
Corrosion	See Chapters 3 and 4 of Ettouney and Alampalli (2012)		

Typical values of some of the parameters of Equation 8.108 are offered by Stein and Sedmera (2006) as shown in Table 8.13.

Equation 8.108 contains costs of reconstruction (first term), detour (second term), and potential fatalities, (third term). We note that the equation is hazard-independent, which makes it fairly suitable for most risk-based computations. We note that the equation do not include some additional costs of bridge failures such as potential business interruptions, permanent business closures or social/political effects. These additional effects are much more difficult to estimate; they are case dependent. Because of this, Equation 8.108 can be considered as a lower estimate limit to the cost of bridge failure $C_{Failure}$. Ettouney and Alampalli (2012) utilized Equation 8.108 to compute the statistical properties of bridge failure. In Chapter 10 of Ettouney and Alampalli (2012) a slightly modified version of Equation 8.108 is used to estimate costs of hazard occurrences.

Upon closer inspection of Equation 8.108, it becomes obvious that most of its independent variables are actually estimates of averages of random variables. This indicates that $C_{Failure}$ itself is a random variable, and Equation 8.108 is an estimate of an average for the cost of bridge failure. In different decision making situations, the uncertain nature of $C_{Failure}$ needs to be considered. Given the complex relationships of Equation 8.108, this can be accomplished in one of two ways: either using a Monte Carlo simulation technique, or use the Taylor series technique. Both techniques are explained elsewhere in this chapter.

8.6.5 Relative Risk: Prioritization

8.6.5.1 General

Relative risk methods utilize the form of Equations 8.105 through 8.107. Definitions of risk components are mostly qualitative that are based on the experience of the professional. This is where SHM can help to a great deal in enhancing the accuracy of any/all of the three components of risk. There are two popular relative risk methods. They are FEMA risk method, and the FMEA Analysis method. The FEMA approach, FEMA (2009) is discussed in Chapter 11 by Ettouney and Alampalli (2012). The FMEA method is discussed next as it applied to bridge scour mitigation risk analysis.

8.6.5.2 Failure Mode and Effect Analysis

The FMEA method produces a qualitative measure of risk. Let us assume that for a particular hazard condition N is the number of possible retrofit options for the system on hand. We observe that for each of the possible retrofit measure there exists M_i set of possible failure modes. Let us define, for each of the ith retrofit measure ($i = 1,2, ..., N$) and j^{th} failure mode ($j = 1,2, ..., M_i$) a consequence of failure rating R_{Cij}, a likelihood of failure occurrence R_{Lij} and an ease (difficulty) of failure detection R_{Dij}. We also define the risk priority number RPN_{ij} as

$$RPN_{ij} = R_{Cij} R_{Lij} R_{Dij} \qquad (8.109)$$

$$0 \le R_{Cij} \le 10 \qquad (8.110)$$

$$0 \le R_{Lij} \le 10 \qquad (8.111$$

$$0 \le R_{Dij} \le 10 \qquad (8.112)$$

There are several ways to continue on the risk ranking. For example, a simple method is to compare qualitatively all RPN_{ij} for a given i (retrofit method). The most preferred ith method is then chosen qualitatively. This approach was used in scour countermeasure risk rating by Johnson and Niezgoda (2003). A perhaps less qualitative approach can be to define a risk priority number for each retrofit measure, $RPNM_i$ such that

$$RPNM_i = \frac{\sum_{j=1}^{j=M_i} RPN_{ij}}{M_i} \qquad (8.113)$$

A final ranking of the set $RPNM_i$ can then be easily performed analytically. The retrofit measure with highest risk is $Max(RPNM_i)$ and the retrofit measure with lowest risk is $Min(RPNM_i)$.

The elegance of the FMEA method is that it is simple, yet it covers many parameters. It is also expandable, since additional retrofit measures, or failure modes can be easily added. On the negative side, the method does not account for cost of retrofit measures, nor does it account for other objective parameters such as ease of installation, time constraints, or experience of contractors. In addition, the method is deterministic in nature, thus the uncertainties in the rankings, and their possible effect on the decision maker, are not evaluated.

Perhaps the most limiting factor in the FMEA method as presented is that it does not recognize the interrelationship between the consequence of failure and the likelihood of failure. It is reasonable to assume that low-level hazard is more likely to occur than high-level hazard, yet it would produce a damage level that is fairly small. The opposite is true for high-level hazard. For example, a magnitude 5 earthquake would produce damage levels in a given structure that are much lower than a magnitude 6 earthquake. Yet, in a given site, the likelihood of a magnitude 5 earthquake is much higher than magnitude 6 earthquake.

One possible way to handle this limitation is to generalize Equation 8.109 from the form of Equation 8.107 to the form of Equation 8.106.

8.6.6   RELATIVE RISK: MATRIX ANALYSIS

8.6.6.1   Introduction

Sometimes it is desired to provide a consistent risk analysis template for bridges that have some generic common feature. Unfortunately, those generically similar bridges (or any other form of structures) have important variability, such as the nature of topology, soil condition, and usage patterns. If the decision maker follows a simple risk method such as the prioritization model of the previous sect, a separate model for each of the generically similar bridges must be built. This will result in an inconsistent modeling, and complexity on the bridge network level. Koller (2000) presented an elegant solution for this situation: the matrix approach, we give an SHM illustration for the use of this approach.

8.6.6.2   Problem Description

Consider the case of a typical simple span post-tensioned reinforced concrete bridge girder. Those types of bridges are fairly common throughout the United States. Let us assume that the decision maker decided that a consistent maintenance prioritization scheme for repairing/replacing post-tensioned tendons is need, given the limited budget that is available for this type of bridges within the network. The decision maker then estimates that there are four generic parameters that affect the repair prioritization of a given bridge: (a) age of construction, (b) traffic patterns, (c) skew angle of the bridge, and (d) repair history. The local factors that can affect each of the generic parameters are (a) seismic zones, (b) substructure conditions, (c) visual inspection, (d) temperature, and (e) humidity. The matrix of generic and local parameters is shown in Table 8.16. *We should emphasize here that the generic and local parameters of this example are hypothetical, and not based on an actual studies: they are used here for demonstration purposes only.*

TABLE 8.16
Matrix of Generic and Local Parameters

Parameters	Seismic Effects	Substructures	Visual Inspection	Temperature	Humidity
Age of construction	Age of construction would affect seismic behavior. This is due to the changes of construction material, construction details, and the structural system of the bridge	Older bridges would be in a more deteriorated condition than newer ones	Older bridges are in more deteriorated state than newer bridges	Cumulative effects of cycles of thermal expansion and freeze and thaw would have more effects on older than newer bridges	Cumulative effects of humidity variations would have more effects on older than newer bridges
Traffic patterns	Traffic patterns have minimal interaction with seismic behavior of a bridge	Designs of substructures are directly related to traffic patterns	Visual inspection would investigate effects of traffic patterns on different bridge components, e.g., collision and/or deck deterioration patterns	Temperature or humidity levels have minimal interaction with traffic patterns	
Skew angle	Skew angle can have an effect on seismic performance	Behavior of connections and other structural details are affected by angle of skew	Components with skew features, such as connections, could experience more signs of deterioration	Repeated cycles of temperature may be important	
Repair history	Seismic events can affect repair activities. Conversely, repairing bridges can affect seismic performance	Repair history has direct effects on substructure behavior	Visual inspection results can lead to repair activity	Temperature (e.g., generating cracks) or humidity (e.g., generating corrosion) effects can lead to repair activities	

In addition to the above matrix, a relationship between damage level and the scores of importance of generic parameters is needed. This is another situation where an SHM database for this class of bridges can be used to generate the required relationship. Table 8.16 shows an example of the desired relationship. The table shows the parameter score range and the corresponding damage level. It should be noted that Table 8.17 can also be generated based on the experience of the analyst; however, an SHM database is preferable, since it is more accurate.

The matrix in Table 8.16 can now be filled on a qualitative manner for each bridge site by a scale from 0 to 10. An entry of 0 indicated that the local parameter do not have any impact on the generic parameter at the same row. A score of 10 indicates a maximum effect of the local condition on the generic parameter. Let us assume that the matrix entries are x_{ij}. Where i and j are the rows and columns of Table 8.16, respectively. Thus $1 \geq x_{ij} \geq 0$.

The values of x_{ij} can be filled by one of three methods

1. Qualitatively: Where each of the bridges of the type under considerations is evaluated by a professional who is familiar with the generic and local conditions of the bridge.
2. Analytically: If there is an analytical relationship between the local and generic parameters, it can be estimated. For example, the importance of the seismic effects and the skew angle of the bridge might be possible to estimate analytically for each bridge under consideration.
3. Experimentally. If there is an SHM activity for a particular bridge, such as records of temperature, humidity, or traffic patterns, they can be used to estimate appropriate x_{ij}.

Of the three methods, it is clear that the use of SHM data would generate the most accurate x_{ij}.

Next, the decision maker will build a histogram of each of the generic parameters using two sources

1. A general histogram that relate the importance of the generic parameter Y_i, to the state of deterioration of the tendons D.
2. A relationship between Y_i and x_{ij}.

Formally

$$D = f(Y_i) \tag{8.114}$$

And

$$Y_i = g(x_{ij}) \tag{8.115}$$

Again, the histograms of Equation 8.114 can be based on SHM data collection, or in lieu of such data, they can be estimated by the official based on "common sense." Similarly, the functional

---

TABLE 8.17
Score Range of Generic Parameters and Damage Levels

Parameters	Damage Level (%)		
	00–20	20–40	40–99
Age of construction	0.00–0.40	0.40–0.80	0.80–1.00
Traffic patterns	0.00–0.70	0.70–0.95	0.95–1.00
Skew angle	0.00–0.60	0.60–0.90	0.90–1.00
Repair history	0.00–0.50	0.50–0.70	0.70–1.00

relationships $g(x_{ij})$ can be estimated using experience of the official. In general, in their simplest form, they can take the form

$$Y_i = \sum_j w_j x_{ij} \qquad (8.116)$$

Where w_j is an adequate weighting factor.

Tables 8.18 and 8.19 show sample values of x_{ij} and w_j for a particular bridge, respectively. Note that x_{ij} is bridge specific, while w_j is generic for the type of bridges under considerations.

The resulting damage score for each of the generic parameters are shown in Table 8.19.

The histograms $D = f(Y_i)$ are generated next. These histograms are the results of several SHM projects that resulted in database for this type of bridges. Tables 8.20, 8.21, 8.22, and 8.23, respectively show sample frequencies of different generic parameters as they are related to degrees of damage of tendons. From the histograms, the cumulative frequency of tendon damage level for the generic parameters. These results are shown in Tables 8.24, 8.25, 8.26, and 8.27, respectively. The histograms and their cumulative distributions are also shown in Figures 8.30, 8.31, 8.32, and 8.33, respectively. The histograms and the cumulative distributions can be used for different decisions regarding probabilities of damage as resulted from different sources (age of construction, traffic patterns, skew angle and repair history in the current example).

TABLE 8.18
Values of X_{ij} for the Matrix of Generic and Local Parameters

Parameters	Seismic Effects	Substructures	Visual Inspection	Temperature	Humidity
Age of construction	1	5	9	8	8
Traffic patterns	0	4	3	0	0
Skew angle	1	7	0	9	5
Repair history	0	3	7	7	7

TABLE 8.19
Values of w_j for the Matrix of Generic and Local Parameters

Parameters	Seismic Effects	Substructures	Visual Inspection	Temperature	Humidity
w_j	7	4	4	8	8

TABLE 8.20
Histograms Relating Bridge Age and Damage Levels

Age of Bridge (Years)	Damage Level		
	00%–20%	20%–40%	40%–99%
00–10	0.13	0.02	0.01
10–20	0.45	0.16	0.03
20–30	0.2	0.41	0.2
30–40	0.11	0.2	0.53
40–50	0.08	0.12	0.15
50–99	0.03	0.09	0.08
Total	1	1	1

TABLE 8.21
Histograms Relating Traffic Patterns and Damage Levels

Traffic Pattern: Number of Trucks per Day	Damage Level		
	00%–20%	20%–40%	40%–99%
00–50	0.003	0.001	0
50–100	0.007	0.009	0.02
100–1000	0.02	0.06	0.11
1000–	0.97	0.93	0.87
Total	1	1	1

TABLE 8.22
Histograms Relating Skew Angle and Damage Levels

Skewness (degrees)	Damage Level		
	00%–20%	20%–40%	40%–99%
00–05	0.02	0.05	0.02
05–15	0.3	0.15	0.08
15–25	0.6	0.25	0.15
25–	0.08	0.55	0.75
Total	1	1	1

TABLE 8.23
Histograms Relating Repair History and Damage Levels

Repair History	Damage Level		
	00%–20%	20%–40%	40%–99%
Past 2 years	0.03	0.01	0.001
Past 10 years	0.07	0.05	0.03
Past 20 years	0.5	0.55	0.45
Never	0.4	0.39	0.519
Total	1	1	1

TABLE 8.24
Histograms Relating Bridge Age and Cumulative Damage Levels

Age of Bridge (Years)	Cumulative Damage Level		
	00%–20%	20%–40%	40%–99%
00–10	0.13	0.02	0.01
10–20	0.58	0.18	0.04
20–30	0.78	0.59	0.24
30–40	0.89	0.79	0.77
40–50	0.97	0.91	0.92
50–99	1	1	1

TABLE 8.25
Histograms Relating Traffic Patterns and Cumulative Damage Levels

Traffic Pattern: Number of Trucks Per Day	Cumulative Damage Level		
	00%–20%	20%–40%	40%–99%
00–50	0.003	0.001	0
50–100	0.01	0.01	0.02
100–1000	0.03	0.07	0.13
1000–	1	1	1

TABLE 8.26
Histograms Relating Skew Angle and Cumulative Damage Levels

Skew Angle (Degrees)	Cumulative Damage Level		
	00%–20%	20%–40%	40%–99%
00–05	0.02	0.05	0.02
05–15	0.32	0.2	0.1
15–25	0.92	0.45	0.25
25–	1	1	1

TABLE 8.27
Histograms Relating Repair History and Cumulative Damage Levels

Repair History	Cumulative Damage Level		
	00%–20%	20%–40%	40%–99%
Past 2 years	0.03	0.01	0.001
Past 10 years	0.1	0.06	0.031
Past 20 years	0.6	0.61	0.481
Never	1	1	1

8.6.7 ABSOLUTE RISK: COST–BENEFIT

8.6.7.1 Introduction

One of the most important findings of a recent workshop on engineering of structural health, Alampalli and Ettouney (2002) was the need for cost–benefit analysis when embarking on an SHM project, or when considering any major improvements to a particular bridge. There are several methods to perform a cost–benefit analysis of an activity. In this section, we present a risk-based approach that uses decision tree topology. This example will show how to apply a risk method to the cost–benefit problem. For our immediate purpose, the example will show (a) how to utilize the results of SHM experiments to produce accurate cost–benefit analysis, and (b) how to compare the value (or cost–benefit) of SHM projects to the more conventional visual inspection projects in a qualitative manner.

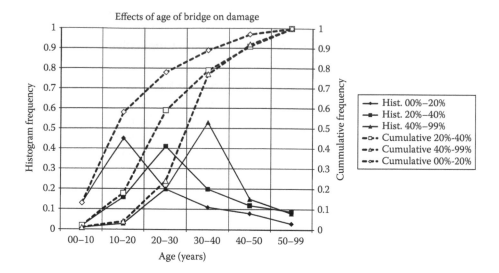

FIGURE 8.30 Frequency and cumulative damage level as related to bridge age.

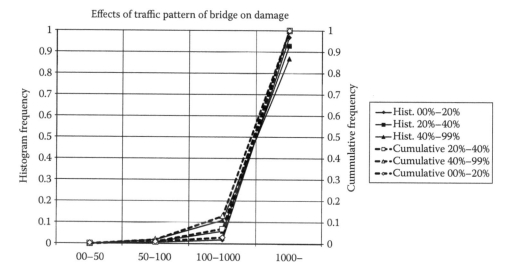

FIGURE 8.31 Frequency and cumulative damage level as related to traffic patterns.

8.6.7.2 Problem Description

The problem that a decision maker has is that it is felt that a monitoring/inspection process for a particular bridge is needed. There are several monitoring/inspection methods that can be followed:

1. Visual/manual inspection
2. Utilize some type of SHM techniques/technologies
3. A combination of the above two methods

The decision maker estimated that after the monitoring/inspection process is finished, there can be three outcomes:

1. There is need for immediate repair
2. There will be some repair needed within 1–2 years
3. There is no obvious need for repair within 2-year period

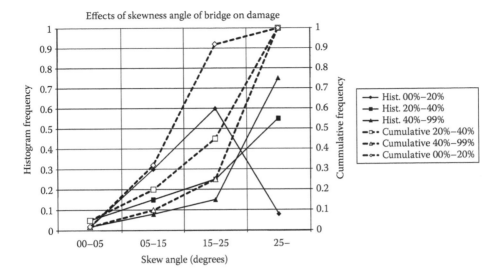

FIGURE 8.32 Frequency and cumulative damage level as related to skew angle.

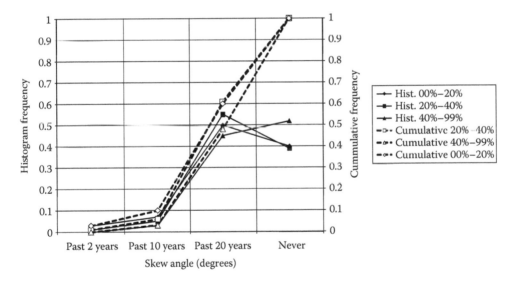

FIGURE 8.33 Frequency and cumulative damage level as related to repair history.

What distinguishes this problem from the other types of problems is that for each possibility there are two basic outcomes:

1. The damage of interest can be discovered by the process
2. The damage can't be discovered by the process

Figure 8.34 shows the generic decision tree for this problem.

On the basis of previous experiences, the decision maker proceeded to assign different probabilities to different potential outcomes. On the basis of past experiences, the probabilities of damage occurrences are estimated in Table 8.28. The potential of observing damage from different methods are estimated in Table 8.29 and the potential monetary consequences of damage observance are estimated in Table 8.30. The table also shows the direct cost of each monitoring method. Clearly,

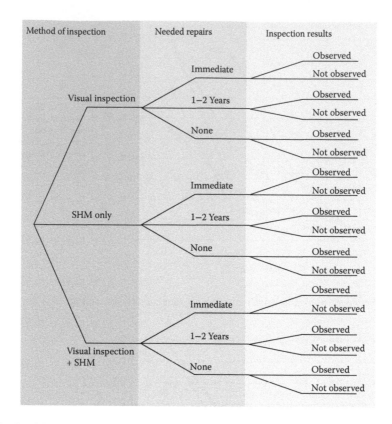

FIGURE 8.34 Decision tree for the inspection methods.

TABLE 8.28
Probability of Damage Condition

Condition of Damage (s)	Frequency (probability) (%)
Needs immediate repair	20
Needs repair within 1–2 years	50
Do not need repair within 2 years	30

TABLE 8.29
Probability of Occurrence of Inspection Results

Method	Condition of Damage	Inspection Results: Probability of Occurrence	
		Damage Observed (%)	Damage Not Observed (%)
Visual inspection	Needs immediate repair	40	60
	Needs repair within 1–2 years	50	50
	Do not need repair within 2 years	60	40
SHM	Needs immediate repair	70	30
	Needs repair within 1–2 years	80	20
	Do not need repair within 2 years	60	40
Both	Needs immediate repair	85	15
	Needs repair within 1–2 years	95	05
	Do not need repair within 2 years	90	10

TABLE 8.30
Costs of Different Outcomes

			Cost (M. $)		
			Direct Cost of Inspection/Monitoring		
Needed Repairs	Inspection Results	General Consequences	Visual Inspection	SHM	Both
Immediate	Damage observed	1.5	0.5	1.0	1.2
	Damage not observed	5.5	0.5	1.0	1.2
1–2 Years	Damage observed	1.0	0.5	1.0	1.2
	Damage not observed	2.3	0.5	1.0	1.2
None	Damage observed	0.0	0.5	1.0	1.2
	Damage not observed	0.0	0.5	1.0	1.2

1. Cost of inspection/monitoring includes direct and indirect costs. Both estimated on a present value basis.
2. General consequences cost includes estimated rehabilitation, social, and long-term costs.

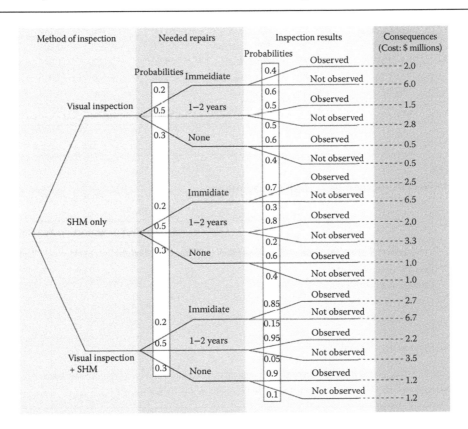

FIGURE 8.35 Decision tree with assigned probabilities and costs.

the direct costs of visual inspection are the least of the three methods. Application of all these estimates to the decision tree branches is shown in Figure 8.35. The total cost of each method can be computed by multiplying each probability along the branch by the total cost of each branch and summing all branches within each method. The process is shown in Table 8.31. The total expected costs for the methods are $2.1 million, $2.17 million, and $2.15 million, for visual inspection, SHM only, and combined methods. Visual inspection method produces the lowest risk while using SHM method would produce the highest risk, when measured in monetary terms.

TABLE 8.31
Deterministic Risk Evaluation

Method	Needed Repairs	Probability (%)	Inspection Results	Probability	Cost	Total Cost	Method Cost
Visual Inspection	Immediate	20.00	Observed	40.00	2.000	0.160	2.10
	Immediate	20.00	Not observed	60.00	6.000	0.720	
	1–2 Years	50.00	Observed	50.00	1.500	0.375	
	1–2 Years	50.00	Not observed	50.00	2.800	0.700	
	None	30.00	Observed	60.00	0.500	0.090	
	None	30.00	Not observed	40.00	0.500	0.060	
SHM Only	Immediate	20.00	Observed	70.00	2.500	0.350	2.17
	Immediate	20.00	Not observed	30.00	6.500	0.390	
	1–2 Years	50.00	Observed	80.00	2.000	0.800	
	1–2 Years	50.00	Not observed	20.00	3.300	0.330	
	None	30.00	Observed	60.00	1.000	0.180	
	None	30.00	Not observed	40.00	1.000	0.120	
Both	Immediate	20.00	Observed	85.00	2.700	0.459	2.15
	Immediate	20.00	Not observed	15.00	6.700	0.201	
	1–2 Years	50.00	Observed	95.00	2.200	1.045	
	1–2 Years	50.00	Not observed	5.00	3.500	0.088	
	None	30.00	Observed	90.00	1.200	0.324	
	None	30.00	Not observed	10.00	1.200	0.036	

TABLE 8.32
Costs of Different Outcomes—High Consequences

Needed Repairs	Inspection Results	General Consequences	Visual Inspection	SHM	Both
			Direct Cost of Inspection/Monitoring		
Immediate	Damage observed	1.5	0.5	1.0	1.2
	Damage not observed	11.0			
1–2 Years	Damage observed	1.0			
	Damage not observed	4.6			
None	Damage observed	0.0			
	Damage not observed	0.0			

Cost (M. $)

1. Cost of inspection/monitoring includes direct and indirect costs. Both estimated on a present value basis.
2. General consequences cost includes estimated rehabilitation, social, and long-term costs.

The example above shows how absolute risk can be used in prioritizing different potential decisions. It is able to do that by using actual monetary values for consequences while using realistic probabilities of different events. To appreciate the power of this approach further, let us increase the cost of not observing the damage for this inspection situation. We can, for example, double such a cost, while keeping all other costs and probabilities intact, as displayed in Table 8.32. The costs of the three methods are shown in Table 8.33. Obviously, there are clear differences between the costs. Now the visual method is having the highest cost. The lowest cost is the combined method. The higher consequences of not observing damage necessitate that the two combined methods of monitoring are used to produce the lowest monetary risk.

TABLE 8.33
Deterministic Risk Evaluation (High Consequences)

Method	Needed Repairs	Probability (%)	Inspection Results	Probability	Cost	Total Cost	Method Cost
Visual	Immediate	20.00	Observed	40.00	2.000	0.160	3.34
Inspection	Immediate	20.00	Not observed	60.00	11.500	1.380	
	1–2 Years	50.00	Observed	50.00	1.500	0.375	
	1–2 Years	50.00	Not observed	50.00	5.100	1.275	
	None	30.00	Observed	60.00	0.500	0.090	
	None	30.00	Not observed	40.00	0.500	0.060	
SHM Only	Immediate	20.00	Observed	70.00	2.500	0.350	2.73
	Immediate	20.00	Not observed	30.00	12.000	0.720	
	1–2 Years	50.00	Observed	80.00	2.000	0.800	
	1–2 Years	50.00	Not observed	20.00	5.600	0.560	
	None	30.00	Observed	60.00	1.000	0.180	
	None	30.00	Not observed	40.00	1.000	0.120	
Both	Immediate	20.00	Observed	85.00	2.700	0.459	2.375
	Immediate	20.00	Not observed	15.00	12.200	0.366	
	1–2 Years	50.00	Observed	95.00	2.200	1.045	
	1–2 Years	50.00	Not observed	5.00	5.800	0.145	
	None	30.00	Observed	90.00	1.200	0.324	
	None	30.00	Not observed	10.00	1.200	0.036	

TABLE 8.34
Probabilistic Risk Results

Method	Cost ($ Million)			
	Mean ($ Million)	Standard Deviation	Sample Minimum	Sample Maximum
Visual inspection	2.10	1.68	0.5	6.0
SHM only	2.17	1.3	1.0	6.5
Both	2.15	.97	1.2	6.7

These results are still an expression of single valued averages. The decision maker still does not have an indication of the level of uncertainty of these numbers. A probabilistic risk method is still needed to gain more feeling of the situation. Using the decision tree of Figures 8.34 and 8.35 it is possible to build three Monte Carlo simulation models for the three observation methods. The number of samples of these models was chosen to be 500 samples. A uniform distribution function was assumed for all events in the tree. A COV of 15% was chosen as a realistic value for each of those distributions. Combined with the information in Figure 8.35, the probabilistic risks can now be evaluated. The resulting mean and standard deviation for the risks of the three observation methods are shown in Figures 8.36 through 8.38. We note that even though the visual inspection method still produces the lowest monetary mean, it also produces the highest standard deviation. The lowest standard deviation is the combined methods. Such results are not entirely surprising. When studying the probabilistic risk for the high consequences case (Table 8.34), the Monte Carlo simulation results produces different prioritization, as in the deterministic risk case as shown in Table 8.35. The PDF and CDF of the high consequences cost situations of the three methods are shown in Figures 8.39 through 8.41. Both the average cost and the standard deviation of the visual inspection method are the highest of the three methods, while the combined method produced the

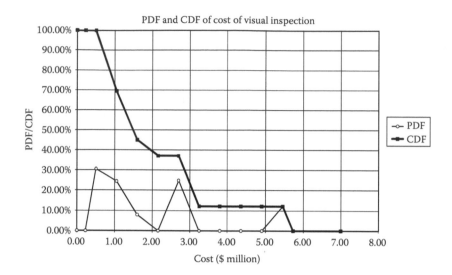

FIGURE 8.36 PDF and CDF of cost of visual inspection.

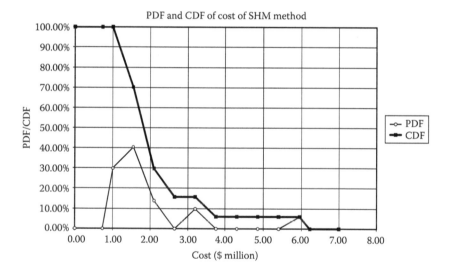

FIGURE 8.37 PDF and CDF of cost of SHM method.

least average cost and standard deviation. Thus, the combined method is the preferred method for this high consequence situation: the decision of which method is preferred is fairly clear.

8.6.7.3 Concluding Remarks

The method just presented utilizes a probabilistic branching of the decision tree, as opposed to the deterministic branching approach. The probabilistic branching versus deterministic branching methods can be compared as follows:

- The probabilistic branching method results in a CDF of the overall costs of different methods, thus making it easy to compare absolute risk of each option. The CDF gives the decision maker the option to use any nonexceedance probability that suites the project's needs. Using deterministic branching offers only a single valued deterministic result, this might not be adequate to make an accurate decision.

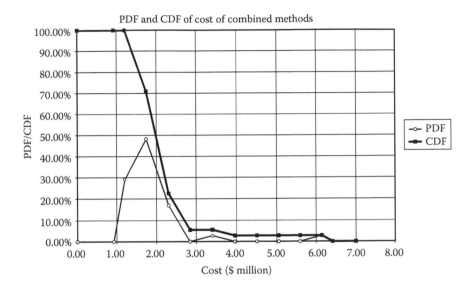

FIGURE 8.38 PDF and CDF of cost of combined (visual and SHM) methods.

TABLE 8.35
Probabilistic Risk Results

	Cost ($ Million)			
Method	**Mean ($ Million)**	**Standard Deviation**	**Sample Minimum**	**Sample Maximum**
Visual inspection	3.28	3.46	0.5	11.5
SHM only	2.75	2.7	1.0	12.0
Both	2.35	1.84	1.2	12.2

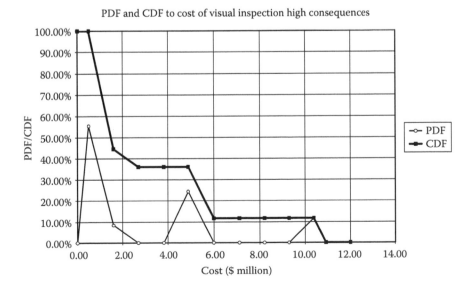

FIGURE 8.39 PDF and CDF of cost of visual inspection (high consequences).

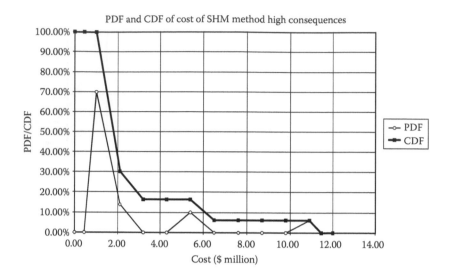

FIGURE 8.40 PDF and CDF of cost of visual inspection (high consequences).

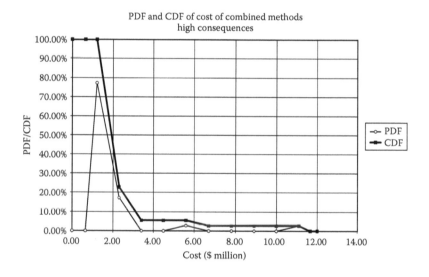

FIGURE 8.41 PDF and CDF of cost of visual inspection (high consequences).

- Advent of powerful computing power, the reliance of Monte Carlo while performing probabilistic branching method is more feasible than ever.
- The probabilistic branching method can utilize the results of an SHM project. If proper coordination and planning are in place, such results can improve the accuracy of the cost–benefit analysis, thus ensuring an optimal financial decision regarding the project on hand.

8.6.8 ABSOLUTE RISK: LIFE CYCLE ANALYSIS

8.6.8.1 Introduction

One implicit assumption that was made for different methods of this chapter so far has been that there are no interdependencies between problem parameters. Witness the many branches of the decision tree of Figure 8.34 are not interconnected, except to the nodes that immediately support them. We also note that Monte Carlo method, which is described in Section 8.7.2, needs special considerations to accommodate such interdependencies between variables. Of course such simplifying assumption is too limiting in practical situations. There are many instances when different problem parameters are interdependent at many levels. Such interdependency can greatly affect the outcome of the problem and the resulting decisions that might be made. We present in this section a fairly simple example of an absolute risk evaluation approach that includes some parameter interdependency.

8.6.8.2 Problem Description

Estimating life cycle costs LCC of a particular system is a recurring dilemma for decision makers. The LCC problem as related to SHM/SHCE is explored in Chapter 10 by Ettouney and Alampalli (2012). For now we present a simple LCC problem with an emphasis on interdependence between parameters. Consider a simple situation of a two parameter cost model: cost of an earthquake hazard and cost of wear and tear (normal deterioration) of a bridge. The decision maker is tasked of estimating the LCC of the bridge in this situation. For simplicity, let us ignore the effects of economic rates (inflation, discount, etc.) over the years. The LCC becomes

$$LCC = \sum_{Years} \left(\sum \left(A_1 C_1 \right) + C_2 \right) \tag{8.117}$$

With A1 is the probability of the annual occurrence of seismic events. The cost of such an earthquake is C_1. The annual cost of normal deterioration of the bridge is C_2. To simplify the problem even further, we assume that there is only one earthquake level that need to be considered the Equation 8.117 becomes

$$LCC = \sum_{Years} \left(A_1 C_1 + C_2 \right) \tag{8.118}$$

Equations 8.117 and 8.118 illustrates the concept of risk. They are similar to the form of Equation 8.107. They account for the effects of hazards, vulnerabilities, and consequences. The risk is expressed in monetary terms. The risk can be evaluated in a deterministic (single valued) manner. Assume, for example, that the earthquake of interest has an annual probability of occurrence of 0.2% (500-years earthquake) and the cost of such an occurrence is estimated at $2.0 million. Furthermore, assume that the cost of annual normal deterioration is $0.5 million. The LCC for 20 years is

$$LCC = 20 \times \left(0.002 \times 2 + 0.5 \right) = \$10.8 \text{ Million}$$

We can, of course, apply a true probabilistic risk analysis to Equation 8.118. If we assume that the two costs are random variables with properties as shown in Table 8.36.

Noting that LCC is now a simple function of random variables, we can use the expressions of Table 8.4 to compute the means and standard deviation or by applying a Monte Carlo simulation analysis, with number of samples of 5000. The mean is μ_{LCC} = $10.11 million. The standard deviation is σ_{LCC} = $2.01 million.

These results offer the decision-maker ability to make more educated decisions than in the deterministic, single valued situation. For example, for conservative estimate of LCC, the decision maker may choose to use the $\mu_{LCC} + \sigma_{LCC}$ in the LCC estimation. Such a value would produce

TABLE 8.36
Statistical Properties of Cost

Parameter	Distribution	Mean $ (Million)	COV $ (Million)
C_1	Uniform	2	0.2
C_2	Uniform	0.5	0.2

a nonexceedance probability level of 84%, if the PDF is normally distributed. Thus estimating the LCC as

$$LCC = \mu_{LCC} + \sigma_{LCC} \qquad (8.119)$$

This would produce $LCC = \$12.12$ million. The difference between the single valued/deterministic, and the more realistic/probabilistic risk is almost 20%.

Let us now turn our attention to the subject of dependency of parameters. Upon further considerations of Equations 8.117 and 8.118, we can only conclude that the two costs C_1 and C_2 must have some kind of dependency. The cost of normal deterioration is usually computed without accounting for any other abnormal rehabilitation costs. In the case of a seismic event, there would be rehabilitation of the bridge that would cost C_1. Such rehabilitation effort would definitely affect the cost of normal deterioration. Such a dependency has not been studied as of the writing of this manuscript. It is clear, however, that the cost of deterioration is C_2 as long as no seismic event occurred. As soon as a seismic event occurs at a particular year, the cost of normal deterioration in the following years will need to be different from C_2. Clearly, Equations 8.117 and 8.118 can't accommodate such a dependency in their current simple forms.

One possible way to handle the dependency is to solve the LCC problem in a true time line fashion, using a Monte Carlo simulation method. Recall that the earthquake hazard was assumed to have an annual probability of occurrence of 0.002%. This can be restated as a uniform PDF in the range of 0–500 years. Figure 8.42 shows both the PDF along with the CDF of this function. We can then build a simulation model as follows:

1. Start the simulation with trial $i = 1$
2. Generate a random number in the range from 0 to 1.0
3. Using the CDF of Figure 8.42, generate the year that an earthquake would happen for that random number N_i
4. If $N_i \geq N_{TOT}C_2$ then there would be no seismic event in this trial
   a. $LCC_i = N_{TOT}$
   b. Go to step 6
5. If $N_i < N_{TOT}$ then there will be an earthquake event in this trial at the year N_i
   a. $LCC_i = C_{21}(N_{TOT} - N_i) + N_iC_2 + C_1$
6. Assign $I = i + 1$
7. Repeat steps 1 through 6 until the desired number of Monte Carlo trials is completed

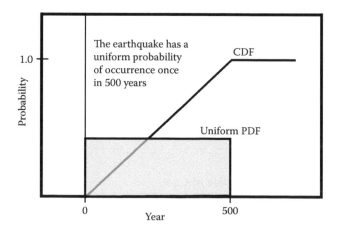

FIGURE 8.42 PDF and CDF of a 500-years earthquake.

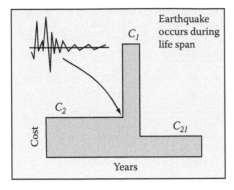
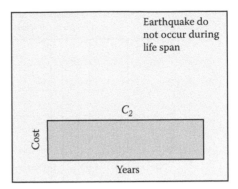

C_2 = Cost of normal deterioration per year
C_1 = Cost of an earthquake event
C_{21} = Cost of normal deterioration per year, after an earthquake event

FIGURE 8.43 Dependency of costs of normal deterioration and earthquake hazard.

TABLE 8.37
LCC for Different Analysis Methods

		Life Cycle Costs $ (Million)						
			Probabilistic					
			Independent			Dependent		
Life Span, Years	Cost of Seismic Retrofit $(M)	Deterministic	Mean	STD	Mean +STD	Mean	STD	Mean +STD
20	2.00	10.08	10.11	2.01	12.13	9.99	1.96	11.95
	20.00	10.80	10.81	2.02	12.83	10.75	4.15	14.90
	40.00	11.60	11.60	2.03	13.63	11.64	8.09	19.73
50	2.00	25.20	25.24	4.99	30.23	24.70	4.95	29.65
	20.00	27.00	26.93	4.95	31.88	26.54	6.87	33.41
	40.00	29.00	28.83	5.03	33.86	28.49	12.00	40.49

The cost C_{21} is the cost or normal deterioration after the earthquake rehabilitation efforts are made. N_{TOT} in the above steps is the required life span of the bridge. The dependency logic for this example is shown in Figure 8.43. Upon finishing steps 1 through 7, the data series LCC_i can be used to generate the CDF of LCC, as well as the mean and standard deviation of LCC.

If we assume that C_{21} = $0.3 million, and performing Monte Carlo simulation process, we can compute the μ_{LCC} and σ_{LCC} for each case of interest. Table 8.37 shows the results for different life spans, seismic retrofit costs, and deterministic, probabilistic independent, and probabilistic dependent cases. The LCC varies considerably; which shows the importance of probabilistic and dependent simulation modeling. The equivalent histograms for life spans of 20 years are shown in Figures 8.44 through 8.49. The CDFs are shown in Figures 8.50 through 8.52.

8.6.8.3 Concluding Remarks

The importance of the dependency between parameters in risk assessment was explored in this example. We would like to mention that the importance of this dependency is not limited to LCC. In most practical situations there will invariably be dependencies between important parameters. It is not an easy task to isolate then quantify those dependencies. Yet, it is essential to accommodate

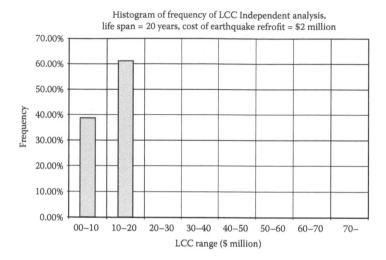

FIGURE 8.44 Histogram: life span = 20, cost of EQ = 20, independent analysis.

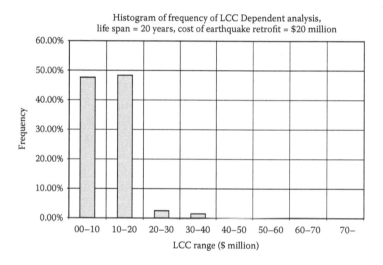

FIGURE 8.45 Histogram: life span = 20, cost of EQ = 20, dependent analysis.

such dependencies in any risk analysis. Short of that, the decision maker might end up with solutions of a problem that is very different from the problem on hand.

One final comment about this example: the reader might ask: what is the role of SHM in this example? We can answer such a question by observing first that to accurately estimate the costs C1 and C2, there is a need for the following:

1. Accurate measurements and estimation of the normal deterioration of the bridge and the different rehabilitation measures that are needed to counter such deterioration
2. Accurate structural analysis and identification to estimate the damaging effects of earthquake on the bridge
3. The above two needs might also include data collection (real time or intermittent)

All of the above are direct and indirect usage of SHM/SHCE tools and methodologies. Without employing some or all of those tools and methodologies, the cost estimations of C1 and C2 will not be more than a theoretical exercise with no great practical value.

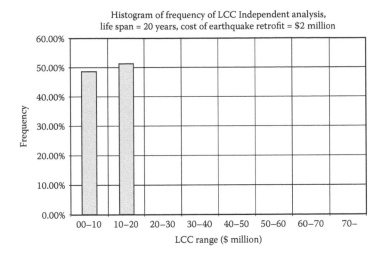

FIGURE 8.46   Histogram: life span = 20, cost of EQ = 2, independent analysis.

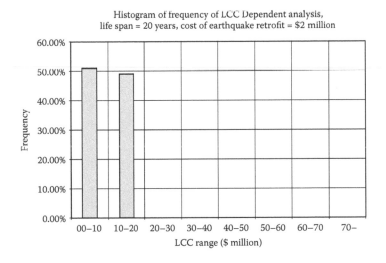

FIGURE 8.47   Histogram: life span = 20, cost of EQ = 2, dependent analysis.

8.7   STOCHASTIC MODELS

Performance of civil infrastructure is a time-dependent issue. The time scale of performance can be extremely short (acoustic emission from cracks, sudden or brittle collapse, etc.), short (earthquake seismic wave effects, hydrodynamics of flash floods, speedy truck, etc.), or long (deterioration of reinforced concrete, corrosion of steel girders, etc.). One of the commonalities between all of these performance measures is that they all exhibit uncertain behavior in one form or another. Such time-dependent uncertain behavior is commonly referred to as stochastic behavior, or stochastic processes. The subject of stochastic processes has been studied by numerous authors; see, for example, Lawler (1995) and Crandal and Mark (1963). Many of the stochastic processes techniques have been used by decision makers for civil infrastructure (see Elishakoff 1983). In this chapter, we review two of the most popular methods in stochastic processes for bridge management systems, namely Markov processes and Monte Carlo methods. We offer some practical examples for each technique.

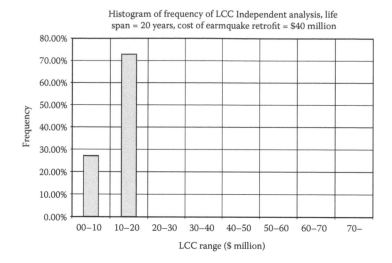

FIGURE 8.48   Histogram: life span = 20, cost of EQ = 40, independent analysis.

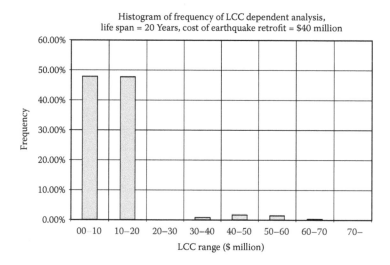

FIGURE 8.49   Histogram: life span = 20, cost of EQ = 40, dependent analysis.

8.7.1 MARKOV PROCESSES

Assume that the is a set of different states θ_i, with i = 1, 2, ..., M. The states can describe states of damage of bridge deck, or a state of deterioration of paint, and so on. Further we assume that in a given time period, the probabilities of the ith state to change to the jth state is constant at p^{ij}. This set of probabilities form the components of an $M \times M$ matrix $[T]$ that is called *transition matrix*. If the initial probability of the formation of each state is defined as $\{p_0\}$, then the probabilities that the subject of interest is at a given state after N periods are the components of the vector $\{p_N\}$ such that

$$\{p_N\} = \{p_0\}[T]^N$$

(8.120)

The above process is called *Markov process* (Figure 8.53). It is used in several applications in civil and structural engineering, especially in the bridge management field. It has the advantage of

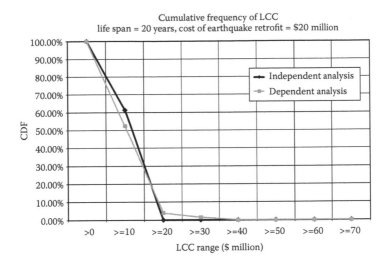

FIGURE 8.50   CDF for life span = 20, cost of EQ = 20.

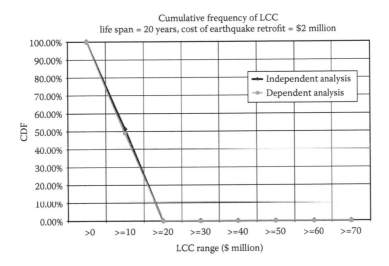

FIGURE 8.51   CDF for life span = 20, cost of EQ = 2.

simplicity. However, one of the most important limitations of the methods is that probabilities in the transition matrix are assumed to be constant.

In addition to the above assumptions, it is clear that the components of the transition matrix [T] must be accurate, otherwise the computed probabilities, and the ensuing decisions, will be faulty. See Benjamin and Cornell (1970) for more on theoretical applications of Markov processes. Also, see Agrawal et al. (2009) for bridge management applications of Markov process.

The field of SHM can take advantage of Markov processes in enhancing decision making for many situations. The above examples of bridge paint, or deck deterioration illustrates such use. Conversely, SHM can also help in producing accurate probabilities for populating the transition matrices. We explore both aspects of SHM/Markov process interrelationship by providing a specific example.

8.7.1.1   Example: Bridge Paint Status

As an example of Markov process use in decision making, let us consider the bridge paint deterioration problem. In our example, the decision maker decides that there are three states

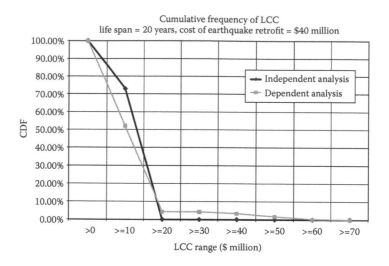

FIGURE 8.52 CDF for life span = 20, cost of EQ = 40.

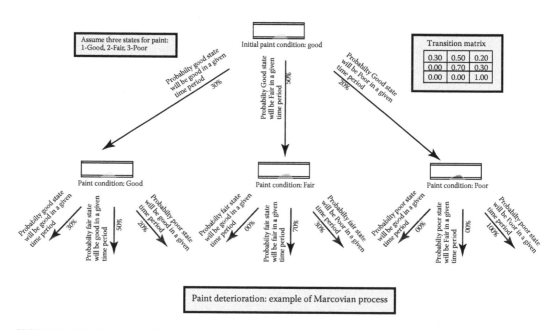

FIGURE 8.53 Example of Markov processes: deterioration of paint.

for paint: good, fair, and poor. Previous experiences and studies have produced a generic paint state transition matrix (based on a 2-year period) similar to that in Table 8.38. Such a generic matrix is used all over network of bridges that the decision maker in our hypothetical example is managing. It is a 3 × 3 matrix. The cells in the matrix are the transition probabilities p_{ij}, which indicates the probability that the state of paint will change from the ith state to the jth state within 2 years. Note that the sum of each of the row in the matrix is 1.0, as expected. Also, note that the probabilities in lower triangle of the matrix are all zeros, which indicates that the state of the paint can't improve during the 2-years period. The transition matrix can be used to estimate the probabilities of a given paint state at any time.

The above process is fairly simple, yet it is very sensitive to the probability values in the transition matrices. SHM can help in producing accurate transition probability matrices, as is discussed next.

TABLE 8.38
Qualitative Estimate of
Transition Matrix of Paint Status

Paint Status	Good	Fair	Poor
Good	0.3	0.5	0.2
Fair	0.0	0.7	0.3
Poor	0.0	0.0	1.0

TABLE 8.39
Initial Paint Status at Different SHM Observation Zones

Bridge	Zone	Initial Paint State
Bridge # 1. Older Bridge; thus has	1	Good
several different initial paint states	2	Fair
	3	Fair
	4	Poor
	5	Poor
Bridge # 2. Newer Bridge, lesser number	6	Good
of different initial paint states	7	Fair
	8	Fair
Bridge # 3. Newest Bridge. Less number	9	Good
of different initial paint states	10	Good

8.7.1.2   Use of SHM in Markov Processes: Paint States of a Bridge

One of the main uses of SHM activities in a Markovian processes is to accurately estimate transition matrices. Generally speaking, transition matrices are estimated using qualitative (best guess) approach. Such an approach is generic and does not account for specific conditions of the process or geometries on hand. For example, the paint state of the bridge that was discussed earlier depend on many general and local conditions. The general conditions, such as type of paint, type of bridge, and age of paint, are applicable to most situations, and can be considered generic in nature. The local conditions, such as temperature variation range, humidity or salt usage (snow removal solution, or sea water) are all local conditions that can affect the deterioration of paint. These local conditions will affect transition matrices and must be accommodated in any Markov process analysis. Let us consider how an SHM project can help in building a transition matrix for the evaluation of the paint deterioration problem in a bridge.

The bridge official is tasked with producing a more realistic transition matrix for bridge paint states. The paint states are kept similar to the ones mentioned earlier. Three bridges that have similar local conditions are chosen for this project. It is decided that manual visual inspection is suited for this project, since the estimation of paint states is mostly qualitative. Each of the three bridges has different paint states at several zones as shown in Table 8.39, for a total of 10 zones. The paint states at each of these zones were observed on a quarterly basis, over 8 years period, thus producing 32 observation points at each zone. The observation results are shown in Table 8.40. The frequencies of the different observations are then tabulated (Table 8.41). These frequencies are then grouped in a suitable form in Table 8.42. Those probabilities are finally arranged in an SHM-based transition matrix as shown in Table 8.43.

8.7.1.3   Value of SHM for a Markov Process: Paint States

We turn our attention now to the value of the SHM project. Recall the earlier qualitative estimate of the paint transition matrix in Table 8.38. We had just evaluated the more accurate transition matrix

TABLE 8.40
SHM Observation Results

Observation	2-Years Increments	Bridge # 1					Bridge # 2			Bridge # 3	
		1	2	3	4	5	6	7	8	9	10
0		Good	Fair	Fair	Poor	Poor	Good	Fair	Fair	Good	Good
1	1	Good	Fair	Fair	Poor	Poor	Good	Fair	Fair	Good	Good
2		Good	Fair	Fair	Poor	Poor	Good	Fair	Fair	Good	Good
3		Good	Fair	Fair	Poor	Poor	Good	Fair	Fair	Good	Good
4		Good	Fair	Fair	Poor	Poor	Good	Fair	Fair	Good	Good
5		Good	Fair	Fair	Poor	Poor	Good	Fair	Fair	Good	Good
6		Fair	Fair	Fair	Poor	Poor	Good	Fair	Fair	Good	Good
7		Fair	Fair	Fair	Poor	Poor	Good	Fair	Fair	Good	Good
8		Fair	Fair	Fair	Poor	Poor	Good	Fair	Fair	Good	Good
9	2	Fair	Fair	Fair	Poor	Poor	Good	Fair	Fair	Good	Good
10		Fair	Fair	Fair	Poor	Poor	Good	Fair	Fair	Good	Good
11		Fair	Fair	Poor	Poor	Poor	Fair	Fair	Fair	Good	Good
12		Fair	Fair	Poor	Poor	Poor	Fair	Fair	Fair	Good	Good
13		Fair	Fair	Poor	Poor	Poor	Fair	Fair	Fair	Good	Good
14		Fair	Poor	Poor	Poor	Poor	Fair	Fair	Fair	Fair	Good
15		Fair	Poor	Poor	Poor	Poor	Fair	Fair	Fair	Fair	Good
16		Fair	Poor	Poor	Poor	Poor	Fair	Fair	Fair	Fair	Good
17	3	Fair	Poor	Poor	Poor	Poor	Fair	Fair	Fair	Fair	Good
18		Fair	Poor	Poor	Poor	Poor	Fair	Fair	Fair	Fair	Good
19		Fair	Poor	Poor	Poor	Poor	Fair	Fair	Poor	Fair	Good
20		Fair	Poor	Poor	Poor	Poor	Fair	Fair	Poor	Fair	Good
21		Fair	Poor	Poor	Poor	Poor	Fair	Fair	Poor	Fair	Good
22		Fair	Poor	Poor	Poor	Poor	Fair	Fair	Poor	Fair	Good
23		Fair	Poor	Poor	Poor	Poor	Fair	Fair	Poor	Fair	Good
24		Fair	Poor	Poor	Poor	Poor	Poor	Fair	Poor	Poor	Good
25	4	Fair	Poor	Poor	Poor	Poor	Poor	Poor	Poor	Poor	Good
26		Fair	Poor	Poor	Poor	Poor	Poor	Poor	Poor	Poor	Good
27		Fair	Poor	Poor	Poor	Poor	Poor	Poor	Poor	Poor	Good
28		Poor	Poor	Poor	Poor	Poor	Poor	Poor	Poor	Poor	Poor
29		Poor	Poor	Poor	Poor	Poor	Poor	Poor	Poor	Poor	Poor
30		Poor	Poor	Poor	Poor	Poor	Poor	Poor	Poor	Poor	Poor
31		Poor	Poor	Poor	Poor	Poor	Poor	Poor	Poor	Poor	Poor
32		Poor	Poor	Poor	Poor	Poor	Poor	Poor	Poor	Poor	Poor

in Table 8.43. It is more accurate, obviously, since it is based on actual observations in more realistic conditions that are similar to the bridge conditions of interest, rather than generic conditions. Qualitatively, it is clear that the components of Table 8.38 and Table 8.43 are considerably different. The bridge official, however, would like to have a more objective procedure for estimating the value of SMH procedures. The first step in estimating the value is to compute the cost of the SHM. Recall that the measurements of the paint deterioration were done by manual observations at three bridge sites, over the span of 8 years, at quarterly intervals, for a total of 32 observations per bridge zone (total of ten zones). The cost of these observations, and the management, archiving, and computations is estimated by the bridge official is C_{SHM}, while the cost of painting the bridge is C_{PNT}.

TABLE 8.41
Frequency of Different Observations

Starting State	Ending State	Bridge # 1					Bridge # 2			Bridge # 3		Total
		1	2	3	4	5	6	7	8	9	10	
Good	Good	1	0	0	0	0	1	0	0	1	3	6
Good	Fair	0	0	0	0	0	1	0	0	1	0	2
Good	Poor	0	0	0	0	0	0	0	0	0	1	1
Fair	Fair	2	1	1	0	0	0	3	2	1	0	10
Fair	Poor	1	1	1	0	0	1	1	1	1	0	7
Poor	Poor	0	2	2	4	4	1	0	1	0	0	14
Total transitions per zone		4	4	4	4	4	4	4	4	4	4	

TABLE 8.42
Consolidated Frequencies of Observations

Total Start States		Total End States		
		Good	Fair	Poor
Good	9	6	2	1
Fair	17	0	10	7
Poor	14	0	0	14

TABLE 8.43
SHM-based Transition Matrix of the Paint Status

Paint Status	Good	Fair	Poor
Good	0.67	0.22	0.11
Fair	0.0	0.59	0.41
Poor	0.0	0.0	1.0

The value of the SHM-based transition matrices is estimated using the potentially saved paint costs. For example, if the initial paint state is "Good," the initial probability vector is

$$\{P_o\} = \{1 \quad 0 \quad 0\} \tag{8.121}$$

After $2N$ years of operations, the paint state probability vector is

$$\{P_N\} = \{P_0\}[T]^N \tag{8.122}$$

Where $[T]$ is a 3 × 3 probability transition matrix. The N power in Equation 8.122 is due to the fact that the transition probabilities are computed for a 2-years time interval.

From Table 8.38 and Equations 8.121 and 8.122, the state probabilities from generic transition matrices after 10 years are

$$\{P_{N1}\} = \{0.00 \quad 0.21 \quad 0.79\} \tag{8.123}$$

From Table 8.43, the SHM-based state probabilities vector after 10 years are

$$\{P_{N1}\} = \{0.14 \quad 0.17 \quad 0.69\} \tag{8.124}$$

To get a wider understanding of the deterioration process as predicted by Markov analysis, we plot the 10 years probabilities of paint being at a given state in Figures 8.54 through 8.56 for good, fair, and poor state, respectively. The differences between the generic and SHM-based probabilities are self-evident. Any management decision that is based on Markov process for paint deterioration would be fairly inaccurate if it is based on generic transition matrices. We can quantify this important conclusion as follows.

If the management rule for that particular office is to repaint if the probabilities of the paint state of "poor" is ≥0.75, then the generic transition matrices based estimations will cause an initiation of repainting. The SHM-based results will not initiate repainting. Continuing the process for the SHM-based computations reveals that 2 more years will pass before an repainting is needed. If we ignore discount and inflation rates, for simplicity, we can deduce that over the course of the life span of the

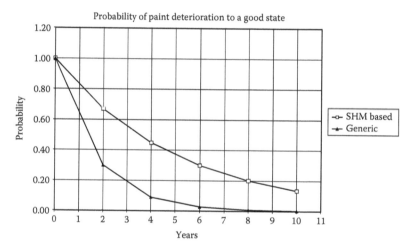

FIGURE 8.54   Probabilities of good paint state.

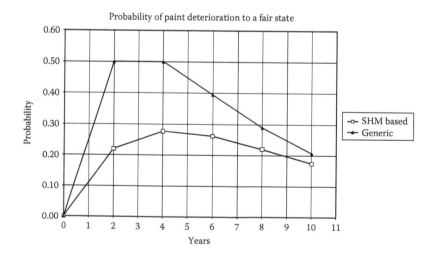

FIGURE 8.55   Probabilities of fair paint state.

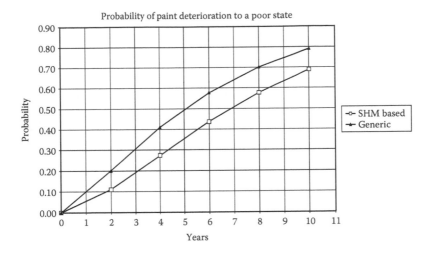

FIGURE 8.56 Probabilities of poor paint state.

bridge the cost savings are C_{PNT}. Further, if we assume that the number of bridges in the network that have similar conditions as the subject bridge is N_{BRIDGE}, the monetary value of the SHM project, over the course of the life span of the bridge is

$$V_{SHM} = N_{BRIDGE} C_{PNT} - C_{SHM} \tag{8.125}$$

The above example showed net monetary value that result from direct cost savings due to later repaint needs. If the SHM-based transition matrices indicate that a sooner repaint is needed, there is still a value for such situation. The value in this case will come from the realized performance of repainting the bridge before its state deteriorate below the stated performance goal (≥ 0.75 probability of a "poor" painting condition). The value is not monetary, but it can be computed using utility theory techniques as mentioned in Section 8.4.2.

One final remark about the above example: it was based on a single paint state throughout the life span of the bridge. More realistic analysis that account for different paint states can be exercised. The computations of value are similar for those more complex examples.

8.7.2 Monte Carlo Method

Monte Carlo simulation method has been used in many fields. Its popularity is attributed to its simplicity. Lately, because of the advances in computing power, Monte Carlo uses have grew even further and faster. For our purposes, we describe Monte Carlo in its simplest form as follows:

1. Input: a single, or a set, of random variables with known probability distributions
2. Using random number generator, generate a single input data set
3. Model: Create a mathematical model that uses the random values in #2 to produces a single output data set. Note that the model itself can contain random properties
4. Repeat #2 and #3 as many times as prescribed, until an ensemble of data sets is formed
5. Form frequency analysis (see Section 8.3.2.2 of this chapter) to produce needed statistical properties, and plot the histograms, or evaluate PDFs, of CDFs, as needed

Figure 8.57 shows a summary of the process. The Monte Carlo simulation method is utilized extensively in this document. For further readings about the basis of the method, or other applications, check Trueman (1977).

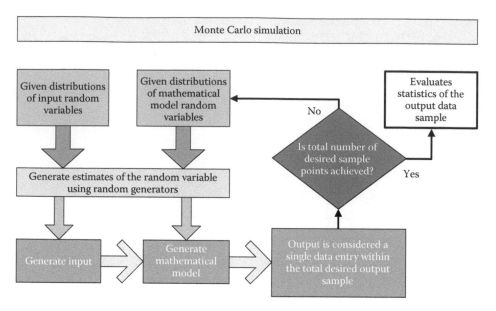

FIGURE 8.57   Monte Carlo simulation technique.

8.8   STRUCTURAL ANALYSIS IN DECISION MAKING

8.8.1   GENERAL

Properties of infrastructure are usually assumed to be single valued, or deterministic. Such an assumption is usually accepted during the course of conventional design procedures. In many decision making situations, such an assumption might not be adequate. Since decision making process involves uncertainties of almost all issues, it is reasonable to accommodate the uncertainties of properties of infrastructure in the decision making process. Ignoring such uncertainties might lead to inadequate decisions. In some situations such as when assessing probabilities and consequences of failure, ignoring uncertainties of infrastructure properties might lead to unsafe decisions.

The problem can be summarized as follows. Find the response of a given structure that has some of properties as random variables. The random input can be described by its PDF, or simply by its mean and variances. The computed output is usually in a form of a mean and variances of output measures, such as displacements, strains, or stresses. There are many methods to this problem. Many solutions address uncertainties in structural components or structural systems; see Elishakoff (1983) or Melchers (2002). Whether the main interest lies in structural components or systems, there is a need for a special probabilistic analysis program. The steps for such an analysis are

1. Model the structure using a specialized probabilistic analysis program
2. Assign random properties as appropriate
3. Assign deterministic properties as appropriate
4. Assign the different design loads
5. Analyze for desired design measures, such as displacements, strains, or stresses. Note that the design measures in this situation will include statistical parameters such as means and standard deviations
6. Use the statistical results of step #5, above to evaluate confidence limits, or probabilistic statements about the safety

7. Use the results from #6 in the remainder of the decision making process. For example, decide if the results of step #6, above, is within acceptable safety limits of the structure on hand

The above steps are applicable if specialized probabilistic analysis tools are available to the official. However, such specialized tools might not be easy to obtain. In the absence of such specialized tools, the following steps might be followed:

1. Model the structure using conventional structural analysis techniques
2. From a random number generator and utilizing the probabilistic statement concerning the uncertain properties, find a random value(s) for those properties
3. Assign the values obtained in #2 to the appropriate locations in the conventional analytical model of step #1
4. Assign deterministic properties to the rest of the model as appropriate
5. Assign different design loads
6. Analyze for desired design measures, such as displacements, strains, or stresses
7. Repeat steps 2 through 6
8. From the information obtained in #6, form the statistical parameters (such as means and standard deviations) for different design measures
9. Use the statistical results of step #8, above to evaluate confidence limits, or probabilistic statements about the safety
10. Use the results from #9 in the remainder of the decision making process. For example, decide if the results of step #9, above are within acceptable safety limits of the structure on hand

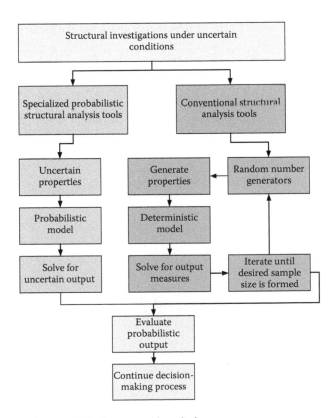

FIGURE 8.58 Methods for probabilistic structural analysis.

The reader would recognize the above steps as a form of Monte Carlo's method.

Both methods would enable the official to account for uncertainties in structural properties. Figure 8.58 shows schematics of the steps of both approaches. There are advantages and limitations of each method:

- The specialized probabilistic method is computationally efficient, but it requires a special probabilistic analytical tools that might not be available to the decision maker
- The Monte Carlo approach is usually available, but it can be computationally inefficient

The appropriate method will then depend on the severity and complexity of the structural system under considerations.

8.8.2 PROBABILISTIC FINITE/BOUNDARY ELEMENT METHODS

8.8.2.1 Theoretical Background

Among specialized probabilistic structural analysis methods for structural systems are probabilistic finite element methods and probabilistic boundary element methods. We present here a brief explanation of probabilistic boundary element method. For complete developments and applications in different fields, the reader is referred to Ettouney et al. (1989a and 1989b) and Daddazio and Ettouney (1990).

Let us assume that for an elastic problem, the discretized boundary element equation can be expressed as

$$M(x_1, x_2) p(x_1, x_2) = N(x_1, x_2) q(x_1, x_2) \tag{8.126}$$

Where $M(x_1, x_2)$ and $N(x_1, x_2)$ are square matrices that include the integrals of the kernel functions of the problem. The vectors $p(x_1, x_2)$ and $q(x_1, x_2)$ include the boundary displacements and traction. We assume that x_1 and x_2 are two uncertain physical properties (e.g., they can represent an uncertain shear modulus and Poisson's ratio). Let us assume that $\bar{x}_1$ and $\bar{x}_2$ are the mean values of x_1 and x_2. Let us define the perturbation values

$$\varepsilon_1 = x_1 - \bar{x}_1 \tag{8.127}$$

$$\varepsilon_2 = x_2 - \bar{x}_2 \tag{8.128}$$

We can now utilize Taylor series to expand Equation 8.126 about the mean values of x_1 and x_2. Utilizing perturbation analysis techniques, Bellman (1964), Equation 8.126 will produce the following recursion equations

$$\bar{M} \bar{p} = \bar{N} \bar{q} \tag{8.129}$$

$$\bar{M} \frac{\partial \bar{p}}{\partial x_1} = \frac{\partial \bar{N}}{\partial x_1} \bar{q} - \frac{\partial \bar{M}}{\partial x_1} \bar{p} \tag{8.130}$$

$$\bar{M} \frac{\partial \bar{p}}{\partial x_2} = \frac{\partial \bar{N}}{\partial x_2} \bar{q} - \frac{\partial \bar{M}}{\partial x_2} \bar{p} \tag{8.131}$$

$$\bar{M} \frac{\partial^2 \bar{p}}{\partial x_1^2} = 2 \frac{\partial \bar{M}}{\partial x_1} \frac{\partial \bar{p}}{\partial x_1} \tag{8.132}$$

Higher order equations can be developed. We now solve the recursive equations for $\bar{p}$, $\partial \bar{p} / \partial x_1$, $\partial \bar{p} / \partial x_2$, and $\partial^2 \bar{p} / \partial x_1^2$ and express the displacements as

$$p(x_1, x_2) = \bar{p} + \varepsilon_1 \frac{\partial \bar{p}}{\partial x_1} + \varepsilon_2 \frac{\partial \bar{p}}{\partial x_2} + \frac{1}{2} \varepsilon_1^2 \frac{\partial^2 \bar{p}}{\partial x_1^2} + \frac{1}{2} \varepsilon_2^2 \frac{\partial^2 \bar{p}}{\partial x_2^2} + \varepsilon_1 \varepsilon_2 \frac{\partial^2 \bar{p}}{\partial x_1 \partial x_2} \tag{8.133}$$

The mean and the mean square of the displacement vector (correct to the second order) can be easily computed by

$$E[p] = \bar{p} + E[\varepsilon_1] \frac{\partial \bar{p}}{\partial x_1} + E[\varepsilon_2] \frac{\partial \bar{p}}{\partial x_2}$$

$$+ \frac{1}{2} E[\varepsilon_1^2] \frac{\partial^2 \bar{p}}{\partial x_1^2} + \frac{1}{2} E[\varepsilon_2^2] \frac{\partial^2 \bar{p}}{\partial x_2^2} + E[\varepsilon_1 \varepsilon_2] \frac{\partial^2 \bar{p}}{\partial x_1 \partial x_2} \qquad (8.134)$$

$$E[p^2] = \bar{p}^2 + E[\varepsilon_1^2] \left(\bar{p} \frac{\partial^2 \bar{p}}{\partial x_1^2} + \left(\frac{\partial \bar{p}}{\partial x_1} \right)^2 \right) + E[\varepsilon_2^2] \left(\bar{p} \frac{\partial^2 \bar{p}}{\partial x_2^2} + \left(\frac{\partial \bar{p}}{\partial x_2} \right)^2 \right) \qquad (8.135)$$

As soon as $E[p]$ and $E[p^2]$ are evaluated, the required mean and variance can be established as in Section 8.3 of this chapter.

Note that Equations 8.126 through 8.135, above can be generalized for more than two uncertain variables. However, the equations can become too complex. In such a situation, the use of other techniques, such as Monte Carlo technique becomes more preferable.

8.8.2.2 Example

Consider the bridge foundation problem of Figure 8.59. The foundation is a single footing that rests directly on a soil layer that rest on a half space. The footing supports a bridge column that supports the bridge superstructure. The foundation-soil system is located in a seismically active region. Because of this, the official desires to compute the equivalent soil springs as detailed by Arya et al. (1979). The elastic properties of the soil layer and the half space are shown in Table 8.44. For our purposes, the foundation has a square footprint, and is assumed to have a stiffness that is much higher than the soil stiffness. By using conventional soil-structure analysis techniques, the official computed the horizontal and vertical soil springs to be as shown in Table 8.45.

Upon further research, the official found out that the soil properties in Table 8.46 were obtained by a soil consultant who provided also the coefficient of variations (COV) of those properties as

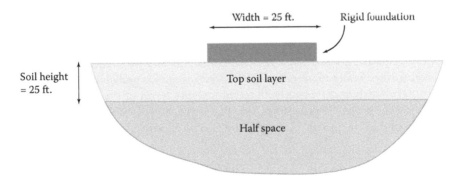

FIGURE 8.59 Soil—structure interaction problem.

TABLE 8.44
Deterministic Input

Soil Layer	Modulus of Rigidity (ksf)	Poisson Ratio
Top layer	2000.00	0.35
Half space	3500.00	0.30

TABLE 8.45
Deterministic Results

Equivalent Spring Type	Deterministic Results (kips/ft)
Vertical	150,366
Horizontal	123,285

TABLE 8.46
Probabilistic Input

Soil Layer	Modulus of Elasticity (ksf)		Poisson Ratio	
	Mean	COV	Mean	COV
Top layer	2000.00	0.21	0.35	0.05
Half space	3500.00	0.26	0.30	0.05

TABLE 8.47
Probabilistic Results

Equivalent Spring Type	Mean μ	Standard Deviation σ	COV	$\mu + \sigma$	$\mu - \sigma$
Vertical	153,461	38,365	0.25	230,192	76,730
Horizontal	119,482	34,649	0.29	188,782	50,182

shown in Table 8.46. The official felt that those COV were higher than normal. It was felt that with high COV, there is a need for a probabilistic analysis. Fortunately, a PBE computer code was available, and the official embarked on a probabilistic analysis that employed an algorithm similar to the one described above. The results are shown in Table 8.47. Note that the means μ of the soil springs have not changed much from the deterministic results of Table 8.45. However, the standard deviations σ and COV produced a $\mu+2\sigma$ and $\mu-2\sigma$ results that are very different from the mean. Those limits of the soil springs would represent a 97% of nonexceedance probability (if the resulting distribution is normal), which is consistent with the probability levels of the whole project. The final decision was to use the probabilistic results. Note that two analyses would be needed for the upper and lower bounds of the soil springs. For more on this type of problems, see Ettouney et al. (1989a, 1989b, and 1993).

8.9 FINANCIAL CONSIDERATIONS

8.9.1 INTRODUCTION

Decision making processes invariably use monetary means. This monetary means can vary from simple present and future value considerations to more complex financial considerations of projects. We describe the basic processes of present and future values in this section. In addition, we show the importance of uncertainty considerations in financial computations. We also introduce the concepts of LCA and its relationship to both SHM and financial analysis concepts. Finally, the role of peer review in the whole process is introduced and discussed. The reader may refer to more advanced engineering economic references, such as Newnan and Lavelle (1988), Park

(1997), Riggs et al. (1996), and Ruegg and Marshall (1990) for more detailed discussions of this all important subject.

8.9.2 PRESENT VALUE

Present value concept recognizes that future costs are worth today less than in the future. This is due to the interest-earning potential of monies at the present. If the yearly discount rate i, is constant, then the present value PV of a future monies FV at n years away is

$$PV = \frac{FV}{(1+i)^n} \tag{8.136}$$

Note that the interest rate is called *discount rate* in the context of present value computations.

As an extension of the present value concept, the present value PV_n of future sums of monies FV_i that are realized at uniform time interval is

$$PV_n = \sum_{j=1}^{j=n} \frac{FV_j}{(1+i)^j} \tag{8.137}$$

If FV_i is constant at FV, then the Equation 8.137 becomes

$$PV_n = FV \sum_{j=1}^{j=n} \frac{1}{(1+i)^j} \tag{8.138}$$

This is a present value of an annuity.

8.9.2.1 Example I: Present Value in Decision Making

Present value is used in several situations in asset management decision making processes. For example, if the decision maker wishes to compare savings from two projects. Project A promises graduated cost savings, while project B promises constant cost savings. The total cost savings for each project is identical for the first 20 years, if discount rates are ignored. By applying the present value Equation 8.137 for each of the projects, it is possible to make a reasonable choice, as detailed in Table 8.48. By comparing the present value of the two projects, it is clear that project B will offer more savings than project A. If the decision maker chooses project B, the total present value savings will be $224.69 thousands, about 4.7%.

8.9.3 FUTURE VALUE

Future value FV is the inverse of present value. It is higher than the present value due to interest effects. Thus, the future value received after n years, with a constant interest i from present monies PV is as follows:

$$FV = PV(1+i)^n \tag{8.139}$$

If there is a recurring cost, or benefit, over n years the future cumulative value is

$$FV_n = \sum_{j=1}^{j=n} P_j (1+i)^j \tag{8.140}$$

With P_j is the cost, or benefit, during the jth year.

8.9.3.1 Example II: Future Value in Decision Making

For example, if a constant yearly inspection cost is $0.5 million. After 20 years, with an interest rate of 7.0%, the compound future value of the inspection costs will be $21.93 million. For a slight

TABLE 8.48
Present Value of Two Projects

Year	Discount Rate (%)	Project A		Project B	
		Cost Savings per Year ($1,000)	Present Value ($1,000)	Cost Savings per Year ($1,000)	Present Value ($1,000)
1	5.00	300	285.71	400	380.95
2	5.00	300	272.11	400	362.81
3	5.00	300	259.15	400	345.54
4	5.00	300	246.81	400	329.08
5	5.00	300	235.06	400	313.41
6	5.00	400	298.49	400	298.49
7	5.00	400	284.27	400	284.27
8	5.00	400	270.74	400	270.74
9	5.00	400	257.84	400	257.84
10	5.00	400	245.57	400	245.57
11	5.00	400	233.87	400	233.87
12	5.00	400	222.73	400	222.73
13	5.00	400	212.13	400	212.13
14	5.00	400	202.03	400	202.03
15	5.00	400	192.41	400	192.41
16	5.00	500	229.06	400	183.24
17	5.00	500	218.15	400	174.52
18	5.00	500	207.76	400	166.21
19	5.00	500	197.87	400	158.29
20	5.00	500	188.44	400	150.76
Total present value			4760.19		4984.88

change of interest rate to 6.5%, the future value of the inspection cost will be $20.65 million. The importance of interest rate is self-evident. Next section we discuss potential fluctuations (uncertainties) in interest rate and their effect on present and future values.

8.9.4 Considerations of Uncertainties

For decisions concerning present of future values, we note that there are uncertainties in all the input parameters. Discount or interest rates i, future values FV_i, or P_j can all be uncertain. Because of this, the resulting present or future value will also be uncertain. Equations 8.136 through 8.140 must be evaluated probabilistically. Monte Carlo simulation method is a common method for computing those values in a probabilistic manner.

8.9.4.1 Example I, Continued: Present Value and Uncertainties

Recall that in example I assumed that the discount rate and the cost savings per year for both potential projects A and B to be deterministic. Such an assumption is not realistic, since it is obvious that those variables are uncertain. Because of this, the decision maker embarks on a probabilistic comparison between the two projects. To facilitate such a comparison, the cost savings per year and the discount rate of both projects, as shown in Table 8.48 were assumed to be the means of both parameters for the next 20 years. The decision maker made some educated assumptions regarding realistic uncertainties for both parameters. First, it is assumed that the uncertainties in both parameters will follow uniform distribution PDF. Second, realistic values of the COV are estimated for that period. The resulting means and COV for both parameters are shown in Table 8.49.

TABLE 8.49
Uncertain Parameters for the Two Projects

Year	Project A				Project B			
	Discount		Cost Savings per Year ($1,000)		Discount		Cost Savings per Year ($1,000)	
	Mean	COV	Mean	COV	Mean	COV	Mean	COV
1	0.05	0.12	300	0.1	0.05	0.12	400	0.1
2	0.05	0.12	300	0.1	0.05	0.12	400	0.1
3	0.05	0.12	300	0.1	0.05	0.12	400	0.1
4	0.05	0.12	300	0.1	0.05	0.12	400	0.1
5	0.05	0.12	300	0.1	0.05	0.12	400	0.1
6	0.05	0.12	400	0.1	0.05	0.12	400	0.1
7	0.05	0.12	400	0.1	0.05	0.12	400	0.1
8	0.05	0.12	400	0.1	0.05	0.12	400	0.1
9	0.05	0.12	400	0.1	0.05	0.12	400	0.1
10	0.05	0.12	400	0.1	0.05	0.12	400	0.1
11	0.05	0.12	400	0.1	0.05	0.12	400	0.1
12	0.05	0.12	400	0.1	0.05	0.12	400	0.1
13	0.05	0.12	400	0.1	0.05	0.12	400	0.1
14	0.05	0.12	400	0.1	0.05	0.12	400	0.1
15	0.05	0.12	400	0.1	0.05	0.12	400	0.1
16	0.05	0.12	500	0.1	0.05	0.12	400	0.1
17	0.05	0.12	500	0.1	0.05	0.12	400	0.1
18	0.05	0.12	500	0.1	0.05	0.12	400	0.1
19	0.05	0.12	500	0.1	0.05	0.12	400	0.1
20	0.05	0.12	500	0.1	0.05	0.12	400	0.1

TABLE 8.50
Present Value of Cost Savings

Project	Mean ($1,000)	STD ($1,000)	COV	Mean – STD ($1,000)
A	4768.2	128.41	0.0269	4511.4
B	4994.9	132.4	0.0265	4730.1

After a Monte Carlo simulation process using Equation 8.137, the statistical results for the present value of the cost savings of the two projects are shown in Table 8.50.

A conservative present value of cost savings would be the mean—two standard deviations. Thus, project B still offer a better solution, with conservative cost savings of $218.7 thousands, or a cost saving of about 4.8%. In this case, considerations of uncertainties provided similar results to deterministic computations: the decision maker can proceed with project B confidently.

8.9.4.2 Example II, Continued: Future Value and Uncertainties

Let us consider the effects of uncertainties in example II. Consider that the yearly inspection costs as a random variable with a mean of $0.5 million. The decision maker feels that such a yearly cost can fluctuate; thus becoming a random variable, with a COV of 0.2. The decision maker assumes that such variability can take a uniform probability distribution. Similarly, the interest rate has uncertainty. The decision maker assumes that the interest rate as a uniformly distributed random variable will have a mean of 7% and COV of 0.2. A simple Monte Carlo simulation algorithm can be performed using Equation 8.139. The results of the simulation show that the average future value of the inspection costs are $22.33 million. The standard deviation is $1.48 million. Thus the two-sigma cost estimation is $25.3 million. Note that such estimation is higher than the earlier deterministic cost of $21.93 million. The cost differential of 25.3-21.93 = $3.37 million is the cost of uncertainty. Such uncertainty should be considered in any decision making process.

8.9.5 LIFE CYCLE COSTS/BENEFITS AND LIFE SPAN

Life cycle analysis is the analysis of different aspects of a particular system during its expected life span. These aspects can include different performance measures such as safety, security, or durability. In practice, LCA refers to estimation of costs and benefits that are expected during the life span of the system. We can state that

> Optimizing life cycle aspects of a particular system, or systems, is perhaps the most important aspect of an asset manager.

As stated above, costs/benefits are popular aspects in the field of LCA. We immediately state that cost and/or benefits need not be of monetary units. Of course, measuring costs/benefits using monetary units is used by many authors in the field.

Accurate LCA can address an important question that faces decision makers often: What is more beneficial: to expend efforts on a short-term project, such as immediately painting a rusted girder, or a long-term project, such as hardening an abutment for an expected 50 years earthquake? In other words, accurate LCA can easily prioritize short-term versus long-term issues. This can be done by incorporating all aspects of the different projects *accurately* within the framework of the LCA. Thus, for the painting versus earthquake hardening issue, the information in Table 8.51 needs to be ascertained.

Clearly, both issues are of importance to the decision maker. Assuming that the monetary metric is of paramount importance to the decision maker in this particular situation, it is clear that considerations of monetary costs and benefits along the life span of the bridge will aid in prioritizing the two options. Note that the three LCA issues, namely hazards, vulnerabilities, and consequences are similar to the main issues involved in risk evaluations, Section 8.6. What differentiate LCA from risk is that it involves considering the passage of time up to a predetermined life span of the system under consideration.

Estimation of LCA involves many of the decision making tools that were presented in this chapter. SHM procedures can aid LCA efforts by providing information regarding performance of the bridge/system. Specific applications of LCA are presented throughout this volume and the companion volume (see Ettouney and Alampalli 2012).

8.9.6 PEER REVIEW

As noted in earlier sections, SHCE project can be a simple, straight forward project or complicated project with significant time and financial commitment depending on project scope, structure involved, and so on. Thus peer review or an independent peer review can be very useful in ensuring successful SHCE project. The primary reason for peer review is to make sure that the SHCE project will indeed produce the value that is anticipated from the project for the money/efforts planned to

---

TABLE 8.51
Information Needed for LCA

Issue for LCA	Painting	Earthquake Hardening
Hazard	Immediate, high probability	Lesser probability, once in every 50 years occurrence
Vulnerability	Rusting, loss of section	Severe damage to bridge
Consequences	Continued deterioration and ultimate failure	Potential bridge failure

be spent. Thus the peer review is to ensure that scientifically the proposed content makes sense, it is practical, has realistic time frame, and, finally, will assist in the decision an owner intends to make using the project results. In general, it is the responsibility of the owner or sponsor of the project to initiate the peer review process.

There are several decisions that have to be made before a peer review is initiated. It is always useful for organizations that will be involved routinely with SHCE projects to develop a set of peer review guidelines. These should include guidelines for project selection, review process, review team composition, compensation for reviewers, guidelines for reviewers, reviewer selection criteria, review time, and conflict resolution (Alampalli 2001).

The project requiring peer review can be based on project cost, duration, or criticality of the decision that will be made on the basis of the project outcome. Review can occur at beginning of the project for small projects or at several stages for big projects covering several years. Normally the initial review is to make sure that the project assumptions, both technical and nontechnical, makes sense, and project results can support the reasons the project was initiated. The number of reviewers and reviewer's qualifications should be decided by the owner/sponsor of the project rather than the entity conducting the project. The compensation for peer reviewers should be included in the project budget. The scope of the review, acceptable time frame for review, compensation schedules, reporting requirements should be detailed well to ensure that there will be no conflicts (or minimal conflicts) in the future and peer review can be done on time and does not impact the project time frame. There should be a mechanism to address the reviewers comments especially if the reviewer opinion differs considerably than the opinions of the proposers. It should also be made clear to everyone in this process that the peer review process is intended to enhance project quality and thus not to serve as watchdog for errors; and review recommendations are not binding on the owner/sponsor.

8.10 APPENDIX: TABLES FOR COMPUTING COST OF BRIDGE FAILURE

Tables 8.52 through 8.60 in this appendix are used in a study by Stein and Sedmera (2006); the study was supported by the National Cooperative Highway Research Program (NCHRP) Project 24-25.

TABLE 8.52
Cost of Bridge Construction

Bridge Superstructure Type, Demolition	Total Cost ($/ft²)
Reinforced concrete flat slab; simple span	50–65[a]
Reinforced concrete flat slab; continuous span	60–80[a]
Steel deck/girder; simple span	62–75[a]
Steel deck/girder; continuous span	70–90[a]
Prestressed concrete deck/girder; simple span	50–70[a]
Prestressed concrete deck/girder; continuous span	65–1 10[a]
Post-tensioned, cast-in-place, concrete box girder cast on scaffolding;	75–110
Span length < = 240 ft	
Steel Box Deck/ Girders:	
Span range from 1 50 ft to 280 ft	76–120
For curvature add a 15% premium segmental concrete box girders; span range from 150 ft to 280 ft	80–100)
Movable bridges; bascule spans & piers	900–1500
Demolition of Existing Bridges:	
Typical	9–15
Bascule spans & piers	63

[a] Increase the cost by 20% for phased construction.

Source: http://www.dot.state.fl.us/structures/Manuals/LRFDSDG2002AugChap11.pdf. Accessed May 26, 2005.

TABLE 8.53
Cost Multiplier for Early Replacement

Average Daily Traffic (ADT)	Cost Multiplier for Early Replacement
ADT < 100	1.0
100 ≤ ADT < 500	1.1
500 ≤ ADT < 1000	1.25
1000 ≤ ADT < 5000	1.5
ADT ≤ 5000	2.0

TABLE 8.54
Comparison of Total and Variable Costs per Mile

Cost Category	Automobiles	Trucks
Total per mile	S0.45	$1.80
Driver costs	–	$0.50
Total vehicle cost per mile	$0.45	$1.30
Variable cost per mile	$0.15	$0.43
Variable as % of total	33%	33%

Source: Minnesota Department of Transportation
(ftp://www. trrb.org/pJf/2003I9.pdf.
Accessed May 26, 2005).

TABLE 8.55
Detour Duration versus ADT

Average Daily Traffic (AUT)	Detour Duration (days)
ADT < 100	1095
100 ≤ ADT < 500	730
500 ≤ Ajyr < 1000	548
1000 ≤ ADT < 5000	365
ADT ≤ 5000	183

TABLE 8.56
Value of Time

State	Mean Wage[a] *($/hour)*	Value of time[b] *($/hour)*
Alabama	15.35	6.29
Alaska	20.27	8.31
Arizona	16.77	6.88
Arkansas	14.21	5.83
California	20.18	8.27
Colorado	19.14	7.85
Connecticut	21.35	8.75
Delaware	18.77	7.70
District of Columbia	27.87	11.43
Florida	16.23	6.65

TABLE 8.56 (continued)
Value of Time

State	Mean Wage[a] ($/hour)	Value of time[b] ($/hour)
Georgia	17.23	7.06
Guam	13.20	5.41
Hawaii	17.67	7.24
Idaho	15.76	6.46
Illinois	18.55	7.61
Indiana	16.26	6.67
Iowa	15.38	6.31
Kansas	16.24	6.66
Kentucky	15.47	6.34
Louisiana	15.02	6.16
Maine	16.09	6.60
Maryland	19.89	8.15
Massachusetts	21.78	8.93
Michigan	19.03	7.80
Minnesota	19.15	7.85
Mississippi	13.77	5.65
Missouri	16.57	6.79
Montana	14.37	5.89
Nebraska	15.89	6.51
Nevada	16.49	6.76
New Hampshire	18.01	7.38
New Jersey	20.69	8.48
New Mexico	15.87	6.51
New York	20.96	8.59
North Carolina	16.40	6.72
North Dakota	14.72	6.04
Ohio	17.26	7.08
Oklahoma	14.97	6.14
Oregon	17.78	7.29
Pennsylvania	17.29	7.09
Puerto Rico	10.61	4.35
Rhode Island	18.38	7.54
South Carolina	15.35	6.29
South Dakota	13.98	5.73
Tennessee	15.74	6.45
Texas	16.98	6.96
Utah	16.40	6.72
Vermont	16.66	6.83
Virgin Islands	13.62	5.58
Virginia	18.81	7.71
Washington	19.65	8.06
West Virginia	14.65	6.01
Wisconsin	16.94	6.95
Wyoming	15.63	6.41

Source: http://www.bls.gov/oes/current/oessrest.htm. Accessed January 12, 2006. The value of time is assumed to be 41% of the mean wage as suggested by José A. Gómez-Ibáñez, William B. Tye, Clifford Winston, *Essays in Transportation Economics and Policy: A Handbook in Honor of John R. Meyer,* 1999.

TABLE 8.57
Occupancy per Vehicle Mile by Daily Trip Purpose

Trip Purpose	Mean	Standard Error
All personal vehicle trips	1.63	0.012
Work	1.14	0.007
Work related	1.22	0.020
Family/personal	1.81	0.016
Church/school	1.76	0.084
Social/recreational	2.05	0.028
Other	2.02	0.130

Source: NHTS, Highlights of the 2001 National Household Travel Survey, Bureau of Transportation Statistics, U.S. Department of Transportation, Washington, DC, 2001.

TABLE 8.58
Value of Time Used in the Derivation of Road User Costs

Vehicle Type	Value of Time from MBC (1990 Dollars)	Value of Time Adjusted (1998 Dollars using CPI)
Small passenger car	$9.75	$12.16
Medium/large passenger car	$9.75	$12.16
Pickup/van	$9.75	$12.16
Bus	$10.64	$13.27
2-axle single unit truck	$13.64	$17.01
3-axle single unit truck	$16.28	$20.30
2-S2 semi truck	$20.30	$25.32
3-S2 semi truck	$22.53	$28.10
2-S1-2 semi truck	$22.53	$28.10
3-S2-2 semi truck	$22.53	$28.10
3-S2-4 semi truck	$22.53	$28.10

Source: http://tti.tamu.edu/documents/407730.pdf. Accessed on May 26, 2005.

TABLE 8.59
Estimates of the Values of Travel Time

Travel Purpose	Automobiles Small	Medium	Trucks 4-Tire	6-Tire
Business Travel				
Value per person[a]	$21.20	$21.20	$21.20	$18.10
Average vehicle occupancy	1.43	1.43	1.43	1.05
Total business	$31.55	$31.96	$32.47	$22.01
Personal Travel				
Value per person[a]	$10.60	$10.60	$10.60	
Average vehicle occupancy	1.67	1.67	1.67	
Total personal	$17.70	$17.70	$17.70	

[a] 2000 Dollars.
Source: FHWA website (http://isdde.dot.gov/olpfiles/fhwa/010617.pdf. Accessed May 26, 2005.)

TABLE 8.60
Assumed Number of Lives Lost in Bridge Failure

Average Daily Traffic (ADT)	Number of Lives Lost
ADT < 100	0
100 < ADT < GOO	1
500 < ADT < 1000	2
1000 < ADT < 5000	1
ADT > 5000 (Not an interstate or arterial)	5
ADT > 5000 (interstate or arterial)	10

REFERENCES

Agrawal, A., Kawaguchi, A., and Qian, G., (2009) "Bridge Element Deterioration Rates: Phase I Reprot," New York State Department of Transportation, Project # C-01-51, Albany, NY.

Alampalli, S., (2001) "Peer Review to Enhance State Transportation Research Program," *Leadership and management in Engineering*, ASCE, *Vol 1*, No. 2, pp. 27–28.

Alampalli, S. and Ettouney, M., (2002) *Proceedings of the Workshop on Engineering Structural Health*, Sponsored by NYS-DOT, Albany, NY.

Arya, S., O'Neill, M., and Pincus, G., (1979), *Design of Structures and Foundations for Vibrating Machines*, Gulf Publishing Company, Houston, TX.

Aven, T., (2003) *Foundations of Risk Analysis*, John Wiley and Sons, West Sussex, England.

Barker, R. and Puckett, J., (2007) *Design of Highway Bridges: An LRFD Approach*, John Wiley Sons, Hoboken, NJ.

Bellman, R., (1964) *Perturbation Techniques in Mathematics, Physics and Engineering*, Rinehart and Winston, New York, NY.

Benjamin, J., and Cornell, A., (1970) *Probability, Statistics, and Decision for Civil Engineers*, McGraw-Hill, New York, NY.

Biggs, J., (1964) *Intorduction to Structural Dynamics*, McGraw Hill, New York, NY.

Chakraborty, S. and DeWolf, J., (2006) "Development and Implementation of a Continuous Strain Monitoring on Multi-Girder Composite Steel Bridge," *Journal of Bridge Engineering*, ASCE, *Vol 11*, No. 6, pp. 753–762.

Crandal, S. and Mark, W., (1963) *Random Vibrations*, Academic Press, New York.

Daddazio, R. and Ettouney, M., (1987) "The Effect of Uncertain Boundary Conditions on the Large Displacement Elasto-Plastic Response of Impulsively Loaded Structures," Proceedings, ASCE Engineering Mechanics 6th Specialty Conference, Buffalo, NY.

Daddazio, R. and Ettouney, M., (1990) "Boundary Element Methods in Probabilistic Acoustic Radiation Problems," *Journal of Vibration and Acoustic, ASME, Vol 112*, pp. 556–560.

Daddazio, R. and Ettouney, M., (1991) "*Probabilistic Acoustic Analysis*," in *Boundary Element Methods in Acoustics*, Editors: Ciskowski, R. D. and Brebbia, C. A. Computational Mechanics Publications, Southampton, UK, pp. 95–108.

Elishakoff, A., (1983) *Probabilistic Methods to the Theory of Structures*, John Wiley & Sons.

Elsayed, A. E. and Ettouney, M. M., (1994) "Perturbation Analysis of Linear Programming Problems with Random Parameters," *Computers and Operation Research, Vol 21*, No 2, pp. 211–224.

Ettouney, M. and Alampalli, S., (2011) *Infrastructure Health in Civil Engineering, Applications and Management*, CRC Press, Boca Raton, FL.

Ettouney, M., Benaroya, H., and Wright, J., (1989a) "Probabilistic Boundary Element Methods (PBEM)," in *Computational Mechanics of Probabilistic and Reliability Analysis*, Editors: Liu, W. K. and Belytschko, T. Elmpress International, Lausanne, Switzerland, pp. 141–166.

Ettouney, M., Benaroya, H., and Wright, J., (1989b) "Boundary Element Methods in Probabilistic Structural Analysis (PBEM)," *Applied Mathematics Modeling, Vol 13*, No. 7, pp. 432–441.

Ettouney, M., Daddazio, R. and Abboud, N. (1993) "*Stochastic Analysis of Soil Dynamics*," in *Computational Stochastic Mechanics*, Editors: Cheng, A. H-D. and Yang, C. Y. Computational Mechanics Publications, Southampton, UK.

FEMA, (2009). "Handbook for Rapid Visual Screening of Buildings to Evaluate Terrorism Risk," FEMA 455, Federal Emergency Management Agency, FEMA Publication, Washington, DC.

Farrar, C. R. and Doebling, S.W., (1998) *Damage Detection - Field Application to a Large Structure*. Proceedings of the Modal Analysis & Testing, NATO-Advanced Study Institute, Sesimbra, Portugal.

Johnson, P. A., and Dock, D. A., (1998) "Probabilistic bridge scour estimates," *Journal of Hydraulic Engineering*, ASCE, 124(7), 750–755.

Johnson, P. A., and Niezgoda, S. L., (2003) "Risk-based method for selecting scour countermeasures," *Journal of Hydraulic Engineering*, ASCE, 130(2), 121–128.

Johnson, P. A., and Simon, A., (1997) "Reliability of bridge foundations in modified channels," *Journal of Hydraulic Engineering*, ASCE, 123(7), 648–651.

Koller, G., (2000) *Risk Modeling for Determining Value and Decision Making*, Chapman & Hall/CRC Press, Boca Raton, FL.

Lawler, G. F., (1995) *Introduction to Stochastic Processes*, Chapman & Hall/CRC, New York.

Melcher, R., (2002) *Structural Reliability Analysis and Prediction*, John Wiley & Sons, New York, NY.

Moses, F., (2001) "Calibration of Load Factors for LRFR Bridge Evaluation," National Cooperative Highway Research Program, NCHRP *Project 454*, Washington, DC.

Murdock, M. B., (2009) "Comparative Load Rating Study Under LRFR And LFR Methodologies For Alabama Highway Bridges," Auburn University, Auburn, AL.

Neubeck, K., (2004) *Practical Reliability Analysis*, Prentice-Hall, NJ.

Newnan, D. G. and Lavelle, J. P., (1988) *Engineering Economic Analysis, 7th Edition*, Engineering Press, Austin, TX

NHTS, (2001), "Highlights of the 2001 National Household Travel Survey," Bureau of Transportation Statistics, U.S. Department of Transportation, Washington, DC.

Park, C. S., (1997) *Contemporary Engineering Economics, 2nd Edition*, Addison- Wesley, Menlo Park, CA.

Pendat, J., and Piersol, A., (2010) *Random Data: Analysis and Measurement Procedures*, Wiley, New York.

Riggs, J. L., Bedworth, D. D., and Randhawa, S. U., (1996) *Engineering Economics, 4th Edition*, McGraw-Hill, New York.

Ruegg, R. T. and Marshall, H. E., (1990) *Building Economics: Theory and Practice*, Van Nostrand Reinhold, New York.

Rausand, M. and Hoyland, A., (2004) *System Reliability Theory: Models, Statistical Methods, and Applications*, Wiley, NJ.

Ravindra, M. and Galambos, T., (1978) "Load and Resistance Factor Design for Steel," *Journal of Structural Division*, ASCE, *Vol 104*, no 9, pp. 1337–1353.

Schlaifer, R. S., (1969) *Analysis of Decision under Uncertainty*, McGraw-Hill, New York, NY.

Stein, S. and Sedmera, K., (2006) "Risk-Based Management Guidelines for Scour at Bridges with Unknown Foundations," National Cooperative Highway Research Program, NCHRP Project 24-25, Washington, DC.

Timoshenko, S. and Gere, J., (1961) *Theory of Elastic Stability*, McGraw-Hill, New York.

Trueman, R. (1977) *An Introduction to Quantitative Methods for Decision Making*, Holt, Rinehart, and Winston, New York, NY.

Von Neumann, J. and Morgenstern, O., (1944) *Theory of Games and Economics*, Princeton University Press, Princeton, NJ.

Appendix

Unit Conversion

Table A.1 contains the units of major engineering metrics used in this volume and the relationships between the units in both SI and US Customary units. The conversion factors are rounded off to a reasonable decimal point.

TABLE A.1
Conversion Table between SI and US Customary Units

Engineering Metric	To Convert From ... SI	to ... US Customary	Divide by
Linear acceleration	m/s^2	in/s^2	0.0254
	m/s^2	ft/s^2	0.3048
Angle	radian	radian	1
Angular acceleration	$radian/s^2$	$radian/s^2$	1
Angular velocity	radian/s	radian/s	1
Area	m^2	in^2	0.000645
	m^2	ft^2	0.0929
	mm^2	in^2	645.2
Energy	J	ft.lb	1.356
Force	N or $kg.m/s^2$	lbf	4.448
	kN	kipf	4.448
Frequency	Hz	Hz	1
Impulse	N.s or kg.m/s	lb . s	4.448
Distance/length	m	in	0.0254
	mm	in	25.40
	m	ft	0.3048
	km	mi	1.609
Mass	kg	lb (mass)	0.4536
	G	oz	28.35
	kg	slug	14.59
	kg	ton	907.2
Moment of a force	N.m	lb.in	0.112867
Power	W or J/s	ft.lbf/s	1.356
	W or J/s	hp	745.7
Pressure or stress	Pa	psi or lb/in^2	6895
	Pa	lb/ft^2	47.88
	kPa	psi or lb/in^2	6.895
Work	J	lb.ft	1.356
Velocity	m/s	in/s	0.0254
	m/s	mi/h	0.447
	km/h	mi/h	1.609
Volume: solid	m^3	in^3	1.64E-05
	m^3	ft^3	0.02832
	cm^3	in^3	16.39

continued

TABLE A.1 (continued)
Conversion Table between SI and US Customary Units

Engineering Metric	To Convert From ...	to ...	Divide by
	SI	US Customary	
Volume: liquid	l	gal (US)	3.785
	l	qt	0.946
Time	s	s	1

Notes: m = meter, s = seconds, kg = kilogram, J = joule, kN = kilo newton, qt = quart, N = newton, Hz = hertz, W = watt, Pa = pascal, l = liter, kipf = kilo pounds (force), in = inches, lb = pounds (force), gal = gallon, ft = feet, mm = millimeter, oz = ounce, mi = mile, km = kilo meter, g = gram.

Index

Note: *Italicized* page references denote figures and tables.

Milton Keynes UK
Ingram Content Group UK Ltd.
UKHW051903071024
449327UK00025B/2073